Saving Food

Saving Food

Production, Supply Chain, Food Waste, and Food Consumption

Edited by

Charis M. Galanakis
Galanakis Laboratories, Chania, Greece

ELSEVIER
ACADEMIC PRESS
An imprint of Elsevier

Academic Press is an imprint of Elsevier
125 London Wall, London EC2Y 5AS, United Kingdom
525 B Street, Suite 1650, San Diego, CA 92101, United States
50 Hampshire Street, 5th Floor, Cambridge, MA 02139, United States
The Boulevard, Langford Lane, Kidlington, Oxford OX5 1GB, United Kingdom

Notices

Knowledge and best practice in this field are constantly changing. As new research and experience broaden our understanding, changes in research methods, professional practices, or medical treatment may become necessary.

Practitioners and researchers must always rely on their own experience and knowledge in evaluating and using any information, methods, compounds, or experiments described herein. In using such information or methods they should be mindful of their own safety and the safety of others, including parties for whom they have a professional responsibility.

To the fullest extent of the law, neither the Publisher nor the authors, contributors, or editors, assume any liability for any injury and/or damage to persons or property as a matter of products liability, negligence or otherwise, or from any use or operation of any methods, products, instructions, or ideas contained in the material herein.

British Library Cataloguing-in-Publication Data

A catalogue record for this book is available from the British Library

Library of Congress Cataloging-in-Publication Data

A catalog record for this book is available from the Library of Congress

ISBN: 978-0-12-815357-4

For Information on all Academic Press publications
visit our website at https://www.elsevier.com/books-and-journals

Publisher: Charlotte Cockle
Acquisition Editor: Megan Ball
Editorial Project Manager: Laura Okidi
Production Project Manager: Nilesh Kumar Shah
Cover Designer: Matthew Limbert

Typeset by MPS Limited, Chennai, India

Contents

List of contributors

Elisabete M.C. Alexandre QOPNA & LAQV-REQUIMTE, Department of Chemistry, University of Aveiro, Aveiro, Portugal; Center for Biotechnology and Fine Chemistry – Associated Laboratory, School of Biotechnology, Catholic University of Portugal, Porto, Portugal

Valérie L. Almli Sensory and Consumer Science Department, Nofima, Ås, Norway

Graciela Alvarez Refrigeration Process Engineering Research Unit, IRSTEA, Antony, France

Jessica Aschemann-Witzel MAPP Centre, Department of Management, Aarhus School of Business and Social Sciences, Aarhus University, Aarhus, Denmark

Carla Caldeira European Commission, Joint Research Centre (JRC), Ispra, Italy

Sara Corrado European Commission, Joint Research Centre (JRC), Ispra, Italy

Christine Costello Assistant Professor, Industrial & Manufacturing Systems Engineering, University of Missouri, Columbia, MO, United States

Ilona E. de Hooge Marketing and Consumer Behaviour group, Wageningen University, Wageningen, The Netherlands

Hans De Steur Department of Agricultural Economics, Faculty of Biosciences Engineering, Ghent University, Ghent, Belgium

Manoj K. Dora College of Business, Arts & Social Sciences, Brunel Business School, Brunel University, London, United Kingdom

Gabriel da Silva Filipini Federal University of Rio Grande, School of Chemistry and Food, Rio Grande, Brazil

Xavier Gellynck Department of Agricultural Economics, Faculty of Biosciences Engineering, Ghent University, Ghent, Belgium

Selale Glaue Efes Vocational School, Dokuz Eylul University, İzmir, Turkey

Nihan Gogus Efes Vocational School, Dokuz Eylul University, İzmir, Turkey

Tiziano Gomiero Independent scholar, Treviso, Italy

Gang Liu SDU Life Cycle Engineering, Department of Chemical Engineering, Biotechnology, and Environmental Technology, University of Southern Denmark, Odense, Denmark

Lara Manzocco Department of Agricultural, Food, Environmental and Animal Sciences, University of Udine, Udine, Italy

Paola Chaves Martins Federal University of Rio Grande, School of Chemistry and Food, Rio Grande, Brazil

Vilásia Guimarães Martins Federal University of Rio Grande, School of Chemistry and Food, Rio Grande, Brazil

Ultan McCarthy School of Science & Computing, Department of Science, Waterford Institute of Technology, Waterford, Ireland

Samuel Mercier Department of Electrical Engineering, University of South Florida, Tampa, FL, United States; Department of Chemical and Biotechnological Engineering, Université de Sherbrooke, Sherbrooke, QC, Canada

Martin Mondor Saint-Hyacinthe Research and Development Centre, Agriculture and Agri-Food Canada, Saint-Hyacinthe, QC, Canada

Sílvia A. Moreira QOPNA & LAQV-REQUIMTE, Department of Chemistry, University of Aveiro, Aveiro, Portugal; Center for Biotechnology and Fine Chemistry – Associated Laboratory, School of Biotechnology, Catholic University of Portugal, Porto, Portugal

Semih Otles Food Engineering Department, Ege University, İzmir, Turkey

Aditya Parmar Natural Resources Institute, University of Greenwich, London, United Kingdom

Darian Pearce Department of Agricultural Economics, Faculty of Biosciences Engineering, Ghent University, Ghent, Belgium

Manuela Pintado Center for Biotechnology and Fine Chemistry – Associated Laboratory, School of Biotechnology, Catholic University of Portugal, Porto, Portugal

Carlos A. Pinto QOPNA & LAQV-REQUIMTE, Department of Chemistry, University of Aveiro, Aveiro, Portugal

Stella Plazzotta Department of Agricultural, Food, Environmental and Animal Sciences, University of Udine, Udine, Italy

Viviane Patrícia Romani Federal University of Rio Grande, School of Chemistry and Food, Rio Grande, Brazil

Serenella Sala European Commission, Joint Research Centre (JRC), Ispra, Italy

Jorge A. Saraiva QOPNA & LAQV-REQUIMTE, Department of Chemistry, University of Aveiro, Aveiro, Portugal

Taija Sinkko European Commission, Joint Research Centre (JRC), Ispra, Italy

Despoudi Stella Aston Business School, Aston University, Birmingham, United Kingdom

Sebnem Tavman Food Engineering Department, Ege University, İzmir, Turkey

Ismail Uysal Department of Electrical Engineering, University of South Florida, Tampa, FL, United States

Sebastien Villeneuve Saint-Hyacinthe Research and Development Centre, Agriculture and Agri-Food Canada, Saint-Hyacinthe, QC, Canada

Joshua Wesana Department of Agricultural Economics, Faculty of Biosciences Engineering, Ghent University, Ghent, Belgium; School of Agricultural and Environmental Sciences, Mountains of the Moon University, Fort Portal, Uganda

Li Xue Institute of Geographic Sciences and Natural Resources Research, Chinese Academy of Sciences, Beijing, P.R. China; SDU Life Cycle Engineering, Department of Chemical Engineering, Biotechnology, and Environmental Technology, University of Southern Denmark, Odense, Denmark; University of Chinese Academy of Sciences, Beijing, P.R. China

Preface

About one-third of the food produced in the world for human consumption gets lost or wasted every year. This quantity is shocking considering that it accounts approximately for 1.3 billion tons of food. As it can be easily understood, the problem of food loss and waste is directly connected to hunger and global sustainability in the 21st century. However, the problem is even bigger than it seems, as food loss also accompanies a major squandering of resources such as water, land, energy, labor, and capital. In addition, it is connected to increased and unwanted greenhouse gas emissions that contribute to global warming and climate change. The problem of food loss and food waste is so big that it cannot be solved with mere activities or simple suggestions. It can be eliminated only by facing challenges and providing continuous solutions, at all levels of food production and consumption for all the involved actors and stakeholders. Correcting the policy framework, optimizing agricultural practices, shaping food production, changing consumers' and companies' attitudes, motivating retailers, promoting packaging and process technologies, valorizing waste streams, and other actions should also be taken into account.

Subsequently, a guide covering the latest developments in this particular direction is required. This book fills these gaps by covering all the aspects of food-loss reduction at all relevant stages and in all possible ways. It provides details about introducing sustainable food production, adapting more sustainable methods for efficient crop cultivation and harvesting, optimizing utilization of resources, eliminating losses in the supply chain, adapting sustainable packaging solutions, appealing enterprises to change consumer behavior, developing food waste valorization strategies, and raising people's awareness of wasted food. The ultimate goal is to support the scientific community, policy makers, professionals, and enterprises, that aspire to set up actions and strategies, to reduce wastage of food. Thereby, the book targets all involved actors and aims to drive innovations, promote interdisciplinary dialogues, and spark debates to generate solutions across the entire value chain from field to fork.

It consists of 13 chapters. Chapter 1 provides an introduction to global food loss and food waste using data for 84 countries and 52 individual years. Chapter 2 reviews soil and crop management practices that may reduce yield loss, or increase yields, while reducing the use of inputs and the environmental impact of agricultural activities. A number of food loss reduction measures (technical and behavioral) are available along the entire value chain, but the motivation to implement them is the one that needs due consideration and action. Further optimization of agricultural practices to save food is described in Chapter 3.

During food production, transport, storage, and final consumption, the food properties may get affected in several ways. To ensure safety and stability of foods and avoid their discharge, effective and economic food preservation methods should be selected. Chapter 4 deals with the conventional and emerging preservation techniques, such as pasteurization, sterilization, cooling, freezing, ohmic heating, microwave, and radio frequency, which are thermal preservation technologies. On the other hand, Chapter 5 deals with the application of nonthermal and eco-friendly emergent processing methodologies such as high pressure processing, pulsed electric fields, and ultrasounds. These modern technologies assure products' safety as well as maintain their original quality, thus contributing to food loss reduction during production.

An efficient way to preserve food is using industrial processes, but it is also possible to use active packaging to extend the shelf life of food products. To this end, Chapter 6 discusses existing and innovative packaging solutions to minimize food waste. Chapter 7 reviews the main stages and technologies used for the preservation of perishable food products along the supply chain, and the amount of food lost or wasted along these stages for the main families of products. It also highlights the need for better refrigeration of food along the last stages of the cold chain (retail and consumer handling) and for better management along the commercial portion of the cold chain in developed countries. Chapter 8 aims to provide an overview on losses in the food industry. At first, food losses in the upstream and downstream supply chain are discussed prior to denoting the different ways to reduce food losses by optimizing supply chains. Solutions at the supply chain entity–level as well as supply chain network–level are provided. Chapter 9 presents mitigating approaches that could be initiated along food supply chains. This is conducted by discussing a case study of measuring food losses in the supply chain through value stream mapping in the dairy sector in Uganda.

Food waste valorization includes different food waste management strategies, whose goal is to turn food waste into value-added derivatives to be used in food or other industrial sectors. These strategies present the advantage of exploiting an always-available and cheap source, such as food waste, for producing derivatives presenting a high potential market value. Chapter 10, discusses the basic definitions and principles at the basis of food waste valorization and presents relevant strategies, with particular emphasis on those in which the great potential of food waste is maximally exploited.

In Chapter 11, the environmental impacts of food production and consumption of an average European citizen are assessed taking the food waste generated along the food supply chain into account. In addition, the impact of food waste reduction and adoption of different diets are estimated. Chapter 12 discusses food waste at the consumer–retailer interface, the so-called "suboptimal food" (reduction of food losses and wastes is one of the agricultural research areas, that has received only limited resources and attention from the public and private sectors in comparison to increased yields per hectare). Finally, Chapter 13 provides an introduction to the concepts of Zero Waste and life-cycle assessment; an overview of the challenges presented by the United States agricultural system as it is today; and a discussion on the food waste management options included in the Environmental Protection Agency's Food Recovery Hierarchy.

Conclusively, the book is a guide for food retailers, supply chain specialists, food scientists, food technologists, food engineers, professionals, agriculturalists, and food producers trying to minimize the food loss and adapt zero waste strategies. It provides critical information in this direction, so that the general public can be aware, the government can set relevant guidelines, and finally the food industry can optimize production lines. It provides an overview and description of the problem from different angles (e.g., environmental impacts, some social and many technological issues) and covering different actors (consumers, producers, processors, industry, policy makers, etc.). This way it can help identify current research gaps and spur more in-depth investigations of certain topics described in the different chapters. It could be of particular interest to food industry stakeholders as it highlights strategies and technologies that could help mitigate food waste. Knowledge of best practices and advanced procedures for the balanced production of agricultural resources and foods, and their redistribution, transportation, and consumption would make it possible to achieve sustainable food systems.

At this point, I would like to express my gratitude to all the authors of the book for their acceptance of my invitation and their participation in this collaborative book that brings together, for the first time, different scientific, technological, and managerial issues of saving food in one comprehensive text. They accepted and followed the editorial guidelines, the book's concept, and the timeline with ultimate attention. All these actions conclude in a great honor for me and are highly appreciated. I consider myself fortunate to have had the opportunity to bring together so many experts from Belgium, Brazil, Canada, China, Denmark, France, Italy, Ireland, Norway, Portugal, The Netherlands, Turkey, Uganda, the United Kingdom, and United States. I would like to thank the acquisition editor Megan Ball, the book manager Katerina Zaliva, and all Elsevier's production staff for their help during the editing and publishing process.

I would also like to thank the Food Waste Recovery Group (www.foodwasterecovery.group) of ISEKI Food Association and its pool of experts that provided us with valuable information about different ways of saving food.

Last but not the least, a message for all the readers: Such collaborative projects of hundreds of thousands of words may contain a few errors and gaps. Any instructive comments or even criticisms are and always will be welcome. Thus, never hesitate to contact me to discuss any issues with the book.

Charis M. Galanakis[1,2]
[1]Food Waste Recovery Group, ISEKI Food Association, Vienna, Austria, [2]Research & Innovation Department, Galanakis Laboratories, Chania, Greece

Introduction to global food losses and food waste

1

Li Xue[1,2,3] *and Gang Liu*[2]

[1]Institute of Geographic Sciences and Natural Resources Research, Chinese Academy of Sciences, Beijing, P.R. China, [2]SDU Life Cycle Engineering, Department of Chemical Engineering, Biotechnology, and Environmental Technology, University of Southern Denmark, Odense, Denmark, [3]University of Chinese Academy of Sciences, Beijing, P.R. China

Chapter Outline

1.1 Introduction

Food losses and food waste (FLW) occur along the whole food supply chain. In recent years, FLW has become a global concern and poses considerable challenges to food security (The Economist Intelligence Unit, 2014), natural resources (FAO, 2013), environment (Katajajuuri et al., 2012), and human health (Pham et al., 2014), and is therefore considered as a key obstacle to sustainable development. Therefore, reducing FLW has been put on the political agenda at the global and national levels. For instance, the United Nations has set a target of halving per capita global food waste at the retail and consumer levels and reducing food losses along production and supply chains by 2030, in the Sustainable Development Goals (SDG) Target 12.3 (United Nations, 2017). The European Union (European Commission Food Safety Home Page, 2017) has taken actions to work towards this

Saving Food. DOI: https://doi.org/10.1016/B978-0-12-815357-4.00001-8

target; in 2015, the United States (United States Department of Agriculture, 2017) also announced its first-ever national goal to reduce food waste by 50% by 2030 to improve food security and protect natural resources; and the African Union also made a commitment to halve postharvest losses by 2025 in the 2014 Malabo Declaration (Lipinski et al., 2016).

Over the past few decades, with growing concerns and attention on FLW from public and political sectors, more and more studies have quantified FLW across the food supply chain at national, regional, and global scales. For example, according to the Food and Agriculture Organization (FAO) of the United Nations, about one-third of food production was lost or wasted worldwide that was meant for human consumption (Gustavsson et al., 2011). This significant amount of FLW would mean 4.4 gigatonnes of CO_2 equivalent (FAO, 2015), 250 km^3 of blue water footprint (FAO, 2013), 28% of the total agriculture land globally during agriculture production, an economic cost of about USD 750 billion (equivalent to the gross domestic product (GDP) of Turkey) (FAO, 2013), and approximately 24% of all food produced when converted into calories (Gustavsson et al., 2011).

Many other studies have also revealed a similar scale of FLW on the regional or country level and its significant impacts on environment, economic development, and food security. For example, it is reported that the EU-28 generate about 100 million tonnes of FLW each year, and the largest contribution is from households (45%) (FUSIONS, 2015). For the member states, households in the United Kingdom wasted approximately 7.2 million tonnes of food in 2012 (WRAP, 2014). The wasted food from households in Finland, Denmark, Norway, and Sweden make up 30%, 23%, 20%, and 10%–20% of food purchased, respectively (Gjerris and Gaiani, 2013). In Switzerland, about one-third of food produced (calorie equivalent) is wasted and households contribute the most (Beretta et al., 2013). Some other developed countries also highlight a similar trend. For example, in the United States, the per capita FLW increased by about 50% between 1979 and 2003 (Hall et al., 2009). In Australia, more than 4.2 million tonnes of FLW goes to landfill per year (Verghese et al., 2013).

In the past decades, some governmental organizations and national agencies have made great effort to quantify FLW. For example, the FAO has issued a number of relevant reports on FLW at a global scale (Gustavsson et al., 2011; FAO, 2014). The United States Department of Agriculture Economic Research Service (USDA-ERS) developed the Loss-Adjusted Food Availability Data Series in 1997, which covers about 200 items for three stages (production to retail, retail, and consumer) of losses in terms of quantities, values, and calories (Buzby et al., 2009; Buzby and Guthrie, 2002). In the United Kingdom, the Waste and Resources Action Programme (WRAP) organization has been set up to reduce food waste, and has released a number of reports on FLW in the food supply chain since 2007 (WRAP, 2008, 2009).

In recent years, relevant stakeholders from academia, industry, and governmental and nongovernmental organizations have participated in research projects and worked on the standardization of quantification and methods of FLW. For example, the project Food Use for Social Innovation by Optimizing Waste Prevention

Strategies (FUSIONS) (2012–16) funded by European Commission has been working towards a more resource efficient Europe, and has issued a number of reports, covering the framework of FLW definition, measurement, and mitigation strategies (Östergen et al., 2014; FUSIONS, 2016). In 2015, the European Commission funded a further project called Resource Efficient Food and dRink for the Entire Supply cHain (REFRESH) (2015–19), which involves 26 partners from 12 European countries and China and focuses on the reduction of avoidable waste and improved valorization of food resources (Refresh Home Page, 2017). In 2016, World Resources Institute, United Nations Environment Programme (UNEP), World Business Council for Sustainable Development, FAO, and WRAP together as a partnership of major international organizations announced the first global standard to quantify FLW (World Resources Institute, 2016).

Though there are continuous efforts on quantifying FLW and some researchers have also stressed the data deficiency and inconsistency and raised concerns on the demand of better measurement of FLW (Parfitt, 2013; Liu, 2014; Shafiee-Jood and Cai, 2016), there are still major gaps in the existing global FLW data as follows:

- The spatial coverage of existing studies is narrow. Most research is carried out in developed countries. For instance, there are plenty of publications drawing out the situation of FLW in the United States (Thyberg et al., 2015; Buzby and Hyman, 2012; Kantor et al., 1997) and Sweden (Bräutigam et al., 2014; Filho and Kovaleva, 2015). In contrast, only a few studies quantified FLW in developing countries, such as Nepal (Choudhury, 2006) and the Philippines (Parfitt et al., 2010) and some countries experiencing a rapid development, such as China and India (Parfitt and Barthel, 2011).
- There is an uneven focus on the different food supply stages. A great many studies have illustrated food waste at retailing and consumption stages (Davies and Konisky, 2000; Stenmarck et al., 2011; Parry et al., 2015), mainly conducted in developed countries, such as the United States. On the other hand, there are few studies revealing the situation of postharvest losses, which are mainly carried out in developing countries, such as India (Gangwar et al., 2014).
- Some existing data are outdated but still in use. Some studies have to depend on the older data due to the lack of updated ones. For example, data on the postharvest losses of fresh fruits and vegetables from one study in the 1980s and 1990s were used in two recent studies (Parfitt et al., 2010; Kader, 2005).
- There is a lack of primary data and a great many studies have to cite data in the existing studies. For example, many researchers have repeatedly cited data from the FAO report issued in 2011 (Oelofse and Nahman, 2013; Lipinski et al., 2013; Nahman and de Lange, 2013). But it may not be representative in terms of time and countries for commodities (Liu, 2014). The data provided by the African Postharvest Loss Information System has been mostly used to address postharvest losses (Prusky, 2011; World Bank, 2011; Segrè et al., 2014).
- The definition of FLW, methods used, and system boundaries are different in existing studies. This makes it difficult to systematically compare and verify FLW data between countries, commodities, and stages. Therefore, it is uncertain to do analysis on the relationship between FLW and social, economic, and environmental factors based on the existing data.

It is particular of importance to clearly and comprehensively understand the existing global FLW data on their quality and availability. First, it is a prerequisite for tracking the progress toward the SDG Target 12.3 and the national FLW reduction goals, and evaluating the effect of relevant policies. Second, it will contribute to raising awareness, informing mitigation strategies, and giving priority to prevent and reduce FLW. Third, better data can been verified and compared among countries, stages, and commodities, helping to distinguish patterns and drivers of FLW generated. Fourth, it can be an essential foundation for further analyzing the social, economic, and environmental impacts of FLW.

In this chapter, a critical overview of all the available FLW data in 202 publications is provided, which could provide a basic database for further analysis of environmental impacts and mitigation strategies of FLW. Bibliometric characteristics of existing literature and methods of measurement (advantages and disadvantages) are assessed, their patterns between countries, food supply chain stages, and food commodities are discussed, and some implications for future work are denoted.

1.2 System definition

1.2.1 Food losses and food waste

FLW occurs across the food supply chain. Some studies have made a difference between the definition of FLW, edible and inedible food waste, avoidable and unavoidable food waste. For example, according to the FAO (FAO, 2014), food loss refers to food that is lost due to quantity or quality reasons, and food waste refers to food that is left to spoil or expire due to carelessness of consumers, which is usually related to discarding deliberately or other use of food (e.g., animal feed). Because of the deficiency of consistencies in the literature reviewed, the distinctions were not considered and we do not differentiate between food loss and food waste in this study, so we define FLW as the combined amount of FLW.

1.2.2 Food supply chain

As shown in Fig. 1.1, FLW involves six major processes. FLW could be further classified into three types: farm losses/waste (during agricultural production and harvesting), postharvest losses/waste (during postharvest handling and storage, manufacturing, distribution, and retailing), and consumer waste (both in household and out-of-home). Agricultural products losses/waste on the farm are mainly caused by insects, diseases, and severe weather. For livestock products, it relates to sickness and death during breeding stage for cattle, pig, and poultry meat, and discarded fish during fishing. Postharvest losses/waste refers to food spoilage and degradation during different stages. It includes postharvest handling and storage (when food is under threshing/shelling or icing and animals transported to slaughtering), manufacturing (when food is processed into various products), distribution (when food is transported, loaded, and uploaded), as well as retailing (includes wholesale, supermarket, and wet market). Consumer food waste occurs both in household and dining out away from home.

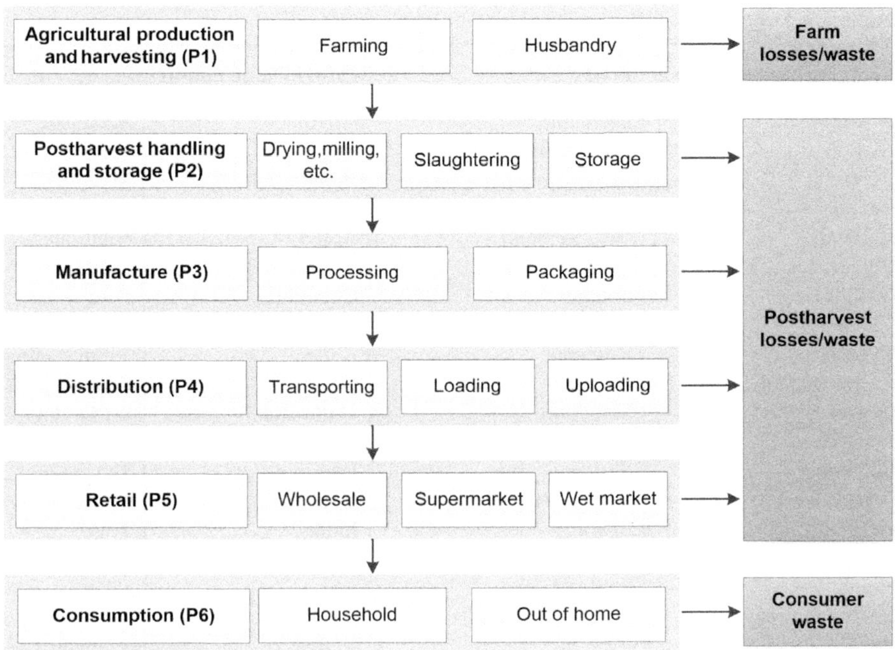

Figure 1.1 Food supply chain for food losses and food waste.

1.2.3 Food commodity groups

The commodities categories were defined based on the classification of FAO and by taking consideration of characteristics of data in the publications. As a result, 10 groups of food commodities were presented:

1. Cereal and cereal products (e.g., wheat, maize, and rice);
2. Roots and tubers (e.g., potatoes and cassava);
3. Oilseeds and pulses (e.g., peanuts and soybeans);
4. Fruits;
5. Vegetables;
6. Meat;
7. Fish and seafood;
8. Dairy products;
9. Eggs;
10. Others or not specified.

1.2.4 Geographical and temporal boundary

The FLW data was collected from as early as possible to 2015 at the global, regional, and national levels. Based on per capita GDP and the classification principles of FAO (Gustavsson et al., 2011), the countries are divided into medium/high-income countries and low-income countries (Table 1.1).

Table 1.1 Grouping of different development levels of countries

Medium/high-income countries		Low-income countries	
Armenia	Lithuania	Angola	Malaysia
Australia	Luxembourg	Argentina	Mexico
Austria	Malta	Bangladesh	Myanmar
Belarus	Netherlands	Benin	Nepal
Belgium	New Zealand	Bolivia	Nigeria
Bulgaria	Norway	Brazil	Pakistan
Canada	Poland	Cambodia	Peru
China	Portugal	Cameroon	Philippines
Cyprus	Romania	Chile	Saudi Arabia
Czech Republic	Russia	Colombia	South Africa
Denmark	Singapore	Costa Rica	Sri Lanka
Estonia	Slovakia	Egypt	Swaziland
Finland	Slovenia	Ethiopia	Tanzania
France	South Korea	Ghana	Thailand
Germany	Spain	India	Togo
Greece	Sweden	Indonesia	Turkey
Hungary	Switzerland	Iran	Uganda
Ireland	United Kingdom	Jamaica	Venezuela
Italy	Ukraine	Kenya	Vietnam
Japan	United States	Laos	Zambia
Latvia		Madagascar	Zimbabwe
		Malawi	

1.3 Food losses and food waste quantification

1.3.1 Bibliometric analysis of literature

1.3.1.1 Type of publications

Web of Science and Google Scholar were the main source for the research, and reports issued by research institutions as well as governmental or nongovernmental organizations were also collected to ensure a wider coverage of available data. Finally, 202 publications were reviewed. They include five types: peer-reviewed journal articles (53.5%), reports (35.6%), PhD and master's theses (5.9%), conference proceedings (3.0%), and book chapters (2.0%). Journal articles were dominant (108) in the reviewed publications, which were published in 69 different journals and covered a wide range of subjects. In total, approximately 45% of them were published in the top 10 journals (Fig. 1.2). The majority of the publications outlets were *Waste Management*, *Waste Management & Research*, *Resources, Conservation and Recycling*, *Food Policy*, and *Journal of Cleaner Production*, representing 15.7%, 7.4%, 5.6%, 4.6%, and 2.8% of the total published articles, respectively.

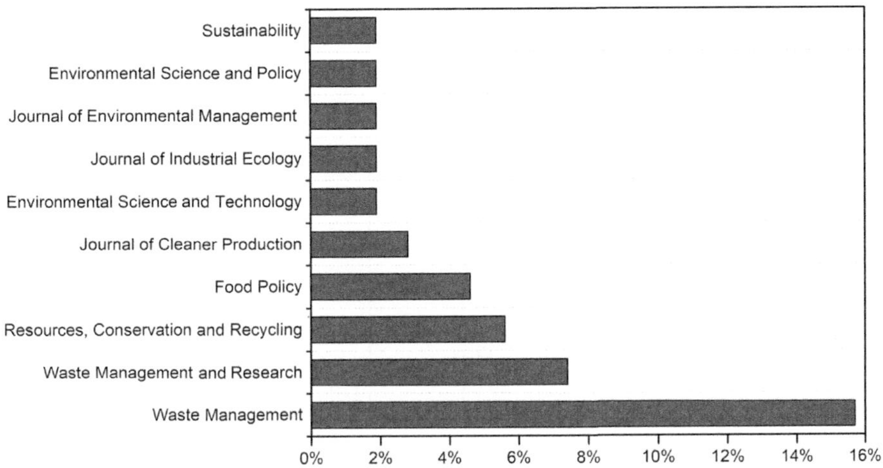

Figure 1.2 The top 10 journals that publishes food loss and food waste data.

1.3.1.2 Temporal trend for year of publications and estimation

Fig. 1.3A shows the number of publications during the 76-year period (1939–2015). In general, the number of publications increased throughout the whole period. It was small and remained stable before 2000. Afterwards, it has seen a gradual increase during 2001–10. In the last five years, the number of studies has grown substantially (137), accounting for 67.8% of the total publications. This means there is an increasing focus on FLW research around the world.

Fig. 1.3B illustrates the time trend of the year of estimation. According to literature, the FLW data was discovered as early as 1933, and the number remained stable and low until 1995. Afterwards, the number has increased significantly by more than 60% over the past 10 years, 38.1% from 2006 to 2010 and 25.1% from 2011 to 2014.

1.3.1.3 Distribution of countries

The 202 publications reported FLW data throughout the food supply chain covering 84 countries (reported 498 times) distributed all over the world. However, the focus on FLW was unbalanced in different regions. Most studies were conducted in the developed areas, such as North America, Northern and Western Europe, whereas little attention was paid to the developing countries, such as India. Fig. 1.4 shows spatial distribution and the top 10 countries have been studied. Most research was conducted in the United Kingdom (Langley et al., 2010; Mena et al., 2014; Vanham et al., 2015; Xu et al., 2015) and United States (Thyberg et al., 2015; Buzby and Hyman, 2012; Kantor et al., 1997), both of which made up more than 10% of the reported times, respectively. Then Sweden (Bräutigam et al., 2014; Filho and Kovaleva, 2015), Germany (Kranert et al., 2012; Jörissen et al., 2015), and Finland (Silvennoinen et al., 2012; Silvennoinen et al., 2015) accounted for 5.4%, 4.4%, and 3.2%, respectively.

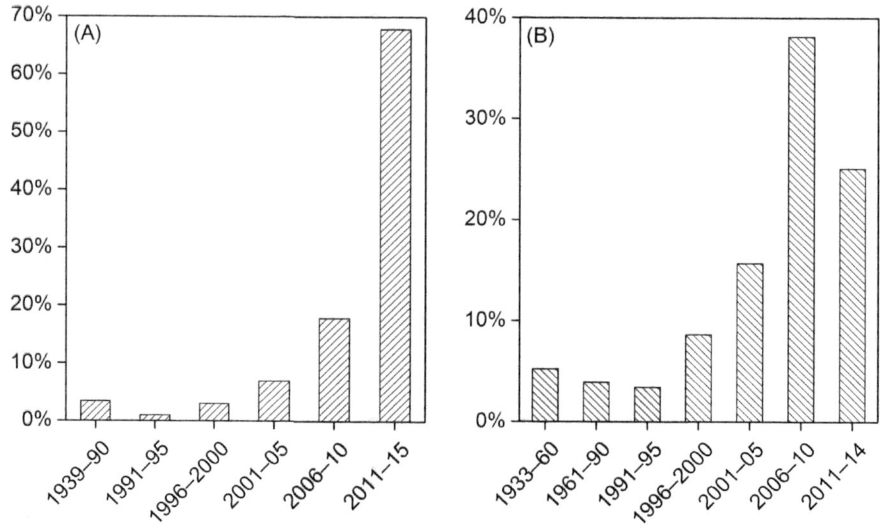

Figure 1.3 (A) Temporal trend of reviewed food losses and food waste (FLW) data in terms of year of publication. (B) Temporal trend of reviewed FLW data in terms of year of estimation.

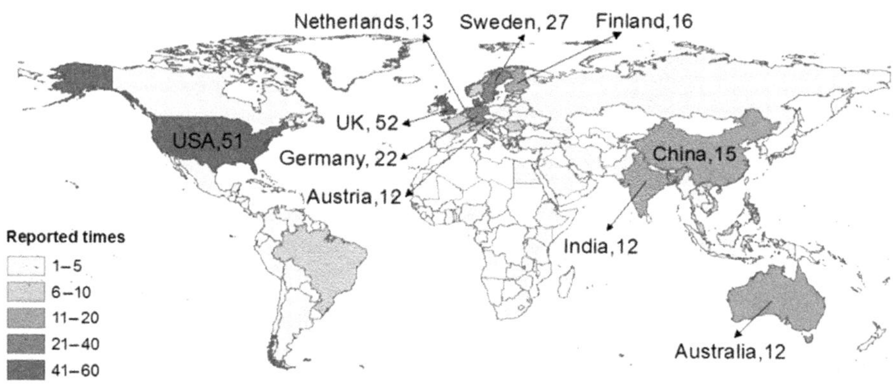

Figure 1.4 Geographical distribution of case countries. The numbers are the reported times of individual countries.

Source: Adopted from Xue, L., Liu, G., Parfitt, J., Liu, X., Van Herpen, E., Stenmarck, Å., et al., 2017. Missing food, missing data? A critical review of global food losses and food waste data. Environ. Sci. Technol. 51 (12), 6618–6633.

1.3.1.4 Food supply chain coverage

According to the publications found, they covered different stages in the food supply chain in terms of medium/high-income countries and low-income countries. Fig. 1.5 shows that most studies covered the retailing and consumption stages. In total, the largest number of studies were carried out in household, accounting for

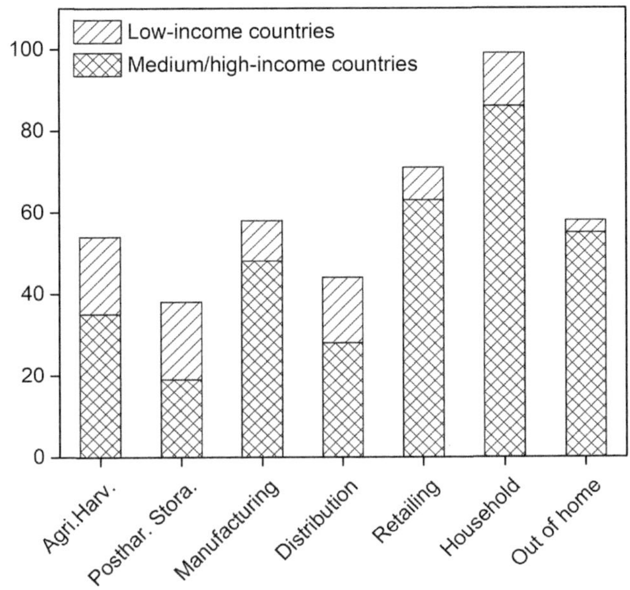

Figure 1.5 The number of publications in terms of different food supply stages and different development levels of countries.

49% of all the publications, which was followed by the retailing stages (35%). However, only a small portion of studies included the stages between agricultural production and distribution. In detail, agricultural production, postharvest handling and storage, manufacturing, and distribution stages accounted for 26.7%, 18.8%, 28.7%, and 21.8%, respectively.

In the case of region studied, the number of publications in medium/high-income countries was much higher than that in low-income countries along the food supply chain, apart from the postharvest handling and storage stage with the same number of publications for both. The majority of studies involving retailing and consumption stages were conducted in medium/high-income countries, occupying 31.2% and 42.6% of all the literature, respectively. On the other hand, low-income countries were targeted mainly in the early and middle stages of the food supply chain, especially for the agricultural production and postharvest handling and storage stages.

1.3.2 Different methods used for food losses and food waste quantification

1.3.2.1 Overview of methods

There were various methods used to measure the quantity of FLW along the food supply chain. Table 1.2 summarizes the methods used to quantify FLW. Two kinds of methodologies have been used to quantify FLW, which can be divided into two

Table 1.2 Description of different methods used for food losses and food waste quantification

	Method	Symbol	Example of case countries/regions	Food supply chain	References
Direct measurement	Weighing	W	Portugal	P6b	Dias-Ferreira et al. (2015)
	Garbage collection	G	Austria	P6a	Dahlén and Lagerkvist (2008)
	Surveys	S	United Kingdom	P1, P2, P3, P5	Mena et al. (2014)
	Diaries	D	United Kingdom	P6a	Langley et al. (2010)
	Records	R	Sweden	P5	Scholz et al. (2015)
	Observation	O	Italy	P6b	Saccares et al. (2014)
Indirect measurement	Modeling	M	United States	P6	Hall et al. (2009)
	Food balance	F	Global	P1, P2, P3, P4, P5, P6	Gustavsson et al. (2011)
	Use of proxy data	P	Singapore	P6a	Grandhi and Appaiah Singh (2016)
	Use of literature data	L	Denmark	P1, P3, P4, P6	Halloran et al. (2014)

Note: P6a = Household, P6b = Out-of-home.

groups: (1) direct measurement or approximation based on first-hand data, and (2) indirect measurement or calculation derived from secondary data. These methods could provide an insight of origins and specific stages in the whole food supply chain of FLW, or an overview of FLW at the regional or global level from a macro-perspective. Detailed information on the methods used is outlined as follows:

Direct measurement involves a variety of methods to quantify or estimate the amount of FLW:

- *Weighing*: It is usually used in restaurants, hospitals, and school via instrument or device to measure the weight of FLW. It may or may not involve weighing each part of FLW for the compositional analysis.
- *Garbage collection*: This involves separation from other types of residual wastes collected to determine the weight or proportion of FLW. It may or may not involve compositional analysis of FLW. It can be collected from households (Gutiérrez-Barba and Ortega-Rubio, 2013).
- *Surveys*: Questionnaires are used to collect information about perceptions and behaviors on FLW answered by a great many individuals, or by face-to-face interviews with major stakeholders in the field. This usually takes place in households, where people can directly estimate the quantity of food waste or the percentage of food purchased that goes to waste in their families (Stefan et al., 2013).
- *Diaries*: It is often used in households and commercial kitchens by recording the quantity of FLW for a certain time, where weighing scales are sometimes used to quantify the amount of the food waste (Rathje and Murphy, 2001).
- *Records*: It is usually used in the retailing and manufacturing stages, especially for supermarkets and large-scale food companies, where regular collection of information (not initially used for FLW record) can determine the quantity of FLW.
- *Observation*: Visually estimating the amount of food left over by using scales with multiple points or assessing the volume of FLW by counting the number of goods.

The other group includes methods based on the existing data from different secondary sources:

- *Modeling*: It uses mathematical models to obtain the amount of FLW on the basis of the factors that affect FLW generation.
- *Food balance*: Using food balance sheet (e.g., FAOSTAT) based on inputs, outputs, and stocks in the food supply chain to calculate FLW, or human metabolism (e.g., the relationship between body weight and the amount of food eaten).
- *Use of proxy data*: Using data from companies or statistical institutions (in an aggregated level) to estimate the amount of FLW.
- *Use of literature data*: Using data from literature directly or estimating quantities of FLW according to the data in other literature.

Fig. 1.6 shows the methods used in the 202 publications. It can be seen that most of the publications depended on the indirect measurement (red-yellow (dark gray in print version) colors in Fig. 1.6). More than 40% of them were only based on literature data, and about one-third used other types of methods with literature data, for instance, modeling (Khan and Burney, 1989; Liu et al., 2013) or proxy data (Gooch, 2012; An et al., 2014) (indirect measurement) or weighing or surveys (Papargyropoulou et al., 2014; Edjabou et al., 2015) (direct measurement).

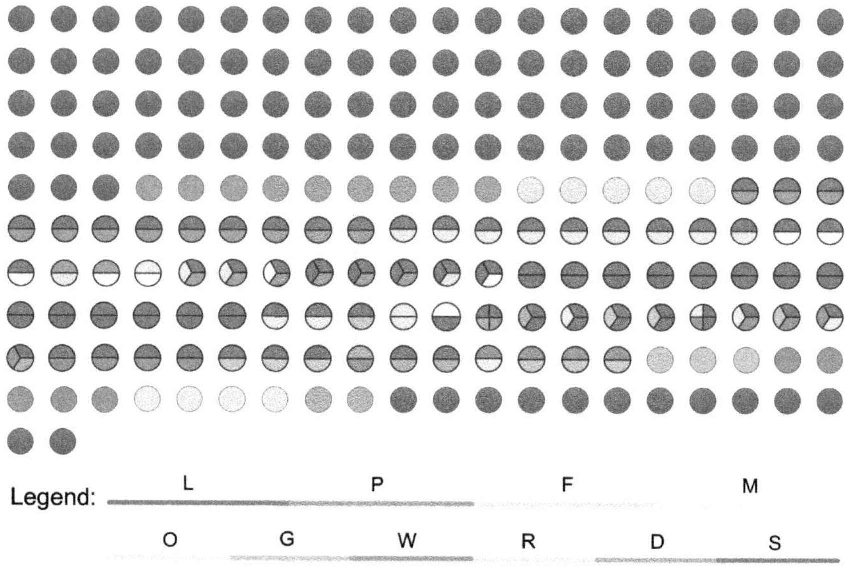

Figure 1.6 An overview of the methods used in the reviewed 202 publications. Each circle indicates one publication, and the colors represent different methods used. Direct measurement includes: weighing (W), garbage collection (G), surveys (S), diaries (D), records (R), and observation (O). Surveys also contain questionnaires, interviews and experts' estimation. Indirect measurement involves: use of literature data (L), use of proxy data (P), food balance (F), and modeling (M).
Source: Adopted from Xue, L., Liu, G., Parfitt, J., Liu, X., Van Herpen, E., Stenmarck, Å., et al., 2017. Missing food, missing data? A critical review of global food losses and food waste data. Environ. Sci. Technol. 51 (12), 6618–6633.

Only a small fraction of the publications depended on the direct measurement. In addition, for the 138 publications using literature data, they often depended on each other and some publications have been highly cited. More than one-fourth of them referred to the data from the top 10 publications cited, and the number of citations has greatly increased since 2008 (Fig. 1.7). The high percentage of using the secondary data may indicate that the available global FLW database has high uncertainties, especially when there is lack of original data for a certain country or a certain year but literature data that are not representative are used.

1.3.2.2 Advantages and disadvantages of methods

Table 1.3 lists the advantages and disadvantages of different methods based on some criteria (e.g., time, cost, and accuracy).

- Weighing and garbage collection can provide relatively detailed, objective, and accurate information of food discarded. These two methods may lead to full quantification of FLW and can produce more detailed data at the food types level. However, they can be

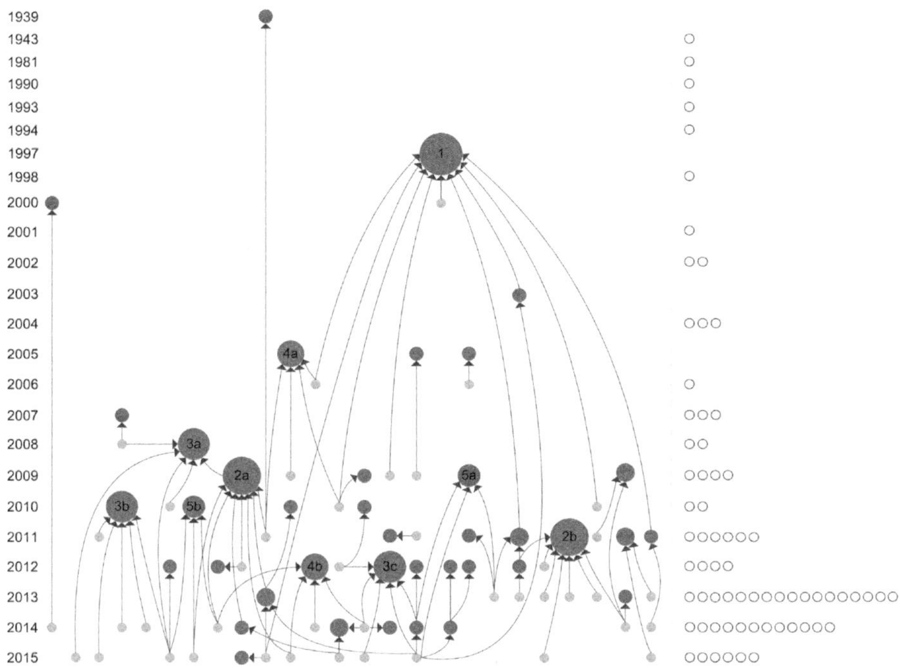

Figure 1.7 The citation network of the 138 publications that used literature data. Each dot indicates one publication. The size of the dot represents the number of citations, and the arrow represents the direction of citation. The dots in white on the right represent publications outside the citation network. The top 10 cited publications are: (1) Kantor et al. (1997); (2a) WRAP (2009); (2b) Gustavsson et al. (2011); (3a) WRAP (2008); (3b) Monier et al. (2010); (3c) Buzby and Hyman (2012); (4a) Kader (2005); (4b) Kranert et al. (2012); (5a) Buzby et al. (2009); (5b) Langley et al. (2010).
Source: Adopted from Xue, L., Liu, G., Parfitt, J., Liu, X., Van Herpen, E., Stenmarck, Å., et al., 2017. Missing food, missing data? A critical review of global food losses and food waste data. Environ. Sci. Technol. 51 (12), 6618–6633.

performed only when space available for classifying food and with device to weigh, and they are also more time-consuming and expensive than other methods. For example, a study on food waste in restaurants was conducted in four Chinese case cities (Beijing, Shanghai, Chengdu, and Lhasa) in 2015, which directly weighed food waste from 3557 tables in 195 restaurants of different categories, including lunch and dinner by individual items. It is estimated that food waste per capita in restaurants (approximately 11 kg/cap) is close to the average level of Western countries. This is a first approximation of the scales and patterns of restaurants food waste in Chinese cities, which can help inform the strategies on food waste reduction (Wang et al., 2017). In addition, the accuracy of waste composition analysis relies on the methods used, and it has identified various sources of error (Lebersorger and Schneider, 2011).

- Surveys, diaries, records, and observation are other ways of direct measuring and approximating FLW data, which consumes less time and costs more than weighing. However, due to some factors such as personal views, the way of raw data collection,

Table 1.3 Advantages and disadvantages of different methods used for food losses and food waste quantification

	Method	Symbol	Time	Cost	Accuracy	Objectivity	Reliability
Direct measurement	Weighing	W	●●●	●●●	●●●	●●●	●●●
	Garbage collection	G	●●●	●●●	●●●	●●●	●●●
	Surveys	S	●●	●●	●●	●●	●●
	Diaries	D	●●●	●●	●●	●●	●●
	Records	R	●	●	●●	●●	●●
	Observation	O	●	●	●	●	●
Indirect measurement	Modeling	M	●●	●	●	●●	●
	Food balance	F	●	●	●●	●●●	●●
	Use of proxy data	P	●	●	●●	●●●	●●
	Use of literature data	L	●	●	●●	●●●	●

and subjectivity of observers, the accuracy of the data collected may be lower. Surveys that include questionnaires can be completed by email or by phone, or by face-to-face interviews and expert estimation. But biases may occur in FLW estimation because this method depends on the memory of people and they may provide answers that the society expects. For example, Naziri et al. (2014) conducted questionnaire surveys, focus group discussion, and key expert interviews on postharvest losses of cassava during July and October 2012 in four individual developing countries (Ghana, Nigeria, Thailand, and Vietnam) to investigate the amount of losses and explore mitigation strategies. Diaries can be a heavy task for participants, and cause gradual decline of participants' enthusiasm (Langley et al., 2010), as well as difficulties in recruitment and high dropout rates (Sharp et al., 2010). In addition, keeping diaries may have influences on changes in awareness and behavior, which will lead to uncertain accuracy of the diaries (Sharp et al., 2010). For example, to analyze the composition of food waste in the United Kingdom households, Langley et al. (2010) asked 13 households to keep a diary for 7 days, recording the information on the type, origin, and weight of food waste. Records often cost less and take little time to get FLW data. Observation is a relatively quick way to estimate FLW, but the accuracy and reliability are questioned.

- Because of low cost and high feasibility, secondary data is widely used to measure the amount of FLW. But there is higher uncertainty among these methods. For modeling, the choice of model parameters and the relationship between these factors and the quantity of FLW would largely affect the results. For food balance method, the accuracy is determined by the quality and comprehensiveness of the food balance sheet data. The most cost-effective and feasible way to obtain data is by using proxy data and literature data, however, their accuracy primarily relies on the quality and representativeness of the source data used. If the data are uncertain and inaccurate, the results would also not be reliable.

In reality, no direct or indirect methods can be satisfactory. Despite the advantages, direct measurement usually involves a limited number of participants in a certain community or city and a certain stage of the food supply chain, which could lead to an unavoidable problem of deficiency of representativeness, especially for the large countries like the United States and China. On the other hand, indirect measurement can provide an overview of the entire country and various stages. A combination of direct and indirect measurement could be a better choice to illustrate the FLW problem. For policy making and mitigation strategies, based on the statistical data at the national or regional level it could determine the severity of the problem. For the design of effective intervention steps, using first-hand data and exploring the driving and influencing factors could be a good approach.

The choice of method has a significant impact on the FLW quantification, which could result in data disparity in the literature examined. For example, it was reported that the food manufacturing industry in Italy produced about 5.7 million tonnes of FLW in 2006 (Monier et al., 2010), while another study based on modeling estimated about 1.9 million tonnes of FLW for this sector (Bräutigam et al., 2014). Such big difference exists between them because they used different data sources and assumptions. The former one included FLW and recycled or reused byproducts, whereas the latter one adopted the loss rate in the manufacturing sector and the method reported by FAO (Gustavsson et al., 2013).

1.3.3 Food losses and food waste in general

1.3.3.1 Farm losses and waste

At the agriculture production stage, the FLW in low-income countries is generally higher than that in medium/high-income countries, because there is more advanced technology and infrastructure for harvesting in rich countries. For example, it is reported that FLW at this stage accounts for the largest portion (26%) of the total FLW in South Africa (Spescha and Reutimann, 2013) whereas it makes up 13% of the overall FLW across the food supply chain in Canada (Nahman and de Lange, 2013).

According to the existing data, there is little information on FLW of food commodities in the agricultural production and harvesting stage. For different food categories, on a per capita level, cereal loss is the largest with a median of roughly 16 kg/cap. For example, it is reported that about 5%–9% of cereal was lost at this stage in China, and a similar trend can be seen in Ghana (World Bank, 2011). Fruits and vegetables are the second largest wasted category at this stage with a value of 13 kg/cap. However, there is a significant difference of fruit and vegetable losses/waste between less developed and industrialized countries. For example, fruit and vegetable FLW made up about 20%–30% of the total production in China (Liu, 2014) while it accounted for only 6%–15% at this stage in Italy (Segrè et al., 2014). The reason for the big difference is that more advanced and newer technologies are used in developed countries. There is a small farm FLW of meat and fish, dairy products, and eggs at the production level (Fig. 1.8).

1.3.3.2 Postharvest losses and waste

Postharvest FLW occurs during the postharvest handling and storage, manufacturing, distribution, and retailing stages, where distinctive characteristics can be seen

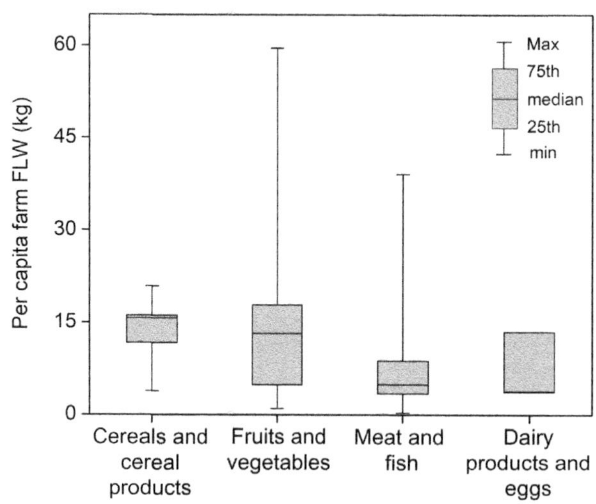

Figure 1.8 Per capita farm food losses and food waste of different food commodities.

for different types of food. Fig. 1.9 shows postharvest FLW for different food commodities based on the existing data.

- As to cereals and cereal products, the postharvest FLW varies largely at different stages. Most FLW occurs at the postharvest handling and storage stage with a median value of over 18 kg/cap in developing countries. For instance, in South and Southeast Asia, cereals have the largest postharvest FLW in all food commodities. In particular, the postharvest FLW rate of rice, which is the staple food in Philippines, was 10% (FAO, 2014). Then the retailing stage is the second highest (over 10 kg/cap), followed by the manufacturing and distribution stages (5 kg/cap).
- Fruits and vegetables have the largest postharvest FLW among all the food groups on a per capita level. For example, the fruits and vegetables FLW at the manufacturing stage was more than 33 kg/cap in South Africa (Nahman and de Lange, 2013), which was far higher than that of all other food types or stages. It should be noted that there is less FLW at the manufacturing stage in developed countries, for example, it is only about 5 kg/cap in Denmark (Smil, 1981). The FLW at the distribution stage is approximately 17 kg/cap. The FLW at the retailing stage is small with a median value of 3 kg/cap.
- Meat and fish products have the least FLW during postharvest stages. There is little information for their postharvest handling and storage FLW. Based on the existing data, the FLW of meat and fish products at this stage is very small (about 0.3 kg/cap). FLW is similar for the manufacturing and retailing stages, both roughly with a median value of 1.3 kg/cap. For example, it is reported that the FLW rates of meat were 0.2%, 5%, and 0.5% for postharvest handling and storage, manufacturing, and distribution stages (Holm, 2013).
- There are also few studies for the quantification of postharvest FLW of dairy products and eggs. For the four substages, the median FLW is at approximately 6, 3, 0.2, and 3.4 kg/cap, respectively. Due to the poor cooling systems, the FLW of milk in the manufacturing and distribution stages was 3%–15% and 8%–11% in Ukraine, respectively (Holm, 2013).

It should be noted that the retailing FLW in the United States attracts a special attention in the literature. It is reported that roughly 2.4 million tonnes of food was lost at the retailing stage in 1995, but it rose to 19.5 million tonnes in 2010, accounting for 10% of the available food supply in the United States (WRAP, 2008). Fig. 1.10 shows that cereals and cereal products, vegetables, and fruits have the greatest contribution to the retailing FLW, with a median value of 10.5, 8, and 6 kg per capita, respectively. For example, some studies point out that the cereal products FLW at the retailing stage makes up 12% of the United States food supply (Buzby and Hyman, 2012). It is also estimated that the FLW of fresh fruits, vegetables, and meat and seafood in the supermarket were 11.4%, 9.7%, and 4.5% on average, respectively (Buzby et al., 2009) in 2005/2006, which is consistent with the estimates of the other developed countries. These data suggest that fresh products dominate in the retailing FLW because of factors like expired shelf dates, overstocking, product damage and quality problems, and inappropriate inventory rotation (Kantor et al., 1997). But FLW at the retailing stage in developed countries, including the United States, mostly takes place in the supermarkets rather than street markets and nonsupermarkets (often found in developing countries).

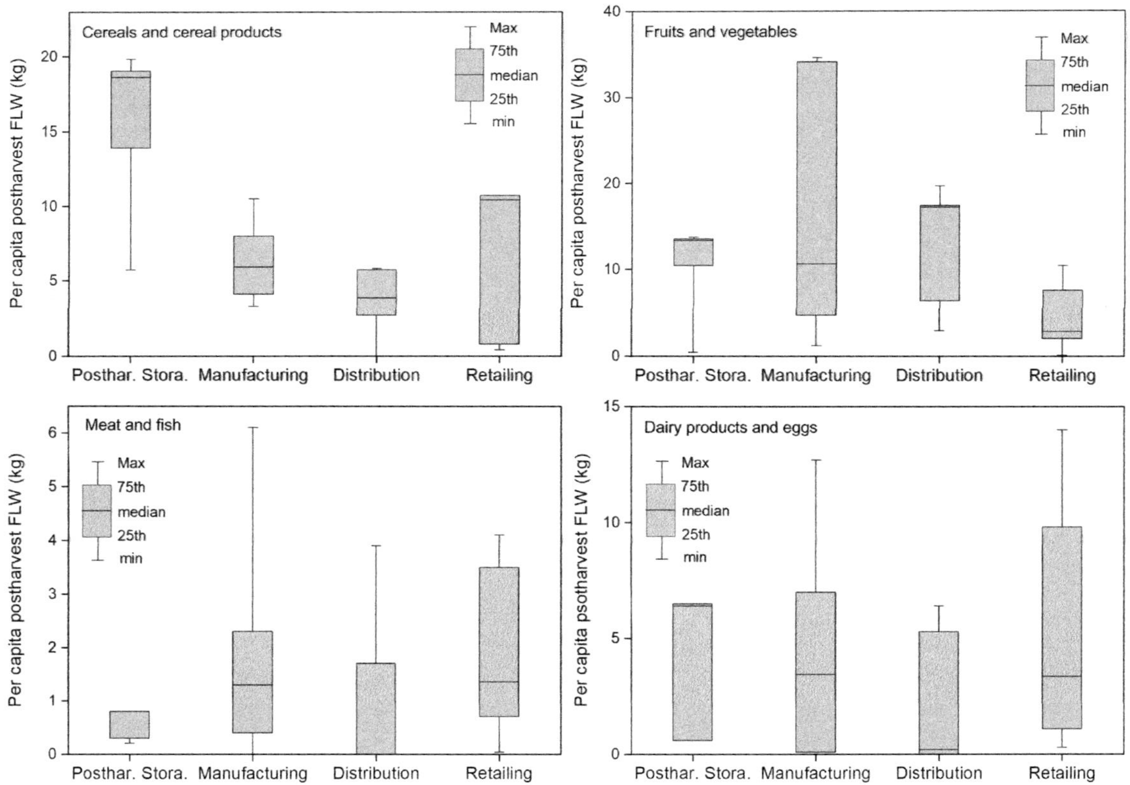

Figure 1.9 Per capita postharvest food losses and food waste at different stages of cereals and cereal products, fruits and vegetables, meat and fish, and dairy products and eggs.

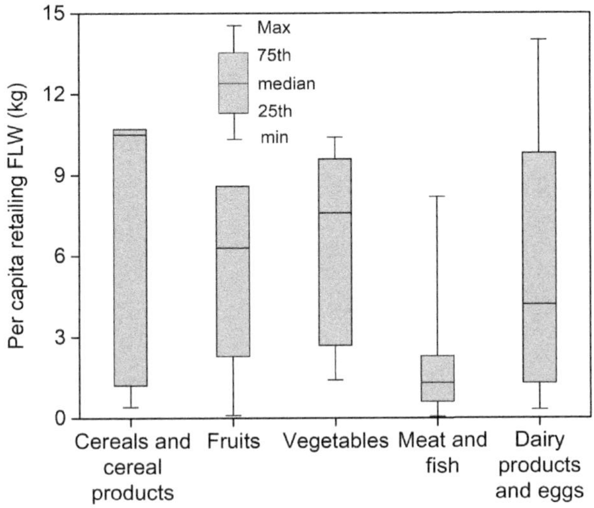

Figure 1.10 Per capita food losses and food waste of different food commodities at retailing stage in the United States.

1.3.3.3 Consumer food waste

- *Household food waste*

Because of limited disposable household income, low-income countries normally waste little food in households. Food waste in households comprises the largest portion of the total FLW in medium/high-income countries, mainly due to poor purchase planning, excessive cooking, overstocking, or misunderstanding the "best before" and "use by" dates (Koivupuro et al., 2012). In the Europe Union, about 38 million tonnes or 42% of the total food was wasted in households, with an average of about 76 kg/cap (FUSIONS, 2015). In Canada, household food waste accounted for 51% of the total FLW along the food supply chain (Gooch et al., 2010). There was about 19% of food and drink bought into households in the United Kingdom, which also represented 70% of the total FLW at postharvest stages and consumption stage (WRAP, 2013). However, according to the existing studies, there is little first-hand data for households in emerging and developing countries. The FLW in households may be much larger than expected, especially in urban areas. Since there is lack of field research in these countries, generalization should be undertaken with caution.

Fig. 1.11 shows a positive relationship between per capita GDP and per capita household food waste. When per capita GDP goes up, the per capita food waste generation from households also sees the same trend. Some previous studies also indicate the same pattern (Holm, 2013). For example, it was estimated that in 2007, food waste produced in South African households was only 7.3 kg per capital (or 0.35 million tonnes in total) (Oelofse and Nahman, 2013) while households in the United Kingdom generated 109.3 kg/cap (or 6.7 million tonnes in total) (Lee et al., 2010).

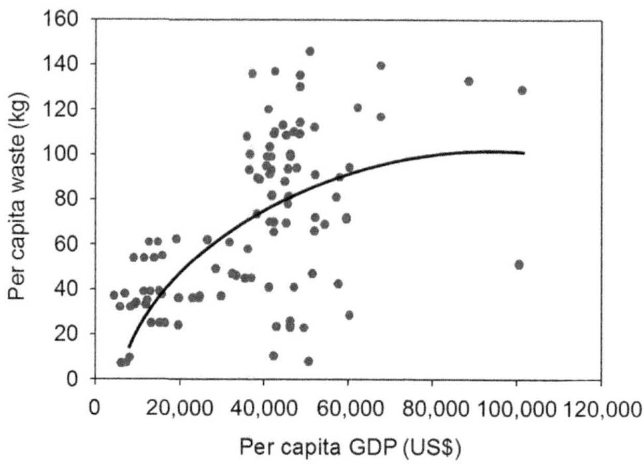

Figure 1.11 Relationship between per capita gross data product and per capita household food waste.
Source: Adopted from Xue, L., Liu, G., Parfitt, J., Liu, X., Van Herpen, E., Stenmarck, Å., et al., 2017. Missing food, missing data? A critical review of global food losses and food waste data. Environ. Sci. Technol. 51 (12), 6618–6633.

However, interestingly, when per capita GDP reaches above a certain level (about USD 50,000), the per capita food waste generation tends to be stable. This might relate to the growing awareness of the public, food waste prevention initiatives, and the impact of market mechanisms (e.g., increasing the price of food and disposal cost of food waste). For example, Australia (Thi et al., 2015) and the United Kingdom (Quested et al., 2011) have effectively taken some campaigns such as "Zero Waste" and "Love Food Hate Waste" against food waste. As a result, FLW from households has reduced by 21% between 2007 and 2012 in the United Kingdom. On the other hand, it may also involve more prepared food consumed and less cooking from scratch in rich countries, which may transfer food waste from household to food industry to some degree.

- *Out-of-home food waste*

Many researchers have studied the situation of food waste outside the home, that is, in the food service industry, including for example restaurants (Papargyropoulou et al., 2014), canteens (Halloran et al., 2014), schools (Okazaki et al., 2008), hospitals (Dias-Ferreira et al., 2015), care centers (Silvennoinen et al., 2015), military institutions (Davies and Konisky, 2000), and in-flight (Li et al., 2003), anywhere responsible for preparing or providing food away from home.

Most research on out-of-home FLW is carried out in the industrialized countries. For example, it was reported that 0.92 million tonnes of food was wasted in this sector every year in the United Kingdom (Parry et al., 2015). In Germany, this sector was the second largest source of food waste, accounting for 17% of the total FLW along the food supply chain (Kranert et al., 2012). In Finland, it was the third largest contributor of FLW (20%) with 0.075–0.085 million tonnes of food wasted, following household (35%) and food industry (27%) (Silvennoinen et al., 2012).

It should be noted that China, as the largest emerging economy in the world, was also facing a high level of food waste in the catering industry, making up about 11%–17% of all food ordered (Liu, 2014). For example, it was reported the total quantity of Horeca (hotels, restaurants, and cafés) food waste in Lhasa (western China, with lower income compared with Western countries) has reached a high level of FLW. However, due to strict regulations issued recently (e.g., the public expense for official extravagance and governmental reception meals), on a per capita level, Horeca food waste generation in Lhasa has decreased from 128 to 98 g/cap/meal during 2011 and 2015 (Wang et al., 2018). Another pilot study focused on the situation of plate waste in school lunch programs in Beijing; it was reported that the average amount of food waste generated by students in Beijing was 130 g/cap/meal in 2014, making up 21% of the total food served. Food supply models, the quality of canteen service, eating habits, and students' knowledge of agricultural production were the main driving factors that influence plate waste (Liu et al., 2016).

In general, food waste per capita out-of-home is lower than that in households (Fig. 1.12). Assuming that as the per capita GDP and living standards increase, people would consume more food out-of-home, it may bring greater food waste for a variety of reasons, such as oversized dishes and taste. However, the relationship between per capita GDP and per capita food waste outside households seems not that significant. The reason can be explained by the fact that the food service sector

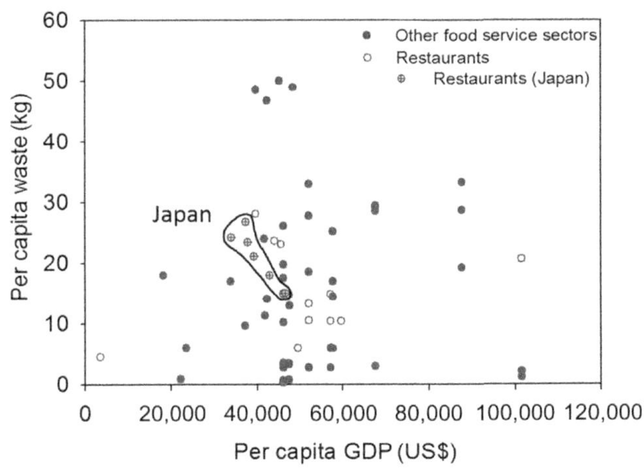

Figure 1.12 Relationship between per capita gross data product and per capita out-of-home food waste. It differentiates restaurants (empty circles) and other food service sectors (e.g., canteens; filled circles), and the circles with a cross enclosed are for restaurants in Japan. Source: Adopted from Xue, L., Liu, G., Parfitt, J., Liu, X., Van Herpen, E., Stenmarck, Å., et al., 2017. Missing food, missing data? A critical review of global food losses and food waste data. Environ. Sci. Technol. 51 (12), 6618–6633.

is diverse, including the "for profit" (e.g., restaurant) and "cost" (e.g., care center) parts, which results in a mixed pattern generation of food waste. It is interesting to point out that food waste in restaurants in Japan shows a downward trend in these years. This may partly relate to the implementation of the Food Recycling Law in Japan in May 2001, which set specific targets for industry sectors to reduce food waste generation. As a result, food waste out-of-home reduced from 3.1 million tonnes in 2007 to 1.92 million tonnes in 2012 (Parry et al., 2015).

1.3.3.4 Comparison of food losses and food waste for different development levels of countries

In Fig. 1.13, we take cereal as an example to show the evolution of food waste at different stages in the supply chain and economies with different development levels, using the United States, China, and South Africa as representative of industrialized, emerging, and average developing countries. Here we take food losses

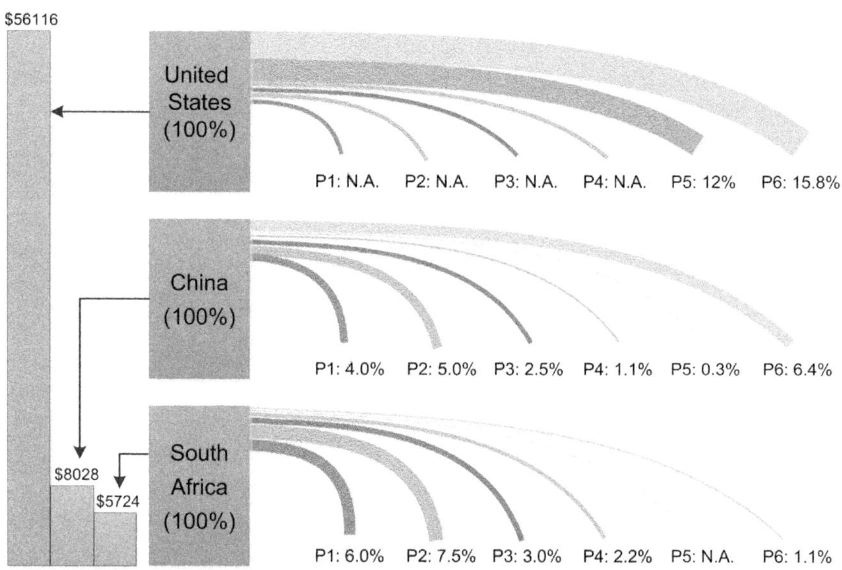

Figure 1.13 FLWR of cereals throughout the food supply chain in the United States, China, and South Africa. The vertical chart on the left represents per capita gross data product in current USD in 2015 for these three countries (according to the World Bank).
P1 = agricultural production and harvesting, P2 = postharvest handling and storage, P3 = manufacturing, P4 = distribution, P5 = retailing, P6 = consumption. N.A. means not available. The reference flow is assumed to be a fictive output of 100% of the amount produced.
Source: Adopted from Xue, L., Liu, G., Parfitt, J., Liu, X., Van Herpen, E., Stenmarck, Å., et al., 2017. Missing food, missing data? A critical review of global food losses and food waste data. Environ. Sci. Technol. 51 (12), 6618–6633.

and food waste rate (FLWR) as an indicator to illustrate the situation, which represents the share of FLW to the total agriculture production.

- As a highly industrialized country, there is little information for cereal losses during postharvest stages in the United States. It could be assumed to be low. For South Africa, the FLWR at agricultural production, postharvest handling and storage, manufacturing, and distribution stages are all higher than those in China. This indicates that as awareness increases and the economy grows, China has adopted more advanced technologies and efficient storage systems in agricultural production, and has largely used improved transportation with large volumes (Liu et al., 2013). This also means it could be an efficient way to reduce FLW by improving the technologies and infrastructure in developing countries such as South Africa.
- With the development of the country and the increase of GDP, cereal waste at the consumption stage also increases. The FLWR of cereals at this stage in the United States is the highest (15.8%) among the three countries, which is around 2.5 and 14.4 times that in China and South Africa, respectively. In China, with the rapid development of economy and the improvement of living standards, the FLWR of cereals has increased to 6.4% in recent years, which is higher than that of all the other stages. For the average developing country, the FLWR of cereals at consumption stage is low in South Africa (1.1%). It should be pointed out that different countries have different production and consumption patterns of cereals, which may contribute to the differences among these countries.

1.4 Implications for future

The study suggests that FLW have attracted more attention, with more than 60% of FLW data reported in recent years. Although they provide an overview of the scale of FLW globally, for a few industrialized countries, and different food supply chain stages such as household, there are still data gaps and deficiencies as to the magnitude of FLW in developing countries (e.g., China and India) that have undergone rapid dietary transformation from starchy staples towards more diverse and fresh food (Parfitt, 2013).

In this case, we list some directions for future study as follows (Fig. 1.14):

First, the systems and methodologies for FLW quantification should be harmonized. It is important to consider these aspects: the definition of FLW [e.g., avoidable and unavoidable food waste (Östergren et al., 2013), food for human consumption vs nonhuman consumption], food supply chain stages (e.g., different segments at consumption), the classification and conversion factors of food commodities (e.g., procedure to convert cooked food items to different categories of raw food products), the treatment of FLW (e.g., donation, incineration, feed production, or landfill), the measurement units (e.g., physical weight, calories, or percentages), and the measurement methods. This would help to compare the available data among countries, food commodities, and stages in the food supply chain, which will further enable exploration of driving factors and patterns of FLW generation. For example, the recent released global Food Loss and Food Waste Protocol (World Resources Institute, 2016) in 2016 is an excellent first step, providing a

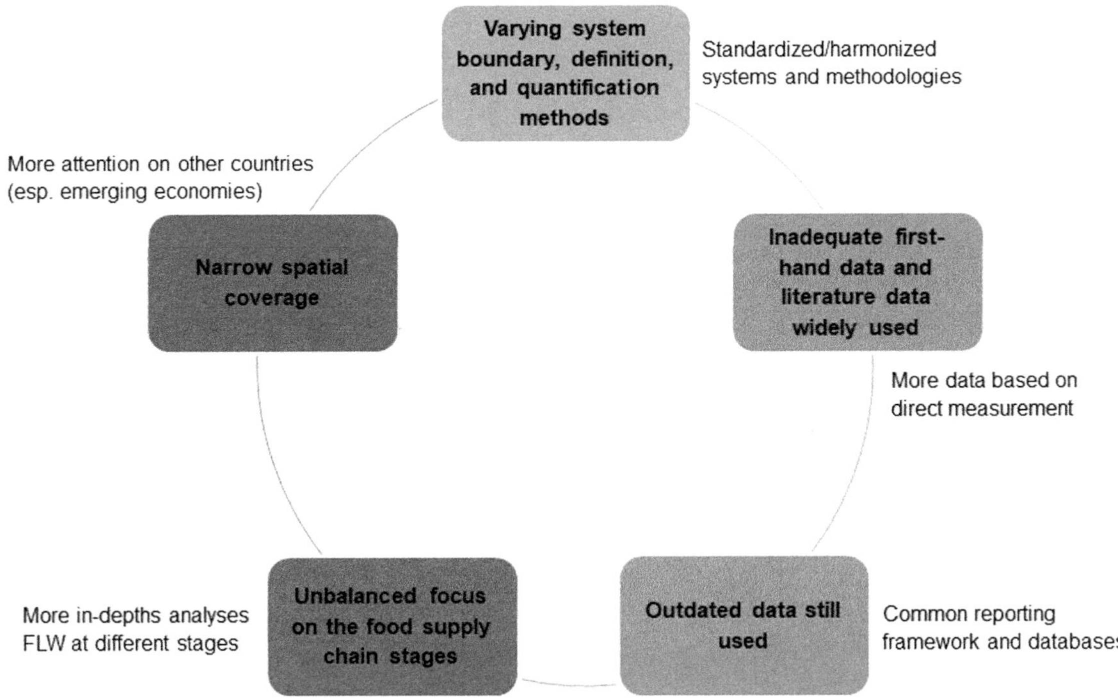

Figure 1.14 Gaps and way forward of the existing global food losses and food waste database.

standard that can be used by any entity and should be promoted more largely. More efforts are needed to further refine and implement these kinds of harmonized methodologies.

Second, there is an urgent need for more data based on first-hand sources. The results show that only about 20% of the existing publications relied on first-hand data. Depending on the unrepresentative data from literature largely may result in high uncertainties. Although the time, labor, and economic cost are high, more field work and collection of first-hand data should be encouraged, which could help verify the existing data, improve the accuracy and reliability of the data, and fill in the gaps where data are not available.

Third, it should focus on the regions that are experiencing rapid development and emerging economies, such as the BRICs, that is, Brazil, Russia, India, China, and South Africa, other than the current hotspot areas (e.g., United States and Europe). There is less information on the quantification of FLW in those developing countries, but the situation may be serious; for example, a report shows that food waste at the consumer level in China is higher than that of the total in EU27 (Liu, 2014). Those countries are also undergoing rapid changes in diet structure, urbanization, and growing household income, which might bring a higher FLW in the future. Relying on the outdated data may result in overestimating FLW at agricultural production stage and underestimating food waste at consumption stage in developing countries (Liu, 2014; Shafiee-Jood and Cai, 2016). In addition, when more data are available for specific countries or cultures, it is better to consider social and cultural background in the FLW quantification and mitigation.

Fourth, deeper analyses should be conducted on FLW at different stages in the food supply chain. The results show that about half of the existing studies focused on the household food waste mainly in developed countries. More attention should be paid to the stages that have less data and are poorly understood, for example, FLW out-of-home (e.g., canteens and restaurants) and postharvest stages in less developed countries. It would help to identify the drivers of FLW at different stages with a more detailed quantification.

Fifth, research should build and maintain consistent databases under a common reporting framework on FLW, and then make the data available to the public through joint efforts from all stakeholders in the whole food supply chain. Those databases would contribute to track the progress towards achieving SDG Target 12.3 and national targets on FLW, as they would provide a benchmark for tracking the progress of FLW reduction. The governmental and nongovernmental organizations such as the UNEP and FAO, as well as national statistical agencies, should play a strong leadership in this area. For example, the data series reported by USDA-ERS and WRAP are good models. It should encourage all related industry or industry associations to report their FLW regularly. In the long term, it is applicable to track FLW reduction by using the "measurable, reportable, and verifiable (MRV)" principle.

Last but not least, quantifying FLW is the foundation for further analysis. Better data measurement would help better understanding of the social, economic, and environmental impacts of FLW, determine hotspots that should be given priority actions, build long-term scenarios to inform the effective policy making and strategies in achieving reduction of FLW, and contribute to improve the efficiency and sustainability of the food system. It should carry out research focusing on these topics at the same time.

1.5 Conclusions

FLW has become a global concern in recent years and has also become a priority in the global and national political agenda. There has been a growing body of literature on FLW quantification in the past decade. However, there are still major gaps, such as various definitions of system boundaries and quantification methods, data deficiencies, narrow geography, and unbalanced food supply chain coverage. Most existing studies are carried out for a few developed countries (e.g., the United Kingdom and United States) and concentrate on the retailing and consumption stages (e.g., household), and more than half of them are based only on secondary data, which signals high uncertainties. The existing data indicates that at farm level, FLW in developing countries is higher than that in developed countries. Per capita fruit and vegetable FLW dominates at postharvest stages. With the increase of per capita GDP, per capita food waste from household also goes up. More standardized systems and methodology, more data based on direct measurement, more in-depth analysis of FLW at different stages, a common reporting framework, and more attention to the emerging economies, are urgently needed to properly inform relevant policy on the reduction of FLW and mitigation strategies on environmental impacts.

References

An, Y., Li, G., Wu, W., Huang, J., He, W., Zhu, H., 2014. Generation, collection and transportation, disposal and recycling of kitchen waste: a case study in Shanghai. Waste Manage. Res. 32 (3), 245–248.

Beretta, C., Stoessel, F., Baier, U., Hellweg, S., 2013. Quantifying food losses and the potential for reduction in Switzerland. Waste Manage. 33 (3), 764–773.

Bräutigam, K.-R., Jörissen, J., Priefer, C., 2014. The extent of food waste generation across EU-27: different calculation methods and the reliability of their results. Waste Manage. Res. 32 (8), 683–694.

Buzby, J.C., Guthrie, J.F., 2002. Plate Waste in School Nutrition Programs: Final Report to Congress. Economic Research Service E FAN-02-009, United States Department of Agriculture, Washington, DC.

Buzby, J.C., Hyman, J., 2012. Total and per capita value of food loss in the United States. Food Policy 37 (5), 561–570.

Buzby, J.C., Wells, H.F., Axtman, B., Mickey, J., 2009. Supermarket Loss Estimates for Fresh Fruit, Vegetables, Meat, Poultry, and Seafood and Their Use in the ERS Loss-Adjusted Food Availability Data; Economic Information Bulletin Number 44. Economic Research Service, United States Deparment of Agriculture, Washington, DC.

Choudhury, M.L., 2006. Recent developments in reducing postharvest losses in the Asia-Pacific region. In: Rolle, R.S. (Ed.), Postharvest Management of Fruit and Vegetables in the Asia-Pacific Region. Asian Productivity Organization, Tokyo, Japan.

Dahlén, L., Lagerkvist, A., 2008. Methods for household waste composition studies. Waste Manage. 28 (7), 1100–1112.

Davies, T., Konisky, D.M., 2000. Environmental Implications of the Foodservice and Food Retail Industries. Resources for the Future, Washington, DC.

Dias-Ferreira, C., Santos, T., Oliveira, V., 2015. Hospital food waste and environmental and economic indicators - a Portuguese case study. Waste Manage. 46, 146–154.

Edjabou, M.E., Jensen, M.B., Götze, R., Pivnenko, K., Petersen, C., Scheutz, C., et al., 2015. Municipal solid waste composition: sampling methodology, statistical analyses, and case study evaluation. Waste Manage. 36, 12–23.

European Commission Food Safety Home Page, <http://ec.europa.eu/food/safety/food_waste/eu_actions_en/accessed> (accessed 20.12.17).

FAO, 2013. Food Wastage Footprint: Impacts on Natural Resources. FAO, Rome, Italy.

FAO, 2014. Mitigation of Food Wastage: Societal Costs and Benefits. FAO, Rome, Italy.

FAO, 2014. Definitional Framework of Food Loss. FAO, Rome, Italy.

FAO, 2015. Food Wastage Footprint & Climate Change. FAO, Rome, Italy.

FUSIONS, 2015. Food Waste Data Set for EU-28. Wageningen University Publishing, Wageningen, The Netherlands.

FUSIONS, 2016. Food Waste Quantification Manual to Monitor Food Waste Amounts and Progression. Wageningen University Publishing, Wageningen, The Netherlands.

Filho, W.L., Kovaleva, M., 2015. Food Waste and Sustainable Food Waste Management in the Baltic Sea Region. Hamburg University of Applied Sciences, Hamburg, Germany.

Gangwar, R.K., Tyagi, S., Kumar, V., Singh, K., Singh, G., 2014. Food production and post harvest losses of food grains in India. Food Sci. Qual. Manage. 31, 48–53.

Gjerris, M., Gaiani, S., 2013. Household food waste in Nordic countries: estimations and ethical implications. Nord. J. Appl. Ethics 7 (1), 6–23.

Gooch, M., Cut Waste, Grow Profit: How to Reduce and Manage Food Waste, Leading to Increased Profitability and Environmental Sustainability; Value Chain Management Centre, George Morris Centre: Oakville, Ontario, 2012. <http://www.valuechains.ca/usercontent/documents/Cut Waste Grow Profit FINAL DOCUMENT Oct3 12.pdf>.

Gooch, M., Felfel, A., Marenick, N., 2010. Food Waste in Canada. Value Chain Management Centre, Oakville, Ontario, Canada.

Grandhi, B., Appaiah Singh, J., 2016. What a waste! A study of food wastage behavior in Singapore. J. Food Prod. Mark. 22, 471–485.

Gustavsson, J., Cederberg, C., Sonesson, U., Otterdijk, R., van; Meybeck, A., 2011. Global Food Losses and Food Waste: Extent, Causes and Prevention. FAO, Rome, Italy.

Gustavsson, J., Cederberg, C., Sonesson, U., Emanuelsson, A., 2013. The Methodology of the FAO Study: "Global Food Losses and Food Waste - Extent, Causes and Prevention"- FAO, 2011. The Swedish Institute for Food and Biotechnology, Gothenburg, Sweden.

Gutiérrez-Barba, B.E., Ortega-Rubio, A., 2013. Household food-waste production and a proposal for its minimization in Mexico. Life Sci. J. 10 (3), 1772–1783.

Hall, K.D., Guo, J., Dore, M., Chow, C.C., 2009. The progressive increase of food waste in America and its environmental impact. PLoS One 4 (11), e7940.

Halloran, A., Clement, J., Kornum, N., Bucatariu, C., Magid, J., 2014. Addressing food waste reduction in Denmark. Food Policy 49, 294–301.

Holm, T., 2013. Reduction of FLW in Europe and Central Asia. Synthesis Report prepared for Food and Agriculture Organization of the United Nations, Regional Office for Europe and Central Asia REU, Budapest, Hungary.

Jörissen, J., Priefer, C., Bräutigam, K.-R., 2015. Food waste generation at household level: results of a survey among employees of two European research centers in Italy and Germany. Sustainability 7 (3), 2695–2715.

Kader, A.A., 2005. Increasing food availability by reducing postharvest losses of fresh produce. ISHS Acta Hortic. 682, 2169–2175.

Kantor, L.S., Lipton, K., Manchester, A., Oliveira, V., 1997. Estimating and addressing America's food losses. Food Rev 20 (1), 2–12.

Katajajuuri, J.M., Silvennoinen, K., Hartikainen, H., Jalkanen, L., Koivupuro, H.K., Reinikainen, A., 2012. Food waste in the food chain and related climate impacts. In: Corson, M.S., van der Werf, H.M.G. (Eds.), Proceedings of the 8th International Conference on Life Cycle Assessment in the Agri-Food Sector LCA Food 2012. INRA, Rennes, France, pp. 627–632.

Khan, M.Z.A., Burney, F.A., 1989. Forecasting solid waste composition — an important consideration in resource recovery and recycling. Resour. Conserv. Recycl. 3 (1), 1–17.

Koivupuro, H.-K., Hartikainen, H., Silvennoinen, K., Katajajuuri, J.-M., Heikintalo, N., Reinikainen, A., et al., 2012. Influence of socio-demographical, behavioural and attitudinal factors on the amount of avoidable food waste generated in Finnish households. Int. J. Consum. Stud. 36 (2), 183–191.

Kranert, M., Hafner, G., Barabosz, J., Schneider, F., Lebersorger, S., Scherhaufer, S., et al., 2012. Determination of Discarded Food and Proposals for a Minimization of Food Wastage in Germany. Institute for Sanitary Engineering, University of Stuttgart, Stuttgart, Germany.

Langley, J., Yoxall, A., Heppell, G., Rodriguez, E.M., Bradbury, S., Lewis, R., et al., 2010. Food for thought?—a UK pilot study testing a methodology for compositional domestic food waste analysis. Waste Manage. Res. 28 (3), 220–227.

Lebersorger, S., Schneider, F., 2011. Discussion on the methodology for determining food waste in household waste composition studies. Waste Manage. 31 (9–10), 1924–1933.

Lee, P., Willis, P., Hollins, O., 2010. Waste Arisings in the Supply of Food and Drink to Households in the UK. Waste and Resources Action Programme WRAP, Banbury, UK.

Li, X.D., Poon, C.S., Lee, S.C., Chung, S.S., Luk, F., 2003. Waste reduction and recycling strategies for the in-flight services in the airline industry. Resour. Conserv. Recycl. 37 (2), 87–99.

Lipinski, B., Hanson, C., Lomax, J., Kitinoja, L., Waite, R., Searchinger, T., 2013. Reducing Food Loss and Waste Creating a Sustainable Food Future, Installment Two. World Resources Institute and United Nations Environment Programme, Washington, DC.

Lipinski, B., O'Connor, C., Hanson, C., 2016. SDG Target 12.3 on Food Loss and Waste: 2016 Progress Report; Champions 12.3: The Hague, The Netherlands. <https://champs123blog.files.wordpress.com/2016/09/sdg-target-12-3-progress-report_2016.pdf>.

Liu, G., 2014. Food Losses and Food Waste in China: A First Estimate. OECD Food, Agriculture and Fisheries Papers, No. 66, OECD Publishing, Paris, France.

Liu, J., Lundqvist, J., Weinberg, J., Gustafsson, J., 2013. Food losses and waste in China and their implication for water and land. Environ. Sci. Technol. 47 (18), 10137–10144.

Liu, Y., Cheng, S., Liu, X., Cao, X., Xue, L., Liu, G., 2016. Plate waste in school lunch programs in Beijing, China. Sustainability 8 (12), 1288.

Mena, C., Terry, L.A., Ellram, L., 2014. Causes of waste across multi-tier supply networks: cases in the UK food sector. Int. J. Prod. Econ. 152, 144–158.

Monier, V., Mudgal, S., Escalon, V., O'Connor, C., Gibon, T., Anderson, G., et al., 2010. Preparatory Study on Food Waste Across EU27. European Commission. DG Environment - Directorate C, Brussels, Belgium.

Nahman, A., de Lange, W., 2013. Costs of food waste along the value chain: evidence from South Africa. Waste Manage. 33 (11), 2493–2500.

Naziri, D., Quaye, W., Siwoku, B., Wanlapatit, S., Viet, T., Bennett, B., 2014. The diversity of postharvest losses in cassava value chains in selected developing countries. J. Agric. Rural Dev. Trop. Subtrop. 115 (2), 111–123.

Oelofse, S.H.H., Nahman, A., 2013. Estimating the magnitude of food waste generated in South Africa. Waste Manage. Res. 31 (1), 80–86.

Okazaki, W.K., Turn, S.Q., Flachsbart, P.G., 2008. Characterization of food waste generators: a Hawaii case study. Waste Manage. 28 (12), 2483–2494.

Östergren, K., Anderson, G., Easteal, S., Gustavsson, J., Hansen, O.J., Moates, G., et al., Food waste prevention: the challenge of making appropriate definitional and methodological choices for quantifying food waste levels. Presented at the 6th International Conference on Life Cycle Management LCM, Gothenburg, Sweden, August 25 – 28, 2013.

Östergen, K., Gustavsson, J., Bos-Brouwers, H., Timmermans, T., Hansen, O.-J., Møller, H., et al., 2014. FUSIONS Definitional Framework for Food Waste. Wageningen University Publishing, Wageningen, The Netherlands.

Papargyropoulou, E., Padfield, R., Rupani, P.F., Zakaria, Z., 2014. Towards sustainable resource and waste management in developing countries : the role of commercial and food waste in Malaysia. Int. J. Waste Resour. 4 (3), 2–7.

Parfitt, J., 2013. Global Food Waste Campaigns Suffer From Data Deficiency. Guardian Professional, London, UK.

Parfitt, J., Barthel, M., 2011. Global Food Waste Reduction: Priorities for a World in Transition. UK Government's Foresight Project on Global Food and Farming Futures, London, UK.

Parfitt, J., Barthel, M., Macnaughton, S., 2010. Food waste within food supply chains: quantification and potential for change to 2050. Philos. Trans. R. Soc. B Biol. Sci. 365 (1554), 3065–3081.

Parry, A., Bleazard, P., Okawa, K., 2015. Preventing Food Waste: Case Studies of Japan and the United Kingdom. OECD Food Agriculture & Fisheries Papers, No. 76, OECD Publishing, Paris, France.

Pham, T.P.T., Kaushik, R., Parshetti, G.K., Mahmood, R., Balasubramanian, R., 2014. Food-waste-to-energy conversion technologies: current status and future directions. Waste Manage. 38, 399–408.

Prusky, D., 2011. Reduction of the incidence of postharvest quality losses, and future prospects. Food Secur. 3 (4), 463–474.

Quested, T.E., Parry, A.D., Easteal, S., Swannell, R., 2011. Food and drink waste from households in the UK. Nutr. Bull. 36 (4), 460–467.

Rathje, W.L., Murphy, C., 2001. Rubbish!: The Archaeology of Garbage. University of Arizona Press, New York.

Refresh Home Page, <http://eu-refresh.org/about-refresh/> (accessed 20.12.17).

Saccares, S., Scognamiglio, U., Moroni, C., Marani, A., Calcaterra, V., Amendola, M., et al., 2014. Evaluation model of plate waste to monitor food consumption in two different catering settings. Ital. J. Food Saf. 3 (2), 1–8.

Scholz, K., Eriksson, M., Strid, I., 2015. Carbon footprint of supermarket food waste. Resour. Conserv. Recycl. 94, 56–65.

Segrè, A., Falasconi, L., Politano, A., Vittuari, M., 2014. Background Paper on the Economics of Food Loss and Waste. FAO, Rome, Italy.

Shafiee-Jood, M., Cai, X., 2016. Reducing food loss and waste to enhance food security and environmental sustainability. Environ. Sci. Technol. 50 (16), 8432–8443.

Sharp, V., Giorgi, S., Wilson, D.C., 2010. Methods to monitor and evaluate household waste prevention. Waste Manage. Res. 28 (3), 269–280.

Silvennoinen, K., Katajajuuri, J.M., Hartikainen, H., Jalkanen, L., Koivupuro, H.K., Reinikainen, A., 2012. Food waste volume and composition in the Finnish supply chain: special focus on food service sector. Fourth International Symposium on Energy from Biomass and Waste. CISA Publisher, Venice, Italy.

Silvennoinen, K., Heikkilä, L., Katajajuuri, J.-M., Reinikainen, A., 2015. Food waste volume and origin: case studies in the Finnish food service sector. Waste Manage. 46, 140–145.

Smil, V., 1981. China's food: availability, requirements, composition, prospects. Food Policy 6 (2), 67–77.

Spescha, G., Reutimann, J., 2013. Reducing Food Waste - A Hidden Opportunity for Investors. <http://www.inrate.com/Inrate/media/Documents/Sustainability Matters/13-01-10-SustainabilityMatters.pdf>.

Stefan, V., van Herpen, E., Tudoran, A.A., Lähteenmäki, L., 2013. Avoiding food waste by Romanian consumers: the importance of planning and shopping routines. Food Qual. Prefer. 28 (1), 375–381.

Stenmarck, Å., Hanssen, O.J., Silvennoinen, K., Katajajuuri, J.-M., Werge, M., 2011. Initiatives on Prevention of Food Waste in the Retail and Wholesale Trades. Nordic Council of Ministers, Copenhagen, Denmark.

The Economist Intelligence Unit, 2014. Food Loss and Its Intersection With Food Security. EIU, London, United Kingdom, <http://www.eiu.com/public/topical_report.aspx?campaignid = foodloss14>.

Thi, N.B.D., Kumar, G., Lin, C.-Y., 2015. An overview of food waste management in developing countries: current status and future perspective. J. Environ. Manage. 157, 220–229.

Thyberg, K.L., Tonjes, D.J., Gurevitch, J., 2015. Quantification of food waste disposal in the United States: a meta-analysis. Environ. Sci. Technol. 49 (24), 13946–13953.

United Nations. United Nations sustainability development goals home page. <http://www.un.org/sustainabledevelopment/sustainableconsumption-production/> (accessed 20.12.17).

United States Department of Agriculture. USDA and EPA Join with Private Sector, Charitable Organizations to Set Nation's First Food Waste Reduction Goals. <https://www.usda.gov/media/press-releases/2015/09/16/usda-and-epa-join-private-sector-charitableorganizations-set> (accessed 20.12.17).

Vanham, D., Bouraoui, F., Leip, A., Grizzetti, B., Bidoglio, G., 2015. Lost water and nitrogen resources due to EU consumer food waste. Environ. Res. Lett. 10 (8), 084008.

Verghese, K., Lewis, H., Lockrey, S., Williams, H., 2013. The Role of Packaging in Minimising Food Waste in the Supply Chain of the Future. RMIT University, Melbourne.

WRAP, 2008. The Food We Waste. Waste and Resources Action Programme WRAP, Banbury, UK.

WRAP, 2009. Household Food and Drink Waste in the UK 2009. Waste and Resources Action Programme WRAP, Banbury, UK.

WRAP, 2013. Household Food and Drink Waste in the United Kingdom 2012. Waste and Resources Action Programme WRAP, Banbury, UK.

WRAP, 2014. Household Food and Drink Waste: A Product Focus. Waste & Resources Action Programme WRAP, Banbury, UK.

Wang, L., Liu, G., Liu, X., Liu, Y., Gao, J., Zhou, B., et al., 2017. The weight of unfinished plate: a survey based characterization of restaurant food waste in Chinese cities. Waste Manage. 66, 3–12.

Wang, L., Xue, L., Li, Y., Liu, X., Cheng, S., Liu, G., 2018. Horeca food waste and its ecological footprint in Lhasa, Tibet, China. Resour. Conserv. Recycl 136, 1–8.

World Bank, 2011. Missing Food: The Case of Postharvest Grain Losses in Sub-Saharan Africa. The International Bank for Reconstruction and Development, The World Bank, Washington, DC.

World Resources Institute, 2016. Food Loss and Waste Accounting and Reporting Standard. WRI, Washington, DC, <http://www.wri.org/sites/default/files/FLW_Standard_final_2016.pdf>.

Xu, Z., Sun, D.-W., Zhang, Z., Zhu, Z., 2015. Research developments in methods to reduce carbon footprint of cooking operations: a review. Trends Food Sci. Technol. 44 (1), 49–57.

Xue, L., Liu, G., Parfitt, J., Liu, X., Van Herpen, E., Stenmarck, Å., et al., 2017. Missing food, missing data? A critical review of global food losses and food waste data. Environ. Sci. Technol. 51 (12), 6618–6633.

Soil and crop management to save food and enhance food security

2

Tiziano Gomiero
Independent scholar, Treviso, Italy

Chapter Outline

2.1 Introduction: enhancing food security by reducing yield loss

The coming decades will present a major challenge for the human population. Managing water, energy, and food procurement to feed the present and future population will call for our utmost ingenuity and wisdom. Although the overall population growth rate is decreasing, the population is still growing, especially in Asia and Africa. The present population of 7.5 billion (of which about 900 million are still undernourished) is expected to reach 8.5 billion in 2030 and 10 billion in 2050 (Alexandratos and Bruinsma, 2012; UN, 2017). Increasing food consumption per capita, and particularly meat intake (Smil, 2013; Godfray et al., 2018), will pose further pressure on natural resources (i.e., water, soil, energy) and exacerbate

Saving Food. DOI: https://doi.org/10.1016/B978-0-12-815357-4.00002-X

human impact on the environment (i.e., agrochemicals, greenhouse gas emissions (GHGs), biodiversity loss) (Pretty, 2008; Godfray et al., 2010, 2018; FAO, 2011a,b; Foley et al., 2011; Gomiero et al., 2011a; Alexandratos and Bruinsma, 2012; Gomiero, 2016; Campbell et al., 2017). It has been argued that, to meet food demand, in 2050 global agricultural production may have to increase by 70%– 110% (Bruinsma, 2003, 2011; Tilman et al., 2011; Alexandratos and Bruinsma, 2012).

Although in the last decades yields have increased dramatically, food loss and waste are still extremely high.

Food losses refer to the decrease in edible food mass throughout the part of the supply chain that specifically leads to edible food for human consumption (Parfitt et al., 2010; FAO, 2011a). FAO (2014a) defined food loss also as "the decrease in **quantity or quality** of food reflected in nutritional value, economic value or food safety of all food produced for human consumption but not eaten by humans" (bold added by the author). Therefore, in addition to yield, expressed in biomass harvested per ha, the nutritional content of produce has also to be addressed. The term *food waste* refers to food losses occurring at the end of the food chain (retail and final consumption), and relates to retailers' and consumers' behavior (Stuart, 2009; Parfitt et al., 2010; FAO, 2011a). Parfitt et al. (2010) stated that addressing moral and economic dimensions of food may lead the following to be included as food loss: crops diverted into feeding livestock, biofuels (see also Gomiero, 2015a), or biomaterials production.

It has been estimated that 30%–40% of all food harvested is lost or wasted each year (Stuart, 2009; Parfitt et al., 2010; FAO, 2011a; Royte, 2016); these estimates may vary greatly depending on the specific crops, locations, and situations involved (Parfitt et al., 2010). The figures are nevertheless indicative of a very significant issue, and reducing food losses is a key step to saving food.

Crop yields lower than potentially achievable can also be considered as food loss. Crops can perform poorly for a number of reasons, for example, weather extremes, pests, and poor agricultural practices. Poor agricultural practices such as monoculture, failing to implement proper crop rotations, intensive use of inputs, and poor water management eventually lead to soil degradation (i.e., reduced fertility and soil erosion), accumulation of toxic compounds in the soil, reduced nutritional content of produce, and a weakening of plant defenses, which in turn facilitates pest attack.

Therefore, to sustain food production in the long run, it is necessary to adopt agricultural practices that preserve soil and crop health. This also in view of the potential effect of climate change, which may dramatically impact on the performance of agriculture systems (affecting both produce yield and quality), as recent work seems to indicate (Medek et al., 2017; Myers et al., 2017; Scheelbeek et al., 2018; Tigchelaar et al., 2018; Zhu et al., 2018). Of course, reducing food losses and improving the sustainability of the food system require a rethinking of the functioning of the whole food system, including the impact of food choices, the alternative use of food such as the production of biofuels, power relations along the food chain, and the impact of the globalization process (Smil, 2000, 2013; Pretty, 2008;

Lang et al., 2009; Perfecto et al., 2009; Stuart, 2009; Conway, 2012; Nestle, 2013; Gomiero, 2015a, 2018a,b).

In this chapter, the relation between soil health, agricultural practices, and yield loss is discussed. I review how unsustainable agricultural practices' effect on soil organic matter (SOM) and soil structure is revised in spite of soil fertility reduction. The potential of agroecological agricultural practices to preserve soil health and increase yields while reducing the use of agrochemicals, as well as their potential limitations, are discussed. The concept of food security is then introduced, followed by a discussion on how unsustainable agricultural practices can reduce yields. Thereafter, soil conservation as an imperative to guarantee food security to the present and future population is denoted. The next section analyzes how unsustainable agricultural practices may impact on crop yield. Some agroecological practices that may help protect soil health and increase yields while reducing the use of inputs are also reviewed, prior to focusing on using crop genetic diversity as a means to enhance crop protection and increase yield. The potential of some technological approaches [namely precision farming and genetically modified (GM) crops] to preserve soil, increase yield, and reduce the environmental impact of food production is discussed, too. Finally, conclusions and other important issues impacting the sustainability of food production (e.g., biofuels, power relations in the food system, and the role of food choices) are presented.

2.2 Yield loss and food security

In the field, crop loss can happen at the time of harvest as edible crops are left in field, ploughed into soil, eaten by birds or rodents, or because timing of harvest is not optimal. Produce may also be damaged while harvesting due to poor harvesting technique (Cassman et al., 2003; Deguine et al., 2009; Parfitt et al., 2010). Yield loss can occur during crop growth due to the combined effect of weeds and pests (insects, rodents, plant diseases caused by bacteria, fungi, or viruses), which reduce yield in the field and may affect produce quality (pests may cause spoilage also during the postharvest phase, i.e., during storage and transportation) (Cassman et al., 2003; Deguine et al., 2009; FAO, 2011a). At the field level, harvest losses have been estimated at around 26%–30% for sugar beet, barley, soya, wheat and cotton, 35% for maize, 39% for potatoes and 40% for rice, with high regional variability (Deguine et al., 2009).

Further to that, yields can be heavily reduced by soil degradation (i.e., loss of soil fertility) (Foley et al., 2005; Montgomery, 2007b; FAO, 2015; FAO and ITPS, 2015; Lal, 2015a; Gomiero, 2016). Panagos et al. (2018) noted that soil erosion, on average, accounts for an 8% yield loss after 25–30 years cropping, notwithstanding the increasing use of inputs to replace nutrient loss due to soil erosion.

Soil compaction is also an important form of soil degradation that greatly affects yield and cost of production (Hamza and Anderson, 2005; USDA, 2008; FAO and ITPS, 2015; Sivarajan et al., 2018). Machines and farm animals are the main cause of soil compaction. Working the soil at the wrong soil water content exacerbates the compaction process. Compaction increases bulk density, and that affects plant health and yield. The more compact the soil (the higher the bulk density), the more energy plants have to spend to root in the soil and to access nutrients and water. Soil compaction is a very serious issue. Once soils undergo compaction it may be difficult to repristinate their previous structure, as such process depends on soil biological activity, which is greatly affected by the compaction process itself.

Deguine et al. (2009) argued that despite the increasing use of pesticides, harvest losses caused by pests have increased from 4% to 10% for wheat, barley, rice, and potatoes, and have remained stable or decreased slightly for maize, soya, cotton, and coffee. It has been estimated that, in the absence of any crop protection measures, about 80% of the world rice harvest, 70% of the potato harvest, and 50% of the wheat harvest might be lost (Deguine et al., 2009).

Nevertheless, den Biggelaar et al. (2004) stated that loss of productivity varies greatly, depending on crop, geographic area and soil type, and that productivity declines may not relate directly to the amount of soil loss but concern a number of erosion-induced changes in the physical, chemical, and biological qualities of soil that influence production (i.e., SOM, water-holding capacity, nutrient contents, bulk density).

Nevertheless, inappropriate agricultural practices, while potentially helping to boost yields in the short term, may expose soil to heavy erosion and put productivity at risk in the long term (under extreme weather, bare soils, low in SOM, may lose several centimeters over a very short space of time) (Morgan, 2005; Montgomery, 2007a,b; Quinton et al., 2010; Gomiero, 2016). As yield reduction tends to be compensated by using an increasing amount of inputs (i.e., fertilizers, pesticides, water), it also leads to increasing the cost of produce, and reducing farmers' profits (Fig. 2.1).

Guaranteeing food security to the world is a major challenge. FAO (2011a) defined food security as a state when "all people, at all times, have physical and

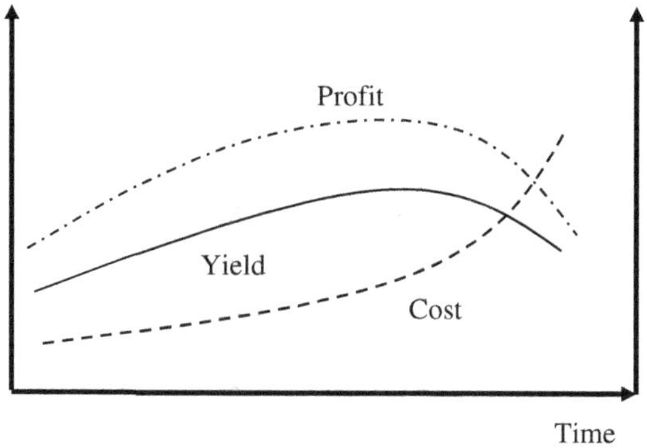

Figure 2.1 Yield loss due to unsustainable agricultural practices drives production costs up and reduces farmers' profits.
Source: Figure by T. Gomiero.

economic access to sufficient, safe and nutritious food for a healthy and active life." It requires maintaining both crop productivity (i.e., yield, which implies also ensuring enough land is available and its quality suitable to sustain production) and the quality of produce, which should be nutritious and free from toxins and other forms of contamination (Fig. 2.2).

Thus, sustainable agricultural practices should aim at reducing yield loss, stabilizing or improving long-term yields, and making the agricultural system resilient to stressors (able to recover from events such as drought and climate extremes or pest attack). At the same time, the environmental impact of agriculture should be reduced and ecosystem services preserved (Foley et al., 2011; Robertson et al., 2014; Hamilton et al., 2015; Gomiero, 2016).

In the long term, adoption of sound agricultural practices by focusing on preserving soil fertility and reducing competition by weeds and pests may help both increase yields and avoid yield loss (Fig. 2.3).

Unsustainable agricultural practices can also affect the nutritional content of produce, by reducing the density of nutrients, such as micronutrients. Nutrient-dense foods are those with a high concentration of nutrients, such as vitamins and minerals, relative to their caloric content (HLPE, 2017). Reducing nutrient density in food may pose a further threat to the health of people, especially in developing countries, where deficiencies in essential vitamins and minerals (also termed "hidden hunger"; Ruel-Bergeron et al., 2015) might affect 2 billion people. Decreasing nutritional content of produce should also be considered as a form of yield loss.

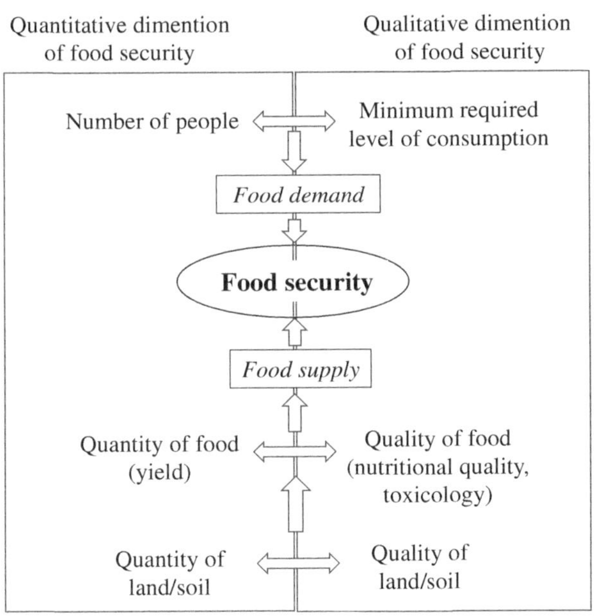

Figure 2.2 Quantitative and qualitative dimensions of food security.
Source: Figure by T. Gomiero.

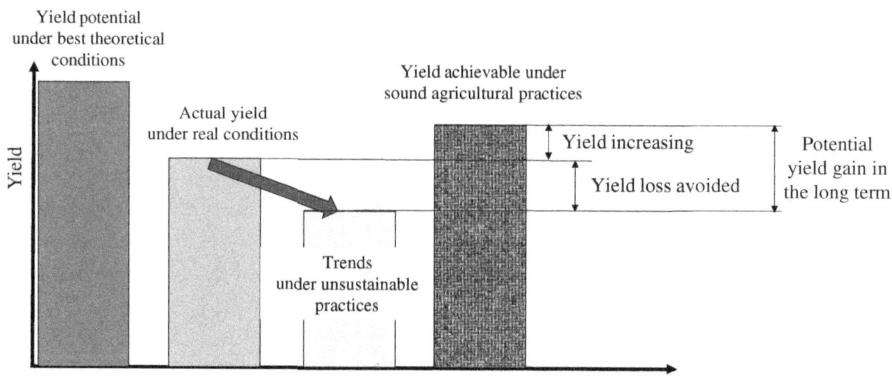

Figure 2.3 Adoption of sound agricultural practices may help both increase yield and prevent yield reduction in the long term. Theoretical yield potential refers to the achievable yield of a genotype under the best possible conditions (solar radiation, temperature, crop canopy, water, nutrients, lack of competing weeds and pests); such conditions are possible only theoretically, as they are never present in reality. Actual yield refers to the productivity of a crop in the field under local environmental conditions. In the long term, intensive or improper practices may lead to yield reduction.
Source: Figure by T. Gomiero.

As for produce quality, scholars highlight that, in the last decades, produce is less nutrient-dense than it used to be, possibly as a result of intensive agricultural practices, new crop genotypes (new high yield cultivars), and soil exhaustion (Mayer, 1997; Fan et al., 2008; Davis, 2009; Blackmore Smith and Hopkins, 2018).

According to some reviews (Mayer, 1997; Davis, 2009), in the United States and United Kingdom, nutrient content in fruit and vegetable decreased by 5%–40% in the last decades. Fan et al. (2008) reported that, from the 1960s to the present, mineral concentration in United States wheat decreased by 20%–30%. Tests reported that increased CO_2 concentration in the atmosphere could enhance yields but at the same time decrease nutritional quality of produce. According to Myers et al. (2017), a CO_2 concentration of 550 ppm can lead to a 5%–10% decrease in mineral content in cereal grains and legumes (e.g., vitB group, iron, zinc, and sulfur). Smith and Myers (2018) estimated that many food crops grown under a CO_2 level of 550 ppm have protein, iron, and zinc contents that are reduced by 3%–17% compared with current conditions. Uddling et al. (2018) claimed that increased CO_2 concentration may dramatically reduce the protein content in produce (with the exception of legumes), also altering amino acid composition.

2.3 Preserving soil health: an imperative if we want to feed the future

So far, according to some analyses, the increase in agricultural productivity (i.e., yield) has been about 70% due to the intensification of agriculture: new high yield varieties (HYV), irrigation, use of agrochemicals, and for the remaining 30% to new land being brought into production (Alexandratos and Bruinsma, 2012; Gibbs et al., 2010; Conway, 2012). Agricultural land has become one of the largest terrestrial biomes on the planet, occupying an estimated 40% of land surface (Tilman et al., 2001; Foley et al., 2005). Doubling of global food production during the past decades has been accompanied by a massive increase in the use of inputs, such as synthetic nitrogen, phosphorus, pesticides, large use of irrigation, and energy (Smil, 2000, 2003; Tilman et al., 2001; Foley et al., 2011; Gomiero et al., 2011a). Agriculture accounts for 70% of all water withdrawn from aquifers, streams, and lakes (Tilman et al., 2001; Molden, 2007; FAO, 2011b).

Since the 1990s there has been a slowdown in the growth of world agricultural production and world cereals output has stagnated and fluctuated widely (Conway, 2012; Grassini et al., 2013; Ray et al., 2012; Alexandratos and Bruinsma, 2012). Recent work by Grassini et al. (2013) seemed to indicate that some physical limits to yield productivity may have already been reached for rice, wheat, and maize, and that further attempts at increasing productivity may result in a decreasing marginal return of investment (see also Cassman et al., 2003).

Recent works (Scheelbeek et al., 2018; Tigchelaar et al., 2018) highlighted that the effects of climate change may greatly impact on crop yields. Tigchelaar et al. (2018) claim that maize yield may decrease by about 10% in Brazil and by up to 50% in the United States. The metaanalysis carried out by Scheelbeek et al. (2018), considering articles published between 1975 and 2016 concerning the effects of ambient temperature, tropospheric carbon dioxide (CO_2), ozone (O_3) concentrations, water availability, and salinization on yields and nutritional quality of vegetables and legumes, reported that in a business-as-usual scenario, predicted changes in environmental exposures would lead to reductions in yields of nonstaple vegetables and legumes.

2.3.1 Land availability and soil quality: undertaking a precautionary approach

In the last few decades, the intensification of agriculture has led to the degradation and exhaustion of soil and land. Foley et al. (2005, pp. 570–571) concluded that "[i]n short, modern agricultural land-use practices may be trading short-term increases in food production for long-term losses, in ecosystem services, including many that are important to agriculture." Soil degradation poses a major threat to food security, especially in poor regions. FAO (2011a) highlighted that there is a strong relation between land degradation and poverty. Later assessments (FAO, 2011a; Bindraban et al., 2012; Gomiero, 2016) estimated 25% of the present agricultural land to be highly degraded, about 44% to be slightly to moderately degraded, and about 10% to be recovering from degradation. Yet, the dire state of soils devoted to agriculture seems to be going unnoticed by policy makers, business, and civil society. It is urgent, therefore, to act to halt soil degradation and adopt agricultural practices that can preserve soil health.

By 2050 the demand for new agricultural land (due to population pressure, diet change, and demand for biofuels) is expected to increase by about 50%. It is very probable that tropical forests will account for that land, therefore, further deforestation is to be expected and soil degradation exacerbated (DeFries et al., 2010; Gibbs et al., 2010; Gomiero, 2015a, 2016). Furthermore, the amount of land is one aspect of the problem, while the other concerns the quality of such land. Soil quality plays a key role in determining yields, production costs, and long-term sustainability of agricultural enterprises. Marginal land, characterized by soil of poor quality, can still be cropped. Nevertheless, yields may be poor and farming may require a high use of inputs (e.g., fertilizers, water), and in the long term such land might be dismissed as it becomes infertile.

Concerning the land available for agriculture expansion, experts have different opinions. According to a review carried out by Gomiero (2016), views on the possibility of expanding agricultural land range from concerned to optimistic.

The concerned. Concerned experts pointed out that soil degradation is of major concern, and argued that, at a global level, there is not much room for the further expansion of agricultural activities and that many densely populated countries are already facing serious problems of land scarcity. They claim that most of the best agricultural land is already cropped. What is left is mostly forested land, where soil may not be very productive (actually, once deforested, such areas are highly prone to soil erosion). It is also pointed out that statistics about yields may not be reliable (an issue recognized by all experts). Some experts denoted that in many developing countries the areas harvested, yields, and production are not accurately measured, and figures may thus be affected by assumptions or political reasons. In addition, they pointed out that soil degradation reduces both actual and potential yields. Conway (2012) pointed out that in the past 50 years the population has grown by 110% and cropland by only 10%, which might be telling figures pointing to the fact that there is not much land left that can be easily cropped. The expansion of soybean (300%) and palm oil (700%) is presumably due to the clearing of the Cerrado in Brazil and of rainforests in many tropical countries (Gibbs et al., 2010; Conway, 2012; Gibbs and Salmon, 2015). It has also been stressed that the Human Appropriation of Net Primary Productivity may have reached 50% and that a further expansion of agricultural activities may erode vital space and resources away from existing biodiversity and ecosystems (Haberl et al., 2014).

The optimistic. Experts holding a more optimistic view, even though in agreement with the call to preserve soil health, argue that there is land available to sustain the further expansion of agriculture. In addition, such experts believe that in many regions of the world productivity is still very low and can be substantially increased with more inputs (i.e., fertilizers) and technology (i.e., irrigation, GM organisms). According to Mauser et al. (2015), improving crop growth management through better technology and knowledge may result in a 39% increase in estimated global production potential, while a further 30% can be achieved by the spatial reallocation of crops to their profit-maximizing locations. According to the authors, the expected yield increase will make cropland expansion redundant, and will it not be necessary to rely on GM crops. Of course, in many developing countries even a minimal investment can lead the average crop yields to rise. However, better technology and knowledge come at a cost: at present, for many developing countries such investments may be out of reach.

Given that food security is at stake, a precautionary approach should be taken, focusing on the adoption of agricultural practices that preserve soil health. Other issues should be analyzed in parallel as well, looking at the functioning of the whole of society. For example, Alexandratos and Bruinsma (2012) stated that, if social policies are not implemented, increasing productivity may spur a population growth trap (as it happened in the case of the Green Revolution), hindering further progress and locking the system into poverty.

2.3.2 The role of soil organic matter in preventing soil degradation and maintaining yields

Agricultural practices adopted in conventional agriculture tend to be poorly concerned with preserving soil health. Soil tillage increases fertility by the mineralization of soil and effectively controls weeds (although it may help the spreading of some weed species). Nevertheless, tillage, ploughing in particular, may trigger soil erosion, and the reduction in SOM that makes soil prone to erosion through the effect of rain and wind reduces soil biodiversity and destroys mycorrhiza (Morgan, 2005; Gliessman, 2014; Lal et al., 2007; Montgomery, 2007a,b; NRC, 2010; Lal, 2002, 2015a,b; Gomiero, 2016). This is exacerbated by practices that leave soil uncovered for long periods. The soil removed by either wind or water erosion is 1.3–5.0 times richer in organic matter than the soil left behind (Montgomery, 2007a). The United States Department of Agriculture (USDA) estimated that it takes 500 years to produce an inch (2.54 cm) of topsoil while it may take a few decades, or just a few years, to erode many centimeters of topsoil (Montgomery, 2007a; Gomiero, 2016).

Resistance of soils to erosion is closely linked to the stabilizing influence of SOM and vegetation cover. High organic matter content inhibits erosion because SOM binds soil particles together, generating an aggregate that resists erosion. In regions such as Asia and Africa, where soil erosion is associated with reduced vegetation cover, loss of soil carbon can trigger catastrophic shifts to severely degraded landscapes.

Soil is an extremely complex ecosystem, still poorly known. It has been estimated that, globally, SOM may contain more than three times as much carbon as either the atmosphere or terrestrial vegetation (Schmidt et al., 2011). SOM found in the topsoil (the upper 15–25 cm soil layer) is of key importance for soil fertility. It was generally believed that most of SOM was found in the topsoil (0-30 cm). Nevertheless, recent analyses proved that SOM at 0.3–1 meter may equal, or more, the amount of SOM found in the upper layer (Schmidt et al., 2011; Gregory et al., 2016). Mineral soils form most of the world's cultivated land and may contain from a trace to 30% organic matter (Bot and Benites, 2005). Fertile agricultural soils can contain up to 100 tons of organic matter per hectare (or 4% of the total soil weight); in the case of most agricultural soils, SOM represents 1%–5% of topsoil (Russell, 1977; Bot and Benites, 2005; FAO and ITPS, 2015). SOM contains roughly 55%–60% carbon by mass (FAO and ITPS, 2015.). About 95% of soil nitrogen and 25%–50% of soil phosphorus are held in the SOM-containing topsoil layer (Lal, 2010). It has been estimated that for every 1% of SOM content, soil can hold 10,000–11,000 L of plant-available water per hectare of soil down to about 30 cm (Sullivan, 2002).

SOM greatly increases the water-holding capacity of soils, which is up to 100% higher in the crop root zone (Lotter et al., 2003; Gomiero, 2016). A number of studies have shown that, under drought conditions, crops in organically managed systems, where soils have higher SOM content compared with conventionally managed fields, produce higher yields than comparable crops managed conventionally (Gomiero et al., 2011b). This advantage can result in organic crops out-yielding conventional crops by 70%–90% under severe drought conditions (Lotter et al., 2003; Pimentel et al., 2005). Other studies have shown that organically managed crop systems have lower long-term yield variability and higher cropping system stability (Pimentel et al., 2005; Reganold and Wachter, 2016).

Unsustainable soil management affects crop yield in the short term by the effect of rill or gulley erosion following intense rainfall, and in the long term by the gradual loss of soil structure and fertility. Yield losses have been reported to range from 10% to 95% per 10 cm of soil loss (Powlson et al., 2011). A review work by den Biggelaar et al. (2004), reported that that average crop yields and effects of past erosion on yields differ greatly by crop, continent and soil order, and that inappropriate soil management may amplify the effect of erosion on productivity by one or several orders of magnitude.

A metaanalysis carried out by Montgomery (2007b) highlighted that the adoption of proper agricultural practices, such as conservation agriculture, greatly reduces soil erosion (Fig. 2.4). Montgomery (2007b) reported that from a database of 39 field tests monitoring the effect of the adoption no-till on soil erosion, no-till practices showed to reduce soil erosion from 2.5 to >1000 times (median and mean values of 20 and 488 times, respectively).

Nevertheless, no-till may also lead to some problems and minimum tillage may have to be preferred (see Section 2.5.1.3). Increasing SOM also contributes to offsetting CO_2. It is generally assumed that 50–70% of soil C stocks have been lost in cultivated soils due to the effects of agricultural activities (Zomer et al., 2017). Zomer et al. (2017) reported that, globally, cropland, may store more than 140 Pg C in the top 30 cm of soil, almost 10% of the global soil organic carbon (SOC) pool. It has been estimated (Smith et al., 2008; Lal, 2015a,b) that, by adopting sustainable practices, agriculture could offset up to about 20% of total global annual CO_2 emissions. It has to be noted that carbon density in soil decreases with temperature, from less than 100 tC/ha in the equatorial belt to 400 tC/ha, or more, in the northern belt (United States, Canada, Europe, Russia) (Lal, 2002; Zomer et al., 2017). According to Zomer et al. (2017), croplands worldwide could sequester between 0.90 and 1.85 Pg C/year, accounting for 26%–53% of the target of the "4 per 1000" initiative (an initiative launched by France in 2015, aiming at achieving an

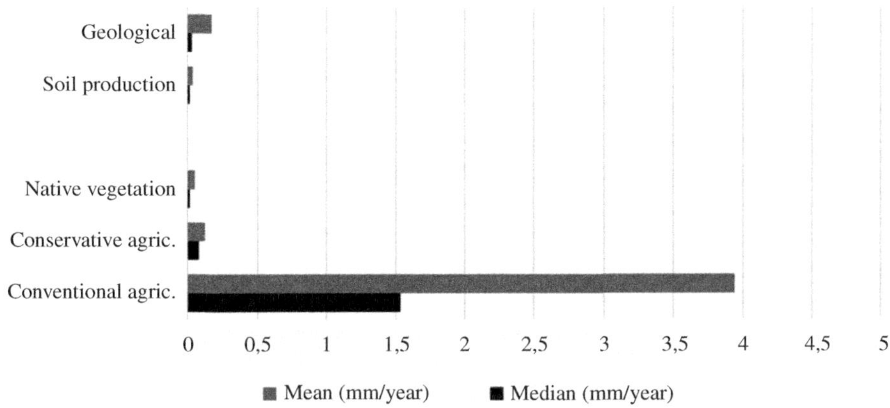

Figure 2.4 Soil erosion rate for managed and natural soils: result from a metaanalysis. Source: Figure from Gomiero, T., 2016. Soil degradation, land scarcity and food security: reviewing a complex challenge. Sustainability 8, 1–41. Available from: http://www.mdpi.com/2071-1050/8/3/281, data after Montgomery, D.R., 2007b. Soil erosion and agricultural sustainability. Proc. Natl. Acad. Sci. 104, 13268–13272, no permission needed for republishing.

annual growth rate of 0.4% in the soil carbon stocks, that would halt the increase in the CO_2 concentration in the atmosphere related to human activities; for details see https://www.4p1000.org/). Although such estimates may be rather optimistic (the authors did not account for differences in climate and important soil process issues, such as nutrient and water limitations), it is clear that increasing SOM in soils represents one of the most effective ways to sequester atmospheric carbon (while benefiting agricultural activities and ecosystem services).

However, it has to be pointed out that SOM can accumulate for some time (20–30 years, Zomer et al., 2017) then it levels off. Therefore, there is a limit to how much carbon the soil can capture acting as a carbon sink and conversion to more sustainable agriculture can only represent a temporary and partial solution to the problem of CO_2 emissions. Long-term solutions concerning GHGs emission abatement should rely, other than preserving C stored in soils and vegetation, also on a more general change of our development path, for instance by reducing overall fossil fuel consumption.

2.4 Unsustainable agricultural practices and their effect on yield loss

2.4.1 "Soil fatigue" and yield decline

Monoculture, poor rotations (i.e., short rotation, wrong species in rotation), and intensive agricultural practices (i.e., use of agrochemicals) can lead to the phenomenon known as "soil fatigue" (or "soil sickness," "yield decline"), an important cause of yield reduction. Soil fatigue is characterized by gradually decreasing yields despite fertilization and other soil preparation efforts, and is related to a complex plant–soil feedbacks (it seems that what is affected is the plants' ability to take up nutrients) (Gamliel et al., 2000; Bennett et al., 2012; van der Putten et al., 2013; Bender et al., 2016; Wolińska et al., 2018). Furthermore, soil fatigue is exacerbated when crops are grown in short rotation (Bennett et al., 2012).

Soil fatigue is a complex phenomenon that may have diverse causes (Gamliel et al., 2000; Jacob et al., 2010; Bennett et al., 2012; van der Putten et al., 2013; Bender et al., 2016; Zhao et al., 2016; Wolińska et al., 2018): (1) soil exhaustion through depletion of some essential plant nutrients, micro and macronutrients by the previous crops; (2) accumulation of soilborne pests and plant pathogens in the soil (e.g., nematodes); (3) accumulation of toxic compounds released from former crops (the effect is known as allelopathy, the production by plants of chemical compounds that affect the germination, growth, survival, and reproduction of other plants), or by soil organisms (i.e., bacteria, fungi, nematodes), affecting the health and growth of other crop species; (4) degradation of soil ecology and soil structure; (5) change in soil pH; and (6) unbalanced soil biodiversity (e.g., bacteria, reduced mycorrhizal fungi).

Soil fatigue has to be managed by restoring soil fertility (supplying micronutrients) when the problem is due to soil exhaustion, introducing proper rotations when allelopathic effects are present, and by soil sterilization in the case of soil toxicity, to eliminate soilborne pathogens. Soilborne plant pathogens have long been fought using soil fumigants, which represent a health hazard, cause environmental pollution, and can cause atmospheric ozone depletion (Gamliel et al., 2000; Dangi et al., 2017). Although some of the most problematic pesticides have been officially banned, such as methyl bromide (within the Montreal Protocol for protection of the ozone layer), they may be still in use in some regions. It has been argued that chemical soil disinfestation (use of fumigants) should be avoided, because, aside from its environmental impact, it leads to eradication of the entire microbial community, thus creating a "microbial vacuum." The latter often leads to a rebounding of pathogens, which can cause even more damage than those originally targeted for control (Gamliel et al., 2000; Dangi et al., 2017). Sound nonchemical methods have been developed that can effectively control soilborne plant pathogens and plant-parasitic nematodes. Some practices are commonly used are:

- sound crop rotations (with species that do not cause allelopathic effects) and use of cover crops (Jacob et al., 2010; Bennett et al., 2012; Zhao et al., 2016; Dangi et al., 2017; Wolińska et al., 2018),
- soil sterilization by steam treatment (180°C–200°C) provides a sound solution for the eradication of soil pests (Johnson, 1946; Gamliel et al., 2000),
- soil flooding (e.g., by introducing paddy rice in rotation) is known to suppress some soilborne pests and has been used in Asia (Momma et al., 2013).

More recently other nonchemical strategies have been developed:

- Anaerobic soil disinfestation (also known as "biological soil disinfestation" or "reductive soil disinfestation") works by creating a temporary anaerobic soil environment to stimulate the growth of facultative and obligate anaerobic microorganisms, which, under anaerobic conditions, decompose the available carbon sources, producing compounds (organic acids, aldehydes, alcohols, ammonia, metal ions that suppress soilborne pests (Blok et al., 2000; Shinmura, 2000; Momma, 2008; Butler et al., 2012; Momma et al., 2013).

- Solarization is accomplished by covering the soil surface with a clear plastic film to trap solar radiation with soil temperature that may rise above 70°C and become lethal to many plant pathogens (Gamliel et al., 2000; Butler et al., 2012; Momma et al., 2013).
- Flame soil disinfestation is a technique by which compressed fuel (i.e., natural gas) is injected into the soil, the flame temperature may reach 1200°C killing soilborne pathogens and weed seeds as well (Gamliel et al., 2000; Mao et al., 2016).
- Increasing SOM and restoring soil biodiversity through more complex cropping patterns and a reduced use of agrochemicals (Bennett et al., 2012; van der Putten et al., 2013; Bender et al., 2016; Wolińska et al., 2018).

The cost of some of these treatments (i.e., soil flooding, solarization) may limit their use to greenhouses or high value crops planted in small plots. Overall, sound preventive measures should rather be implemented.

It has to be highlighted that, although allelopathy may be a cause of soil fatigue, when properly managed it may represent an effective means to cope with weeds, as it has been the case for some rice varieties (Kong et al., 2008; Pheng et al., 2010).

2.4.2 The effect of synthetic fertilizers on pests and soil health

Intensive agriculture greatly relies on the use of synthetic fertilizers to spur plant growth. Often, fertilizers are used far beyond the real needs (Good and Beatty, 2011). Indeed, HYV was created to take full advantage of the high supply of synthetic nitrogen. Nevertheless, although a high amount of input greatly stimulates plant growth, such strategy has a number of drawbacks. It is estimated that only 30%–50% of the nitrogen applied is taken up by plants (Good and Beatty, 2011), while the rest is lost to the environment. The amount of reactive nitrogen (N compounds that support plant growth directly or indirectly) used to produce food is on average about 10-fold higher than its consumption by plants, the rest being a major cause of environmental pollution (i.e., eutrophication, water and air contamination by N-based toxic compounds) (Erisman et al., 2013).

Furthermore, plants growing on synthetic fertilizers are reported to be more prone to pest attack and to have weaker defenses. Since the 1950s, numerous studies reported that heavily fertilized crops were two to three times more prone to be attacked by pests (Altieri et al., 2012). This is possibly due to:

1. physiological changes induced in the plants by synthetic fertilizers (e.g., growth rate),
2. an altered balance between protein content (supplying high-quality food for pests) and secondary metabolite concentrations (many acting as defense compounds),
3. changes in soil ecology that affect plant nutrition (Altieri et al., 2012).

More recent studies have proven that pests prefer plants grown with synthetic fertilizer rather than those growing in organically managed soil (Phelan et al., 1995,

1996, 2009; Alyokhin et al., 2005; Hsu et al., 2009). This is explained by the "mineral balance hypothesis" (Phelan et al., 1996), which states that organic matter and microbial activity associated with organically managed soils allow enhancement of the nutrient balance in plants, which in turn can better respond to pest attack. Under greenhouse-controlled experiments, females of European corn borer (*Ostrinia nubilalis*) were found to lay consistently fewer eggs in maize on organic soil than on conventional soil (Phelan et al., 1995, 1996; Phelan, 2009). The butterfly *Pieris rapae crucivora*, a cabbage pest, prefers to lay eggs on foliage of synthetically fertilized plants (Hsu et al., 2009). Densities of Colorado potato beetle (*Leptinotarsa decemlineata*) have been reported as generally lower in plots receiving manure and soil amendments, in combination with reduced amounts of synthetic fertilizers, compared with plots receiving full rates of synthetic fertilizers, but no manure (Alyokhin et al., 2005). Staley et al. (2010) reported the case of two aphid species presenting a different response to fertilizers. *Brassica* specialist *Brevicoryne brassicae* was found to be more abundant on organically fertilized plants, while the generalist *Myzus persicae* was found more abundant on synthetically fertilized plants. Staley et al. (2010) also reported that the diamondback moth *Plutella xylostella* (a *Brassicaceae* specialist) was more abundant on synthetically fertilized plants and preferred to oviposit on these plants. The authors found that glucosinolate concentrations (a plant defense compound, widely present in *Brassicaceae*) were up to three times greater on plants grown organically, while nitrogen content was maximized on plant foliage under higher or synthetic fertilizer treatments. In China, the great population increases of major insect pests of rice were closely related to the long-term excessive application of nitrogen fertilizers (Lu et al., 2007).

A better management of inputs that avoids an overuse of synthetic fertilizers may greatly benefit crops and reduce the impact of pests (Lu et al., 2007; Good and Beatty, 2011; Erisman et al., 2013), in turn increasing farmers' profit and limiting the environmental impact caused by agrochemicals.

2.5 Agricultural practices for a more sustainable agriculture

It is urgent to work out agricultural practices that guarantee food production while preserving soil health, reducing water consumption and the use of agrochemicals. Such practices should also make crops able to withstand the potential effects of climate change (drought in particular). A number of approaches and agriculture practices have been proposed, such as no-till, minimum tillage, conservation agriculture (CA), agroecology, integrated pest management (IPM or integrated pest control), and organic agriculture (Altieri, 1987; Lampkin, 2002; Cassman et al., 2003; Altieri and Nicholls, 2004; Hobbs, 2007; Gliessman, 2014; Deguine et al., 2009; Perfecto et al., 2009; Glover et al., 2010a,b; Gomiero et al., 2011a; Lal, 2015a,b: Furlan et al., 2017).

Such practices differ in their focus and scale, some addressing specific issues. CA and no-till focus mainly on preventing soil erosion, agroecology is concerned with a more ecological management of the whole farm, which is seen as an integrated system, while organic agriculture, in addition to embracing agroecology, is also regulated by law and bans the use of synthetic agrochemicals and of GM organisms[1] (Table 2.1).

2.5.1 Conservation agriculture

The dramatic effects of the "dust bowl" that hit the US plains in the 1930s (Worster, 2004; Montgomery, 2007a; Kassam et al., 2014) forced farmers and agronomists to reflect on the use of ploughing (inversion tillage), where the soil is turned upside down by moldboard ploughing followed by disking one or more times (also referred to as "conventional tillage") (Phillips et al., 1980; Lal et al., 2007; Kassam et al., 2014; Islam and Reeder, 2014; Lal, 2015a,b). Experimentation with no-till farming practices, then defined as CA (also referred to as direct seeding, zero tillage, conservation tillage), began in the 1960s in the United States, at Ohio University (Kassam et al., 2014; Islam and Reeder, 2014). In no-till farming, soil is completely undisturbed prior to planting, except for a narrow slot used for seeding, and weed control is achieved by herbicides. In the United States, no-till began to be widely adopted in the 1980s, with the availability of better planters and cheaper herbicides.

2.5.1.1 Principles of conservation agriculture

Phillips et al. (1980, p. 1108) defined the no-till system as "one in which the crop is planted either entirely without tillage or with just sufficient tillage to allow placement and coverage of the seed with soil to allow it to germinate and emerge." Early tests provided evidence that no-till practices reduced the use of energy, labor, and machinery inputs, and provided effective soil erosion control (reducing soil erosion next to zero even in sloping land), improved soil water retention and fertilizer use efficiency, while crop yields were as high as or higher than yields from crops produced by conventional tillage (Phillips et al., 1980; Islam and Reeder, 2014). Nevertheless, drawbacks were also reported, such as a great increase in the use of herbicides (50% more for maize were reported), increase in pests resulting in crop damage, possibly higher than in the conventional tillage system, because of a more favorable habitat and lower soil temperature (Phillips et al., 1980). Early conservation agriculture tests in Ohio were based on the adoption of no-till, crops rotation, and cover crops. Most of the time farmers plant just one or two species of cover crops together, but a "cocktail" of cover crops is also used. Such cocktail consists of 5–10 species with differences in type and architecture

[1] The term "organic agriculture" defines products that are produced according to standards established by international and national institutional bodies. Standards concerned mostly with the ban of agrochemicals (synthetic fertilizers, herbicides and pesticides), the strictly regulated use of drugs in animal rearing, and the prohibition of use of GMOs. The certification includes production, handling and processing (Codex Alimentarius 2004; EC, 2018; IFOAM, 2015; USDA, 2018).

Table 2.1 Agricultural practices, at different scales, that may help to preserve soil health, reduce the use of agrochemicals, and enhance crop production

Soil management		
Soil protection	***Reduce the impact of tillage***	***Reduce the impact of field operations and irrigation***
• Maintaining soil cover (cover crops, mulching) • Enhance soil organic matter • Enhance soil biodiversity • Sound livestock density	• Avoid soil compaction • Minimum tillage • Deep cultivation may be needed, but it has to be done in dry conditions (also to avoid compaction) • Contour tillage • Ridge tillage • Avoid turning the lower layers	• Avoid bare land • Reduce soil compaction • Prevent salinization • Drip irrigation (reduce water use)
Crop management		
Cropping pattern	***Cropping biodiversity***	***Reducing inputs***
• Avoid monoculture • Long rotations • Polyculture/ multiple crops • Intercropping • Agroforestry	• Suitable landraces • Varietal mixture • Preserving agrodiversity • Preserving ecological structures	• Minimum use of agrochemicals • Integrated pest management • Managing the supply of nutrients to crops • Precision agriculture • Agroecological practices • Organic agriculture
Agroecological landscape management		
Preserving/enhancing ecological structures	***Cropping pattern***	***Land works***
• Grass strips • Hedgerows • Woodlot/forest • Wild vegetation	• Complex cropping pattern at landscape level • Integrating crops with wild vegetation	• Contour tillage • Terracing, stone walls • Irrigation/waterways • Repristinating ecological structures

(C3 vs C4), plant height and growth pattern, root distribution, nutrient and allelopathic chemical content, and adaptability. It has been reported that cover crop cocktails significantly improved soil properties and reduced soil compaction (Islam and Reeder, 2014).

Other conservation practices were subsequently tested (Lal, 2015a,b). In the 1970s, reduced tillage (mulch tillage/minimum tillage) practices were implemented (using chisels, field cultivators, discs, sweeps, blades), relying on herbicides for weed control. In the 1980s, ridge tillage was introduced: 10–15 cm high ridges are made either during the previous season during cultivation or at planting time, and crop residues are removed from ridge tops and put into the adjacent furrow.

Since the 1980s, CA has been promoted by international organizations (i.e., FAO), donors, farms, and nongovernmental organizations as a means to halt soil degradation and overcome food insecurity. FAO (2018a) defined CA as "a farming system that promotes maintenance of a permanent soil cover, minimum soil disturbance (i.e., no tillage), and diversification of plant species. It enhances biodiversity and natural biological processes above and below the ground surface, which contribute to increased water and nutrient-use efficiency and to improved and sustained crop production." CA practices are based on:

1. continuous minimum mechanical soil disturbance,
2. permanent organic soil cover (at least 30% with crop residues), and
3. diversification of crop species grown in sequences and/or associations (rotation).

Reducing soil disturbance and maintaining crop residues near the surface contributes to biological diversity both above and below the soil surface, while crop rotation reduces the risk of pest outbreaks and improves soil health (Bot and Benites, 2005; Hobbs et al., 2008; Friedrich and Kassam, 2012; Friedrich et al., 2012; Kassam et al., 2014; FAO, 2015, 2018a; Lal, 2015a,b).

It has been suggested that CA should also include integrated nutrient management (Lal, 2015a,b; Wu and Ma, 2015), that is, integrate old and modern nutrient management methods, drawing from all of their strengths to achieve an ecologically sound and economically optimal farming system. Integrated nutrient management practices, apart from increasing nutrient-use efficiency, should also aim at reducing losses incurred through leaching, runoff, volatilization, and GHGs emissions. A review of the literature indicated that adoption of integrated nutrient management practices improved yields by 8%–150% (Wu and Ma, 2015). Therefore, policies should contemplate subsidies for use of organic manures, ensuring a better balance between inorganic and organic fertilizers (Wu and Ma, 2015).

Environmental advantages of CA include soil and water conservation, carbon sequestration in the soil, landscape protection, flood mitigation, reduced pollution of waterways arising from sediments and in particular from bound phosphorus, and improved drought proofing (Friedrich and Kassam, 2012; Friedrich et al., 2012; Kassam et al., 2014; Kertész and Madarász, 2014; Wezel et al., 2014; Busari et al., 2015; Lal, 2015a,b). It has been claimed that CA is also less costly, and economically, environmentally, and socially beneficial (Kassam et al., 2014), when properly implemented. That is to say, when other than no-till, also the other two CA principles are also implemented (Hobbs et al., 2008; Friedrich and Kassam, 2012; Friedrich et al., 2012; Lal, 2015a,b). In rainfed experiments, no-till plots with residue retention resulted in higher and more stable yields than conventionally tilled plots with residues incorporated (Hobbs et al., 2008). Nevertheless, no-till farming may also cause some important problems (which I will discuss in Section 2.5.1.3).

2.5.1.2 The adoption of conservation agriculture

CA (with no-till) is practiced on more than 125 Mha around the world, covering approximately 9%–10% of the global arable land surface (Kassam et al., 2014; Kertész and Madarász, 2014). Adoption of CA practices varies among regions. According to FAO (in Kassam et al., 2014), CA accounts for about 60% of arable cropland in South America [also thanks to the introduction of herbicide-resistant (HR) crops, which are genetically engineered to resist herbicides such as glyphosate], 60% in Australia and New Zealand, 15% in the United States, 3% in Russia and Ukraine, about 1% in Asia, 0.5% in Europe, and 0.3% in Africa. According to Eurostat (2013), conventional tillage is the most widespread tillage practice in EU-27, where almost two-thirds of arable land is tilled with conventional tillage practices, about a fifth is tilled with conservation practices adopting minimum tillage, while no-till is rarely practiced.

It has been argued that CA, in particular no-till, is not equally suitable for all European agroecosystems (e.g., soil erosion risk is low in northern, cool, and temperate regions when compared with the semiarid Mediterranean regions), and that the ban on HR crops made no-till practices of little interest (Lahma, 2010; Kassam et al., 2014; Kertész and Madarász, 2014; Wezel et al., 2014; Zikeli and Gruber, 2017). Reduction in operating costs has been reported to be a major consideration in farmers' decisions to adopt conservation practices (Kertész and Madarász, 2014), even if environmental awareness is becoming an important issue for farmers. The European Common Agricultural Policy and the system of financial and institutional supports may significantly impact on farmers' decisions to adopt CA practices.

Thus, adoption of CA seems highly dependent on local societal context and farming activities. In general, CA seems to perform better in certain contexts, such as in rainfed agroecosystems in dry climates (Kertész and Madarász, 2014; Pittelkow et al., 2015a,b). Some authors (Corbeels et al., 2014; Kirkegaard et al., 2014) claimed that the existence of different biophysical and socioeconomic contexts requires a pragmatic approach to CA; for example, when mixed crop–livestock systems are widely in place, such as in Australia, Africa, and South America (where livestock are used to graze crop residues after harvest, reducing soil cover, and impacting on soil structure), or when farmers have a diverse set of objectives (i.e., protecting soil, saving time, increasing yields, increasing overall income). Nevertheless, in the case of Australia, Kirkegaard et al. (2014) reported that good livestock management within mixed crop–livestock systems can provide the same, if not greater, soil benefits as ideal CA practices.

In Africa, CA began in the 1970s in Nigeria (Kassam et al., 2014; Corbeels et al., 2014), and since then it has been increasingly promoted to preserve soil and sustain yields (both urgent priorities). Nevertheless, despite more than two decades of research and development investments, and even though CA can potentially lead to increased crop yields in the long term as a result of a gradual increase in overall soil quality, success on farms has been limited (Corbeels et al., 2014; Wezel et al., 2014; Pittelkow et al., 2015a,b), or adoption has been limited to some of the farming strategies suggested by CA (Brown et al., 2018). According to Corbeels et al. (2014) the reasons are:

1. lack of an immediate increase in farm income, deterring farmers from adopting the CA package, as, for smallholders, future benefits do not outweigh their immediate need for an income;
2. farmers owning livestock use crop harvest residues as fodder for livestock rather than as soil cover; and
3. markets for purchase of inputs and sale of produce are still lacking.

In the case of China, Zhao et al. (2017) reported that the slight decline in agronomic yield per unit area and time has deterred Chinese farmers from implementing the CA package (although, they noted that in the long term, yields may be comparable with tilled systems).

Proper assessment of local constraints to the adoption of CA is required to better meet the characteristics and needs of local farmers. In some regions it may be challenging to leave crop residues in the field due to strong pressure for residues to be used for livestock or other purposes. In the case of resource-poor and vulnerable smallholder farmers, yields should be monitored and support provided in case of yield reduction during the transition period. CA should be integrated into ad hoc conservation practices, rather than offered as a package that may fail to respond to local needs.

2.5.1.3 No-till farming: assessing the drawbacks

No-till stands at the core of CA. Nevertheless, without implementing other conservation actions (i.e., long rotation, leaving residues in the field) no-till may actually cause important problems to soil conservation, and in the long run may reduce soil health and affect yields (Bot and Benites, 2005; Friedrich and Kassam, 2012; Friedrich et al., 2012; Wezel et al., 2014; Lal, 2015a,b). Furthermore, even when properly implemented, no-till may have some important drawbacks. Therefore, CA needs to be carefully monitored and the practice might need to be adapted to different biophysical and socioeconomic contexts. Hereafter, a review of the main problems related to no-till practice follows, organized by theme.

Overall environmental benefits. In some regions and agricultural systems, energy use (i.e., fuel, use of inputs), GHG emission, and C sink in soil may not differ between conservation and conventional farming practices (Luo et al., 2010; Kirkegaard et al., 2014; Powlson et al., 2014; Lal, 2015b).

Soil compaction. No-till practice, in the long run, may exacerbate the problem due to the additive effects of equipment traffic, especially when the soil is wet or poor rotation practices are in place (Hamza and Anderson, 2005; USDA, 2008; FAO and ITPS, 2015). Nevertheless, soil compaction has been reported under minimum tillage (5–7 cm) (e.g., Peigné et al., 2018), and can happen also when the soil is tilled (the "plow pan").

Impact on soil ecology. In Argentina, negative effects of no-till on soil macrofauna and litter decomposition, as compared with natural grasslands, have been reported (Domínguez et al., 2010; Álvarez et al., 2014; Domínguez and Bedano, 2016).

Weeds ecology and use of herbicides. Although tillage (i.e., ploughing) is used to fight weeds, tillage may also incorporate weed seeds into the soil, where they can be protected and conserved for many years, and may also spread perennial weeds by cutting and distributing rhizomes and other propagating parts (Friedrich and Kassam, 2012). It has been claimed that in mature and well managed no-till systems, weeds are reduced and the use of herbicides can decrease as well, helping also to reduce management costs (Friedrich and Kassam, 2012). Nevertheless, moving from ploughing to the use of herbicides affects the weed ecology of farmed land. A new weed community could develop composed by plants that are more tolerant to herbicides (Kirkegaard et al., 2014) (tolerant plants are plant that are affected but not killed the chemicals; resistant plants are plants that are not affected by the chemicals). When no-till is poorly implemented (i.e., monocropping, or short-term rotation), weeds may quickly develop resistance to herbicides, forcing farmers to increase the dose of herbicide, eventually ending up contaminating soil and water (Chhokar and Sharma, 2008; Kirkegaard et al., 2014). This seems to be the case in the introduction of herbicide-resistant GM crops (i.e., soybean, maize, canola, cotton). Continuous use of the same product (glyphosate- and glufosinate-based herbicides) led weeds to develop resistance to the chemicals, forcing farmers to use more and more herbicides (the *herbicide treadmill*). Cases have been documented where eventually hand-weeding and deep ploughing had to be used to fight resistant weeds (Powles, 2008; Binimelis et al., 2009; Benbrook, 2016; Bonny, 2016).

Pests and fungal spores may accumulate in soil and infect the following crops. No-till may increase the number of pests that affect crops. The accumulation of residues on soil represents a microhabitat for many pests that cannot be reached by pesticides (Altieri, 1987; Altieri and Nicholls, 2004). An important issue concerns the concentration of fungal spores in the soil (most importantly molds of the genus *Aspergillus* and *Fusarium*), which may infect crops the following year. The case of genus *Aspergillus* and its toxins (aflatoxins) is highly relevant because such mold can infect both conventional and Bt maize. Bt maize was engineered to produce a toxin derived from *Bacillus thuringiensis* that is able to kill some insects (i.e., coleoptera and lepidoptera) that are maize pests, such as the corn borer (*O. nubilalis*), a moth. Such pests, when attacking maize, make it prone to be colonized by molds and then infected by the highly toxic compounds they produce. It has been known since the late 1990s that Bt maize is more resistant than conventional maize to fungi from the genus *Fusarium*, which produces a class of toxins known as fumosins. Such fungi colonize maize through the damage done by some arthropods (i.e., insects, mites). Therefore, by reducing the attack of some parasites (e.g., moths such as the European corn borer), Bt maize can reduce the presence of *Fusarium* and related toxins in maize. The presence of fumosins is reduced (by about 30%) but not eliminated. It is also well known that fungi of the genus *Aspergillus*, which produce the highly dangerous aflatoxins, colonize maize even without the damage caused by parasites. Indeed, with respect to aflatoxins there are no differences between conventional and Bt maize (Hammond et al., 2004; Williams et al., 2010; Reay-Jones and e Reisig, 2014; Abbass et al., 2016; Mitchella et al., 2017).

Limited soil organic carbon accumulation. Some studies pointed out that, although no-till increases SOC concentration in the upper layers of some soils, it does not store it more than conventional tillage for the whole soil profile (Baker, 2017; Blanco-Canqui and Lal, 2008; Powlson et al., 2014; Pittelkow et al., 2015a). Therefore, agricultural land under no-till may not represent as important a C sink as previously believed (Baker, 2017; Kirkegaard et al., 2014; Powlsen et al., 2015a,b).

Yield may be reduced and is context dependent. Although it was believed that yield was higher in no-till than in conventional tillage systems (Friedrich and Kassam, 2012), yield reduction has been reported in no-till systems, on average from 2% to 6% (but much higher in some cases, see Kirkegaard et al., 2014; Pittelkow et al., 2015a), especially during the initial stages of its implementation (Kirkegaard et al., 2014; Lal, 2015b; Pittelkow et al., 2015a; Zhao et al., 2017). This has been attributed to insufficient seed–soil contact, poor seeding equipment, and stunted seedling growth because of suboptimal soil temperatures (Lal, 2015b; Zhao et al., 2017). A global metaanalysis produced by Pittelkow et al. (2015a,b) showed that no-till performance is lower (about 5%) for most crops, and highly context dependent. For example, no-till resulted in maize yield declines at tropical latitudes, but in increased yields, relative to conventional tillage systems, in arid regions, where there is restricted water availability for crop growth (the authors did not report on profitability, which might be higher for no-till due to its potential for reducing costs, e.g., saving on fuels). Grassini et al. (2015) reported that, in the case of irrigated soybean, yields were not higher in no-till fields, and that a yield reduction was observed in no-till fields compared with minimum tillage fields, especially in regions/years with cooler early-season temperatures. Nevertheless, other authors (Hussain et al., 1999; Lal, 2015b; Zhao et al., 2017) noted that in the long term (about 10 years) yields under no-till tend to stabilize and to be comparable to crop yields under tillage.

No-till may reduce the speed of residues decomposition. No-till greatly reduces the speed at which residues are decomposed, and that may constitute a problem. Minimum tillage (or reduced tillage), where soil is tilled at depths of at most 10–20 cm, is a better means to integrate crop residues in soil and can fight weeds without the use of herbicides. Therefore, minimum tillage may substitute no-till, because when sound crop management practices are implemented, the former can still represent an effective way to reduce soil erosion.

In light of what has been discussed above, it is clear that when addressing no-till we have to address the whole management system, since its impact can be very different depending on the biophysical and social context where it is applied.

Practicing sound reduced tillage could potentially represent a more sustainable soil management practice in many contexts, while helping to reduce the use of herbicides (Kirkegaard et al., 2014; Powlson et al., 2014; Pittelkow et al., 2015a,b; Cooper et al., 2016). Ridge tillage (a technique that consists of preparing a seedbed that is elevated above the mean land surface of the field) has also been suggested as a better alternative to no-till, as it enhances soil fertility, improves water management, reduces water and wind erosion control (compared with conventional tillage), facilitates multiple cropping, enhances rooting depth, and improves pest management (Hatfield et al., 1998; Liu et al., 2018).

2.5.1.4 No-till and organic agriculture

In organic agriculture, where the use of synthetic agrochemicals is not allowed, CA, in particular no-till, is hardly practicable, due to the difficulties in controlling weeds (Delate et al., 2011; Carr et al., 2013; Wezel et al., 2014; Vincent-Caboud et al., 2017; Zikeli and Gruber, 2017). Tests on organic CA carried out in France reported a yield reduction of 25% for soybean and 75% for maize compared to ploughing (Vincent-Caboud et al., 2017). In Germany, long-term tests reported organic wheat under tillage yielding 70%–100% more than under no-till (Zikeli and Gruber, 2017). Tests carried out in the United States reported soybean under CA yielding about 10% less then under tillage, while, in the case of maize, yields in no-till fields were about 40% to 90% lower (Delate et al., 2011). Delate et al. (2011) reported that, in the case of soybean, economic return was nevertheless similar, or even higher, in no-till fields compared with tilled fields. Reduced yield in no-till tests have been attributed to weed infestation, cover crop regrowth competing with main crops, and N immobilization during cover crop decomposition (Delate et al., 2011; Vincent-Caboud et al., 2017; Zikeli and Gruber, 2017). Delate et al. (2011) pointed out that reducing tillage in organic crop production may be enhanced by "green payments" for soil conservation, which can compensate the effort needed to offset yield and economic losses. Cooper et al. (2016) noted that reduced tillage (less the 25 cm with no inversion) can represent a valid alternative to no-till to control soil erosion while preserving yield. Some authors (Carr et al., 2013; Wezel et al., 2014) did not exclude that no-till practice can be implemented in organic agriculture and point out that there are encouraging results in this sense. Nevertheless, additional research about conservation tillage effects on weed communities and on the biological, chemical, and physical properties of soils should be conducted under organic management conditions.

2.5.2 The agroecological approach

According to Wezel et al. (2009), the term *agroecology* was firstly used by Bensin, a Russian agronomist, in his work in the late 1920s: he suggested the term *agroecology* to describe the use of ecological methods to conduct research on commercial crop plants. In the 1940s and 1950s, the term *agroecology* was independently reinvented by other scholars and was adopted in both in the United States and Europe (Dalgaard et al., 2003; Francis et al., 2003; Wezel et al., 2009; Gliessman, 2014). By the 1980s, the concept of agroecology had reached a broad diffusion worldwide (Altieri, 1987, 2002; Wezel et al., 2009; Wezel and Soldat, 2009; Gliessman, 2014).

Slightly different definitions of agroecology have been proposed (Altieri, 1987, 2002; Wezel et al., 2009; Gliessman, 2014). Altieri (1987, p. 8, bold in original) defines agroecosystems as "communities of plants and animals interacting with their physical and chemical environments that have been modified by people to produce food, fiber, fuel and other products for human consumption and processing. *Agroecology* is the holistic study of agroecosystems including all the environmental and human elements. It focuses on the form, dynamics and functions of their interrelationship and the processes in which they are involved."

Gliessman (2014, p. 345) defines *agroecosystem* as "an agricultural system understood as an ecosystem" and *agroecology* as "the science of applying ecological concepts and principles to the design and management of sustainable food systems." Gliessman (2014, p. 1) claimed that "[i]n agroecology, we move from a narrow concern with farming practices to the whole universe of interactions among crop plants, soil, soil organisms, insects, insect enemies, environmental conditions, and management actions and beyond that to the effects of farming systems on surrounding natural ecosystems."

While CA aims basically at protecting the soil, the agroecological approach aims at redesigning cropping systems and the agroecological landscape according to ecological principles, to achieve multiple objectives: protect the soil, protect crops from pests, reduce the use of inputs, increase efficiency, preserve farm and landscape biodiversity, preserve crops biodiversity, and guarantee yields and profits to farmers. In this sense, it may better respond to the call for sustainable agriculture (Altieri, 1987, 2018; Gliessman, 2014; Wezel et al., 2014). The scales and dimensions of agroecological investigations changed over the past decades, moving from addressing the field alone to integrating farm and agroecosystem (Dalgaard et al., 2003; Wezel et al., 2009, 2014). Agroecology, because of its attempt to stabilize yields while minimizing the use of inputs, has been indicated as a sound agricultural practice for smallholders and poor farmers (Altieri, 1987, 2002; Altieri et al., 2012).

To properly study the functioning and management of agroecosystems, the multiple scales and dimensions of agroecosystems have to be addressed. The relation between agroecosystems and the structure and functioning of the agrofood system and society need also to be addressed to assess the feasibility and viability of alternative production strategies (Giampietro, 2004; Giampietro et al., 2014; Gomiero, 2016, 2017, 2018c).

DeLonge et al. (2016) pointed out that in the United States, notwithstanding the alarming impact of industrial agriculture and the investments in assessing such an impact, only at best 10% of all public funds devoted to agricultural research concern projects with an emphasis on agroecology. The authors claim that there is an urgent need for additional public funding for systems-based agroecology and sustainable agriculture research. As for Europe, to my knowledge figures are not available, but the situation might not differ much from those of the United States.

(At the global level, concerning research expenditure in organic agriculture, Niggli et al. (2017) estimated that they amount at 0.5% of the total investment in all agricultural research and development).

2.5.2.1 Agroecological practices

Agroecological practices concern a more ecological management of soil, crops, farms, and landscape (Altieri, 1987, 1999, 2018; Francis, 1989; Altieri and Nicholls, 2004; Zehnder et al., 2007; Deguine et al., 2009; Glover et al., 2012; Gliessman, 2014; Wezel et al., 2014; Wojtkowski, 2016; Furlan et al., 2017):

- Reduced tillage is preferred to no-till, as it reduces the use of herbicides, avoids the insurgence of pests that find refuge in crop residues, and better integrates residues in the soil.
- Care is taken to maintain the soil covered all the time, by using cover crops, or leaving residues on the field.
- Crops are managed through long rotations, polyculture, intercropping, relay cropping (the maturing annual crop is interplanted with seedlings or seeds of the following crop), or agroforestry (the practice of including trees or shrubs in crop or animal production agroecosystems).
- Synergies among crops are used to reduce pest insurgence and dealing with weeds, as, for example, is the case for intercropping cereals with legumes, or for the *milpa* system, widespread in central America, where maize, beans, and squash are intercropped to improve soil fertility and shadowing weeds (many more species can be grown in the *milpa* system).
- Use of locally adapted landraces should be preferred when possible, as they may be more resistant to pests, and better adapted to local conditions.
- Agricultural systems should be designed so as to make effective use of sunlight, soil nutrients, rainfall, and biological resources. Fields should be reorganized in mosaics within the farm leaving nontreated strips at the field margins, or embedded within the field.
- Ecological structures such as grass strips, flowering plant corridors, hedgerows, and woodlots are preserved or created to increase biodiversity and its services in pest control (creating suitable habitats for predators and a more complex agroecological mosaic).
- In the case of conventional agriculture, any plant that does not belong to the crop is perceived as a competing weed that has to be eliminated. The agroecological approach distinguishes between species that may harm crops, and should be taken care of, and species that may actually benefit crops, and therefore can be left in the field or at its margins.
- The use of synthetic agrochemicals, although not forbidden (as in organic agriculture), is, nevertheless, reduced to a minimum. Green manure (vegetable biomass) or animal manures should be used when possible.
- Pest control should rely on IPM. IPM was developed in the 1950s, after findings higher levels of pest control in a crop of alfalfa where lower doses of insecticide were used. The lower dose of pesticide sufficed to effectively eliminate part of the pest population, while the rest was eliminated by beneficial species that had survived the treatment thanks to the low application. The term IPM was first used by Ray Smith and Robert van den Bosch in 1967 (Flint and van den Bosch, 1981, p. 6). IPM is "an ecologically based pest control strategy that relies heavily on natural mortality factors such as natural enemies and weather and seeks out control tactics that disrupt these factors as little as possible. IPM uses pesticides, but only after systematic monitoring of pest populations and natural

control factors indicates a need." (Flint and van den Bosch, 1981, p. 6). Strategies for pest control are based on a number of different techniques, including management of the crops in the agroecosystem, management of wild vegetation, biological control, adoption of proper agricultural practices and use of resistant varieties, while pesticides are used only as last resort, at minimum level, and according to established guidelines (Flint and van den Bosch, 1981; Hoy, 1998; Altieri and Nicholls, 2004; Deguine et al., 2009; Wezel et al., 2014; Furlan and Kreutzweiser, 2015; Furlan et al., 2017). IPM should rely on:

1. cultural practices compatible with natural processes (i.e., crop rotation, soil management);
2. vegetation management to enhance natural enemy impact and exert direct effects on pest populations;
3. use of trap crops for pests and host plants for indigenous natural enemies, inundative and inoculative releases of biological control agents;
4. use of mating disruption, insecticides of biological and mineral origin (as in organic agriculture), and IPM.

Monitoring should be carried out to identify pests, weeds, and other potential diseases and ecological practices that exploit the characteristics of cropped and wild species (Hoy 1998; Altieri and Nicholls, 2004; Gurr 2016; Furlan et al., 2017). A very successful practice is the "pull and push" system, that consists in intercropping species that repel pests, "push species," and plants and attract and trap them "pull species." The function of push components of the push–pull strategy is to make the protected resource hard to locate, unattractive, or unsuitable to the pest. The function of the push component is to concentrate pests in a predetermined site, so that they can be efficiently controlled (preferably through highly selective natural pesticides, which are preferred to broad spectrum, synthetic insecticides) (Altieri and Nicholls, 2004; Cook et al., 2007; Glover et al., 2012; Gliessman, 2014).

- Implementing agroecological practices at the landscape level (agricultural landscape) in order optimize the benefits of environmental services provided by the resulting extended agroecosystem (synergic effects of scale).

The adoption of agroecological practices may present some drawbacks, which may prevent farmers from their adoption. Agroecological practices require farmers to be more knowledgeable and skilled, and farmers may prefer simpler conventional practices. Poor farmers, then, may avoid taking risks involved in the adoption of more complex practices (Brookfield, 2001; Carlisle, 2016). Some agroecological practices may limit mechanization of production and may require more labor (Gliessman, 2014). Carlisle (2016) claimed that, to facilitate the adoption of agroecological practices it is necessary to carry on education and research programs, implement sound policies, adopt measures to overcome equipment barriers, and better address the complex relation between farmers and food systems.

2.5.2.2 Crop management

Other than on soil protection, the agroecological approach focuses on sound crops and whole farm management, which should also be extended to the landscape level, to better take advantage of the agroecosystem's environmental services. Crop management aims at increasing the agroecological complexity of fields by rotating crops, or by adoption of a polyculture cropping system, whereby two or more crops are grown in the same productive season. Hereafter, a brief review of the main characteristics of these cropping strategies follows.

Rotation. Crop rotation is an ancient and essential agricultural practice to restore soil fertility (e.g., rotations with legumes such as alfalfa or fava beans), and it also contributes to weed and pest control (by breaking the life cycles of the organisms), thereby increasing yields (Francis, 1989; Bennett et al., 2012; Gliessman, 2014; Wezel et al., 2014; Brankatschk and Finkbeiner, 2015). Conventional agriculture, heavily reliant on agrochemicals (and also due to the novel characteristics of the food system), greatly simplifies the rotation pattern, often reducing it to two crops (i.e., maize, soybean) or even carrying on with monoculture for many years in a row (i.e., soybean in Latin America). This greatly affects soil health and crop yields. For example, for the case of a continuous cultivation of wheat (*Triticum aestivum*), it is well known that the second crop yields about 10% less than the first crop, and third crop yields can be 10%–15% lower than the second (Bennett et al., 2012). Yield decline due to short rotations can range from 10%–20% for maize and wheat (but higher losses have been reported) to 20%–40% for rice, and up to 50% for sweet potatoes and sugar cane (Bennett et al., 2012). Agroecological practices adopt long-term rotation to avoid the accumulation of pests and weeds and soil overexploitation and degradation. Furthermore, sound rotation may increase crop yields and/or reduce the amount of inputs (i.e., N-based fertilizers). For example, maize yield increases when maize is rotated with a legume crop, compared with a continuous maize monoculture system, or a rotation with other cereals (Gentry et al., 2013). Nevertheless, rotations have to be carefully planned to ensure that crop features are appropriate for the rotation cycle, taking advantage of synergistic effects (Wezel et al., 2014).

Multicropping (multiple cropping). Multicropping refers to growing two or more crops on the same field in the same growing seasons. Crops may be planted one after another, in temporal succession, or in different plots within the same field, or intercropped.

Intercropping. Intercropping is the practice of growing two or more crops or genotypes together for part of or the whole growing season (Fig. 2.5).

For intercropping to be successful, the majority of interactions that occur among crop species (i.e., the effect on soil, light, water, nutrients, pests) should be beneficial and/or complementary (i.e., facilitative interactions – Brooker et al., 2016, p. 99, define "*facilitative plant–plant interactions are 'positive, non-trophic interactions that occur between physiologically independent plants and that are mediated through changes in the abiotic environment or through other organisms*") (Francis, 1989; Hainzelin, 2013; Li et al., 2013; Gliessman, 2014; Wezel et al., 2014; Brooker et al., 2016; Wojtkowski, 2016; Altieri, 2018).

Intercropping presents also some drawbacks: it may limit mechanization of production, it may require more labor, the use of herbicides may be constrained, or when not properly chosen, a secondary crop may compete with the main crop reducing its yield and economic performance (Gliessman, 2014).

Some traditional intercropping systems include maize/bean, sorghum/pigeon pea, banana/coffee, and maize/cassava, and involve intercrops of plants with dissimilar size

Figure 2.5 Intercropping of vegetables and potatoes (on the right) in a plot of an organic farm in Padova, Northeast Italy. Note the massive hedgerow that surrounds the farm, and the grass strip at the bottom.
Source: Photo T. Gomiero.

and growth cycle in the field, so as to also have better vertical distribution of leaves in the total canopy (Francis, 1989; Hainzelin, 2013; Gliessman, 2014; Altieri, 2018).

The *milpa* system, the maize–beans–squash intercrop, is a very old and well-known traditional practice in Mesoamerica. In the *milpa* system, beans fix nitrogen, which is then made available to maize through mycorrhizal fungal connections between root systems. The squash, providing shadow to the soil, helps control weeds. Tests carried out in Mexico reported maize yields in the *milpa* system could achieve yields as high as 50% above monoculture yields (planting on land that had only been managed using local traditional practices, and making use of these practices) (Altieri, 1987, 2018; Altieri and Nicholls, 2004; Gliessman, 2014).

Traditional, multiple-cropping systems may provide about 15% to 20% of the world's food supplies. In Latin America, farmers grow 70% to 90% of their beans in combination with maize, potatoes, and other crops. Sixty percent of maize grown in this region is intercropped (Altieri and Nicholls, 2004). Traditional cropping systems are also genetically diverse, containing numerous varieties of domesticated crop species as well as their wild relatives. In the Andes, farmers cultivate as many as 50 potato varieties in their fields (Altieri and Nicholls, 2004; Altieri, 2018), and up to 80 landraces in some Andean valleys of Peru and Bolivia (Brookfield, 2001). Genetic diversity confers at least partial resistance to diseases that are specific to particular crop strains and allows farmers to exploit different soil types and microclimates for a variety of nutritional and other uses (Brookfield, 2001; Altieri and Nicholls, 2004; Brookfield and Padoch, 2007; Jarvis et al., 2007; Altieri, 2018).

Traditional agroforestry systems throughout the tropics commonly contain well over 100 annual and perennial plant species per field, species used for construction materials, firewood, tools, medicine, livestock feed, and human food (Altieri and Nicholls, 2004; Atangana et al., 2014; Farrell and Altieri, 2018).

Combining two or more crops within a mixture can sometimes increase total crop productivity because facilitative interactions among the crop species result in greater total resource utilization compared with growing the component crops as monocultures (Francis, 1989; Li et al., 2004; Gliessman, 2014; Li et al., 2013; Brooker et al., 2016; Gurr et al., 2016; Martin-Guaya et al., 2018; Reiss and Drinkwater, 2018). In some crops, yield increases have been reported to reach 90%, by reducing limitations to crop growth imposed by nitrogen/phosphorus availability and/or the presence of disease (Li et al., 2007). Intercropping has been practiced by farmers in China for more than 2000 years (Li et al., 2007). Knörzer et al. (2009) reported that in China's northeast 300,000 ha of maize fields have been converted to intercropping with sweet clover (*Melilotus officinalis*), resulting in maize yields about the same as those from monoculture, but in an additional 15 t/ha of sweet clover, which can be used to feed three cows a year. Li et al. (2007) reported increased yield for maize and wheat when intercropped with a legume like soybean or faba bean, due to complementary N use (i.e., wheat is much better at extracting soil-available N than legumes are, and thus legumes are forced to get nitrogen from atmospheric N fixation). Intercropping is known to be more efficient in poorer soil and poorer environmental conditions, because of higher nutrient uptake, improved resource utilization, and low-input cultivation, but it loses this advantage if combined with high-input cultivation (Knörzer et al., 2009).

Nevertheless, in some cases intercropping can decrease yields by interspecific competition (Li et al., 2007).

A key indicator to assess the performance of intercropping is the land equivalent ratio (LER), a measure of the yield advantage obtained by growing two or more crops as an intercrop compared with growing the same crops as a collection of separate monocultures (Vandeermer, 2011; Gliessman, 2014; Wojtkowski, 2016). For example, crops A and B are intercropped. The yield of crop A in the intercropped system is 10 t/ha and in monoculture is 8 t/ha. The yield of crop B in the intercropped system is 5 t/ha and in monoculture is 3 t/ha. The LER is calculated as $8/10 + 3/5 = 1.4$, a figure indicating that to achieve the same yield from the crops A and B under monoculture we would require 40% more land than as having A and B intercropped. LER is, therefore, an indicator of the intensification achieved by the intercrop system.

Recent meta-analyses of intercropped systems reported LERs of 1.28 (Martin-Guaya et al., 2018) and 1.17 (Yu et al., 2015). The authors noted that intercropping could potentially be a sound means to supply more food to feed people in the future. Nevertheless, it has to be stressed that LER does not provide an economic assessment of production, while being an important indicator in assessing the overall sustainability of an enterprise. Therefore, in some cases farmers may require some economic incentives for adopting this more complex practice (Martin-Guaya et al., 2018). Intercropping systems are usually less affected by pests, but the effect depends on the choice of the crops intercropped, as in some cases crop mixtures may, for example, change the microclimate, increasing humidity and favoring the presence of some pests (Francis, 1989; Gliessman, 2014; Wezel et al., 2014). Intercropping can also reduce weeds. Intercropping maize with legumes has shown to reduce weeds and increase yields by 37% (Verret et al., 2017). Therefore, intercropping can increase outputs while decreasing management costs (Fig. 2.6).

The integration of crops and small livestock, complex crop associations and rotations, agroforestry, and remarkable tropical home garden systems characterize traditional agriculture, which has fed people well for a very long time (Altieri, 1987; Gliessman, 2014). Nevertheless, it has to be highlighted that adoption of agroecological practices may be more labor intensive. Therefore, in societies where the cost of labor is high (i.e., Europe, United States), farmers may have limited interest in such practices. Such an issue directly concerns how a society accounts for externalities. If the cost of externalities caused by unsustainable agricultural practices were to be internalized, adopting agroecological practices may prove much more cost effective (Gomiero, 2016; Wojtkowski, 2016).

Agroforestry. Agroforestry refers to land-use systems and technologies where woody perennials (trees, shrubs, etc.) are grown on the same land-management units as agricultural crops and/or animals, in some form of spatial arrangement or temporal sequence. Agroforestry allows to diversify production, protect soil, reduce pest pressure, and increase social, economic, and environmental benefits for farmers

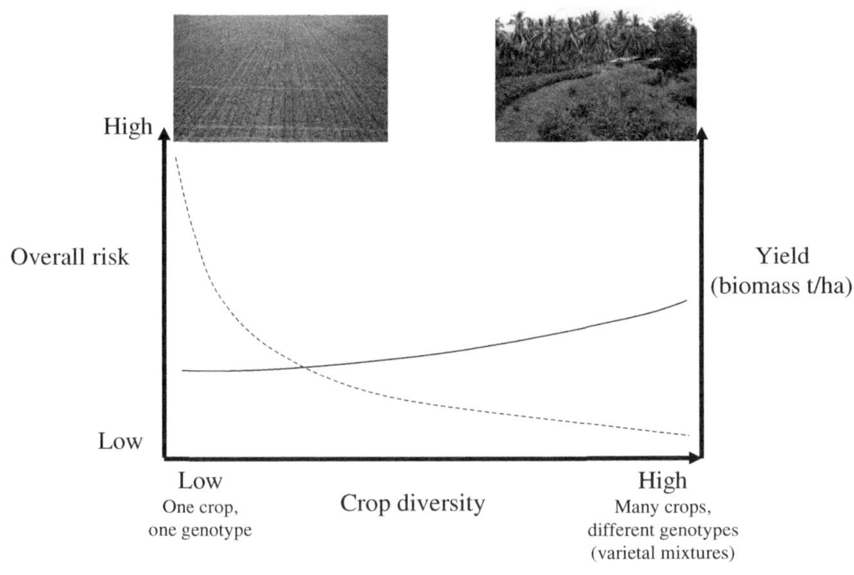

Figure 2.6 Increasing crop biodiversity (both as number of species and varieties) greatly reduces risks and enhances total yield per hectare.
Source: Photo on the left (maize monoculture) from FAO (http://www.fao.org/docrep/006/x8234e/x8234e08.htm); photo on the right (complex cropping system) from FAO (http://ref.data.fao.org/photo?entryId = 3f405bf8–54e2–4be1–834b-bc7ba2c3636a).

and for society at large (Altieri and Nicholls, 2004; Perfecto et al., 2009; Gliessman, 2014; Atangana et al., 2014; Wezel et al., 2014; Wojtkowski, 2016; Farrell and Altieri, 2018; FAO, 2018b). Agroforestry is a key feature of tropical home gardens, where tens of different species can be cultivated on the same plots, on three to four different layers (Atangana et al., 2014; Farrell and Altieri, 2018; FAO, 2018b). Complex tropical agroforest systems can achieve high agricultural yields, make systems more resilient and preserve high levels of biodiversity, both agricultural and wild (Perfecto et al., 2009; Clough et al., 2011; Atangana et al., 2014; Farrell and Altieri, 2018). Agroforestry is spreading also within Europe, especially in the southern part, as a way to diversify cultures, promote soil and environmental conservation, increase the carbon sink, reduce the use of agrochemicals, and as a means to respond to the effects of climate change (Wezel et al., 2014; Torralba et al., 2016; Mosquera-Losada et al., 2018).

2.5.2.3 The importance of adopting agroecological management at the landscape level

Agricultural intensification results in a dramatic simplification of landscape composition and in a sharp decline of biodiversity, which affects the functioning of natural pest control, as natural habitats provide shelter for a broad spectrum of natural species that operate as pest control for all crops. Research demonstrates that pesticides disrupt the communities of pests' natural enemies, in turn leading to increased pest damage in crops (van den Bosh, 1989; Winston, 1997; Crowder et al., 2010; Bommarco et al., 2011; Hamilton et al., 2015; Gurr et al., 2016). Hoy (1998) pointed out that effective resistance mitigation requires a holistic approach to pest management.

Landscape heterogeneity is a key factor in promoting biodiversity in the agricultural landscape. A mosaic landscape may support a larger number of species in a given area, simply because the landscape contains a larger number of habitats. Properly preserving a healthy agroecological landscape and landscape-ecological structures (i.e., hedgerows, herbaceous strips, and woodlot) helps protect crops by relying on helpful organisms that predate on pests (e.g., ladybird beetles, aphids, parasitoid wasps, aphids, and caterpillars) (Thies and Tscharntke, 1999; Bianchi et al., 2003; Altieri and Nicholls, 2004; Perfecto et al., 2009; Macfadyen et al., 2009; Crowder et al., 2010; Hobbs et al., 2014; Gurr et al., 2016; Wezel et al., 2014; Hamilton et al., 2015; Gomiero, 2015b; Wojtkowski, 2016). Mols and Visser (2007), for example, found that the great tit (*Parus major* L.), a European cavity-nesting bird, reduces the abundance of harmful caterpillars in apple orchards by as much as 50%–99%. In the Netherlands, the foraging of *P. major* increased apple yields by 4.7–7.8 kg per tree. Bianchi and Van Der Werf (2003), found that landscapes with 9%–16% noncrop habitat provided enough resources for local populations of ladybird beetles to control aphid outbreaks. On the contrary, reducing ecological structures and causing habitat fragmentation results in a significant reduction in local biodiversity and this may impact on the biological control of pests (Thies and Tscharntke, 1999; Altieri and Nicholls, 2004; Bianchi et al., 2006; Gardiner et al., 2009; Wezel et al., 2014; Hamilton et al., 2015). It has also been suggested that biodiversity conservation, by retaining local food web complexity, can also represent an effective management strategy against the spread of invasive species that often act as pests in new environments (Kennedy et al., 2002). This may help avoid the downside of using exotic natural enemies to fight novel invasive species, as species introduced for biocontrol can act as invasive species in their own right (Thomas and Reid, 2007). Perfecto et al. (2009) studied the effect of trees in shadow coffees in Mexico, and found that biodiversity harbored by trees allows for the existence of a complex web of relations among ants, ladybird beetles, birds, spiders, and parasitic wasps contributing to effectively control about a hundred potential coffee pests. Castelan et al. (2018) carried out a test in Brazil to assess the role of natural forests on banana plantations. The authors demonstrated that preserving natural ecosystems near plantations reduced pest attack on banana plants (increasing yields) and improved the nutritional quality of produce. A multi-site field studies carried out in Asia, on rice fields, by Gurr et al., (2016), seems confirming that increasing biodiversity promotes ecological intensification of agriculture. The authors report that inexpensive intervention aiming at increasing nectar-producing plants around rice fields, significantly reduced populations of two key pests, reduced insecticide applications by 70%, increased grain yields by 5% and delivered an economic advantage of 7.5%.

2.6 Cropping biodiversity to reduce losses and increase yields

2.6.1 The potential benefits of varietal mixture to cope with pest and increase yields

Before the green revolution, farmers selected local crop varieties (landraces) aiming at obtaining crops resistant to local pests and local environmental conditions (which is also what many small subsistence farmers do today). Since the process of industrialization of agriculture started and agrochemicals became available, crops have been selected mainly with the aim to increase yields and to fit in with extensive monotypic monoculture. This has led to selection of genetically uniform HYV that have become increasingly vulnerable to pests and dependent on human management. In the process, thousands of landraces have been lost all over the world (Marshall, 1977; Wolfe,1985; Brookfield, 2001; Heal et al., 2004; Fowler and Hodgkin, 2005; Jarvis et al., 2007, 2008, 2011; Frison et al., 2011; Hainzelin, 2013; Wojtkowski, 2016).

The risks arising from reduced crop genetic diversity have been discussed by scholars since the 1950s, when such new genetically homogeneous varieties began to enter the market (Marshall, 1977; Heal et al., 2004). Scholars have argued that the impoverished germplasm of HYV would have weakened the resistance of crops in front of pest attack. Browning (in: Heal et al., 2004) pointed out that diversity was the only defense against the unknown (such as the presence of novel pests or environmental conditions).

The continuous and ever-increasing use of pesticides required to protect new varieties eventually caused major environmental pollution, while pesticide residues became common on food. Furthermore, pesticides did not solve the problems faced by crops, as pests quickly became resistant to them. This led to a process known as the "pesticide treadmill" (Flint and van den Bosch, 1981; van den Bosh, 1989; Vandeermer, 2011), whereby an increasing quantity and expanding range of chemicals have to be used (Flint and van den Bosch, 1981; Hoy, 1998; Altieri and Nicholls, 2004; Vandeermer, 2011).

A promising strategy to control pests and hinder pathogen adaptation to varietal resistance is the use of mixtures of varieties such that the mix will form a heterogeneous environment for the parasite (Marshall, 1977; Wolfe,1985; Zhu et al., 2000, 2007; McDonald and Linde, 2002; Brookfield, 2001; Brookfield et al., 2002; Altieri and Nicholls, 2004; Brookfield and Padoch, 2007; Hainzelin, 2013; Han et al., 2016). The main purpose of genetic mixtures (crop variety mixtures) for pest and disease management is to slow down pest and pathogen spread. In variety mixtures (also mixture of landraces), two or more component varieties are grown concurrently within the same field, introducing diversity to the crop stand. Cultivation of variety mixtures is a characteristic trait of subsistence agriculture, nevertheless, due to the benefits provided, this strategy is also gaining increasing attention in industrialized countries (Brookfield, 2001; Altieri and Nicholls, 2004; Jarvis et al., 2007; Kiær et al., 2009; Newton et al., 2010; Frison et al., 2011; Hainzelin, 2013; Brooker et al., 2016; Dwivedi at al., 2016; Wojtkowski, 2016). Dwivedi et al. (2016) stressed that landraces, given their more than millennial evolutionary history and adaptation to stressful environments, can represent an ideal resource to explore novel genetic variation that overcomes challenges to crop production, enhancing the yield (through, for example, a process of facilitation), and stability of staple crops in vulnerable environments.

It has been reported that cropping varietal mixtures allows for better pest management, provides buffering against variation in environmental factors, and guarantees more stable and potentially increased crop yields (Zhu et al., 2000; Kiær et al., 2009; Newton et al., 2010; Mulumba et al., 2012; Hainzelin, 2013; Li et al., 2013; Zhang et al., 2013; Yu et al., 2016; Wojtkowski, 2016). Döringa et al. (2015) reported that increasing plant diversity in the field raised wheat yields by 2%–4% over monocultures. A metaanalysis conducted by Kiær et al. (2009) confirms the potentials of seed mixtures of wheat and barley to provide increased grain yields and improve its stability over time. Although the overall yield increase found for cultivar mixtures compared with the expected yield from their component monocultures was just 2.2%, this increase is comparable to the average annual rate of yield gain due to plant breeding improvements (between 1% and 3%) (Reiss and Drinkwater, 2018). In China, Zhu et al. (2000, 2007) conducted a series of tests on rice, in areas heavily affected by the rice blast fungus (*Pyricularia oryzae*, one of the mayor epidemic diseases limiting rice production in southwest China), and on wheat fields (intercropped with broad bean, *Vicia faba*) affected by wheat rust. In both cases, a number of varietal mixtures were interplanted. The authors claim that results were impressive. In the case of rice, pests were reduced by 94%, yields were 85% higher than usual monoculture, and farmers' average income increased by US \$150 ha^{-1}. In the case of wheat, wheat rust was reduced by 25%, wheat yield remained the same, while the yield of broad beans increased. Positive effects induced by the adoption of varietal mixtures are also reported for other crops in Asia (Zhu et al., 2000; Li et al., 2013; Reiss and Drinkwater, 2018). Further to that, such an approach helps foster conservation of agrobiodiversity, crop varieties, and landraces in situ, reducing the risk of further biodiversity losses (Zhu et al., 2003).

Nevertheless, issues associated with this agricultural practice have also been documented. Reiss and Drinkwater (2018) argued that, in specific cases, negative mixing effects have been observed, and at times both positive and negative mixing effects are observed in the same trial. The authors highlighted that the mixing effect of a specific variety mixture may be difficult to predict, therefore results of individual trials may not apply to other mixtures and other growing conditions. For example, tests carried out by Han et al. (2016) proved that only some rice combinations grown in a mixture showed effective control of rice blast, while using other varietal mixtures did not achieve the same results. Developing pest-resistant varieties (landraces) has been indicated as a possible solution to cope with pests and reduce the use of agrochemicals. Nevertheless, if crops are not properly managed, pests can eventually overcome crops' resistance.

2.6.2 Cropping perennial crops

Since the 1980s, in the United States, due to the dramatic consequences of ploughing on soil conservation (such as the "dust bowl," Worster, 2004; Montgomery, 2007a), some authors began to suggest moving from an agriculture based on annual crops to an agriculture relying on the cultivation of perennial crops, so that the detrimental effect of soil tillage and agrochemical usage could be avoided, or at least greatly reduced (Jackson, 1980, 2002; Soule and Piper, 1991).

Although conservation tillage and the adoption of cover crops can improve SOM, in general, they cannot accumulate as much SOM stock as in grasslands, forests, or the native ecosystems that agriculture replaces. Developing perennial grain agroecosystems may greatly benefit agriculture and help accumulate SOM in fields and reduce nitrogen loss. Field tests carried out in the central USA by Glover et al. (2010a) reported better performance of perennial grassland (mixtures of about 30 species) compared with wheat fields for all agronomic (i.e., biomass harvested, use of inputs, management costs) and environmental performance parameters (i.e., soil structure, biodiversity). Management costs are reduced because perennial crops do not need to be replanted every year, so they also require fewer passes of farm machinery and fewer inputs of pesticides and fertilizers, which reduces fossil fuel use. Glover et al. (2007) reported that herbicide costs for annual crop production may be 4–8.5 times the herbicide costs for perennial crop production, so fewer inputs in perennial systems mean lower cash expenditures for farmers.

Perennial crops, with their roots exceeding depths of two meters, can improve ecosystem functions such as water conservation, nitrogen cycling, and carbon sequestration by more than 50% when compared with conventional crops. Perennial crops are reported to be 50 times more effective than annual crops in maintaining topsoil, to reduce N losses 30- to 50-fold, and to store around 300–1100 kg carbon/ha per year compared with the 0–400 kg carbon/ha per year of annual crops (Cox et al., 2006; Glover et al., 2007, 2010b; Crews and Rumsey, 2017). Due to their effect on soil carbon and their lower inputs requirements, some authors claim that perennials could help slow down climate change (Cassman et al., 2003; Cox et al., 2006, 2010; Glover et al., 2007, 2010a,b; Powlson et al., 2011; Crews and Rumsey, 2017; Baker, 2017). Perennial crops are predicted to better adapt to temperature increases of the magnitude predicted by most climate-change models. Cassman et al. (2003) reported that increase of 3°C–8°C are predicted to increase yields of switchgrass (*Panicum virgatum*), a perennial forage and energy crop, by 5000 kg/ha, whereas annual species yields are predicted to decline (e.g., maize, −1500 kg per ha; soybean, −800 kg per ha; sorghum, −1000 kg/ha). Perennial cereals can also be intercropped with legume forages, which offers some important benefits, such as providing nitrogen to the grain crops, facilitating the accrual of SOM, increasing forage quality, and helping support pollinators (Hayes et al., 2016; Ryan et al., 2018).

In recent years, there has been an increasing focus on perennial breeding programs in all continents (Cox et al., 2010; Glover et al., 2012; Pimentel et al., 2012; Baker, 2017). Glover et al. (2012) reported the increasing adoption of perennial species in Africa, as such species are better suited to the poor African soils and can gain access to more of the soil's nutrients and water, and for a longer time, compared with annual crops. In 2014, an international workshop on the topic was held in Rome, Italy, at the FAO headquarters (FAO, 2014b).

Novel perennial crops have been developed both by hybridizing high-performing domestic annual species with closely related wild perennials (wheat, rye, sorghum, rice), and by the domestication of wild perennial plant species with the potential to serve as new grain crops (e.g., *Helianthus maximiliani* and *Silphium integrifolium*, which are related to common sunflower *Helianthus annuus*) (Cox et al., 2010; Baker, 2017; Ryan et al., 2018). Some authors reported that intermediate wheatgrass (*Thinopyrum intermedium*) is perhaps the most advanced example of a recently domesticated perennial grain crop, and that grain from improved lines of this crop is marketed as Kernza and is now being used in restaurants, bakeries, and commercial products (Baker, 2017; Ryan et al., 2018).

However, perennial crops also present some problems. Even if perennials yield more aboveground biomass, edible yield of perennial species is lower than conventional species (species that went thought long-term domestication). Nevertheless, it is believed that artificial selection in a properly managed agricultural environment could increase seed yield while maintaining perenniality

(Cox et al., 2006; Glover et al., 2010b; Crews and Rumsey, 2017; Baker, 2017; Ryan et al., 2018). Perennials' longevity may lead plants to allocate resources to belowground biomass, reducing seed productivity. Experts claim that this might not preclude selection of perennials that are high-yielding and economically viable (Glover et al., 2010b; Ryan et al., 2018). Perennial crops may be slow growing but then stay in the ground for multiple years. Such characteristics made fields cropped with perennial crops more susceptible to weed invasion than those cropped with annual crops, being thus at risk of poor establishment and crop failure (Pimentel et al., 2012). It has been argued that in some cases (i.e., perennial wild rice in the United States) perennial crops may become invasive (Pimentel et al., 2012).

Farmers should be aware of such potential problems and preempt them. For example, interplanting perennial grain crops with legume crops can reduce potential weed problems (Pimentel et al., 2005; Hayes et al., 2016; Ryan et al., 2018). The authors claimed that investing in proper breeding programs may overcome the present problems, focusing in particular on:

1. producing perennials with reliable regrowth and high grain yield and quality over multiple years,
2. making them adapted to abiotic stresses (i.e., water and nutrient deficiencies),
3. making them more resistant to pests and diseases (Glover et al., 2010b).

Profitability is a crucial factor in farmers' decision-making. Some authors (Baker, 2017; Ryan et al., 2018) argued that perennials help save on annual sowing and production costs (e.g., fuel, fertilizers, pesticides) compared with annual grain crop production, and that cost reduction, along with price premium for grain quality, may make perennials interesting for farmers. Ryan et al. (2018) claimed that the value of ecosystem services provided by perennial grain crops should be recognized if we want to develop a true multifunctional agriculture, and that information about this agricultural practice should be made known to farmers.

2.7 Technological approaches

In this section, I will briefly review the potential role of technology in enhancing the performance and sustainability of agricultural practices, namely the use of precision farming and transgenic crops.

2.7.1 Precision agriculture

Precision agriculture (PA) (also known as "precision farming," "site-specific crop management," "prescription farming," and "variable rate technology") has been developing since the 1990s, and refers to agricultural management systems carefully tailoring soil and crop management to fit the different conditions found in each field. PA is an information and technology-based agricultural management system (e.g., using remote sensing, geographic information systems, global positioning systems, and robotics) to identify, analyze, and manage soil spatial and temporal variability within fields for optimum profitability, sustainability, and protection of the environment (Bongiovanni and Lowenberg-Deboer, 2004; NRC, 1997; Gebbers and Adamchuk, 2010; Schrijver, 2016). PA is believed to be able to reduce the amount of inputs required, and better protect crops and soil.

Performances of PA are still debated, as comprehensive research is lacking (Yost et al., 2017). Paustian and Theuvsen (2017) noted that the advantages of PA adoption by farmers have been demonstrated by numerous expost studies, but most existing studies concentrate on only a few aspects of PA adoption. Paustian and Theuvsen (2017) noted that, in Germany, farming a large amount of arable land has a significant effect on PA adoption by farmers. However, the authors do not provide information about any changes in performance since PA adoption. Yost et al. (2017) analyzed a long-term dataset from Missouri on wheat, maize, and soybean cultivation, comparing yields before and after adoption of PA practices. The authors concluded that the greatest production advantage of a decade of PA lay in reducing temporal yield variation but did not concern yields. Nevertheless, they claimed that reducing yield variation was a positive outcome, as it leads to greater yield stability and resilience to a changing climate. Unfortunately, the work does not deal with the economic issues, and the economic sustainability of the enterprise is unclear. Robertson et al. (2009) analyzed the economic performance of six large Australian farms (1250–5800 ha cropping program) and found a benefit of PA adoption ranging from \$1 to \$22 ha^{-1} across the six farms, with the initial capital outlay recovered within 2–5 years. Due to the characteristics of PA, where data concerning farm and management practices are stored in databases, it has been highlighted that an issue might be posed by the future ownership of data (Schrijver, 2016).

2.7.2 Genetically modified crops

Adoption of herbicide-resistant GM crops (Roundup/glyphosate, or gluphosinate) allowed the adoption of no-till practices. Nevertheless, due to agricultural policies (e.g., heavy subsidies on maize in the United States) and global markets (e.g., increasing demand for soybean in Asia), herbicide-resistant crops have been eventually cropped continuously (maize after maize, or soybean after soybean), or, at best, in a very simple rotation (such as maize alternated to soybean in the United States), without adoption of long rotations, as recommended by the CA. Such practices led to weeds becoming resistant to herbicides (to date, about 220 weed species have evolved resistance to one or more herbicides, Heap, 2014; Bonny, 2016). Herbicides were thus used in higher amounts, dramatically increasing environmental pollution and causing human health concerns (Binimelis et al., 2009; Powles and Yu, 2010; Cerdeira et al., 2011; Benbrook, 2016; Bonny, 2016). Producers of GM crops are counteracting the issue by making HR crops resistant to other herbicides such as Dicamba and 2,4-D, which will further increase herbicide use and contamination, and are already causing dramatic conflicts among farmers. Dicamba, for example, is highly volatile and can drift away for many kilometers, affecting non-HT crops in the surrounding area and causing enormous economic damage. Damage for which, it seems, nobody is held accountable. Some farmers' associations sued Monsanto over the release of drifting herbicides (Charlier, 2017; Hall and Lokai-Minnich, 2018; Beck, 2018), and claimed that use of such compounds may be part of a strategy to force all farmers to plant HT crops, which, at present, are produced in a regime of near-monopoly. It is difficult to find any rationale in this approach to weed control, as it is clearly going to fail (and cause increased toxic compound accumulation in GM crops). Actually, in the United States, resistance to Dicamba in noxious weeds such as Burning bush (*Kochia scoparia*) and Pigweed (*Amaranthus* spp.) have been reported since the early 1990s (Cranston et al., 2001; Harwood et al., 2001; Steckel, 2017; LeClere et al., 2018). Therefore, widespread use of said herbicide on GM crops would quickly increase the problem.

GM crops engineered to produce Bt toxins went through a similar path, requiring agrochemical companies to increase the number of Bt toxins produced by plants (Tabashnik et al., 2013; Gassmann et al., 2014; Bøhn and Lövei, 2017), as well as to adopt other techniques of pest control at the same time (e.g., fungicides, hormonal traps).

It has been claimed that GM crops helped reduce the use of agrochemicals and increase farmers' profits, due to the lower amount of inputs and labor required by GM crops [for a later review see Klümper and Qaim (2014), although the choice of literature, including many references, may not meet rigorous scientific standards]. Positive outcomes may have been achieved in the early years following introduction of GM crops. However, as soon as weeds and pests developed resistance to glyphosate and Bt, use of agrochemicals skyrocketed. Further to that, along with the increasing cost of GM seeds (in the last decade, in the United States, the price of GM seeds increased by 300%–400%), farmers' profits were quickly eroded, and, as a matter of fact, many farmers, to save money, have been reported to be turning to conventional seeds (Bunge, 2016a,b; Hakim, 2016).

Adoption of GM crops has raised a number of concerns. Genetic contamination of wild plants and weeds are widely reported (Andersson and de Vicente, 2010; Bauer-Panskus et al., 2013). In the mid-2000s, nontransgenic canola fields were reported to have become contaminated with glyphosate-resistant canola (Cerdeira and Duke, 2006). Although herbicide-resistant GM crops are not affected by the herbicide, they do accumulate the herbicide in the plant, including the edible parts (Bøhn et al., 2014). That means that animals and humans are exposed to the compounds present in the herbicides: active principles, adjuvants (which may be much more toxic than the active principles, Mesnage et al., 2014), and their degradation products (which may be highly toxic too). Due to pest resistance to the first Bt toxins, GM crops are being engineered to produce many different Bt toxins (Bt maize producing 4–5 different toxins are on the market). Although Bt is considered unable to affect human health, and producers claim that GM Bt crops are safe (Koch et al., 2015), the increasing quantity of Bt toxins (of different types) present in GM crops, and the possible interactions between Bt toxins and pesticide residues and other compounds present in plants should be a matter of concern (Then, 2010; Mesnage et al., 2013; Then and Bauer-Panskus, 2017).

Introduction of GM crops did not result in increased crop yields. Although in the last few decades the yield of soybean, maize, and other crops has risen substantially, that was due to the improvement of varieties (achieved by traditional breeding) or improvement of agricultural practices (Grassini et al., 2013). NAS (2016, p. 7) states that "...there is no evidence from USDA data that they have substantially increased the rate at which of US agriculture is increasing yield." Recently, Nilsen (2017) reported that the rate of yield increase of US maize has been the same for the last 50 years, and that introduction of GM maize has not resulted in a noticeably increased yield growth trend.

It has been claimed that adoption of GM crops may help African countries to better feed themselves. Some experts highlight that very low amounts of inputs are used in Africa, therefore higher yields can be achieved by simply helping farmers buy fertilizers, or adopt more suitable crop varieties and agricultural practices (Sanchez, 2010). The IAASTD report[2] (IAASTD, 2009) on GM crops, which involved more than 400 experts, concluded that they may not represent a suitable tool for reducing hunger and poverty, improving nutrition, providing health and rural livelihoods, and facilitating social and environmental sustainability.

2.8 Conclusion

A number of issues are of major concern for the sustainability of our food system. A larger population has to be fed, and economic development is driving hundreds of millions of people to shift from a diet based mostly on vegetables to one rich in animal products. This is leading to an increasing cost of food production, both in terms of resource use and environmental impact of agriculture. In many regions of the globe, especially in those more densely populated, we are already experiencing dramatic problems concerning soil degradation, water shortages, energy supply, and environmental contamination by agrochemicals. Intensive and inappropriate agricultural practices, while boosting yield and profits in the short term, may put long-term productivity and food security at risk. Therefore, adoption of sound agricultural practices is of primary importance to preserve soil health, reduce the environmental impact of agriculture, and reduce yield loss.

In this chapter, a number of agricultural practices proposed as sustainable alternatives to conventional agriculture have been reviewed and assessed. Such practices have different goals. Some of them (i.e., CA), aim at improving soil conservation, in particular at reducing soil erosion. Others (i.e., those proposed by the agroecology movement) are more concerned with an ecological management of crops, integrated with the local landscape, that can protect soil, prevent the insurgence of pests, and reduce the use of agrochemicals.

[2] IAASTD was a 3-year project promoted by the United Nations, the World Bank and the World Health Organization, aiming at assessing agricultural knowledge and science and technology in relation to reducing hunger and poverty, improving nutrition, providing health and rural livelihoods, and facilitating social and environmental sustainability.

Upon reviewing the above, it appears that CA, while offering important benefits in view of soil protection, may nevertheless present some important drawbacks. Therefore, the impact of its adoption has been carefully monitored, and depending on the specific case, it may be useful to favor minimum tillage in place of continuous no-till. It has to be stressed that CA relies on the use of herbicides, which are required for weed control. In the long term, such practice leads to soil and water contamination by chemicals, and when badly implemented (as it is often the case, especially since GM crops were widely cropped), causes weeds to become resistant to herbicides, forcing farmers to apply more and more herbicides. The latter issue is of great concern, with particular reference to the sustainability of GM crops (another highly relevant issue concerns the fact that GM crops also accumulate herbicides in the edible parts). Agroecology offers a range of practices that seem able to provide multiple benefits: preventing yield loss, protecting the soil, reducing the use of inputs, preserving crop genetic biodiversity, and preserving the agroecological landscape. Also, in this case, monitoring is required to assess how well a practice may fit into the specific features of the local agroecological and socioeconomic systems, and pros and cons have to be carefully weighed.

Further research is needed to explore the potential of low-impact agroecological practices, to further improve them and make them available to an ever-greater number of farmers.

In parallel, sound agricultural policies have to be developed (fostering collaboration among the different actors of the food system) to facilitate adoption of those practices by farmers. Eventually, functioning of the whole food system should be addressed, including critical issues such as postharvest food waste, the overall impact of food choices, the alternative use of food such as the production of biofuels, power relations along the food chain, and the impact of the globalization process.

The rapid changes we are experiencing both socially (e.g., population growth) and environmentally (e.g., the potential impact of climate change) provide an urgent warning signal to policy makers, researchers, and society as a whole to address such issues promptly to better cope with the major challenges waiting ahead.

Acknowledgments

I wish to thank Dr. Charis Michel Galanakis, for inviting me to join this project. I wish to thank Dr. Lucio Marcello, researcher at the Rivers and Lochs Institute, University of the Highlands and Islands, UK, for editing the manuscript.

References

Abbas, H.K., Bellaloui, N., Bruns, H.A., 2016. Investigating transgenic corn hybrids as a method for mycotoxin control. Food Nutr. Sci. 7, 44–54.

Alexandratos, N., Bruinsma, J., 2012. World Agriculture: Towards 2030/2050: The 2012 Revision. FAO, Rome, Italy. Available online at: <http://www.fao.org/docrep/016/ap106e/ap106e.pdf> (accessed 10.10.15.).

Altieri, M., 1999. The ecological role of biodiversity in agroecosystems. Agric. Ecosyst. Environ. 74, 19–31.
Altieri, M., 1987. Agroecology: The Science of Sustainable Agriculture. Westview Press, Boulder, NY, USA.
Altieri, M.A., 2018. Agroecology the Science of Sustainable Agriculture, second ed. CRC press, Boca Raton, FL, USA.
Altieri, M.A., 2002. Agroecology: The science of natural resource management for poor farmers in marginal environments. Agric. Ecosys. & Environ. 93, 1–24.
Altieri, M.A., Nicholls, C.I., 2004. Biodiversity and Pest Management in Agroecosystems, second ed. CRC Press, Boca Raton, FL, USA.
Altieri, M.A., Ponti, L., Nicholls, C.I., 2012. Soil fertility, biodiversity and pest management. In: Gurr, G.M., Wratten, S.D., Snyder, W.E., Read, D.M.Y. (Eds.), Biodiversity and Insect Pests: Key Issues for Sustainable Management. John Wiley & Sons, Chichester, West Sussex, UK, pp. 72–84.
Alyokhin, A., Porter, G., Groden, E., Drummond, F., 2005. Colorado potato beetle response to soil amendments: a case in support of the mineral balance hypothesis? Agric. Ecosyst. Environ. 109, 234–244.
Álvarez, C.R., Taboada, M.A., Perelman, S.B., Morrás, H.J., 2014. Topsoil structure in no-tilled soils in the Rolling Pampas, Argentina. Soil Research 52, 533–542.
Andersson, M.S., de Vicente, M.C., 2010. Gene Flow Between Crops and Their Wild Relatives. Johns Hopkins University Press, Baltimore, Canada.
Atangana, A., Khasa, D., Chang, S., Degrande, A., 2014. Tropical Agroforestry. Springer, Dordrecht, The Netherlands.
Baker, B., 2017. Can modern agriculture be sustainable? BioScience 67, 325–331.
Bauer-Panskus, A., Hamberger, S., Then, C., 2013. Transgene escape: global atlas of uncontrolled spread of genetically engineered plants. Testbiotech. Institute for Independent Impact Assessment in Biotechnology, Munich, Germany. Available from: http://www.testbiotech.org/en/node/944.
Beck, M., 2018. Federal suit alleges companies knew dicamba would drift, Monsanto created monopoly. Available from: http://www.kunc.org/post/federal-suit-alleges-companies-knew-dicamba-would-drift-monsanto-created-monopoly
Benbrook, C.M., 2016. Trends in glyphosate herbicide use in the United States and globally. Environ. Sci. Eur. 28, 3. Available from: https://doi.org/10.1186/s12302-016-0070-0.
Bender, S.F., Wagg, C., van der Heijden, M.G.A., 2016. An underground revolution: biodiversity and soil ecological engineering for agricultural sustainability. Trends Ecol. Evol. 31, 440–452.
Bennett, A.J., Bending, G.D., Chandler, D., Hilton, S., Mills, P., 2012. Meeting the demand for crop production: the challenge of yield decline in crops grown in short rotations. Biol. Rev. 87, 52–71.
Bianchi, F., Van Der Werf, W., 2003. The effect of the area and configuration of hibernation sites on the control of aphids by *Coccinella septempunctata* (Coleoptera: Coccinellidae) in agricultural landscapes: a simulation study. Environ. Entomol. 32, 1290–1304.
Bindraban, P.S., van der Velde, M., Ye, L., van den Berg, M., Materechera, S., Innocent Kiba, D., et al., 2012. Assessing the impact of soil degradation on food production. Curr. Opin. Environ. Sustain. 4, 478–488.
Binimelis, R., Pengue, W., Monterroso, I., 2009. "Transgenic treadmill": responses to the emergence and spread of glyphosate-resistant johnsongrass in Argentina. Geoforum 40, 623–633.

Blackmore Smith, C., Hopkins, E., 2018. Produce is less healthy than it was 70 years ago. USA today. https://eu.usatoday.com/story/money/nation-now/2018/07/05/cover-crops-they-future-farming-united-states/759371002/.

Blanco-Canqui, H., Lal, R., 2008. No-tillage and carbon sequestration: An on-farm assessment. Soil Sci. Soc. Am. J. 72, 693–701.

Blok, W.J., Lamers, J.G., Termorshuizen, A.J., Bollen, G.J., 2000. Control of soilborne plant pathogens by incorporating fresh organic amendments followed by tarping. Phytopathology 90, 253–259.

Bøhn, T., Lövei, G.L., 2017. Complex outcomes from insect and weed control with transgenic plants: ecological surprises? Front. Environ. Sci. 5. Available from: https://doi.org/10.3389/fenvs.2017.00060.

Bøhn, T., Cuhra, M., Traavik, T., Sanden, M., Fagan, J., Primicerio, R., 2014. Compositional differences in soybeans on the market: glyphosate accumulates in Roundup Ready GM soybeans. Food Chem. 153, 207–215.

Bommarco, R., Miranda, F., Bylund, H., Björkman, C., 2011. Insecticides suppress natural enemies and increase pest damage in cabbage. J. Econ. Entoml. 104, 782–791.

Bongiovanni, R., Lowenberg-Deboer, J., 2004. Precision Agriculture and Sustainability. Precision Agriculture 5, 359–387.

Bonny, S., 2016. Genetically modified herbicide-tolerant crops, weeds, and herbicides: overview and impact. Environ. Manage. 57, 31–48.

Bot, A., Benites, J., 2005. The Importance of Soil Organic Matter Key to Drought-Resistant Soil and Sustained Food Production. FAO Soils Bulletin 80, FAO, Rome, Italy. Available from: http://www.fao.org/docrep/009/a0100e/a0100e00.htm.

Brankatschk, G., Finkbeiner, M., 2015. Modeling crop rotation in agricultural LCAs—challenges and potential solutions. Agric. Syst. 138, 66–76.

Brooker, R.W., Karley, A.J., Newton, A.C., Pakeman, R.J., Schöb, C., 2016. Facilitation and sustainable agriculture: a mechanistic approach to reconciling crop production and Conservation. Functional Ecology 30, 98–107.

Brookfield, H.C., 2001. Exploring Agrodiversity. Columbia University Press, New York, USA.

Brookfield, H.C., Padoch, C., Parsons, H., Stocking, M., 2002. Cultivating Biodiversity: Understanding, Analysing and Using Agricultural Diversity. ITDG Publications, London, UK.

Brookfield, H., Padoch, C., 2007. Managing biodiversity in agricultural ecosystemsIn: Jarvis, D.I., Padoch, C., Cooper, H.D. (Eds.), Columbia University Press, New York and Bioversity International, Rome, Italy, pp 321–338.

Brown, B., Nuberg, I., Llewellyn, R., 2018. Constraints to the utilisation of conservation agriculture in Africa as perceived by agricultural extension service providers. Land Use Policy 73, 331–340.

Bruinsma, J., 2011. The Resources Outlook: By how much do land, water and crop yields need to increase by 2050? FAO, Rome, Italy. Available online at: <ftp://ftp.fao.org/agl/aglw/docs/ResourceOutlookto2050.pdf> (accessed 10.10.15.).

Bruinsma, J. (Ed.), 2003. World Agriculture: Towards 2015/2030. An FAO Perspective. Earthscan, London, UK. Available online at: <ftp://ftp.fao.org/docrep/fao/005/y4252e/y4252e.pdf> (accessed 10.10.15.).

Bunge, J., 2016a. Farmers' illegal use of herbicide takes toll on neighboring crops. Wall Street Journal . Available from: https://www.wsj.com/articles/farmers-illegal-use-of-herbicide-takes-toll-on-neighboring-crops-1470130201.

Bunge, J., 2016b. Behind the Monsanto deal, doubts about the GMO revolution. The Wall Street. Journal, Sept. 14, 2016. Available from: https://www.wsj.com/articles/behind-the-monsanto-deal-doubts-about-the-gmo-revolution-1473880429.

Busari, M.A., Singh Kukal, S.S., Kaur, A., Bhatt, R., Dulazi, A.A., 2015. Conservation tillage impacts on soil, crop and the environment. Int. Soil Water Conserv. Res. 3, 119−129.

Butler, D.M., Kokalis-Burelle, N., Muramoto, J., Shennan, C., McCollum, T.G., Rosskopf, E. N., 2012. Impact of anaerobic soil disinfestation combined with soil solarization on plant−parasitic nematodes and introduced inoculum of soilborne plant pathogens in raised-bed vegetable production. Crop Prot. 39, 33−40.

Campbell, B.M., Beare, D.J., Bennett, E.M., Hall-Spencer, J.M., Ingram, J.S.I., Jaramillo, F., et al., 2017. Agriculture production as a major driver of the Earth system exceeding planetary boundaries. Ecol. Soc. 22 (4), 8. Available from: https://doi.org/10.5751/ES-09595-220408.

Carlisle, L., 2016. Factors influencing farmer adoption of soil health practices in the United States: a narrative review. Agroecol. Sustain. Food Syst. 40, 583−613.

Carr, P.M., Gramig, G.G., Liebig, M.A., 2013. Impacts of organic zero tillage systems on crops, weeds, and soil quality. Sustainability 5, 3172−3201. Available from: https://doi.org/10.3390/su5073172.

Cassman, K.G., Dobermann, A., Walters, D.T., Yang, H., 2003. Meeting cereal demand while protecting natural resources and improving environmental quality. Annu. Rev. Environ. Resour. 28, 315−358.

Castelan, F.P., Castro-Alves, V.C., Saraiva, L.A., Nascimento, T.P., Cálhau, M.F.N.S., Dias, C.T.S., et al., 2018. Natural ecosystem surrounding a conventional banana crop improves plant health and fruit quality. Front. Plant Sci. 9, . Available from: https://doi.org/10.3389/fpls.2018.00759Article 759.

Cerdeira, A.L., Duke, S.O., 2006. The current status and environmental impacts of glyphosate-resistant crops: a review. J. Environ. Qual. 35, 1633−1658.

Cerdeira, A.L., Gazziero, D.L., Duke, S.O., Matallo, M.B., 2011. Agricultural impacts of glyphosate-resistant soybean cultivation in South America. J. Agric. Food Chem. 59, 5799−5807.

Charlier, T., 2017. Farmers sue Monsanto, other producers over drifting herbicide. USA today network, July 28, 2017. https://eu.commercialappeal.com/story/news/2017/07/28/farmers-sue-monsanto-other-producers-over-drifting-herbicide/504656001/.

Chhokar, R.S., Sharma, R.K., 2008. Multiple herbicide resistance in little seed canary grass (*Phalaris minor*): a threat to wheat production in India. Weed Biol. Manage. 8, 112−123.

Clough, Y., Barkmann, J., Juhrbandt, J., Kessler, M., Wanger, T.C., Anshary, A., et al., 2011. Combining high biodiversity with high yields in tropical agroforests. Proc. Natl. Acad. Sci. 108, 8311−8316.

Codex Alimentarius, 2004. Guidelines for the production, processing, labelling and marketing of organically produced foods (GL 32−1999, Rev. 1−2001). Internet <http://www.codexalimentarius.net/web/standard_list.do?lang = en> (accessed 11.12.).

Conway, G., 2012. One Billion Hungry: Can We Feed the World? Cornell University Press, Ithaca, USA.

Cook, S.M., Khan, Z.R., Pickett, J.R., 2007. The use of push-pull strategies in integrated pest management. Annu. Rev. Entomol. 52, 375−400.

Cooper, J., Baranski, M., Stewart, G., Nobel-de Lange, M., Bàrberi, P., et al., 2016. Shallow non-inversion tillage in organic farming maintains crop yields and increases soil C stocks: a meta-analysis. Agron. Sustain. Dev. 36 (1), 22.

Corbeels, M., de Graaff, J., Ndah, T.H., Penot, E., Baudron, F., Naudin, K., et al., 2014. Understanding the impact and adoption of conservation agriculture in Africa: a multi-scale analysis. Agric., Ecosyst. Environ. 187, 155–170.

Cox, T.S., Glover, J.D., Van Tassel, D.L., Cox, C.M., DeHaan, L.R., 2006. Prospects for developing perennial grain crops. Bioscience 56, 649–659.

Cox, T.S., Van Tassel, D.L., Cox, C.M., Dehaan, L.R., 2010. Progress in breeding perennial grains. Crop Pasture Sci. 61, 513–521.

Cranston, H.J., Kern, A.J., Hackett, J.L., Miller, E.K., Maxwell, B.D., Dyer, W.E., 2001. Dicamba resistance in kochia. Weed Science 49, 164.

Crews, T.E., Rumsey, B.E., 2017. What agriculture can learn from native ecosystems in building soil organic matter: a review. Sustainability 9, 578. Available from: https://doi.org/10.3390/su9040578.

Crowder, D.W., Northfield, T.D., Strand, M.R., Snyder, W.E., 2010. Organic agriculture promotes evenness and natural pest control. Nature 466, 109–112.

Dalgaard, T., Hutchings, N.J., Porter, J.R., 2003. Agroecology, scaling and interdisciplinarity. Agr. Ecosyst. Environ. 100, 39–51.

Dangi, S.R., Tirado-Corbalá, R., Gerik, J., Hanson, B.D., 2017. Effect of long-term continuous fumigation on soil microbial communities. Agronomy 7 (2), 37. Available from: https://doi.org/10.3390/agronomy7020037.

Davis, D.R., 2009. Declining fruit and vegetable nutrient composition: what is the evidence? Hortscience 44, 15–19.

DeFries, R.S., Rudel, T., Uriarte, M., Hansen, M., 2010. Deforestation driven by urban population growth and agricultural trade in the twenty-first century. Nat. Geosci. 3, 178–181.

Deguine, J.-P., Ferron, P., Russell, D., 2009. Crop Protection From Agrochemistry to Agroecology. CRC Press Taylor & Francis Group, Boca Raton, FL, USA.

Delate, K., Cwach, D., Chase, C., 2011. Organic no-tillage system effects on soybean, corn and irrigated tomato production and economic performance in Iowa, USA. Renew. Agric. Food Syst. 27 (1), 49–59.

DeLonge, M.S., Miles, A., Carlisle, L., 2016. Investing in the transition to sustainable agriculture. Environ Sci Policy. 55, 266–273.

den Biggelaar, C., Lal, R., Wiebe, K., Breneman, V., 2004. The global impact of soil erosion on productivity. I: absolute and relative erosion-induced yield losses. Adv. Agron. 81, 1–48.

Domínguez, A., Bedano, J.C., Becker, A.R., 2010. Negative effects of no-till on soil macrofauna and litter decomposition in Argentina as compared with natural grasslands. Soil Tillage Res. 110, 51–59.

Domínguez, A., Bedano, J.C., 2016. The adoption of no-till instead of reduced tillage does not improve some soil quality parameters in Argentinean Pampas. Appl. Soil Ecol. 98, 166–176.

Döringa, T.F., Annicchiarico, P., Clarke, S., Haigh, Z., Jones, H.E., Pearce, H., et al., 2015. Comparative analysis of performance and stability among composite cross populations, variety mixtures and pure lines of winter wheat in organic and conventional cropping systems. Field Crops Res. 183, 235–245.

Dwivedi, S.L., Ceccarelli, S., Blair, M.W., Upadhyaya, H.D., Are, A.K., Ortiz, R., 2016. Landrace germplasm for improving yield and abiotic stress adaptation. Trends Plant Sci. 21, 31–42.

EC (European Commission), 2018. Organic production and labelling of organic products. European Parliament legislative resolution of 19 April 2018. <http://www.europarl.europa.eu/sides/getDoc.do?pubRef = -//EP//TEXT + TA + P8-TA-2018-0180 + 0 + DOC + XML + V0//EN>.

Erisman, J.W., Galloway, J.N., Seitzinger, S., Bleeker, A., Dise, N.B., Petrescu, A.M.R., et al., 2013. Consequences of human modification of the global nitrogen cycle. Phil. Trans. R. Soc. B 368, 20130116.

Eurostat, 2013. Agri-environmental indicator - tillage practices. <http://ec.europa.eu/eurostat/statistics-explained/index.php?title = Agri-environmental_indicator_-_tillage_practices>.

Fan, M.-S., Zhao, F.J., Fairweather-Tait, S.J., Poulton, P.R., Dunham, S.J., McGrath, S.P., 2008. Evidence of decreasing mineral density in wheat grain over the last 160 years. J Trace Elem Med Bio. 22, 315–324.

FAO, 2011a. Global Food Losses and Food Waste – Extent, Causes and Prevention. FAO, Rome, <http://www.fao.org/docrep/014/mb060e/mb060e00.pdf>.

FAO, 2011b. The State of the World's Land and Water Resources for Food and Agriculture (SOLAW) – Managing Systems at Risk. Food and Agriculture Organization of the United Nations, Rome and Earthscan, London, UK. Available online at: <http://www.fao.org/docrep/017/i1688e/i1688e.pdf> (accessed 15.09.15.).

FAO, 2014a. Technical Platform on the Measurement and Reduction of Food Loss and Waste. FAO, Roma. Available from: http://www.fao.org/platform-food-loss-waste/en/.

FAO, 2014b. Perennial Crops for Food Security. Proceedings of the FAO Expert Workshop August 28–30, 2013, Rome, Italy. http://www.fao.org/3/a-i3495e.pdf.

FAO (The Food and Agriculture Organization of the United Nations), 2015. Soil Degradation. FAO, Rome, Italy, <http://www.fao.org/soils-portal/soil-degradation-restoration/it/> (accessed 10.10.15.).

FAO and ITPS, 2015. Status of the World's Soil Resources (SWSR) – Main Report. Food and Agriculture Organization of the United Nations and Intergovernmental Technical Panel on Soils, Rome, Italy, 2015. Available online at <http://www.fao.org/3/a-i5199e.pdf> (accessed 20.03.18.).

FAO, 2018a. Conservation Agriculture. <http://www.fao.org/conservation-agriculture/en/>.

FAO, 2018b. Agroforestry. <http://www.fao.org/forestry/agroforestry/en/>.

Farrell, J.G., Altieri, M.A., 2018. Agroforestry systems. In: Altieri, M.A. (Ed.), Agroecology the Science of Sustainable Agriculture, second ed. CRC Press, Boca Raton, FL, USA, pp. 247–263.

Flint, M.L., van den Bosch, R., 1981. Introduction to Integrated Pest Management. Plenum Press, New York, USA.

Fowler, C., Hodgkin, T., 2005. Plant genetic resources for food and agriculture: assessing global availability. Annu. Rev. Environ. Resour. 29, 143–179.

Foley, J.A., Defries, R., Asner, G.P., Barford, C., Bonan, G., Carpenter, S.R., et al., 2005. Global consequences of land use. Science 309, 570–574.

Foley, J.A., Ramankutty, N., Brauman, K.A., Cassidy, M.S., Gerber, J.S., Johnston, M., et al., 2011. Solutions for a cultivated planet. Nature 478, 337–342.

Friedrich, T., Kassam, A., 2012. No-till farming and the environment: do no-till systems require more chemicals? Outlooks Pest Manage. 23, 153–157.

Friedrich, T., Derpsch, R., Kassam, A., 2012. Overview of the global spread of conservation agriculture. <http://factsreports.revues.org/1941>.

Francis, C.A., 1989. Biological efficiencies in multiple-cropping systems. Advances in Agronomy 42, 1–42.

Frison, A.E., Cherfas, J., Hodgkin, T., 2011. Agricultural biodiversity is essential for a sustainable improvement in food and nutrition security. Sustainability 3, 238–253.

Francis, C.A., Lieblein, G., Gliessman, S., Breland, T.A., Creamer, N., Harwood, R., et al., 2003. Agroecology: The ecology of food systems. J. Sustain. Agric. 22, 99–118.

Furlan, L., Kreutzweiser, D., 2015. Alternatives to neonicotinoid insecticides for pest control: case studies in agriculture and forestry. Environ. Sci. Pollut. Res. 22, 135–147.

Furlan, L., Contiero, B., Chiarini, F., Colauzzi, M., Sartori, E., Benvegnù, I., et al., 2017. Risk assessment of maize damage by wireworms (Coleoptera: Elateridae) as the first step in implementing IPM and in reducing the environmental impact of soil insecticides. Environ. Sci. Pollut. Res. 24, 236–251.

Gamliel, A., Austerweil, M., Kritzman, G., 2000. Non-chemical approach to soilborne pest management—organic amendments. Crop Prot. 19, 847–853.

Gardiner, M.M., Landis, D.A., Gratton, C., DiFonzo, C.D., O'Neal, M., Chacon, J.M., et al., 2009. Landscape diversity enhances biological control of an introduced crop pest in the north-central USA. Ecol. Appl. 19, 143–154.

Gassmann, A.J., Petzold-Maxwel, J.L., Clifton, E.H., Dunbar, M.W., Hoffmann, A.M., Ingber, D.A., et al., 2014. Field-evolved resistance by western corn rootworm to multiple *Bacillus thuringiensis* toxins in transgenic maize. Proc. Natl. Acad. Sci. 111, 5141–5146.

Gentry, L.F., Ruffo, M.L., Below, F.E., 2013. Identifying factors controlling the continuous corn yield penalty. Agron. J. 105, 295–303.

Gebbers, R., Adamchuk, V.I., 2010. Precision agriculture and food security. Science 327, 828–831.

Giampietro, M., 2004. Multi-Scale Integrated Analysis of Agro-ecosystems. CRC Press, Boca Raton, FL.

Giampietro, M., Aspinall, R.J., Ramos-Martin, J., Bukkens, S.G.F., 2014. Resource Accounting for Sustainability Assessment: The Nexus Between Energy, Food, Water and Land Use. Routledge, New York, USA.

Gibbs, H.K., Salmon, J.M., 2015. Mapping the world's degraded lands. Appl. Geogr. 57, 12–21.

Gibbs, H.K., Ruesch, A.S., Achard, F., Clayton, M.K., Holmgren, P., Ramankutty, N., et al., 2010. Tropical forests were the primary sources of new agricultural land in the 1980s and 1990s. Proc. Natl. Acad. Sci. 107, 16732–16737.

Gliessman, S.R., 2014. Agroecology. The Ecology of Sustainable Food System. CRC Press, Boca Raton, FL, USA.

Glover, J.D., Cox, C.M., Reganold, J.P., 2007. Future farming: a return to roots? Sci. Am. 83–89. August.

Glover, J.D., Reganold, J.P., Cox, C.M., 2012. Plant perennials to save Africa's soils. Nature 489, 359–361.

Glover, J.D., Culman, S.W., DuPont, T., Broussard, W., et al., 2010a. Harvested perennial grasslands provide ecological benchmarks for agricultural sustainability. Agric. Ecosyst. Environ. 137, 3–12.

Glover, J.D., Reganold, J.P., Bell, L.W., Borevitz, J., et al., 2010b. Increased food and ecosystem security via perennial grains. Science 328, 1638–1639.

Godfray, C., Beddington, J., Crute, I., Haddad, L., Lawrence, D., Muir, J., et al., 2010. Food security: the challenge of feeding nine billion people. Science 27, 812–818.

Godfray, H.C.J., Aveyard, P., Garnett, T., Hall, J.W., Key, T.J., Lorimer, J., et al., 2018. Meat consumption, health, and the environment. Science 361, eaam5324. Available from: https://doi.org/10.1126/science.aam5324.

Gomiero, T., 2015a. Are biofuels an effective and viable energy strategy for industrialized societies? A reasoned overview of potentials and limits. Sustainability 7, 8491–8521. Available from: https://doi.org/10.3390/su7078491.

Gomiero, T., 2015b. Effects of agricultural activities on biodiversity and ecosystems: organic vs. conventional farming. In: Robinson, G., Schmallegger, D. (Eds.), The Globalisation of Agriculture. Edward Elgar Handbook, pp. 77–105. Chapter 3.

Gomiero, T., 2016. Soil degradation, land scarcity and food security: reviewing a complex challenge. Sustainability 8, 1–41. Available from: http://www.mdpi.com/2071-1050/8/3/281.

Gomiero, T., 2017. Biophysical analysis of agri-food systems: scales, energy efficiency, power and metabolism of society. In: Fraňková, E., Haas, W., Singh, S.J. (Eds.), Sociometabolic Perspectives on Sustainability of Local Food Systems: New Insights for Science, Policy and Practice. Springer, pp. 69–102. Chapter 3.

Gomiero, T., 2018a. Large-scale biofuels production: a possible threat to soil conservation and environmental services. Appl. Soil Ecol. 123, 729–736.

Gomiero, T., 2018b. Food quality assessment in organic vs. conventional agricultural produce: Findings and issues. Applied Soil Ecology 123 (2018), 714–728.

Gomiero, T., 2018c. Agriculture and degrowth: state of the art and assessment of organic and biotech-based agriculture from a degrowth perspective. J. Cleaner Prod. 197, 1823–1839.

Gomiero, T., Pimentel, D., Paoletti, M.G., 2011a. Is there a need for a more sustainable agriculture? Crit. Rev. Plant Sci. 30, 6–23.

Gomiero, T., Pimentel, D., Paoletti, M.G., 2011b. Environmental impact of different agricultural management practices: conventional vs. organic agriculture. Crit. Rev. Plant Sci. 30, 95–124.

Good, A.G., Beatty, P.H., 2011. Fertilizing nature: a tragedy of excess in the commons. PLoS Biol. 9 (8), e1001124. Available from: https://doi.org/10.1371/journal.pbio.1001124.

Grassini, P., Eskridge, K.M., Cassman, K.G., 2013. Distinguishing between yield advances and yield plateaus in historical crop production trends. Nat. Commun. 4, 2918. Available from: https://doi.org/10.1038/ncomms3918.

Grassini, P., Torrion, J.A., Yang, H.S., Rees, J., Andersen, D., Cassman, K.G., et al., 2015. Soybean yield gaps and water productivity in the western U.S. Corn Belt. Field Crops Res. 179, 150–163.

Gregory, A.S., Dungait, J.A.J., Watts, C.W., Bol, R., Dixon, E.R., White, R.P., et al., 2016. Long-term management changes topsoil and subsoil organic carbon and nitrogen dynamics in a temperate agricultural system. Eur J Soil Sci. 67, 421–430.

Gurr, G.M., Lu, Z., Zheng, X., Xu, H., Zhu, P., Chen, G., et al., 2016. Multi-country evidence that crop diversification promotes ecological intensification of agriculture. Nature Plants 2, 16014.

Haberl, H., Erb, K.-H., Krausmann, F., 2014. Human appropriation of net primary production: patterns, trends, and planetary boundaries. Annu. Rev. Environ. Resour. 39, 363–391.

Hakim, D., 2016. Doubts about the promised bounty of genetically modified crops. The New York Times, October 29, 2016. <https://www.nytimes.com/2016/10/30/business/gmo-promise-falls-short.html>.

Hainzelin, É., 2013. Cultivating Biodiversity to Transform Agriculture. Springer, Dordrecht, The Netherlands.

Hall, P.K., Lokai-Minnich, C., 2018. Documentation will be crucial in 2018 dicamba drift situations. <https://www.ohiofarmer.com/crops/documentation-will-be-crucial-2018-dicamba-drift-situations>.

Hamilton, S.K., Doll, J.E., Robertson, J.P., 2015. The Ecology of Agricultural Landscapes: Long-Term Research on the Path to Sustainability. Oxford University Press, New York, NY, USA.

Hammond, B.G., Campbell, K.W., Pilcher, C.D., Degooyer, T.A., Robinson, A.E., McMillen, B.L., et al., 2004. Lower fumonisin mycotoxin levels in the grain of Bt corn grown in the United States in 2000 – 2002. J. Agric. Food Chem. 52 (5), 1390–1397.

Hamza, M.A., Anderson, W.K., 2005. Soil compaction in cropping systems: a review of the nature, causes and possible solutions. Soil Tillage Res. 82, 121–145.

Han, G.Y., Lang, J., Sun, Y., Wang, Y.Y., Zhu, Y.Y., Lu, B.R., 2016. Intercropping of rice varieties increases the efficiency of blast control through reduced disease occurrence and variability. J. Integr. Agric. 15, 795–802.

Harwood, J.C., Kern, A.J., Hackett, J.L., Miller, E.K., Maxwell, B.D., Dyer, W.E., 2001. Dicamba resistance in kochia. Weed Sci. 49, 164–170.

Hatfield, J.L., Allmaras, R.R., Rehm, G.W., Lowery, B., 1998. Ridge tillage for corn and soybean production: environmental quality impacts. Soil Tillage Res. 48, 145–154.

Hayes, R.C., Newell, M.T., Crews, T.E., Peoples, M.B., 2016. Perennial cereal crops: an initial evaluation of wheat derivatives grown in mixtures with a regenerating annual legume. Renew. Agric. Food Syst. 32, 276–290.

Heal, G., Walker, B., Levin, S., Arrowg, K., Dasgupta, P., Daily, G., et al., 2004. Genetic diversity and interdependent crop choices in agriculture. Resour. Energy Econ. 26, 175–184.

Heap, I., 2014. Global perspective of herbicide-resistant weeds. Pest Manage. Sci. 70, 1306–1315.

HLPE (High Level Panel of Experts on Food Security and Nutrition), 2017. Nutrition and food systems. A Report by the High Level Panel of Experts on Food Security and Nutrition of the Committee on World Food Security. FAO, Rome, <http://www.fao.org/3/a-i7846e.pdf>.

Hobbs, P.R., 2007. Conservation agriculture: what is it and why is it important for future sustainable food production? J. Agr. Sci. 145, 127–137.

Hobbs, P.R., Sayre, K., Gupta, R., 2008. The role of conservation agriculture in sustainable agriculture. Phil. Trans. R. Soc. B 363, 543–555.

Hobbs, R.J., Higgs, E., Hall, C.M., Bridgewater, P., Chapin III, F.S., Ellis, E.C., 2014. Managing the whole landscape: historical, hybrid, and novel ecosystems. Front. Ecol. Environ. 12, 557–564. Available from: https://doi.org/10.1890/130300.

Hoy, M.A., 1998. Myths, models and mitigation of resistance to pesticides. Phil. Trans. R. Soc. B 353, 787–1795.

Hsu, Y.-T., Shen, T.-C., Hwang, S.-Y., 2009. Soil fertility management and pest responses: a comparison of organic and synthetic fertilization. J. Econ. Entomol. 102, 160–169.

Hussain, I., Olson, K.R., Ebelhar, S.A., 1999. Impact of tillage and no-till on production of maize and soybean on an eroded Illinois silt loam soil. Soil Tillage Res. 52, 37–49.

IAASTD (International Assessment of Agricultural Knowledge, Science and Technology for Development), 2009. Agriculture at the crossroad. Synthesis report. Island Press, Washington, USA. Available from: http://apps.unep.org/publications/pmtdocuments/-Agriculture%20at%20a%20crossroads%20-%20Synthesis%20report-2009Agriculture_at_Crossroads_Synthesis_Report.pdf.

IFOAM (International Movement of Organic Agriculture Movements), 2015. Principles of organic agriculture. <http://www.ifoam.bio/sites/default/files/poa_english_web.pdf>.

Islam, R., Reeder, R., 2014. No-till and conservation agriculture in the United States: an example from the David Brandt farm, Carroll, Ohio. Int. Soil Water Conserv. Res. 2, 97–107.

Jackson, W., 1980. New Roots for Agriculture. University of Nebraska Press.

Jackson, W., 2002. Natural systems agriculture: a truly radical alternative. Agric. Ecosyst. Environ. 88, 111–117.

Jacob, J., Shirmila, J., Sarada, S., Anu, S., 2010. Role of allelopathy in vegetables crops production. Allelopathy J. 25 (2), 275–312.

Jarvis, D.I., Brown, A.H.D., Cuong, P.H., Panduro, L.C., Moreno, L.L., Gyawali, S., et al., 2008. A global perspective of the richness and evenness of traditional crop-variety diversity maintained by farming communities. Proc. Natl. Acad. Sci. USA 105, 5326–5331.

Jarvis, D.I., Hodgkin, T., Sthapit, B.R., Fadda, C., Lopez-Noriega, I., 2011. An heuristic framework for identifying multiple ways of supporting the conservation and use of traditional crop varieties within the agricultural production system. Crit. Rev. Plant Sci. 30, 125–176.

Johnson, J., 1946. Soil steaming for disease control. Soil Sci. 61, 83–92.

Kassam, A., Derpsch, R., Friedrich, T., 2014. Global achievements in soil and water conservation: the case of conservation agriculture. Int. Soil Water Conserv. Res. 2, 5–13.

Kennedy, T.A., Naeem, S., Howe, K.M., Knops, J.M.H., Tilman, D., Reich, P., 2002. Biodiversity as a barrier to ecological invasion. Nature 417, 636–638.

Kertész, Á., Madarász, B., 2014. Conservation agriculture in Europe. Int. Soil Water Conserv. Res. 2, 91–96.

Kiær, L.P., Skovgaard, M., Østergård, H., 2009. Grain yield increase in cereal variety mixtures: a meta-analysis of field trials. Field Crops Res. 114, 361–373.

Kirkegaard, J.A., Conyers, M.K., Hunt, J.R., Kirkby, C.A., Watt, M., Rebetzke, G.J., 2014. Sense and nonsense in conservation agriculture: principles, pragmatism and productivity in Australian mixed farming systems. Agric. Ecosyst. Environ. 187, 133–145.

Klümper, W., Qaim, M., 2014. A meta-analysis of the impacts of genetically modified crops. PLoS One 9 (11), e111629.

Knörzer, H., Graeff-Hönninger, S., Guo, B., Wang, P., Claupein, W., 2009. The Rediscovery of Intercropping in China: A Traditional Cropping System for Future Chinese Agriculture – A Review. In: Lichtfouse, E. (Ed.), Climate Change, Intercropping, Pest Control and Beneficial. Springer, Dordrecht, The Netherland, pp. 13–44. Microorganisms, Sustainable Agriculture Reviews Vol. 2, 2009.

Koch, M.S., Ward, J.M., Levine, S.L., Baum, J.A., Vicini, J.L., Hammond, B.G., 2015. The food and environmental safety of Bt crops. Front. Plant Sci. 6, . Available from: https://doi.org/10.3389/fpls.2015.00283article 283.

Kong, C.H., Hu, F., Wang, P., Wu, J.L., 2008. Effect of allelopathic rice varieties combined with cultural management options on paddy field weeds. Pest Manage. Sci. 64, 276–282.

Lahma, R., 2010. Adoption of conservation agriculture in Europe: lessons of the KASSA project. Land Use Policy 27, 4–10.

Lal, R., 2002. Soil carbon dynamics in cropland and rangeland. Environ. Pollut. 116, 353–362.

Lal, R., 2010. Enhancing eco-efficiency in agro-ecosystems through soil carbon sequestration. Crop Sci. 50, 120–131.

Lal, R., 2015a. Restoring soil quality to mitigate soil degradation. Sustainability 7, 5875–5895. Available from: https://doi.org/10.3390/su7055875.

Lal, R., 2015b. On sequestering carbon and increasing productivity by conservation agriculture. J. Soil Water Conserv. 70, 55–62.

Lal, R., Reicosky, D.C., Hanson, J.D., 2007. Evolution of the plow over 10,000 years and the rationale for no-till farming. Soil Tillage Res. 93, 1–12.

Lampkin, N., 2002. Organic Farming, revised ed. Old Pond Publishing, Suffolk, UK.

Lang, T., Barling, D., Caraher, M., 2009. Food Policy: Integrating Health, Environment and Society. Earthscan, London, UK.

LeClere, S., Wu, C., Westra, P., Sammons, R.D., 2018. Cross-resistance to dicamba, 2,4-D, and fluroxypyr in *Kochia scoparia* is endowed by a mutation in an AUX/IAA gene. Proc. Natl. Acad. Sci. 115 (13), e2911–e2920.

Li, L., Li, S.-M., Sun, J.-H., Zhou, L.-L., Bao, X.-G., Zhang, H.-G., et al., 2007. Diversity enhances agricultural productivity via rhizosphere phosphorus facilitation on phosphorus-deficient soils. PNAS 204, 1192–11196.

Li, L., Zhang, L.Z., Zhang, F., 2013. Crop Mixtures and the Mechanisms of Overyielding. In: second edition Levin, S.A. (Ed.), Encyclopedia of Biodiversity, Volume 2. Academic Press, Usa, Waltham, MA, pp. 382–395.

Liu, S., Zhang, X., Liang, A., Zhang, J., Müller, C., Cai, Z., 2018. Ridge tillage is likely better than no tillage for 14-year field experiment in black soils: insights from a 15N-tracing study. Soil Tillage Res. 179, 38–46.

Lotter, D.W., Seidel, R., Liebhart, W., 2003. The performance of organic and conventional cropping systems in an extreme climate year. Am. J. Altern. Agr. 2003 (18), 146–154.

Lu, Z.X., Yu, X.-P., Heong, K.-L., Hu, C., 2007. Effect of nitrogen fertilizer on herbivores and its stimulation to major insect pests in rice. Rice Sci. 14, 56–66.

Luo, Z., Wang, E., Sun, O.J., 2010. Can no-tillage stimulate carbon sequestration in agricultural soils? A meta-analysis of paired experiments. Agric. Ecosyst. Environ. 139, 224–231.

Macfadyen, S., Gibson, R., Raso, L., Sint, L., Traugott, M., Memmott, J., 2009. Parasitoid control of aphids in organic and conventional farming systems. Agric. Ecosyst. Environ. 133, 14–18.

Mao, L., Wang, Q., Yan, D., Li, Y., Ouyang, C., Guo, M., et al., 2016. Flame soil disinfestation: a novel, promising, non-chemical method to control soilborne nematodes, fungal and bacterial pathogens in China. Crop Prot. 83, 90–94.

Marshall, D.R., 1977. The advantages and hazards of genetic homogeneity. Ann. NY Acad. Sci. 287, 1–20.

Martin-Guaya, M.-O., Paquette, A., Dupras, J., Rivest, D., 2018. The new Green Revolution: Sustainable intensification of agriculture by intercropping. Science of The Total Environment 615, 767–772.

Mauser, W., Klepper, G., Zabel, F., Delzeit, R., Hank, T., Putzenlechner, B., et al., 2015. Global biomass production potentials exceed expected future demand without the need for cropland expansion. Nat. Commun. 6, 8946. Available from: https://doi.org/10.1038/ncomms994.

Mayer, A.-M., 1997. Historical changes in the mineral content of fruits and vegetables. Br. Food J. 99, 207–211.

McDonald, B.A., Linde, C., 2002. Pathogen population genetics, evolutionary potential, and durable resistance. Annu. Rev. Phytopathol. 40, 349–379.

Medek, D.E., Schwartz, J., Myers, S.S., 2017. Estimated effects of future atmospheric CO_2 concentrations on protein intake and the risk of protein deficiency by country and region. Environ. Health Perspect. Available from: https://doi.org/10.1289/EHP41.

Mesnage, R., Clair, E., Gress, S., Then, C., Székács, A., Séralini, G.E., 2013. Cytotoxicity on human cells of Cry1Ab and Cry1Ac Bt insecticidal toxins alone or with a glyphosate-based herbicide. J. Appl. Toxicol. 33 (7), 695–699.

Mesnage, R., Defarge, N., Spiroux de Vendômois, J., Séralini, G.E., 2014. Major pesticides are more toxic to human cells than their declared active principles. BioMed Res. Int. Available from: https://doi.org/10.1155/2014/179691.

Mitchella, N.J., Bowersb, E., Hurburghb, C., Wu, F., 2017. Potential economic losses to the USA corn industry from aflatoxin contamination. Food Addit. Contam. Part A Chem. Anal. Control Expo. Risk Assess. 33 (3), 540–550.

Molden, D. (Ed.), 2007. Water for Food, Water for Life. A Comprehensive Assessment of Water Management in Agriculture, Earthscan, London, UK. Available at <http://www.iwmi.cgiar.org/assessment/Publications/books.htm> (accessed 10.10.15.).

Mols, C.M.M., Visser, M.E., 2007. Great tits (*Parus major*) reduce caterpillar damage in commercial apple orchards. PLoS One 2, e202. Available from: https://doi.org/10.1371/journal.pone.0000202.

Momma, N., 2008. Biological soil disinfestation (BSD) of soilborne pathogens and its possible mechanisms. Jpn. Agric. Res. Quart. 42, 7–12.

Momma, N., Kobara, Y., Uematsu, S., Kita, N., Shinmura, A., 2013. Development of biological soil disinfestations in Japan. Appl. Microbiol. Biotechnol. 97, 3801–3809.

Montgomery, D.R., 2007a. Dirt: The Erosion of Civilization. University of California Press, Berkeley, USA.

Montgomery, D.R., 2007b. Soil erosion and agricultural sustainability. Proc. Natl. Acad. Sci. 104, 13268–13272.

Morgan, R.P.C., 2005. Soil Erosion and Conservation, third ed. Blackwell, Oxford, UK.

Mosquera-Losada, M.R., Santiago-Freijanesa, J.J., Rois-Díaza, M., Moreno, M., den Herder, M., Aldrey-Vázquez, J.A., et al., 2018. Agroforestry in Europe: a land management policy tool to combat climate change. Land Use Policy 78, 603–613.

Mulumba, W., Nankya, R., Adokorach, J., Kiwuka, C., Fadda, C., De Santis, P., et al., 2012. A risk-minimizing argument for traditional crop varietal diversity use to reduce pest and disease damage in agricultural ecosystem of Uganda. Agric. Ecosyst. Environ. 157, 70–86.

Myers, S.S., Smith, M.R., Guth, S., Golden, C.D., Vaitla, B., Mueller, N.D., et al., 2017. Climate change and global food systems: potential impacts on food security and undernutrition. Annu. Rev. Public Health 38, 259–277.

Newton, A.C., Akar, T., Baresel, J.P., Bebeli, P.J., Bettencourt, E., Bladenopoulos, K.V., et al., 2010. Cereal landraces for sustainable agriculture. a review. Agron. Sustain. Dev. 30, 237–269.

NAS (National Academies of Sciences, Engineering, and Medicine), 2016. Genetically Engineered Crops: Experiences and Prospects. The National Academies Press, Washington, DC. Available from: https://www.nap.edu/catalog/23395/genetically-engineered-crops-experiences-and-prospects.

Nestle, M., 2013. Food Politics: How the Food Industry Influences Nutrition and Health. University of California Press, USA.

Niggli, U., Andres, C., Willer, H., Bak, B.P., 2017. Building a global platform for organic farming research, innovation and technology transfer. Organic Agriculture 7, 209–224.

Nilsen, R.L., 2017. Historical corn grain yields for the U.S. <http://www.kingcorn.org/news/timeless/YieldTrends.html>.

NRC (National Research Council), 1997. Precision Agriculture in the 21st Century: Geospatial and Information Technologies in Crop Management. The National Academies Press, USA, Washington, DC. Available from: https://doi.org/10.17226/5491.

NRC (National Research Council), 2010. Toward Sustainable Agricultural Systems in the 21st Century. National Academies Press, Washington, DC, USA. Available online: <http://www.nap.edu/catalog/12832.html> (accessed 03.03.12.).
Panagos, P., Standardi, G., Borrelli, P., Lugato, E., Montanarella, L., Bosello, F., 2018. Cost of agricultural productivity loss due to soil erosion in the European Union: from direct cost evaluation approaches to the use of macroeconomic models. Land Degrad. Dev. 29, 471–484.
Parfitt, J., Barthel, M., Macnaughton, S., 2010. Food waste within food supply chains: quantification and potential for change to 2050. Phil. Trans. R. Soc. B 365, 3065–3081.
Paustian, M., Theuvsen, L., 2017. Adoption of precision agriculture technologies by German crop farmers. Precis. Agric.s 18, 701–716.
Peigné, J., Vian, J.-F., Payet, V., Saby, N.P.A., 2018. Soil fertility after 10 years of conservation tillage in organic farming. Soil Tillage Res. 175, 194–204.
Perfecto, I., Vandermeer, J., Wright, A., 2009. Nature's Matrix: Linking Agriculture, Conservation and Food Sovereignty. Earthscan, London, UK.
Pheng, S., Olofsdotter, M., Jahn, G., Adkins, S., 2010. Use of phytotoxic rice crop residues for weed management. Weed Biol. Manage. 10, 176–184.
Phelan, P.L., 2009. Ecology-based agriculture and the next green revolution. In: Bohlen, P.J., House, G. (Eds.), Sustainable Agroecosystem Management. CRC Press, Boca Raton, FL, pp. 97–135.
Phelan, P.L., Mason, J.F., Stinner, B.R., 1995. Soil-fertility management and host preference by European corn borer, *Ostrinia nubilalis* (Hübner), on *Zea mays* L.: a comparison of organic and conventional chemical farming. Agric. Ecosyst. Environ. 56, 1–8.
Phelan, P.L., Norris, K.H., Mason, J.F., 1996. Soil-management history and host preference by *Ostrinia nubilalis*: evidence for plant mineral balance mediating insect-plant interactions. Environ. Entomol. 25, 1329–1336.
Phillips, R.E., Blevins, R.L., Thomas, G.W., Frye, W.W., Phillips, S.H., 1980. No-tillage agriculture. Science 208, 1108–1113.
Pimentel, D., Hepperly, P., Hanson, J., Douds, D., Seidel, R., 2005. Environmental, energetic, and economic comparisons of organic and conventional farming systems. Bioscience 55, 573–582.
Pimentel, D., Cerasale, D., Stanley, R.C., Perlman, R., Newman, E.M., Brent, L.C., et al., 2012. Annual vs perennial grain production. Agric. Ecosyst. Environ. 161, 1–9.
Pittelkow, C.M., Liang, X., Linquist, B.A., van Groenigen, K.J., Lee, J., Lundy, M.E., et al., 2015a. Productivity limits and potentials of the principles of conservation agriculture. Nature 517, 365–368.
Pittelkow, C.M., Linquist, B.A., Lundy, M.E., Liang, X., van Groenigen, K.J., Lee, J., et al., 2015b. When does no-till yield more? A global meta-analysis. Field Crop. Res. 183, 156–168.
Powles, S.B., 2008. Evolved glyphosate-resistant weeds around the world: lessons to be learnt. Pest Manage. Sci. 64, 360–365.
Powles, S.B., Yu, Q., 2010. Evolution in action: plants resistant to herbicides. Annu. Rev. Plant Biol. 61, 317–347.
Powlson, D.S., Gregory, P.J., Whalley, W.R., Quinton, J.N., Hopkins, D.W., Whitmore, A.P., et al., 2011. Soil management in relation to sustainable agriculture and ecosystem services. Food Policy 36, 72–87.
Powlson, D.S., Stirling, C.M., Jat, M.L., Gerard, B.G., Palm, C.A., Sanchez, P.A., et al., 2014. Limited potential of no-till agriculture for climate change mitigation. Nat. Clim. Change 4, 678–683.

Pretty, J.N., 2008. Agricultural sustainability: concepts, principles and evidence. Philos. Trans. R. Soc. Lond. 363, 447–465.

Quinton, J.N., Govers, G., van Oost, C., Bardgett, R.D., 2010. The impact of agricultural soil erosion on biogeochemical cycling. Nat. Geosci. 3, 311–314.

Ray, D.K., Ramankutty, N., Mueller, N.D., West, P.C., Foley, J.A., 2012. Recent patterns of crop yield growth and stagnation. Nat. Commun. 1–19.

Reay-Jones, F.P.F., e Reisig, D.D., 2014. Impact of corn earworm injury on yield of transgenic corn producing Bt toxins in the carolinas. J. Econ. Entomol. 107 (3), 1101–1109.

Reganold, J.P., Wachter, J.M., 2016. Organic agriculture in the twenty-first century. Nat. Plant 1–8.

Reiss, E.R., Drinkwater, L.E., 2018. Cultivar mixtures: a meta-analysis of the effect of intraspecific diversity on crop yield. Ecol. Appl. 28, 62–77.

Robertson, M., Carberry, P., Brennan, L., 2009. Economic benefits of variable rate technology: case studies from Australian grain farms. Crop Pasture Sci. 60, 799–807.

Robertson, G.P., Gross, K.L., Hamilton, S.K., Landis, D.A., Schmidt, T.M., Snapp, S.S., et al., 2014. Farming for ecosystem services: an ecological approach to production agriculture. Bioscience 64, 404–415.

Royte, E., 2016. How 'Ugly' fruits and vegetables can help solve world hunger. National Geographic, March 2016. <https://www.nationalgeographic.com/magazine/2016/03/global-food-waste-statistics/>.

Ruel-Bergeron, J.C., Stevens, G.A., Sugimoto, J.D., Roos, F.F., Ezzati, M., Black, R.E., et al., 2015. Global update and trends of hidden hunger, 1995-2011: the hidden hunger index. PLoS ONE 10 (12), e0143497. Available from: https://doi.org/10.1371/journal.pone.0143497.

Russell, E.W., 1977. The role of organic matter in soil fertility. Philos. Trans. R. Soc. Lond. B 281, 209–219.

Ryan, M.R., Crews, T.E., Culman, S.W., Dehaan, L.R., Hayes, R.C., Jungers, J.M., et al., 2018. Managing for multifunctionality in perennial grain crops. BioScience 68, 294–304.

Sanchez, P.A., 2010. Tripling crop yields in tropical Africa. Nat. Geosci. 3, 299–300.

Schmidt, M.W.I., Torn, M.S., Abiven, S., Dittmar, T., Guggenberger, G., Janssens, I.A., et al., 2011. Persistence of soil organic matter as an ecosystem property. Nature 478 (2011), 49–56.

Schrijver, R., 2016. Precision agriculture and the future of farming in Europe. Scientific Foresight Study. European Union, Brussels, <http://www.europarl.europa.eu/RegData/etudes/STUD/2016/581892/EPRS_STU(2016)581892_EN.pdf>.

Scheelbeek, P.F.D., Bird, F.A., Tuomisto, H.A., Green, R., Harris, F.B., Joy, E.J.M., et al., 2018. Effect of environmental changes on vegetable and legume yields and nutritional quality. Proc. Natl. Acad. Sci. 115, 6804–6809.

Shinmura, A., 2000. Causal agent and control of root rot of Welsh onion (in Japanese). PSJ Soilborn Dis. Work Rep. 20, 133–143.

Sivarajan, S., Maharlooei, M., Bajwa, S.G., Nowatzki, G., 2018. Impact of soil compaction due to wheel traffic on corn and soybean growth, development and yield. Soil Tillage Res. 175, 234–243.

Smith, M.R., Myers, S.S., 2018. The risk of increased atmospheric CO_2 on human nutritional adequacy. Nat. Clim. Change 8, 834–839.

Smith, P., Martino, D., Cai, Z., Gwary, D., Janzen, H., Kumar, P., et al., 2008. Greenhouse gas mitigation in agriculture. Philos. Trans. R. Soc. B 363, 789–813.

Smil, V., 2000. Feeding the World: A Challenge for the Twenty-first Century. The MIT Press, Cambridge, MA, USA.

Smil, V., 2003. Energy at the Crossroads. The MIT Press, Cambridge, USA.

Smil, V., 2013. Should We Eat Meat? Evolution and Consequences of Modern Carnivory. Wiley-Blackwell.

Soule, J., Piper, J., 1991. Farming in Nature's Image: An Ecological Approach to Agriculture. Island Press, Washington D.C., USA.

Staley, J.T., Stewart-Jones, A., Pope, T.W., Wright, D.J., Leather, R.S., Hadley, P., et al., 2010. Varying responses of insect herbivores to altered plant chemistry under organic and conventional treatments. Proc. R. Soc. B 277, 779–786.

Steckel, L., 2017. Reports of poor Palmer Amaranth control with Dicamba. <http://news.utcrops.com/2017/05/reports-poor-palmer-amaranth-control-dicamba/>.

Stuart, T., 2009. Waste: Uncovering the Global Food Scandal. Penguin Books, London.

Sullivan, P., 2002. Drought Resistant Soil. ATTRA, National Center for Appropriate Technology USDA, Washington, DC, USA. Available online: <http://attra.ncat.org/attra-pub/PDF/drought.pdf> (accessed 24.11.09.).

Tabashnik, B.E., Brévault, T., Carrière, Y., 2013. Insect resistance to Bt crops: lessons from the first billion acres. Nat. Biotechnol. 31, 510–521.

Then, C., 2010. Risk assessment of toxins derived from *Bacillus thuringiensis*-synergism, efficacy, and selectivity. Environ. Sci. Pollut. Res. Int. 17, 791–797.

Then, C., Bauer-Panskus, A., 2017. Possible health impacts of Bt toxins and residues from spraying with complementary herbicides in genetically engineered soybeans and risk assessment as performed by the European Food Safety Authority EFSA. Environ. Sci. Eur. 29 (1), 1. Available from: https://doi.org/10.1186/s12302-016-0099-0.

Thies, C., Tscharntke, T., 1999. Landscape structure and biological control in agroecosystems. Science 285, 893–895.

Tigchelaar, M., Battisti, D.S., Naylor, R.L., Ray, D.K., 2018. Future warming increases probability of globally synchronized maize production shocks. Proc. Natl. Acad. Sci. Available from: https://doi.org/10.1073/pnas.1718031115.

Tilman, D., Fargione, J., Wolff, B., D'Antonio, C., Dobson, A., Howarth, R., et al., 2001. Forecasting agriculturally driven global environmental change. Science 292, 281–284.

Tilman, D., Balzer, C., Hill, J., Befort, B.L., 2011. Global food demand and the sustainable intensification of agriculture. Proc. Natl. Acad. Sci. 108, 20260–20264.

Torralba, M., Fagerholma, N., Burgess, P.J., Moreno, G., Plieninger, T., 2016. Do European agroforestry systems enhance biodiversity and ecosystem services? A meta-analysis. Agric. Ecosyst. Environ. 230, 150–161.

Thomas, M.B., Reid, A.M., 2007. Are exotic natural enemies an effective way of controlling invasive plants? TREE 22, 447–453.

UN (United Nations), 2017. World Population Prospects: The 2017 Revision, Key Findings and Advance Tables. United Nations, Department of Economic and Social Affairs, Population Division (2017). Working Paper No. ESA/P/WP/248. Available online: <https://esa.un.org/unpd/wpp/Publications/Files/WPP2017_KeyFindings.pdf>. (accessed 20.08.17.).

Uddling, J., Broberg, M.C., Feng, Z., Pleijel, H., 2018. Crop quality under rising atmospheric CO_2. Curr. Opin. Plant Biol. 45 (Pt B), 262–267.

USDA (The United States Department of Agriculture), 2018. Organic regulations. <https://www.ams.usda.gov/rules-regulations/organic>.

USDA (The United States Department of Agriculture), 2008. Soil Quality Indicators. USDA Natural Resources Conservation Service, <https://www.nrcs.usda.gov/Internet/FSE_DOCUMENTS/nrcs142p2_053256.pdf>.

van der Putten, W.H., Bardgett, R.D., Bever, J.D., Bezemer, T.M., Casper, B.B., Fukami, T., et al., 2013. Plant–soil feedbacks: the past, the present and future challenges. J. Ecol. 101, 265–276.

Vandeermer, J.H., 2011. The Ecology of Agroecosystems. Jones and Barlett Publishers, Boston, Canada.

van den Bosh, R., 1989. The Pesticide Conspiracy. University of California Press, Berkeley, USA.

Verret, V., Gardarin, A., Pelzer, E., Médiène, S., Makowski, D., Valantin-Morison, M., 2017. Can legume companion plants control weeds without decreasing crop yield? A meta-analysis. Field Crop Res. 204, 158–168.

Vincent-Caboud, L., Peigné, J., Casagrande, M., Silva, E.M., 2017. Overview of organic cover crop-based no-tillage technique in Europe: farmers' practices and research challenges. Agriculture 7, 42. Available from: https://doi.org/10.3390/agriculture7050042.

Wezel, A., Soldat, V., 2009. A quantitative and qualitative historical analysis of the scientific discipline of agroecology. Int. J. Agric. Sustain. 1, 3–18.

Wezel, A., Bellon, S., Doré, T., Francis, C., Vallod, D., David, C., 2009. Agroecology as a science, a movement or a practice. a review. Agron. Sustain. Dev. 29, 503–515.

Wezel, A., Casagrande, M., Celette, F., Vian, J.-F., Ferrer, A., Peigné, J., 2014. Agroecological practices for sustainable agriculture. a review. Agron. Sustain. Dev. 34, 1–20.

Williams, W.P., Windham, G.L., Krakowsky, M.D., Scully, B.T., Ni, X., 2010. Aflatoxin accumulation in BT and Non-BT maize testcrosses. J. Crop Improv. 24, 392–399.

Winston, K.M., 1997. Nature wars: People vs. pests. Harvard University Press, Cambridge, USA.

Wojtkowski, P.A., 2016. Agroecology: The Universal Equations. CRC Press Taylor & Francis Group, Boca Raton, FL, USA.

Wolfe, M.S., 1985. The current status and prospects of multiline cultivars and variety mixtures for disease resistance. Ann. Rev. Phytopathol. 23, 251–273.

Wolińska, A., Kuźniar, A., Zielenkiewicz, U., Banach, A., Błaszczyk, A., 2018. Indicators of arable soils fatigue – bacterial families and genera: a metagenomic approach. Ecol. Indic. 93, 490–500.

Worster, D., 2004. Dust Bowl. The Southern Plains in the 1930s. Oxford University Press, New York, USA.

Wu, W., Ma, B., 2015. Integrated nutrient management (INM) for sustaining crop productivity and reducing environmental impact: a review. Sci. Total Environ. 512–513, 415–427.

Yost, M.A., Kitchen, N.R., Sudduth, K.A., Sadler, E.J., Drummond, S.T., Volkmann, M.R., 2017. Long-term impact of a precision agriculture system on grain crop production. Precis. Agric. 18, 823–842.

Yu, Y., Stomph, T.J., Makowski, D., van derWerf, W., 2015. Temporal niche differentiation increases the land equivalent ratio of annual intercrops: a meta-analysis. Field Crop Res. 184, 133–144.

Yu, Y., Stomph, T.-J., Makowski, D., Zhang, L., van der Werf, W., 2016. A meta-analysis of relative crop yields in cereal/legume mixtures suggests options for management. Field Crop Res. 198, 269–279.

Zehnder, G., Gurr, G.M., Kühne, S., Wade, M.R., Wratten, S.D., Wyss, E., 2007. Arthropod pest management in organic crops. Annu. Rev. Entomol. 52, 57–80.

Zhao, Y.-P., Lin, S., Chu, L., Gao, J.T., Azeem, S., Lin, W., 2016. Insight into structure dynamics of soil microbiota mediated by the richness of replanted *Pseudostellaria heterophylla*. Sci. Rep. 6, 26175.

Zhao, X., Liu, S.L., Pu, C., Zhang, X.Q., Xue, J.-F., Ren, Y.-X., et al., 2017. Crop yields under no-till farming in China: a meta-analysis. Eur. J. Agron. 84, 67–75.

Zhu, Y.Y., Chen, H., Fan, J., Wang, Y., Li, Y., Chen, J., et al., 2000. Genetic diversity and disease control in rice. Nature 406, 718–722.

Zhu, Y.Y., Wang, Y.Y., Chen, H.R., Lu, B.-R., 2003. Conserving traditional rice varieties through management for crop diversity. BioScience 53, 158–162.

Zhu, Y.Y., Wang, Y.Y., Zhou, J.H., 2007. Crop variety diversification for disease control. In: Jarvis, D.I., Padoch, C., Cooper, H.D. (Eds.), Managing Biodiversity in Agricultural Ecosystems. Columbia University Press, New York and Bioversity International, Rome, Italy, pp. 320–337.

Zhu, C., Kobayashi, K., Loladze, I., Zhu, J., Jiang, Q., Xu, X., et al., 2018. Carbon dioxide (CO_2) levels this century will alter the protein, micronutrients and vitamin content of rice grains with potential health consequences for the poorest rice-dependent countries. Sci. Adv. 4 (5), eaaq1012. Available from: https://doi.org/10.1126/sciadv.aaq1012.

Zikeli, S., Gruber, S., 2017. Reduced tillage and no-till in organic farming systems, Germany —status quo, potentials and challenges. Agriculture 7, 35. Available from: https://doi.org/10.3390/agriculture7040035.

Zomer, R.J., Bossio, D.A., Sommer, R., Verchot, L.V., 2017. Global sequestration potential of increased organic carbon in cropland soils. Sci. Rep. 7, 15554.

Optimizing agricultural practices

3

Aditya Parmar
Natural Resources Institute, University of Greenwich, London, United Kingdom

Chapter Outline

3.1 Introduction

For a long time, insufficient regard was given to food loss and waste, as the major emphasis of agricultural research and development was on crop production and breeding improved cultivars (Snowdon, 2010). However, in recent times interest to reduce food loss and waste has gained momentum. Food losses are not only an important driver of food insecurity in developing countries, but they also represent a gross wastage of the natural resources such as freshwater, cropland, and energy utilized in the production of the lost food (Smil, 2000; Kummu et al., 2012). Food losses are commonly referred as the physical loss of agricultural produce, which was intended for human consumption (i.e., food) along the food value chain (Gustavsson et al., 2011), with an exception to consumption or distribution stages, where the losses are instead termed as food waste. Fewer food losses and waste

Saving Food. DOI: https://doi.org/10.1016/B978-0-12-815357-4.00003-1

would also mean that there would be reduced pressure on cropland expansion and intensifying agriculture to increase production, which will conserve biodiversity and forests (Sharma and Wightman, 2016). The definitive causes of food losses and wastes vary with specific products and prevailing socioeconomic conditions in each country or location. Past research has revealed that food losses are a significant problem in developing countries due to lack of infrastructure and capacity of the food supply systems, whereas food waste is a phenomenon of the developed world (Hodges et al., 2011; Parfitt et al., 2010; Gustavsson et al., 2011; Kummu et al., 2012). The food loss figures around the world demonstrate a significant spatial and temporal variability, due to the complex and individual characteristics of crop and location. However, general estimates made by the Food and Agricultural Organization of the United Nations (FAO, 2012) suggest that 30% of the global cereal production and 45% of fruits, vegetables, roots, and tubers are lost at harvest and postharvest.

Typical agricultural commodities and product value chain can be broken down into five system boundaries, namely (1) agricultural production, (2) postharvest handling and storage, (3) processing, (4) distribution, and (5) consumption. These agricultural commodities can be further categorized into cereals and pulses, fruits and vegetables, roots and tubers, and livestock and dairy products. Identification of causes and drivers of food losses and waste along these system boundaries and commodities is critical to pinpoint and implement the prevention and reduction solutions and practices (Gustavsson et al., 2011; FLWP, 2016). The agricultural produce is exposed to various postharvest deteriorations, which in a broad sense can be categorized into physiological degradation (due to the natural processes such as climacteric rise in climacteric fruits, etc.), physical damages (increased heat output due to higher respiration, moisture loss, and entry points for microbial infections), chemical injuries (caused by agrochemicals), and pathological decay (infection by fungi and bacteria) (Snowdon, 2010). Hence, the causes of food losses can range from biological, mechanical, to behavioral and organizational issues in a food supply system. Table 3.1 lists some of the common causes and drivers of food losses in production and postharvest systems of agricultural commodities and products particularly in the context of developing countries. In the process of food loss reduction or prevention, the first step is to identify the nature of the deterioration (immediate cause of loss) that is causing rejection or discarding of the food, for example, if there are any microbiological organisms (bacterium, fungus) or insects involved. Secondly, it is required to understand the possible contributory causes (drivers) such as preharvest factors (e.g., weather conditions during production), harvest stage/maturity and method, or postharvest handling and storage.

The focus of this chapter will be to optimize various practices in the agricultural production system boundary (constituting preharvest and harvest subcategories of the food supply system) for *durables* (cereals and pulses) and *perishables* (horticultural products (i.e., fruits and vegetables and roots and tubers), to reduce food losses in developing countries. *Durables* are characterized by high dry matter content, hard texture, and small and homogenous in size.

Table 3.1 Common food loss causes in agricultural production and postharvest management in developing countries

System boundary	Agricultural production		Postharvest	
Subcategory	**Preharvest**	**Harvest**	**Handling and storage**	**Processing and distribution**
Causes and drivers	• Genetic factors ◦ Cultivar selection ◦ Pest and disease resistant cultivars ◦ Reduced physiological disorders • Environment and cultural practices ◦ Pest and natural calamity ◦ Poor agronomic and cultural practices (soil type, mulching, fertilization, irrigation, pruning and thinning of fruits crops) • Poor market access • Poor organization among producers	• Maturity at harvest ◦ Early or delayed harvesting is having an impact on postharvest shelf life. • Harvesting method or technique ◦ Physical injuries (mechanical damages) • Lack of on-farm packaging and storage facilities	• Inadequate drying of grains and cereals ◦ Temperature and humidity maintenance ◦ Curing of roots and tubers • Improper containers and packaging • Improper use of agrochemicals ◦ Various treatments (waxing, irradiation, fumigation, fungicide treatments, ethylene) • Inadequate information and knowledge on proper postharvest practices. • Poor roads, transportation, and market access • Lack or poor storage conditions and facilities	• Lack or inadequate processing facilities • Inadequate processing capacities • Delays in distribution • Poor storage conditions during distribution • Inappropriate packaging

Perishables, on the contrary, have a higher moisture content (MC), soft texture, and large and irregular shape and size (making them susceptible to desiccation (wilting, shriveling) and mechanical injury during harvest and handling) (Rees, 2012; Kader, 2013). Pest and disease are the major concern for the *durables*, whereas maintenance of physiological processes (internal processes such as respiration and ethylene production) is crucial in case of *perishables*. Grains such as wheat, rice, and maize are primary staples around the globe providing the major part of the calorie intake, however in various parts of the world other crops like cooking banana in East Africa, breadfruit in the South Pacific Islands, yams and cassava in West Africa, and potatoes in South America are important staples (Snowdon, 2010). Moreover, legumes and pulses are important sources of protein in various parts of the world (e.g., in India), whereas fruits and vegetables provide essential vitamins and minerals for a balanced diet. Apart from food, horticultural products contribute significantly to the economies of various countries regarding exports.

Several variables in the agricultural production stage (before farm gate or at farm gate, Table 3.1) determine the extent of food losses, for example, method of harvesting, variety of the crop, timing of the harvest (ripeness of the grains or fruits) (Smil, 2000; Brasil and Siddiqui, 2018). Hodges et al. (2011) stated in a review of food losses from developed and developing countries that it is critical for the developing countries to look before farm gate to reduce food losses. Some of the critical drivers mentioned in this review were educating farmers on postharvest management and causes of food losses, improved infrastructure to access markets, emphasis on building smallholder organizations, and access to postharvest technology investment microcredits. The World Bank (2011) in a report on "missing food" reiterated that most of the food losses in developing countries occur near to the farm, where the choices of crop, cultivar, harvesting technique and initial handling are made. Siddiqui (2018) restated that among all the factors responsible for the overall postharvest quality of the perishable products (especially fruits and vegetables), more than 70% lies in the preharvest and harvesting conditions. For example, for roots and tubers globally about 20% of the product is lost at agricultural production stage (Gustavsson et al., 2011), whereas close to 6% of the cereals are lost in African countries in agricultural production stage during harvest and field drying. African postharvest loss information systems (APHLIS, 2017) recorded losses (dry weight basis) for wheat, maize, rice, sorghum, and barley of about 3.5%, 6.2%, 4.4%, 4.4%, and 3.5% respectively during harvesting and field drying.

It is known that overall state and quality of agricultural products (both perishables and durables) cannot be further improved once harvested, which depends on the producer's choices of crop species, cultivar (or variety), time of planting, and following cultural and harvest practices. Moreover, market factors were also considered as one of the preharvest factor affecting postharvest gains or losses; the market has a significant influence (when the majority of the crop is intended for sale) on the producer's decisions and the required quality criterion for the intended consumers (Simson and Straus, 2010).

3.2 Preharvest systems

Preharvest practices and treatments play a key role in enhancing the quality and shelf life of the agricultural products (Sidiqqui, 2018). In the postharvest phase, one is only limited to keeping the quality obtained at harvest by decreasing the rate of deterioration but this does not necessarily improve on what has already been harvested. Some of the important preharvest factors that influence the postharvest quality (sensory and keeping/shelf life) are variety (genetic/cultivar), climatic (or weather) conditions, irrigation (water supply), soil fertility, fertilizer application, use of agrochemicals, and last but not the least insect pest and diseases (FAO, 1989; Fallik, 2008; Snowdon, 2010). These factors can also be categorized into *abiotic* and *biotic* stresses (Kays, 1999; Kader, 2000; El-Ramady et al., 2015). The typical *abiotic* stresses are water stress (deficit or access), temperature (too hot or too cold), and soil salinity. Whereas, *biotic* stress would be the stress caused by other living organisms such as bacteria, fungi, virus, parasites, and weeds. Examples of various preharvest factors affecting postharvest quality are listed in Table 3.2. Sams (1999) in a similar review listed several examples of how various abiotic and biotic factors would affect the texture of horticultural products. Texture traits (defined as a feeling of touch, deformation, disintegration, and flow, which can be measured objectively with a function of force, time, and distance) are very important in determining the market acceptability. The textural properties could also relate to mechanical, geometrical, and chemical characteristics, for example, hardness, shape, and MC respectively. The products of poor texture may be rejected at one or the other stage of the value chain resulting in significant food losses and wastes. The texture is particularly influenced by changes in cell organelles and biochemistry, MC, and composition of the cell wall. Any of the biotic or abiotic factors affecting these properties can change the texture leading to the changes in the product quality.

Physical injuries are a major problem in almost all the agricultural products from cereals and grains to fruits and vegetables. However, the extent and impact of this problem are much more in the horticultural sector. One of the most common types of injuries is bruise damage, which can occur in preharvest, harvest, handling, and transportation stages, and are known to cause considerable economic and physical losses along the product value chain. Hussein et al. (2018), lists various genotypic, environmental, seasonal, and management related factors during the production of the horticultural crops, which significantly affects the postharvest susceptibility to injuries like bruises.

3.2.1 Cultivar and environment

Every agricultural commodity has a range of genotypic variations in composition, quality, and postharvest shelf life potential (Kader, 2000). Classical/traditional breeding and modern-day molecular based genetic manipulation have the potential to maintain flavor and nutritional qualities and the introduction of resistance to

Table 3.2 The broad category of preharvest factors affecting the postharvest quality of agricultural produce

Factors	Type of causes	Examples
Biological	Pathological, entomological, animal	• Bacterial spot on tomato fruit. • Fungal—white mold on the blossom end of cucurbits • Virus—zucchini yellow mosaic virus of squash • Nematodes—northern root-knot nematode • Cabbage loopers (*Trichoplusia ni Hüber*) • Tomato horn • Worms (*Manduca* spp.) • Flower thrips (*Frankliniella occidentalis* Pergande) • Sweet potato weevil (*Cylas* spp.)
Physiological	Physiological disorders, nutritional imbalance	• Tomato blossom-end rot (a calcium deficiency related problem) • Cracking in tomato and cherry fruit is a nutrition and water-related problem • Nitrogen deficiency in leafy vegetables • High nitrogen is resulting in poor coloration in apple, cranberry
Environmental/ cultural factors	Climate, weather, soils, water relations, light intensity	• Chilling injuries • Freezing/frost damage • High-temperature stress • Wind damage • Soil textural properties (for root crops)
Extraneous matter	Growing medium, vegetable matter, chemical residues	• Ions of heavy metals, (Ag, Cd, Co, Mg, Mn, Ni, and Zn) • Ozone causes surface blistering of spinach (*Spinacia oleracea* L.) leaves
Genetic variations	Cultivar tolerance to pest and disease, physiological stress, and injuries (physicomechanical properties)	• Shape, size, and color • Desired phenotype • Heterozygous

Source: Adapted from Kays, S.J., 1999. Preharvest factors affecting appearance.pdf. Postharvest Biol. Technol., 15 (June 1998), 233–247. http://dx.doi.org/10.1016/S0925-5214(98)00088-X; Sams, C.E., 1999. Preharvest factors affecting postharvest texture. Postharvest Biol. Technol., 15 (3), 249–254. https://doi.org/10.1016/S0925-5214(98)00098-2; Mattheis, J.P., Fellman, J.K., 1999. Preharvest factors influencing flavor of fresh fruit and vegetables. Postharvest Biol. Technol. 15 (3), 227–232. https://doi.org/10.1016/S0925-5214(98)00087-8; Hussein et al., 2018.

postharvest physiological disorder and pests and diseases. Farmers and development agents must consider the choice of the right varieties for a particular location and target markets. Use of cultivars, resistant to disease and pest, having a longer postharvest shelf life and integrated crop management systems can maximize the yields and quality of the product at harvest (Kader, 2013). FAO (1989) gave an example of mangoes in Indonesia where 242 mango varieties are available, but only 7 of them have commercial potential. Hence local growers shall choose what cultivar they want to grow to enhance their income. For example, cultivars with thicker skin and firm texture (particularly for fresh fruits and vegetables) tend to have a longer shelf life (Snowdon, 2010).

Environmental or climatic variations (climatic conditions) are the second most important factor after cultivar determining the flavor and ratio of different constituents of various fruits and vegetables. For example, acid and sugar ratio, and pungency were significantly affected by season in tangerine (*Camellia reticulata*), grapefruit (*Citrus paradise*), and onions respectively (Mattheis and Fellman, 1999). Excessive exposure to sunlight (light above the photosynthetic saturation) can increase the temperature of the fruit resulting in damage and loss of firmness; an example of such damage is sunscald, which is common in apples and tomatoes (Sams, 1999). Due to higher tissue density, firmness of most nonchilling sensitive fruits is higher at low temperatures.

3.2.2 Integrated pest and disease management

Pest control protocols need to be implemented and followed along the entire food value chain to ensure efficient control. Use of clean (disease and pest free) planting material (true seeds, seed tubers, and cuttings) is a prerequisite to control disease in the field and postharvest. For example, onion seeds are very often infected with molds (Snowdon, 2010), requiring a preplanting fungicide treatment. Establishing seed certification schemes for important crops in the developing countries is critical to obtain healthy planting material in preharvest for better postharvest quality and shelf life. Reducing mycotoxigenic fungi load during postharvest stages of various crops (not only for cereals and nuts/peanuts but also apples and grapes) has emerged as a global challenge. Some of the most potent mycotoxigenic plant pathogens are *Fusarium*, *Aspergillus*, and *Penicillium* spp. One of the potential solutions for this problem is mycoparasitism where the preharvest application of beneficial fungi such as filamentous fungi and yeasts is incorporated (Sarrocco and Vannacci, 2017) to reduce the infection of pathogenic fungi. Recovering and reimplementing the indigenous knowledge in reducing the losses at farm due to insect and pests is critical, particularly in the developing world where landholdings are small, and cost of modern pesticides and insecticides are high. Sharma and Wightman (2016), quoted a wonderful example where scientists tested indigenous knowledge (shakedown approach following with hen who feeds on the larvae) to control pod-borers in pigeon pea, and the results were fantastic; the simple shakedown approach reduced the losses by 85%. In a recent survey Parmar et al. (2017a), along with various other prospective control measures to reduce sweet potato postharvest losses in

Figure 3.1 Sweet potato weevil (*Cylas puncticollis* and *Chorthippus brunneus*) damaged root at harvest in southern Ethiopia.
Source: Picture by Aditya Parmar.

Ethiopia, mentioned the preharvest practices such as using early maturing and deep-rooted varieties, crop rotation, clean planting material, sex pheromone traps, and hilling-up as some of the measures to control sweet potato weevil, which is a major production and postproduction problem in East Africa. A picture of a sweet potato weevil–damaged root is presented in Fig. 3.1.

3.2.3 Diversifying the crop production

Researchers at the Royal Botanical Garden have estimated that close to 200,000 plant species are edible and safe for human consumption out of more than 400,000 presently alive plants on Earth. Given this vast number of plant species available for human consumption, the current human diet is highly conservative; we rely on fewer than 150 plant species for our food supply, and close to three quarters of the global grain market comprises of wheat, rice, maize, and barley (Sharma and Wightman, 2016; Charrondière et al., 2013). Presently 95% of the global carbohydrates supply is provided by a mere 30 species, which has significant impact not only biodiversity, food losses, plant disease, and pest infestation but also on human nutrition and health (WHO, 2015; Charrondiére et al., 2013). Rosner (2014) mentioned that about 18,000 legume species are available around the world; these species are nutritious, resistant to pest, diseases and adverse climate and weather conditions, yet fewer than 50 are domesticated for human food. Some of the examples of these underutilized legumes are potato beans (*Apios americana*), Marama beans (*Tylosema esculentum*), Yehub nut (*Cordeauxia edulis*), and beans of genus *Lupinus*. Diversifying the crop mix in agricultural systems of the developing countries can be one of the coping strategies against food insecurity and can enable smallholders to harvest at maturity (Hodges et al., 2011). For example, resource poor smallholders in low-income countries may harvest their crops prematurely, due to food shortage or urgent need for cash, particularly around the second half of the agricultural season. This premature harvesting results in loss of nutrition, financial value, and unsuitability for processing or certain consumptions (Gustavsson et al., 2011). One way to avoid such a situation and associated food loss is by diversifying the production of resource poor smallholders into a variety of cash crops and animal products. By this way they are also more likely to obtain credits from microfinancing institutions or advances from wholesalers or processors (Gustavsson et al., 2011). The crop diversification not only leads to lower chances of food waste at farm and market but also leads to a more nutritious diet for the community. In recent times there has been a significant thrust on improving the diversity of human diets, as it is a common belief among natural scientists that intensive industrial agriculture is not sustainable and does not lead to a healthy human diet (Dwivedi et al., 2017). In a mega study from 150 countries Khoury et al. (2014) highlighted that total crop diversity has shrunken significantly in the last 50 years, and diets have become more homogeneous (depending on the regionally important crops). Such a finding put additional pressure on diversifying farm production and providing market access to a host of diversified crops, particularly on the smallholder subsistence farmers (Dwivedi et al., 2017). Another indicator of global loss of crop diversity is the production data from FAOSTAT (2015), which shows that from 1961 to 2013 land area cultivated for wheat, rice, and maize has increased up to 79%, whereas for barley, millet, oats, rye, sorghum it has reduced by 19% to 33%.

3.2.4 Improved agronomic and cultural practices

Appropriate agronomic and cultural practices ensure the produced agricultural commodity is of higher quality, which reduces losses from rejections and discards along the value chain. Various abiotic and biotic stresses in preharvest can have a negative impact on the quality of the produce postharvest. For example, preharvest water stress can affect quality and shelf life of the product postharvest. Water stress during production is known to be connected with higher weight loss during postharvest storage in various fruits and vegetables (El-Ramady et al., 2015). Another important factor related to the quality of the fruits and vegetables is moisture or water loss in (pre- and postharvest stages of the food supply chain), a water loss of as little as 3%–5% can reduce the marketability of many leafy vegetables and fruits. Plant nutritional factors during the growth and maturation stages have again a significant effect on the postharvest quality. Some of the more important nutrients that fall into this category are nitrogen, phosphorus, and potassium (NPK) and calcium (Ca). Applying more than the recommended amount of nitrogen fertilizer preharvest can result in discoloration of cabbage (due to high accumulation of Zn and Al and deficiency of Mn) and potato (black spots due to bruising) during postharvest storage (El-Ramady et al., 2015). Moreover, excess N and P are known to decrease firmness in many crops (Sams, 1999). Ca is another element that is directly related to fruit and vegetable firmness and textural properties as it is essential to strengthen the plant organ cell wall, which is very important from a postharvest point of view. Soil that is deficient in Ca and K resulting in deficiencies in fruits and vegetable is linked to various postharvest disorders (Kader, 2000). In some cases, preharvest sprays of Ca on fruits and vegetables are required to improve the postharvest shelf life and rigidity of these products to mechanical injuries during postharvest handling.

Some of the important aspects of good cultivation practices that are essential for improved quantity and quality of the fresh produce are better weed control (which competes with crops for nutrient and water) and crop hygiene or sanitation in terms of using clean planting material and removing potential source of infections such as infected parts of the crop from previous harvest before planting new crop. Nowadays, use of agrochemicals (pesticides, herbicides, and growth regulators) are common in most of the production systems around the world, and care must be taken in their use as overuse and choosing inappropriate chemicals for weed and pest control can affect the physical (e.g., spray-burns), sensory, and safety quality (toxic residues) of the product (Snowdon, 2010; Simson and Straus, 2010). For certain roots crops (potato, cassava, sweet ptoato), dehaulming or canopy removal a few days before harvest can lead to improved skin-setting resulting in reduced moisture loss and injuries during postharvest handling and transportation. Parmar et al. (2017b), through an empirical study, demonstrated how preharvest dehaulming (sometimes called as preharvest curing) of sweet potatoes in southern Ethiopia resulted in enhanced skin strength or adhesion.

3.3 Harvest systems

The harvest is the operation of collecting the mature, useful, or economic part of the plant from the fields (Simson and Straus, 2010). In developing countries where the majority of the crop production is done by smallholders on small farms mechanization is limited (due to economies of scale issues), and harvesting time of the crop cultivation is the most labor incentive activity. Bad or extreme weather conditions during harvesting (e.g., frost, rain, storm, and unusually warm or cold conditions) can affect the quality and yields of the crop. Hence, the timing and an unlikely weather event can be damaging for the mature crop ready to be harvested.

The harvest of the crop is one of the most crucial transitions of the agricultural commodities. This transition is much more stressful for perishables (horticultural crops) rather than durable crops such as cereals and grains. Good harvesting practices and knowledge of proper harvest maturity indices is critical for ensuring longer postharvest shelf life and improved quality of the product. Moreover, scheduling of harvesting times with target markets can improve the income of the smallholders and reduce the losses due to oversupply (market gluts). Maturity, internal characteristics at harvest, stage, and proper method of harvesting help retain the quality of horticultural produce and reduce postharvest losses (Prasad et al., 2018).

3.3.1 Harvest and handling techniques

Harvesting technique can be subdivided into two broad categories (Fig. 3.2): complete hand harvesting, and mechanical or mechanical assisted harvesting (which is a combination of manual and machine, e.g., loosening of soil in case of root crops with a mechanical digger and manual collection of roots). In general, it can be said that hand harvesting may result in the harvest of more undamaged and horticultural matured fruits and vegetables. However, it may depend on the training, management, and contractual arrangement with the involved manual labor. Experience from past studies has shown that when the contractual agreements are short-term and based on per box or sack, or per hectare; there is lack of teamwork; working hours are long without breaks; weather conditions are bad; and laborers lack training; this can result in careless harvesting, causing significant mechanical injuries and poor quality of horticultural commodities (Simson and Straus, 2010).

For fruits and vegetables especially, harvesting should take place during the coldest time of the day (early morning), and the product should be placed in the shade immediately to avoid direct sunlight (especially in tropical conditions). Moreover, the harvested crop should have minimum levels of physical damage (skinning, bruising, cuts, and breakages), mainly when the commodity has to travel to a distant market or needs to be stored for a longer duration. El-Ramady et al. (2015) and Sharma and Singh (2012) distinguished the mechanical injuries (or lesions) for horticultural crops into two broader types:

Cuts and punctures: Loss of tissue integrity, surface injuries. Such injuries can lead to increased rates of respiration and ethylene evolution leading to a rapid deterioration of the product, whereas they are mostly caused by harvesting tools. The product becomes vulnerable to pathogen infections, most predominantly rotting

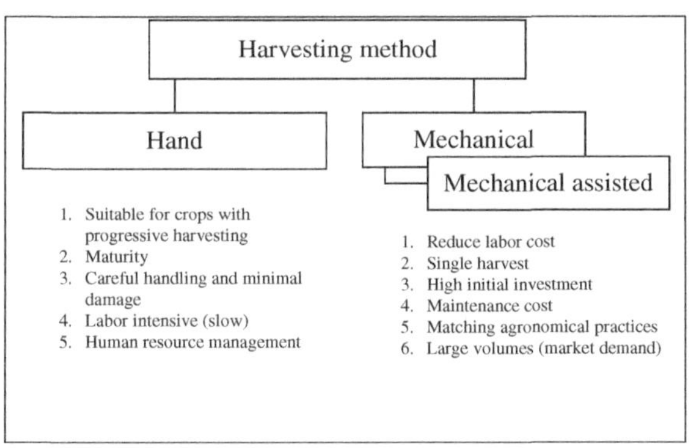

Figure 3.2 Harvesting methods.
Source: Adapted from Sharma, R.M., Singh, R.R., 2012. Harvesting, postharvest handling and physiology of fruits and vegetables. In: Verma, L.R., Joshi, V.K. (Eds.), Postharvest Technology of Fruits and Vegetables (fourth ed.). Indus Publishing Company, New Delhi, pp. 1–484.

fungi and bacteria. An example of cuts and punctures is demonstrated in Fig. 3.1, where harvesting tool–associated cuts and punctures in sweet potato are shown.

Bruises (Impact, compression, abrasion injuries): Such injuries may depend on maturity, water potential, tissue firmness, and temperature of the product. They are subsurface, which may only be detected several days after when the product has already reached the final consumer.

Moreover, during harvest or immediately after presorting, removal of inedible parts such as leaves and folios and packaging takes places to prepare the product for market. During these activities (handling) the fresh horticultural produce can sustain various injuries resulting in a cumulative effect on the marketability and shelf-life of the product (Simson and Straus, 2010).

Harvesting tools and containers (particularly for horticultural products) play an important role in reducing the extent and number of these injuries. Care must be taken in choosing an appropriate container for harvesting to avoid injuries and contamination of the freshly harvested product. Fig. 3.3 provides an example of sweet potato roots being extremely vulnerable to rotting if they are damaged during harvesting and handling. As for the time of the day when the harvest should take place,

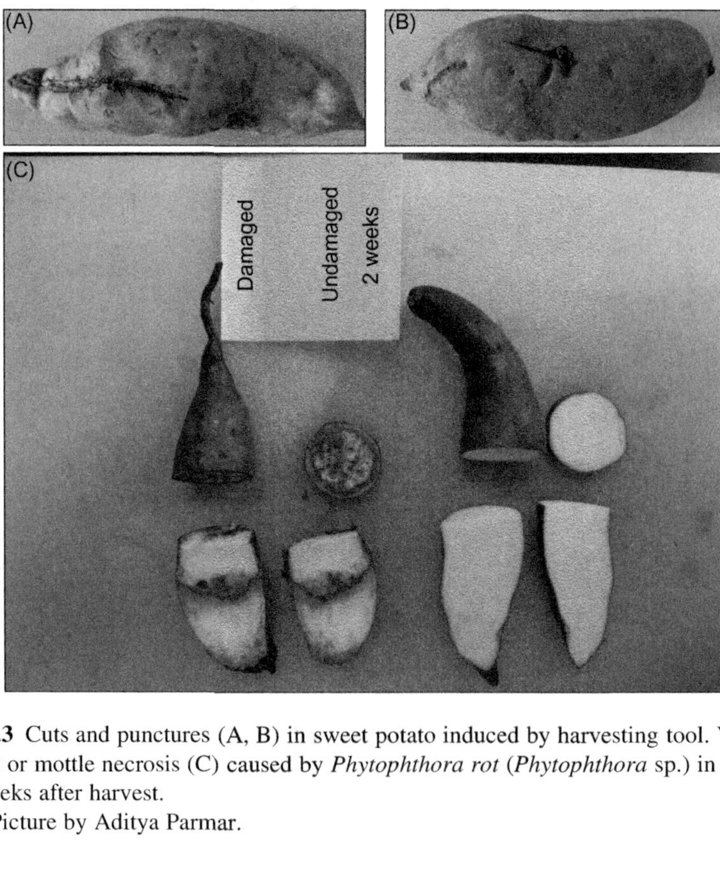

Figure 3.3 Cuts and punctures (A, B) in sweet potato induced by harvesting tool. Vascular browning or mottle necrosis (C) caused by *Phytophthora rot* (*Phytophthora* sp.) in damaged root 2 weeks after harvest.
Source: Picture by Aditya Parmar.

this should be during cool morning hours, as firmness is higher during cool periods and less energy is required for precooling (Simson and Straus, 2010). Keeping the freshly harvested horticultural product in the shade, protecting it from direct sunlight in tropical climates is important. Lastly, training of the harvest labor is the most important factor; the training should focus on how to handle the product gently and with care, and we aware of the postharvest losses that may be caused by injuries induced during harvest and handling.

3.3.2 Harvesting maturity

Harvesting at optimum maturity is critical to determine the final consumer quality of the horticultural products; immature fruits may sustain shriveling, internal and external injuries and poor sensory attributes, whereas overly matured may become too soft, mealy, and insipid before reaching the consumer (Fallik, 2008; Simson and Straus, 2010). The maturity of the agricultural commodities (fruits and vegetables) can be categorized into two types:

Physiological maturity: When translocation of photosynthates stops, and no further increase in dry matter content of the product. For fruits and vegetables, it is the stage of development when these plant parts continue to ripen after harvest.

Horticultural (or commercial) maturity: When the product is harvested at a stage for a specific purpose, for example, for storage or prerequisite usage such as processing into juice.

All agricultural produce may show one or more distinguishable signs or characteristics of its physiological (or biological) maturity. Tomatoes, for example, at physiological maturity, fill their internal locules with gelatinous mass, and when the matured fruit is cut with a sharp knife seeds remains intact (Simson and Straus, 2010). However, the physiological maturity may not always coincide with commercial or horticultural maturity, which is important from a transportation and market point of view. Delayed or too early harvest of agricultural commodities can limit the shelf life and economic value of the product. The maturity of the produce at harvest is one of the most important factors that defines the postharvest shelf life and quality (e.g., regarding appearance, aroma, texture, flavor, nutrition). Harvesting at proper or appropriate maturity (depending on the specific crop and end consumer use) can facilitate more efficient packaging, transportation, and distribution (El-Ramady et al., 2015).

All agricultural produce (grains/seeds, fruits, vegetable, roots and tubers) is the living organs of the plant, which are metabolically active, undergoing respiration (carbohydrates such as starch and sugars are degraded in the presence of oxygen are converted into carbon dioxide, water and energy), ripening, and senescence (for fruits and vegetables). The generic equation for aerobic respiration is presented in Eq. (3.1). The rate of respiration (mL CO_2/kg · h) would depend on the ambient temperature and nature of the product.

$$C_6H_{12}O_6 + 6O_2 = 6CO_2 + 6H_2O + \text{heat} \tag{3.1}$$

To calculate heat production multiply mL CO_2/kg · h by 440 to get Btu/ton/day or by 122 to get kcal/metric ton/day (UCDAVIS, 2018).

Fruits, depending on their biological control of ripening [the production of ripening hormone ethylene (C_2H_4)] and respiratory rates, can be classified into two groups (Rees, 2012; Sharma and Singh, 2012; Prasad et al., 2018):

Climacteric: The fruits belonging to this category possess autonomous ripening behavior (when harvested as physiological mature but not ripe or horticulturally mature), meaning they are capable of developing changes in taste, aroma, color, and texture after being detached from the mother plant. The ripening is associated with increased rate of respiration (also known as climacteric rise) and production of ripening hormone ethylene. Examples of climacteric fruits are apple, avocado, mango, banana, papaya, peach, tomato, etc.

Nonclimacteric: The fruits are incapable (or have the very limited ability) of developing a horticultural maturity after harvest. They do not show an increase in respiration and ethylene production rates, but a constant decline during the ripening process. Examples of nonclimacteric fruits are cherry, grape, lemon, lychee, mandarin, pineapple, strawberry, watermelon, etc.

Hence, the climacteric fruits can be harvested at their physiological maturity for the distant markets as the transportation time can be utilized to achieve horticultural maturity. However, it is more problematic for nonclimacteric fruits where horticultural and physiological maturity are very close, and requiring rapid transportation and better cold chain conditions to reduce losses.

Management of these metabolic and physiological processes at harvest and postharvest is crucial to reduce physical and nutritional losses along the value chain. Postharvest environmental conditions (temperature and humidity in particular) have a major impact on the quantitative (physical) and qualitative (nutritional) losses (Brasil and Siddiqui, 2018). For example, in case of fruits and vegetables, an increase of 10°C in the storage temperature could double, triple, or quadruple the rates of the respiration leading to weight and nutrient loss (Snowdon, 2010). Moreover, for grains and cereals, apart from temperature, MC (or water activity) can play an important role in determining the respiration rates (Coleman et al., 1928; Raudienė et al., 2017). For example, Huang et al. (2013) reported that the respiration rate of corn (maize) increased 100 times when the MC of the corn increased from 14% to 22%. Whereas the reparation rates doubled with every 10°C increase in the ambient temperatures. Raudienė et al. (2017) recorded that the respiration rates of wheat increased from approximately 5 mg CO_2/kg/h with 13% MC and 20°C ambient temperature to 30 mg CO_2/kg/h for 19% MC, and at higher temperatures the respiration rates of wheat showed increment; the increment was starker when the wheat had higher MC. A similar increase in rice and soybean respiration rates with increased MC and ambient temperatures has been reported by Dillahunty et al. (2000) and Sorour and Uchino (2004) respectively. This is a typical problem in tropical countries where ambient temperatures are high; the problem is further compounded due to lack of money to provide controlled atmospheric storage and transportation facilities. Therefore, cost-effective technical innovations to reduce food losses in developing countries are one of the major focuses of postharvest research and development. The respiration rate is the main factor determining the shelf life (or perishability) of the agricultural produce. Table 3.3 provides the respiration rates of some of the important cereals, fruits, and vegetables.

In cereals and grain, harvest takes place in general after 10–15 days of attaining physiological maturity. Moreover, the MC is the most important indicator for harvest maturity, for much of cereals crops, optimum MC (percentage-wet basis) at harvest shall range from 10% to 23%. Providing specific examples, for example, wheat shall be harvested at 16%–18% MC, paddy at 20%–22%, maize at 20%–23%, and groundnuts at 10%–12%. For storage of these cereals, they need to be

Table 3.3 Respiration rates of various agricultural product groups

Agricultural product group	Respiration rate (mL CO_2/kg · h) at 20°C	Rate of ethylene (μL C2H4/kg · h) production at 20°C
Durables[a]		
Wheat, 13% MC, 20°C (Raudienė et al., 2017)	~5 mg CO_2/kg/h	–
Rice, 15% MC, 20°C (Dillahunty et al., 2000)	>1 mg CO_2/kg-h	–
Soybean, 14% MC, 20°C (Sorour and Uchino, 2004)	>1 mg CO_2/kg dry matter.h	–
Corn, 14% MC, 20°C (Huang et al., 2013)	0.31 mg CO_2/kg/h	–
Semiperishables (roots and tubers) (UCDAVIS, 2018)		
Potato	9–23	Very low
Sweet potato	27–35	< 0.1
Onion	27–29	–
Jicama	3–4	Very low
Perishables		
Climacteric [lower value for preclimacteric and higher for climacteric peak] (UCDAVIS, 2018)		
Apple (red delicious)	12–25	20–125
Avocado	40–150	>100
Banana	20–70	0.3–10
Breadfruit	38–178	0.1–1.6
Mango	35–80	0.5–8
Nonclimacteric (UCDAVIS, 2018)		
Grapes	12–15	–
Grapefruit	7–12	<0.1
Lychee	25–40	<0.5
	15–20	<0.2

MC, moisture content.
[a]The respiration rate for cereals and grains would depend significantly on the MC of the grain seeds and the ambient temperature.

dried further postharvest to attain option storage MC that in most cases lies between 10% and 12% (Sahay and Singh, 2016). It is always recommended to use several maturity indices as environmental factors can influence individual maturity indices so that no mistake in harvesting time is made and the crop is always harvested at its optimum desired quality and postharvest life.

Assessment of maturity at harvest is a problem for both climacteric and nonclimacteric fruits (Table 3.4). Climacteric fruits, if harvested at an immature stage, may not ripen properly, and if harvested at an advanced stage of maturity may

Table 3.4 Various harvest maturity indices

Methods	Indices	
	Fruits and vegetables (horticultural crops)	**Cereals and pulses (grains)**
Computational	1. Calendar date 2. Days from full bloom 3. Mean heat units 4. T-stage	1. Days after flowering (*anthesis*) 2. Mean heat units
Physical	1. Fruit retention strength 2. Fruit size and surface morphology 3. Weight 4. Specific gravity 5. Color 6. Firmness 7. TSS	1. Color of the glumes, peduncle and pods (e.g., yellowing of the leaves, pods turning brown, stems turn to straw color) 2. Drying of the pods, ears, leaves, or other parts of the crop 3. Kernel hardness 4. Seed development 5. Shedding of lower older leaves
Chemical	1. Titratable acidity 2. TSS/acid ratio 3. Sugar content (Brix readings) 4. Sugar/acid ratio 5. Bioelectric conductance 6. Starch content 7. Starch-iodine test 8. Tannin content 9. Juice or oil content	1. Moisture or dry matter content 2. Oil content (e.g., in oil crops) 3. Starch content 4. Sugar content
Physiological	1. Respiration rate 2. Ethylene evolution rate	1. Photosynthesis assimilates supply stops (no more increase in dry matter content) 2. Seed black layer and milk line development (for maize) 3. Hard dough stage

TSS, Total soluble solids.
Source: Adapted from Sharma, R.M., Singh, R.R., 2012. Harvesting, postharvest handling and physiology of fruits and vegetables. In: Verma, L.R., Joshi, V.K. (Eds.), Postharvest Technology of Fruits and Vegetables (fourth ed.). Indus Publishing Company, New Delhi, pp. 1–484 and Sahay, K.M., Sing, K., 2016. Unit Operations of Agricultural Processing. Vikas Publishing House Pvt Ltd., Noida, India.

already reach its climacteric peak in the transitory period before it is consumed. As after the climacteric peak, the rate of respiration declines and senescence and decay of the product quality initiates. Moreover, the problem is more complex in the case of nonclimacteric fruits, which do not undergo a ripening process when detached from the plant. Whereas vegetables (e.g., green leafy ones) start to decay as soon as they are harvested; the most prominent physiological process in these vegetables is the loss of chlorophyll (senescence resulting in the yellowing of the tissue) (Snowdon, 2010).

For practical purposes at the farm level, commercial and smallholder farmers employ color and degree of developing as the two most common indices to determine harvest time. In certain commercial or industrial farming days from flowering and accumulation of heat, units may be calculated to determine optimum maturity time.

3.4 On-farm postharvest systems

Postharvest handling takes place at the stage of crop production immediately after harvest. The typical steps in a postharvest operation may include cooling and sorting of horticultural crops and drying and on-farm storage of the cereals. The harvest (the useful part of the plant) is the living organs or seeds of the plant, which continue to respire and metabolize. The focus of this chapter is on the optimization of the handling and storage practices on the farm to reduce on-farm losses, which are a major concern in developing countries. As soon as the crop is uprooted from the ground or detached from the mother plant, it starts to deteriorate. The postharvest management of the crop will ultimately decide if the product would finally reach in its intended quality (or harvest quality) to the final consumer or not. Hence, the initial few hours or days after harvest are critical for defining the shelf life or long-term storage potential of a crop.

For certain roots and tubers (such as cassava, garlic, potato, onions, sweet potato, and yams) immediate postharvest curing (in most cases at the farm, for 3 up to 7 days) may be essential to improve postharvest shelf life and overall quality of the product. The drying out and thickening of skin tissue as a barrier to moisture loss and pathogenic infections are the primary goals of curing. In the bulb crops such as onions and garlic, the outer shell or scalers are dried to ensure longer shelf life. Whereas for potatoes and sweet potatoes wound healing and skin setting (lignification of periderm) to reduce skinning and bruising injuries during successive handling and storage is attained. In most of the cases curing required hot and humid conditions (with good ventilation) ranging from 15°C–20°C to 30°C–32°C and 85%–90% Relative Humidity (RH) for potatoes and sweet potatoes, and 32°C–40°C and 90%–100% RH for tropical roots and tubers such as cassava and yams. Whereas, bulb crops like onions and garlic may be cured 33°C–45°C at 60%–75% RH.

For smallholders in developing countries where controlled atmospheric curing chambers may not be available due to economic reasons, curing of crops like yams, sweet potato, and cassava can be obtained by piling the roots and tubers in the shade (heaping under a tree for example) and covering them with straw and grass to increase relative humidity, the tropical temperatures may be just enough to get the desired results.

In postharvest, temperature and humidity are the two most important factors that influence the decay and deterioration of the horticultural commodities (perishables, fruits and vegetables, and roots and tubers). Heat injuries, particularly in the tropical environments, can be harmful to perishable products causing localized necrosis in the form of sunscalds and general decay. Every commodity has an optimum condition at which it will maintain quality and improved shelf life. When not stored in option temperature and humidity decay rate may increase two to three times for every 10°C increase in the temperature (Kader, 2013). Hence, rapid removal of field heat is crucial to arrest the physiological deterioration, however, apart from this initial cooling, a continuous cool chain would still require along the entire value chain to minimize deterioration and subsequent discard. Management of relative humidity on the other hand (which is a measure of the amount of water that can be held by the atmosphere in the form of water vapor at a temperature and pressure) is particularly important to reduce moisture loss (evaporation losses). The optimum temperature for many of the fruits and vegetables lies close to 0°C (Kader, 2013); the fruits and vegetables which are susceptible to chilling injuries should be stored at relatively higher temperature, for example, ranging from 5°C to 15°C. Similarly, various fresh perishable commodities have their optimum relative humidity, however, in general, all fresh fruits and vegetables require high relative humidity (ranging from 80% up to 100% in some cases) to control water loss and achieving ripening in certain fruits. The temperature management during long-term storage is the key to slow down the physiological and pathological decay of the agricultural produce. From an "on-farm postharvest activities" point of view, precooling to remove the field heat in horticultural produce is an important aspect. However, to ensure the long-term storage and maintenance of desired quality (by retarding the ripening and senescence) it is essential to keep the cool chain conditions all along the food supply chain. Evaporative cooling (zero energy cooling chamber) and night ventilation are some of the effective cooling techniques for smallholder farmers in developing countries who may not have facilities to mechanically precool the produce.

For grain crops such as wheat, maize, rice etc., and dehydrated fruits and vegetables, the water activity is an important parameter. The simplest way of expressing water activity is as follows:

$$Aw = \frac{p}{P^0}$$

Aw = Water activity
p = Partial vapor pressure of the water in the material being measured
P^0 = Vapor pressure of the pure water at the same temperature

Water activity is an expression of the amount of water in processed food products or raw materials that is available to support the growth of microorganism. Water activity of food products is directly proportional to the moisture content of

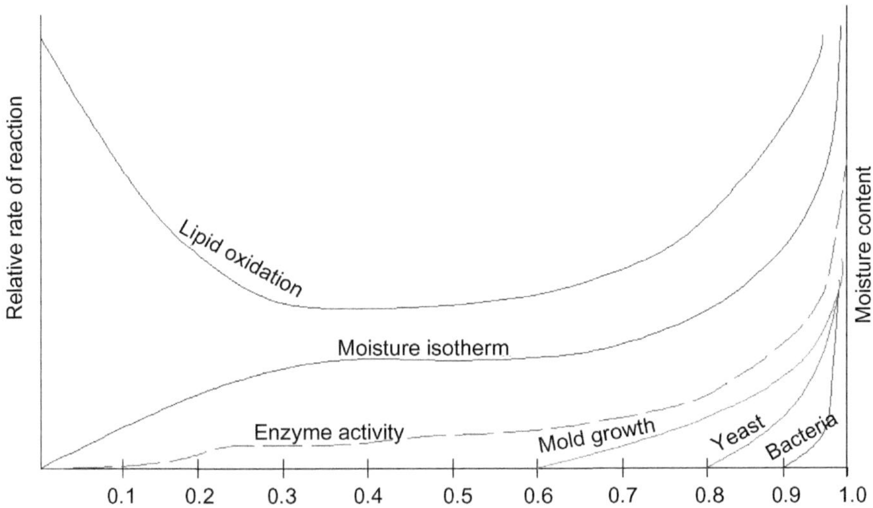

Figure 3.4 A simplified water activity stability diagram for microbial growth and rate of other reactive degradations.
Source: Adapted from Aqualab, 2018. Water activity. Retrieved July 31, 2018, from: http://aqualab.decagon.com.br/educacao/measurement-of-water-activity-for-product-quality/.

the crop, when moisture content increases water activity rises, which encourages fungal and insect problems, and respiration. A simplified water activity stability diagram is presented in Fig. 3.4.

3.4.1 On-farm handling and storage

Improving the storage facility for perishables at farm level is lacking in the developing countries. A significant amount of product is lost due to improper storage at farm and delays in transportation of the product to the market. The contributing factors to deterioration such as temperature, ethylene production, microbial load, pest infestation, moisture content, etc. need to be controlled at the farm gate to ensure better postharvest life.

In relation to perishables (horticultural crops), the durables (cereals and grains) are less susceptible to losses during harvesting and handling. However, the African Postharvest Loss Information System (https://www.aphlis.net/en#/) reports on-farm losses (including harvesting, threshing, winnowing, field drying, on-farm storage) can be in the range of 5%–10% for cereal crops such as wheat, maize, barley, rice, and sorghum (Table 3.5). Storage pests and moisture may make their initial attack at the farm, and the infestation and damage may be aggravated along the value chain as the product travel from farm to consumer.

Condensation in the storage of grains may be caused by temperature fluctuations, and infestation with pests (due to respiratory activity the temperature increases and the environment capacity to hold water increases; as the ambient environment cools

Table 3.5 Safe moisture content of the stored produce for long-term storage (up to a year)

Produce	Conventional storage (in sack or silos) %	Hermetic storage %
Maize	13	<10
Wheat	13	<10
Millet	16	12
Sorghum	12.5	<10
Paddy	14	<10
Rice	13	<10
Cowpeas, beans	15	<10
Groundnuts	7	<5
Cocoa	7	<5
Copra	7	<5
Palm kernels	5	<5
Coffee	13	<10

down the condensation occurs). Hermetic storage may require even lower moisture contents.

Many farmers in the developing regions store their grains in unthreshed form, where the storage period can last from 6 to 12 months. The grains are stored in a variety of traditional storage structures, which are based on socioeconomic, climatic conditions, and availability of local raw materials. The two typical forms of storage systems in developing tropical countries are defined by Gwinner et al. (1990):

1. Open and semiopen storage systems

   An open storage system is suitable for unfavorable hot and humid climates as the freshly harvested grain products may still have higher moisture content, whereas semiopen stores are common in semiarid regions. Raised wooden platforms with a straw roof (protection from rain) on which cobs or panicles are stacked in layers are widespread in the humid tropics. These are simple constructions and have poor store hygiene as insects, birds, and rodents have direct access to the crop; however natural ventilation enables the crop to dry and hinders the development of mold.
2. Closed storage systems

   Closed storage systems are widespread in arid regions and primarily constructed from locally available material such as mud mixed with straw, where crops are stored after threshing. This provides good protection against pests, and has a cool and dry microclimate (especially in mud construction).

3.5 Farmer organization, value addition, training, and access to market

The global agricultural economy is rapidly changing due to increased urbanization and globalization. The product supply chains are becoming complex, and there are various opportunities as well as risks that primary producers face due to these trends. The price of common commodities remains static or has even declined in certain cases, so it is desirable that farmers invest in high value crops and take part in value addition by agricultural processing of their produce. There have been several debates whether a public or private sector approach would support the smallholders in achieving these endeavors. Capacity building of the smallholders in the least developed countries is the key to improve the postharvest systems and in turn, reduce food losses and improve livelihoods. Strengthening the smallholder extension services to disseminate knowledge on preharvest and postharvest factors to improve crop shelf life and quality is paramount to achieving sustainable agricultural systems.

Improving the linkages vertically and horizontally along the entire value chain is important to minimize inefficiencies and reduce the risk of market failures (oversupply, price crashing, or gluts). At the same time, encouraging public/private partnerships is necessary to improve the product quality, add value, and enhance the shelf life of perishable crops. It is a common phenomenon (in particular in industrialized countries, but not limited to them) that overproduction leads to higher food losses, as having produced more than needed the excess harvest is sold at lower prices as animal feed or other nonfood purposes or processing (biofuel). Such a situation leads to the economic losses to farmers, which is another way of translating physical or weight losses of food. The prevention of such circumstances can be achieved by enhancing the communication and cooperation among farmers' groups.

Market access is one of the significant problems smallholders in developing regions face, due to disaggregate nature, small production volumes, lack of connectivity to urban markets, and other logistic issues. Farmer groups and cooperatives/associations can help link the smallholders to the market, sometimes with contractual farming for big traders and processors. Hellin et al. (2007) presented a case study from Central America on maize and high-value vegetables and concluded that farmers' organizations were much more effective in the vegetable sector, which is known for its high transaction cost to access markets. Linking producers directly to the market was a key element of this success; it was demonstrated in this case study that when high value vegetable farmers' organizations were directly linked to supermarkets in urban centers, the impact on financial sustainability of these organizations was immense. However, the weakness or the risk that was highlighted was the lack of business skills and nonreplicable structure of these linkages.

In another example from Malawi, Kachule et al. (2005) reported from a semistructured survey of 12 farmers' organizations the importance of involvement in agricultural processing, food chain management, and development of human capital. Other factors that were a barrier in realizing the full potential of these organizations were lack of financial resources, lack of assets, and limited networking and coordination among them. Critical areas that needed attention were capacity building, networking and linkages, and design of governance systems. Ferris et al. (2014) stated that modernizing the extension services in developing countries can have an impact on millions of smallholders by improving productivity at harvest and postharvest stages. However, the challenges that cannot be addressed by extension alone are the micro- and macroeconomic issues, such as local government structures, high internal transport costs, poor access to inputs, and uncoordinated domestic and international trade policies. Engagement of the private sector along with a stable government and effective policy reforms are some of the solutions that can improve the opportunities to millions of smallholders.

3.6 Climate changes and potential impacts on crop postharvest

It is very clear that in the near future and over the long term, our climate is going to change eventually. Some of the Intergovernmental Panel on Climate Change global warming scenarios and increased amounts of various greenhouse gases are illustrated in Figs. 3.5 and 3.6 respectively. Changing climate in the form of increased temperature, increased levels of carbon dioxide and other gases in the atmosphere, rainfall, humidity, and extreme events, not only has an impact on the primary production stages of the agriculture, but also during postharvest. In addition, postharvest losses in themselves are a big contributor to climate change, which is why it is important to reduce losses, otherwise we can be stuck in a vicious cycle of climate change and associated food losses. Estimates from the World Resources Institute suggest that food loss and waste accounts for 4.4 gigatonnes of greenhouse gases. From the food systems perspective, this calls for an improved cooperation between all the value chain stakeholders to develop more diversified cultivars that are more nutritious, reduce soil degradation, have lower or no dependence on unsuitable external inputs (fertilizers, pesticides), have resistance to current and emerging pests, and have increased adaption to the changing climate (Chegere, 2017; Vermeulen et al., 1958). Dwivedi et al. (2017) in a recent review emphasized that climate change is going to pose a serious threat to crop production and protection globally, particularly in the areas of the world that need to significantly

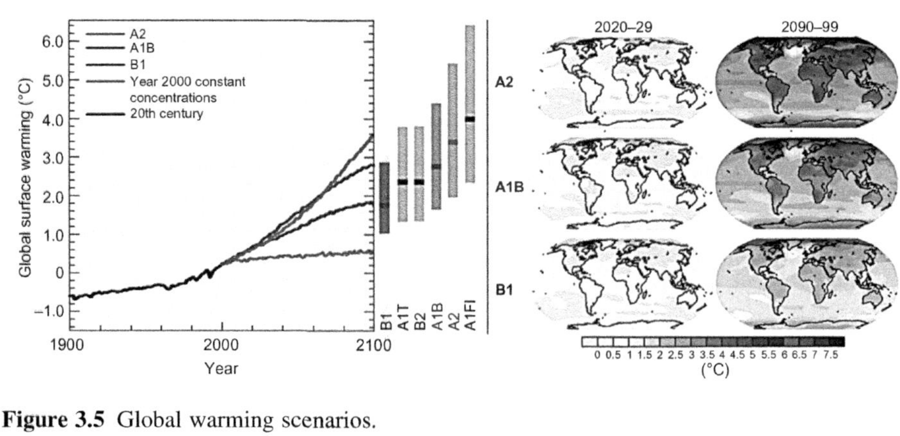

Figure 3.5 Global warming scenarios.
Source: IPCC, 2007a, AR4-synthesis report, Chapter 3, https://www.ipcc.ch/. Courtesy IPCC.

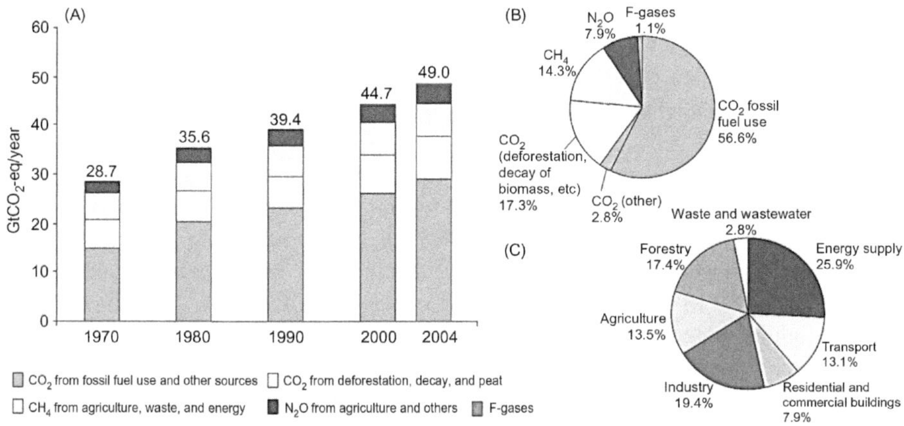

Figure 3.6 Greenhouse gases and their emissions.(A) Global annual emissions of anthropogenic GHGs, (B) Share of different anthropogenic GHGs in total emissions in 2004 in terms of carbon dioxide equivalents (CO2 -eq), (C) The share of different sectors in total anthropogenic GHG emissions in 2004 in terms of CO2 -eq.
Source: IPCC, 2007b, AR4-synthesis report, Chapter 2, https://www.ipcc.ch/. Courtesy IPCC.

increase their production due to rising population and urbanization. Some of these areas that are going to have severe effects of climate change are Africa, Mesoamerica, the Andes, and South-Central Asia. A large proportion of populations in these areas have to adapt and adopt mitigation strategies, which are limited in these regions. They will need new crop cultivars that can sustain changing climate during production and postproduction stages, and in such a situation, reduction of food losses will play a significant part. An example of one of the most popular staple crops is maize in Africa; it has been identified that hybrid high yielding varieties tend to be susceptible to insect and pest damage as well as more tolerant to erratic weather conditions than the indigenous crops such as sorghum and millet (Kossou et al., 1993).

Climate change is a key factor that will affect the postharvest systems including changes in temperature, rainfall, humidity, and events of extreme weather (or worst case, natural disasters). Stathers et al. (2013) provided an extensive review paper on the effects of climate change trends on grain postharvest systems of eastern and southern Africa. The postharvest elements that according to this review paper can be significantly affected by climate change are viability of seeds, survival and reproduction of storage insects and pests, performance of storage protectants, and finally the shelf-life of the perishable and durable crops. For example, in a particular example for cereals and grains storage, if moisture, temperature, and gas composition of the storage environment changes this can have considerable effect on increased infestation of fungi, insects, mites, and rodents, which will worsen the postharvest loss situation in already vulnerable regions. One important thing in this respect is that climate change will have different effects in different areas, so the adaptation strategies need to be very regional and local in their nature. Higher temperatures for the storage of grains and other dehydrated food products can mean more rapid growth and infestation of storage insect pests and growth of fungal organisms particularly those that are associated with the production of myco- and aflatoxins. For perishable products, high temperature means shorter shelf life and more energy demands for cooling, which may not be sustainable in many developing countries. Also, insects/pests and crop disease pathogens (in most cases insects and pests are the vectors of these diseases) migrate and expand their territories with global warming, and more land mass would be become suitable in terms of temperature for these insect and pests. Moreover, the increase in temperature and subsequent occurrences of the extreme weather conditions (droughts, flooding, storms, etc.) can induce water stress (excess or deficit) for the crops in preharvest affecting the crop quality postharvest. El-Ramady et al. (2015) stated that in tree crops, which in most cases are not irrigated but depend on the rainfall, the water deficit for tree crops not only would reduce the productivity but also accelerate the ripening process, which influences the postharvest shelf life. Some of the adaptation strategies to increase the resilience of postharvest systems of cereals and grain in eastern and southern Africa to climate change, which were identified by Stathers et al. (2013), are as follows:

1. Improvement and modifications in grain drying and storage management, which will reduce the mycotoxin contamination and storage pest infestation.
2. Appropriate value addition/food processing opportunities based on the local or global (for export for example) demands and food systems.
3. While selecting the crop cultivars for primary production look at the properties for both the pre- and postharvest stages and activities.
4. Improving the market information systems and networks, which will improve the market linkages and capture new market opportunities.
5. Training and learning of extension agents into climate change and its effects on postharvest systems.
6. Investing in more accurate and responsive weather forecast and early warning systems, which will inform various stakeholders of an oncoming weather event that can negatively affect their crop in pre- and postharvest stage.

Coming from the example to cereals and grain crops to fruits and vegetables, studies have shown that an increase in temperature and exposure to increased levels of gases such as CO_2 and O_3 can directly or indirectly have an effect on the postharvest quality of fresh fruits and vegetables. In an elaborative review Moretti et al. (2010) talked about how increased temperature can lead to changes in photosynthesis and changes in sugars, organic acids, flavonoids, firmness, and antioxidant levels. Whereas higher levels of CO_2 in the atmosphere, for example, in potatoes, can cause tuber malformation during production and in postharvest stages leading to problems such as scab disorder and changes in composition of reducing sugars. Moreover, there are several examples where extreme unpredicted hot and wet spells at harvest of vegetables and sugarcane in Vietnam, Indonesia, and Australia resulted in loss of multimillions of dollars (Vermeulen et al., 1958).

Hence it can be said that climate change can seriously affect various stages of the food chain, from the more obvious link to primary production, to various direct and indirect linkages to postharvest problems leading to food losses. The impact of this is expected to be widespread and complex in nature. More emphasis on research and studies is required to gather more evidence regarding the impact of climate on the postharvest food chain, as the present data is limited and scattered.

References

APHLIS, 2017. African postharvest loss information systems. Retrieved January 1, 2018, from: <https://www.aphlis.net/index.php/en?form = country_narratives#/>.

Aqualab, 2018. Water activity. Retrieved July 31, 2018, from: <http://aqualab.decagon.com.br/educacao/measurement-of-water-activity-for-product-quality/>.

Brasil, I.M., Suddiqui, M., 2018. Preharvest modulations of postharvest fruit and vegetable quality. In: Siddiqui, M.W. (Ed.), Postharvest Quality of Fruits and Vegetables: An Overview. Academic Press, New York, pp. 1–40.

Chegere, M.J., 2017. Climate change and post-harvest agriculture. Request PDF. Available from: <https://www.researchgate.net/publication/324861568_Climate_change_and_post-harvest_agriculture> (accessed 11.09.18.).

Charrondiére, U.R., et al., 2013. FAO/INFOODS food consumption database for biodiversity. Food Chem. 140, 408–412.

Coleman, D.A., Rothgeb, B.E., Fellows, H.C., 1928. Respiration of Sorghum Grains. USDA Technical Bulletin.

Dillahunty, a L., Siebenmorgen, T.J., Buescher, R.W., Smith, D.E., Mauromoustakos, A., 2000. Effect of moisture content and temperature on respiration rate of rice 1. Cereal Chem. (C), 3–5.

Dwivedi, S.L., Lammerts van Bueren, E.T., Ceccarelli, S., Grando, S., Upadhyaya, H.D., Ortiz, R., 2017. Diversifying food systems in the pursuit of sustainable food production and healthy diets. Trends Plant Sci. 22 (10), 842–856. Available from: https://doi.org/10.1016/j.tplants.2017.06.011.

El-Ramady, H.R., Domokos-Szabolcsy, É., Abdalla, N.A., Taha, H.S., Fári, M., 2015. Postharvest management of fruits and vegetables storage. In: Lichtfouse, E. (Ed.), Sustainable Agriculture Reviews. Sustainable Agriculture Reviews, vol. 15. Springer, Cham.

Fallik, E., 2008. Postharvest treatments affecting sensory quality of fresh and fresh-cut products. In: Paliyath, G., Murr, D.P., Handa, A.K., Lurie, S. (Eds.), Postharvest Biology and Technology of Fruits, Vegetables, and Flowers. Wiley-Blackwell, p. 497.

FAO, 1989. Prevention of Post-Harvest Food Losses: Fruits, Vegetables and Root Crops. Food and Agriculture Organization of the United States, pp. 1–159.

FAO, 2012. Save food. Retrieved December 18, 2017, from: <http://www.fao.org/save-food/en/>.

FAO, 2015. FAOSTAT, Food and Agriculture Organization of the United Nations.

Ferris, B.S., Robbins, P., Best, R., Seville, D., Buxton, A., Shriver, J., et al., 2014. Linking Smallholder Farmers to Markets and the Implications for Extension and Advisory Services. Modernizing Extension and Advisory Services, pp. 1–52.

FLWP, 2016. Food Loss and Waste Accounting and Reporting Standard. Washington, DC.

Gustavsson, J., Cederberg, C., Sonesson, U., van Otterdijk, R., Meybeck, A., 2011. Global Food Losses and Food Waste: Extent, Causes and Prevention. International Congress: Save Food!, Düsseldorf, Germany. https://doi.org/10.1098/rstb.2010.0126.

Gwinner, J., Harnisch, R., Mück, O., 1990. Manual on the Prevention of Post-Harvest Grain Losses. GTZ, Hamburg, Germany.

Hellin, J., Lundy, M., Meijer, M., 2007. Farmer organization and market access. LEISA Mag. 23 1 (67), 26–27. Available from: https://doi.org/10.2499/CAPRiWP67.

Hodges, R.J., Buzby, J.C., Bennett, B., 2011. Postharvest losses and waste in developed and less developed countries: opportunities to improve resource use. J. Agric. Sci. 149, 37–45. Available from: https://doi.org/10.1017/S0021859610000936.

Huang, H., Danao, M.G.C., Rausch, K.D., Singh, V., 2013. Diffusion and production of carbon dioxide in bulk corn at various temperatures and moisture contents. J. Stored Prod. Res. 55, 21–26. Available from: https://doi.org/10.1016/j.jspr.2013.07.002.

Hussein, Z., Fawole, O.A., Opara, U.L., 2018. Preharvest factors influencing bruise damage of fresh fruits – a review. Sci. Hortic. (Amsterdam). 229, 45–58.

IPCC, 2007a. *Climate Change 2007: Synthesis Report. Contribution of Working Groups I, II and III to the Fourth Assessment Report of the Intergovernmental Panel on Climate Change* [Core Writing Team, Pachauri, R.K and Reisinger, A. (eds.)]. IPCC, Geneva, Switzerland, 104 pp.

IPCC, 2007b. *Climate Change 2007: Synthesis Report. Contribution of Working Groups I, II and III to the Fourth Assessment Report of the Intergovernmental Panel on Climate Change* [Core Writing Team, Pachauri, R.K and Reisinger, A. (eds.)]. IPCC, Geneva, Switzerland, 104 pp.

Kader, A.A., 2000. Pre- and postharvest factors affecting fresh produce quality, nutritional value, and implications for human health. In: Proceedings of the International Congress Food Production and the Quality of Life, September 4–8, 2000, vol. 1, Sassari, Italy, pp. 109–119.

Kader, A.A., 2013. Postharvest technology of horticultural crops - an overview from farm to fork. J. Appl. Sci. Technol. 1 (1), 1–8.

Kays, S.J., 1999. Preharvest factors affecting appearance.pdfPostharvest Biol. Technol. 15, 233–247 (June 1998). Available from: https://doi.org/10.1016/S0925-5214(98)00088-X.

Kachule, R., Poole, N., Dorward, A., 2005. Farmer Organisations for Market Access: Farmers Organisations in Malawi, vol. 8275.

Khoury, C.K., Bjorkman, A.D., Dempewolf, H., Ramirez-Villegas, J., Guarino, L., Jarvis, A., et al., 2014. Increasing homogeneity in global food supplies and the implications for food security. Proc. Natl. Acad. Sci. 111 (11), 4001–4006. Available from: https://doi.org/10.1073/pnas.1313490111.

Kossou, D.K., Mareck, J.H., Bosque-Perez, N.A., 1993. Comparison of improved and local maize varieties in the Republic of Benin with emphasis on susceptibility to *Sitophilus zeamais* Motschulsky. J. Stored Prod. Res. 29, 333–343.

Kummu, M., de Moel, H., Porkka, M., Siebert, S., Varis, O., Ward, P.J., 2012. Lost food, wasted resources: global food supply chain losses and their impacts on freshwater, cropland, and fertiliser use. Sci. Total Environ. 438, 477–489. Available from: https://doi.org/10.1016/j.scitotenv.2012.08.092.

Mattheis, J.P., Fellman, J.K., 1999. Preharvest factors influencing flavor of fresh fruit and vegetables. Postharvest Biol. Technol. 15 (3), 227–232. Available from: https://doi.org/10.1016/S0925-5214(98)00087-8.

Moretti, C.L., Mattos, L.M., Calbo, A.G., Sargent, S.A., 2010. Climate changes and potential impacts on postharvest quality of fruit and vegetable crops: a review. Food Res. Int. 43 (7), 1824–1832. Available from: https://doi.org/10.1016/j.foodres.2009.10.013.

Parfitt, J., Barthel, M., Macnaughton, S., 2010. Food waste within food supply chains: quantification and potential for change to 2050. Philos. Trans. R. Soc. Lond. Ser. B, Biol. Sci. 365 (1554), 3065–3081. Available from: https://doi.org/10.1098/rstb.2010.0126.

Parmar, A., Hensel, O., Sturm, B., 2017a. Post-harvest handling practices and associated food losses and limitations in the sweet potato value chain of southern Ethiopia. NJAS - Wageningen J. Life Sci. 80, 65–74. Available from: https://doi.org/10.1016/j.njas.2016.12.002.

Parmar, A., Kirchner, S.M., Sturm, B., Hensel, O., 2017b. Pre-harvest curing: effects on skin adhesion, chemical composition and shelf-life of sweet potato roots under tropical conditions. East Afr. Agric. For. J. 0 (0), 1–14. Available from: https://doi.org/10.1080/00128325.2017.1340141.

Prasad, K., Jacob, S., Siddiqui, M.W., 2018. Fruit Maturity, Harvesting, and Quality Standards. In: Mohammed Wasim Siddiqui (Ed.),Preharvest Modulation of Postharvest Fruit and Vegetable Quality (Chapter 2). 41–69. Available from: https://doi.org/10.1016/B978-0-12-809807-3.00002-0.

Raudienė, E., Rušinskas, D., Balčiūnas, G., Juodeikienė, G., Gailius, D., 2017. Carbon dioxide respiration rates in wheat at various temperatures and moisture contents. Mapan – J. Metrol. Soc. India 32 (1), 51–58. Available from: https://doi.org/10.1007/s12647-016-0202-4.

Rees, D., 2012. Introduction. In: Rees, D., Farrell, G., Orchard, J. (Eds.), Crop Post-Harvest: Science and Technology, vol. 1. Blackwell Publication Company, Oxford.

Rosner H., 2014. How we can tame overlooked wild plants to feed the world. <http://www.wiredcom/2014/06/potato-bean/> (accessed 11.06.18.).

Sahay, K.M., Sing, K., 2016. Unit Operations of Agricultural Processing. Vikas Publishing House Pvt Ltd, Noida, India.

Sams, C.E., 1999. Preharvest factors affecting postharvest texture. Postharvest Biol. Technol. 15 (3), 249–254. Available from: https://doi.org/10.1016/S0925-5214(98)00098-2.

Sarrocco, S., Vannacci, G., 2017. Preharvest application of beneficial fungi as a strategy to prevent postharvest mycotoxin contamination: a review. Crop Protect. Available from: https://doi.org/10.1016/j.cropro.2017.11.013.

Sharma, R.M., Singh, R.R., 2012. Harvesting, postharvest handling and physiology of fruits and vegetables. In: Verma, L.R., Joshi, V.K. (Eds.), Postharvest Technology of Fruits and Vegetables, fourth ed. Indus Publishing Company, New Delhi, pp. 1–484.

Sharma, B.S., Wightman, J.A., 2016. Vision Infinity to Food Security, Some Whys, Why Nots and Hows!. Springer Briefs on Agriculture.

Sidiqqui, M.W., 2018. Preface. In: Sidiqqui, M.W. (Ed.), Preharvet Modulation of Postharvest Fruits and Vegetable Quality. Academic Press, New York.

Simson, S.P., Straus, M.C., 2010. Post-Harvest Technology of Horticultural Crops. Oxford Book Company, Jaipur, India.

Smil, V., 2000. Feeding the world: A challenge for the twenty-first century. The MIT Press, Cambridge, Massachusetts, London.

Snowden, A.L., 2010. *Post-Harvest Diseases and Disorders of Fruits and Vegetables: Volume 1*. Manson Publishing, London, UK.

Stathers, T., Lamboll, R., Mvumi, B.M., 2013. Postharvest agriculture in changing climates: its importance to African smallholder farmers. Food Secur. 5 (3), 361–392. Available from: https://doi.org/10.1007/s12571-013-0262-z.

Sorour, H., Uchino, T., 2004. The effect of storage condition on the respiration of soybean. J. Jpn. Soc. Agric. Mach. 66 (1), 66–74. Available from: https://doi.org/10.5772/27025.

UCDAVIS, 2018. Commodity fact sheet. Retrieved January 1, 2018, from: <http://postharvest.ucdavis.edu/Commodity_Resources/Fact_Sheets/>.

Vermeulen, S.J., Campbell, B.M., Ingram, J.S.I., 1958. Climate Change and Food Systems. Annual Review of Environment and Resources. Available from: https://doi.org/10.1146/annurev-environ-020411-130608.

World Bank, 2011. Missing Food : The Case of Postharvest Grain Losses in Sub-Saharan Africa. The World Bank, Washington, DC.

WHO, 2015. World Health Organization and Secretariat of the Convention of Biological Diversity. Connecting Global Priorities: Biodiversity and Human Health: A Status of Knowledge Review. WHO.

Further reading

Arah, I., Arah, I.K., Amaglo, H., Kumah, E.K., Ofori, H., 2015. Preharvest and postharvest factors affecting the quality and shelf life of harvested tomatoes: a mini review. Int. J. Agron. 2015 (6), 1–6. Available from: https://doi.org/10.1155/2015/478041.

HLPE, 2014. Food losses and waste in the context of sustainable food systems. A Report by the High Level Panel of Experts on Food Security and Nutrition of the Committee on World Food Security. HLPE Report, Rome. Available from: https://doi.org/65842315.

Lundqvist, J., Fraiture, C., De, Molden, D., 2008. Saving water: from field to fork curbing losses and wastage in the food chain. SIWI Policy Brief 5–29.

Prasad, K., Jacob, S., Sidiqqui, M.W., n.d. Fruit Maturity, Harvesting and Quality Standards.

Food preservation technologies

4

Sebnem Tavman[1], *Semih Otles*[1], *Selale Glaue*[2] *and Nihan Gogus*[2]
[1]Food Engineering Department, Ege University, İzmir, Turkey, [2]Efes Vocational School, Dokuz Eylul University, İzmir, Turkey

Chapter Outline

4.1 Introduction

According to the United Nations Economic and Social Affairs Department, the world's population was 7.3 billion in 2015 and will reach 8.5 billion by 2030, 9.7 billion by 2050, and 11.2 billion in 2100. Global action against hunger predicts that we must increase food production by 70% to feed this rapidly growing world population. However, increasing farming practices will not be enough to solve global food demand. The development and global implementation of more efficient, energy- and resource-saving technologies helps the industry respond to increasing food demand while reducing environmental impact and providing sustainability. Reducing food waste must also represent an important part of the equation. The reports of the United Nations estimate that 30%–50% of the globally produced food—about 2 billion tons—is lost or wasted. Food may be lost due to some

Saving Food. DOI: https://doi.org/10.1016/B978-0-12-815357-4.00004-3

inadequacies during production and processing, as well as market and consumer trends. They range from poor storage and inadequate transportation logistics to customer purchases of extra products. The establishment of efficient, sustainable food preservation technologies also plays a vital role in preventing wastage associated with food spoilage. The rapidly increasing world population necessitates that the amount of food wasted due to spoilage is kept to a minimum. Food production is only one part of the process to ensure continuous, diverse, safe, food supplies to meet the consumer demands. Lack of food safety systems costs the food industry millions of dollars annually through waste, reprocessing, recalls, and the resulting loss of sales. Foodborne diseases are no longer limited to developing countries. Unsafe food is not fit for human consumption and therefore is wasted. Failure to comply with minimum food safety standards can cause food losses. Several factors can lead to food being unsafe, such as naturally occurring toxins in food itself, microorganisms, contaminated water, unsafe use of pesticides, and veterinary drug residues. Poor and unhygienic handling and storage conditions, and lack of adequate temperature control, can also cause unsafe food and should be prevented by a food preservation technique (FAO, 2011). In the food supply chain of developing and developed countries consumers are responsible for most of the food waste where a recent European food waste program has defined consumer food waste as a major problem. The amount of consumer food waste has attracted the attention of The COST Network, EU network on food waste evaluation, in terms of solving this challenge through technological and political prevention. A safer global food system minimizing all food losses is important, where it also shows us how consumers can reduce food waste in their houses. At this point food preservation methods play an important role in waste reduction by ensuring safer food usage with minimizing food loss (Martindale and Schibel, 2017).

Since ancient times preservation of food has played an important role in human life. The first preservation methods used by early humans were sun drying, salting, and fermentation, which were used to provide food in periods when fresh foods were not available. The need of greater quantities and better quality of processed food increased and continued to increase with the development of civilization. Therefore, this caused increasing interest to the large food preservation industry, which attempts to supply food that is economical, nutritious, and satisfying (Karel and Lund, 2003a). Moreover, food preservation decreases food degradation and enhances the utilization of food by several conventional and innovative preservation methods, which in turn minimizes waste production, saves food, and promotes a sustainable food industry (Martindale and Schiebel, 2017; Chemat et al., 2017). The global value of food lost or wasted is estimated at US$1 trillion per year by the FAO ("Global Initiative on Food Loss and Waste Reduction," 2015), representing a loss of economic value. Therefore, food technology under severe or nonclassical conditions is a currently developing area in applied research and industry to preserve foods and reduce waste. Innovative processing, preservation, and extraction procedures may improve production efficiency and contribute to environmental preservation by reducing the use of water and solvents, fossil energy, prevention of wastewater, and production of hazardous substances (Chemat et al., 2017).

Thermal processing of foods, which is also known as the conventional preservation technique, forms the large part of this food processing industry. Also, autoxidation in food and biological systems is responsible for many adverse effects and implications in food stability and preservation as well as human health. Human health is now assumed to be affected by oxidative damage of foods, which causes the occurrence of some important diseases such as cardiovascular diseases, diabetes, hypertension, metabolic syndrome, cancers, etc. in humans. Therefore, it is essential to preserve foods in terms of waste reduction and human health maintenance by using different commercial and innovative preservation techniques or their combination.

4.2 Thermal food preservation

Food preservation by thermal treatment is one of the most often used and known methods for the inactivation of microorganisms and enzymes, both to prevent a risk to public health and to achieve a commercially reliable shelf life for foods. By using thermal food preservation, long shelf-life foods that do not require refrigeration can be produced. It is also easy to control process conditions during the treatment. Other advantages are the destruction of some antinutritional items that may be present in the food (e.g., destruction of trypsin inhibitor in some legumes) and the ease of digestion and absorption of some of the food items found in the food (e.g., facilitation of digestion of proteins, gelatinization of starch) (Awuahet al., 2007; Augusto et al., 2018).

Thermal processing is based on the use of thermal energy (heat) where the food is heated by a hot fluid to a specific temperature, kept for a certain time that has been previously calculated to optimize the product characteristics, called processing time, and then cooled by a cold fluid to interrupt the thermal actions (Augusto et al., 2018; Karel and Lund, 2003b).

However, temperature speeds up degradation reactions, so that heat treatments adversely affect nutritional compounds and sensorial properties. Therefore, the thermal process should be designed to balance the needs of commercial sterility with the commercial demand to submit a high-quality product. New technologies in food processing operations aim to reduce damage to nutrients and sensory elements by reducing heating times and optimizing heating temperatures. Besides inactivation of pathogens, thermal treatment can also form some other desirable changes, such as protein coagulation, texture softening, and formation of aromatic components.

Thermal processing effectiveness is affected by the characteristics of the product and microorganisms, and by the processing conditions. Fat level, composition, pH, size and shape, used preservatives, and water activity are the product characteristics that influence thermal processing. The microorganism characteristics include the strain, growth conditions, and resistance to stress such as acid and/or heat. The processing conditions are the heating source, heating rate, processing type, and environmental conditions during processing (Osaili, 2012).

The thermal processing is designed according to the process target. It can be a vegetative cell, a microbial spore, or an enzyme that must be chosen to ensure safety and quality of the processed food (Augusto et al., 2018; Karel and Lund, 2003b).

Thermal destruction of microorganisms is a time/temperature process. It can be explained by two notions: decimal reduction time (D value) and thermal resistance constant (z value). The D value is the heating time required to kill 90% of a certain number of microorganisms at a given temperature. It indicates the tolerance of the microorganism to the increase in heating time at a given temperature. As D value increases, thermal resistance of microorganisms increases. D value is calculated at each temperature from the linear regression model between $\log_{10}$ of the bacterial survivors and heating time. The D value is the negative inverse slope of the survivor curve and can be expressed mathematically as follows:

$$D = \frac{t_2 - t_1}{\log_{10}(A) - \log_{10}(B)} \tag{4.1}$$

where A and B represent the survivor counts following heating for times t_1 and t_2 minutes.

The z value is the temperature difference required for the thermal inactivation curve to cause a 1 $\log_{10}$ reduction. It indicates the tolerance of a specific pathogen to the temperature changes in the product. The greater the z value, the less the microorganisms are affected by the temperature rise. The z value is calculated by determining the linear regression between $\log_{10}$ of D values and their corresponding temperature. The z value is the negative inverse slope of the thermal resistance curve and can be expressed mathematically as follows:

$$z = \frac{T_2 - T_1}{\log_{10}(D_1) - \log_{10}(D_2)} \tag{4.2}$$

where D_1 and D_2 are D values at temperatures T_1 and T_2, respectively (Osaili, 2012).

The major conventional thermal processes applied in food preservation are sterilization and pasteurization. Cooling and freezing are also basic methods that are often used. Recently, electroheating technologies such as microwave heating, ohmic heating, and radio frequency (RF) have emerged to resolve the problems with conventional preservation processes. These technologies have been used to inactivate microorganisms in different types of food (Osaili, 2012).

4.2.1 Pasteurization

Pasteurization is the most important preservation method and is essential for food safety. It kills all the disease-causing and most other bacteria that might cause deterioration with minimal changes in sensory and nutritional properties. It is the milder

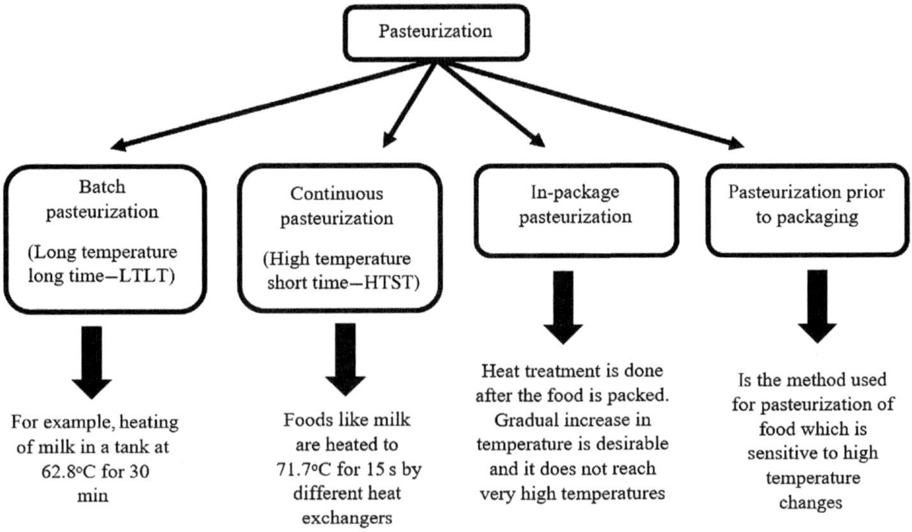

Figure 4.1 Types of pasteurization.

thermal process and generally used to extend the shelf life of food varying from few days (low-acid products with high a_w) up to months (high-acid products and/or with low a_w) at low temperatures, usually at 4°C (Ramesh, 2007a; Augusto et al., 2018).

The intensity of the heat treatment is specified mostly by the pH value of the food. At pH values greater than 4.5 (low-acid food) extermination of pathogenic bacteria, whereas below 4.5 extermination of spoilage microorganisms or inactivation of enzyme is significant. In high-acid foods ($pH < 3.9$) generally non–spore-forming bacteria, yeast, and molds cause spoilage, while in acidic foods (pH 4.0–4.4) yeasts, molds, and both thermophilic and mesophilic bacteria grow. High-acid fruits also have enzymes such as catalase, peroxidase, polyphenol oxidase, pectin esterase, etc. that must additionally be inactivated (Ramesh, 2007a). Pasteurization is generally applied under atmospheric pressure at temperatures about 60°C–100°C for adequate time to preserve foods like milks, fruit juices, beer, and fermented drinks. This method needs to be used with another preservation method to assure stability because pasteurized products have low stability and variable shelf-life (Augusto et al., 2018).

There are four types of pasteurization as seen at Fig. 4.1 (Ramesh, 2007a):

4.2.2 Sterilization

Sterilization is another thermal process used in food preservation. This is an intensive heat treatment applied at temperatures above 100°C (generally 115°C–130°C) for inactivation of microorganisms. The final product should have no viable organisms. It destroys molds, yeasts, vegetative bacteria, and spores. It enables the stability of the product at ambient temperatures and extends the shelf life. It is applied to low-acid foods (pH > 4.6) that may be contaminated with *Clostridium botulinum* spores during storage and therefore should be treated at least 121.1°C for 3 minutes. to achieve a 12*D* reduction of microorganism. All the process conditions should be designed according to the "cold point", which is the slowest heating part of the product. If this point is sterilized, the rest of the product is sterilized.

During sterilization, first the product is heated up a temperature of 110°C–125°C. Then, the product needs a few minutes to equilibrate and reduce the temperature gradient between the surface and center. After the equilibrium, the product should stay at this temperature for a certain time to ensure the sterilization value, which is determined by F_o value. (If the sterilization temperature is 121°C and the z value is 10°C, the sterilization time required for this temperature is indicated as the F_o value.) Finally, the product should be cooled.

There are two types of sterilization. The first is complete sterilization, which means there should not be any living microorganisms. But this type of sterilization leads to reduced quality and nutritional value. The other type is commercial sterilization, which is intended to destroy all pathogens and the microorganisms that may degrade food under normal storage conditions. Some microorganisms that do not deteriorate and have high thermal resistance may survive. The purpose here is to protect the food quality (Ramesh, 2007b).

The target microorganism at sterilization is *C. botulinum*, which is known to be the most resistant pathogen in low-acidity foods. It is mesophilic anaerobic bacteria and has very heat resistant spores. Botulinum toxins are the most potent biological toxins known.

There are two food sterilization methods: thermal and nonthermal processing. Thermal processing is also divided into in-container sterilization (bulk canning) and aseptic sterilization (processing) (Ramesh, 2007b).

4.3 Developments in cooling and freezing technology

4.3.1 Cooling/chilling

Cooling can be described as the storage of food products at temperatures above freezing and below 15°C, keeping the water in liquid phase. It is widely used for short-term preservation and prolongs shelf life with less damage to sensory and nutritional properties. It slows the growth of microorganisms and retards the chemical reactions like enzyme-catalyzed oxidative browning or oxidation, chemical changes that cause color degradation, moisture loss, postharvest and postslaughter metabolic activities of plant and animal tissues. Except raw materials it is generally used with other preservation processes because the physicochemical, microbiological, and biochemical reactions continue to occur at these temperatures (Augusto et al., 2018; Karel and Lund, 2003b).

The reduction in most reaction rates can be explained by the Q_{10} notion (Eq. (4.3)). It can be defined as the variation in the rate of the reactions by changing the temperature at 10°C, describing an exponential reduction in the reactions with decreasing temperature (Fig. 4.2).

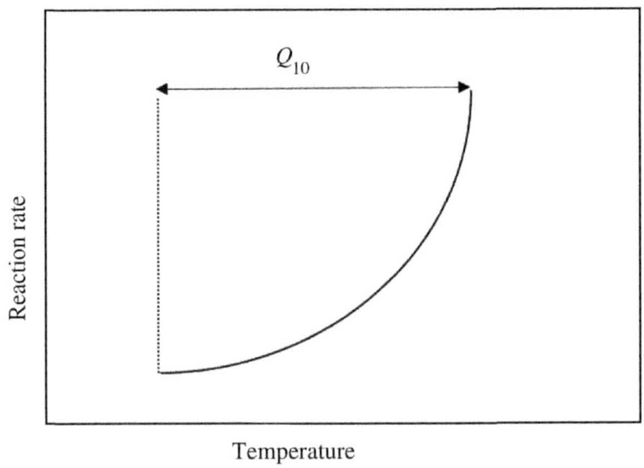

Figure 4.2 Q_{10} concept to define the rate of reaction as a function of temperature.

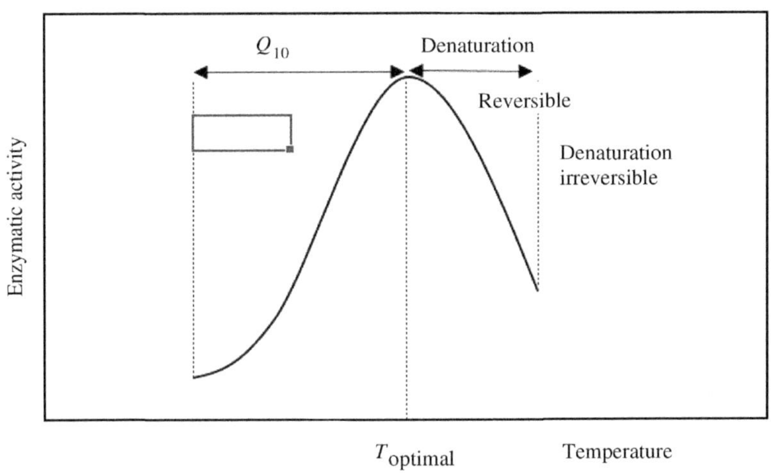

Figure 4.3 Enzymatic activity in relation to temperature.

Generally, for the food products Q_{10} value changes between 2 and 5, which means reducing the temperature by 10°C, reducing the rate of the reactions approximately 2 and 5 times. The longer shelf life of refrigerated products is explained by this notion:

$$Q_{10} = \frac{\text{Reaction rate in } T = T_i + 10^{\circ}C}{\text{Reaction rate in } T = T_i} \tag{4.3}$$

However, the actions of enzymatic and microbial reactions are more complicated and cannot be expressed only by the Q_{10} concept.

Fig. 4.3 shows the activity of an enzyme as a function of temperature. Even though the phenomenon of Q_{10} may explain its activity in a certain temperature range, the denaturation occurs due to maintenance at higher temperatures than the optimum result in reduced enzymatic activity (Augusto et al., 2018).

Parallel activity can be seen for microbial growth with one important difference: the minimum development temperature of each microorganism, below which it remains in a latent state. The minimum and optimum growth temperature ranges of four main types of microorganisms are as follows, respectively: psychrophiles (0°C–5°C), (12°C–18°C); psychrotrophs (0°C–5°C), (20°C–30°C); mesophiles (5°C–10°C), (30°C–40°C); and thermophiles (30°C–40°C), (55°C–65°C).

Consequently, cooling can reduce the microbial growth and inhibit the growth of microorganisms but does not ensure microbial inactivation. Therefore, it must be used with another preservation method (Augusto et al., 2018).

4.3.2 Freezing

Freezing is one of the best ways to extend shelf-life even though several new preservation techniques are acquiring popularity and importance. Freezing is a conventional and simply applied method with a major advantage being its capability to achieve stability without loss of initial quality (Karel and Lund, 2003b).

Freezing influences the physical condition of the substance due to the transformation of water into ice. This occurs when the energy is removed as cooling under the freezing temperature. The general practice for freezing requires a temperature drop to the storage level of around −18°C. All the freezing processes like precooling, supercooling, freezing, tempering, eutectic, ice nucleation, and glass transition points of the products can be clearly demonstrated using freezing or cooling curves and phase diagrams (Tavman and Tuncay, 2018; Rahman and Velez-Ruiz, 2007; Cheng et al., 2017).

Freezing has three stages. First is the cooling of the product to its freezing point (precooling or chilling stage). Here, only sensible heat is removed, and temperature is reduced to simplify ice crystallization of free water. Second stage is moving away the latent heat of crystallization (phase transition stage). This stage comprises the conversion of the water into ice during the crystallization process. So, it is the fundamental step that determines the quality of the frozen product and freezing efficiency. The final stage is cooling the product to the final storage temperature (tempering stage) (Kiani and Sun, 2011).

The breakdown of food by biochemical and physicochemical reactions slows down by freezing but does not stop. The most important factors that prevent quality losses in frozen products are storage temperature, time, and thawing procedures. At temperatures below −18°C, all the microbial activity stops, but enzymatic and nonenzymatic reactions continue slowly during frozen storage. Freezing also inhibits random motion of water molecules.

It is crucial to form small ice crystals in freezing to get the least tissue damage and drip loss in thawing. In general, freezing is faster than thawing and during thawing, damage occurs by microorganisms, chemical, and physical changes (Li and Sun, 2002).

By controlling the freezing process together with storage and preparation conditions, the quality of the frozen food can be increased. Usually, fast freezing is preferred rather than slow freezing. Whereas, some products can be damaged if they are submitted directly to extremely low temperature for a long time. Volume expansion and contraction expansion can express this freeze cracking damage and volume expansion occurs during ice formation. Ice crystal structure has a great importance to protect the quality of frozen products. It is desirable to form fine ice crystals distributed inside and outside the cell to obtain better quality and better-preserved foods, because of less damage to the tissue. On the other hand, large crystals are more acceptable in processes like freeze-drying and freeze concentration. So, control of the water crystallization during the freezing process is very important and recently, new fast freezing technologies to control the crystallization of water have been applied (Kiani and Sun, 2011).

When compared with canning and drying, freezing is the cheapest method of food preservation. The crucial point of freezing is the starting temperature. The material should begin at a temperature near to the freezing point to have preferable crystallization (Tavman and Tuncay, 2018).

The freezing method should be selected according to the type of product, sanitation requirements, reliability, and economic reasons. The common methods used in the food industry include plate contact, air blast, fluidized bed, immersion, and cryogenic freezing.

In the plate freezing method, the product is compressed between metal plates with pressure for good contact pressure. This method can only be used for regular shaped materials. After freezing is completed, hot liquid is used to dissolve the ice.

In immersion freezing, product is immersed in a brine at low temperature to obtain rapid temperature drop by direct heat exchange.

In the air-blast freezing method, food is subjected to cold air flowing at a high speed (2.5–5 m/s) in a tray. At this method velocity selection is important. If the speed is low, the product will freeze slowly, which is not desirable. Even if it is fast, unit-freezing costs increase. The types of air-blast freezing are fluidized bed freezing, belt freezing, spiral freezing, and tunnel freezing.

In cryogenic freezing, food is contacted with liquid gases (liquid nitrogen, liquid carbon dioxide, or their vapor) where the temperatures are below −60°C. This freezing method is very fast, can be used for lots of products, capital cost is low, and high-quality products can be obtained.

4.3.2.1 New freezing technologies

The new freezing technologies include high pressure freezing, ice nucleating proteins, antifreeze proteins, ultrasound-assisted freezing, magnetic resonance freezing, and microwave assisted freezing (Rahman and Velez-Ruiz, 2007).

4.3.2.1.1 High pressure freezing

When the water is frozen at atmospheric pressure, the resulting ice will damage the tissue by increasing the volume. But under high pressure, different kinds of ices are formed and during phase transition, they do not increase volume and they do not harm tissue. As a result, high pressure helps the freezing process and improves product quality.

We can change the physical state of food by using external temperature or pressure. Under high pressure, the freezing point of liquid water can be reduced to below 0°C. When pressure is released, a high supercooling can be obtained, and as a result the ice-nucleation rate is greatly increased.

The use of high pressure simplifies supercooling and it is convenient where uniform and small sized ice crystal distribution is required (Cheng et al., 2017; Li and Sun, 2002).

4.3.2.1.2 Dehydrofreezing

At this type of freezing, food is first dehydrated to a desired moisture and then frozen. Dehydrofreezing ensures the preservation of fruits and vegetables by removing part of the water from food materials prior to freezing. This dehydration process, by removing a part of cellular water, decreases freezing point and freezing time. Thus, as an advantage, improved food quality can be achieved. Also, low energy consumption, as well as low packaging, distribution, and storage costs are the other advantages (Cheng et al., 2017; Li and Sun, 2002).

4.3.2.1.3 Antifreeze protein and nucleation protein

Antifreeze protein and ice-nucleation protein are inserted directly to the food and interact with ice. They are two opposite groups of protein and have different functions. They affect the ice crystal size and crystal structure in food.

The function of the antifreeze protein is to lower the freezing temperature and to restrict the ice formation and change the ice nature by suppressing the growth of ice nuclei. It also delays the recrystallization on frozen storage. Ice nucleating proteins increase ice nucleating temperatures and reduce the degree of supercooling (Li and Sun, 2002).

4.3.2.1.4 Ultrasound-assisted freezing

The application of power ultrasound in food freezing has various aspects, which include initiation of nucleation control of size of ice crystals, speeding up the freezing rate, and improvement of quality of frozen foods (Cheng et al., 2015).

In addition to its traditional application in accelerating the ice-nucleation process, it can also be applied to freeze concentration and freeze-drying processes to control crystal size distribution in the frozen products. If it is applied to the process of freezing fresh foodstuffs, ultrasound can not only increase the freezing rate, but also improve the quality of the frozen products. The ability of power ultrasound in performing these functions is affected by a variety of parameters, such as the duration, intensity, or frequency of ultrasonic waves, etc. (Zheng and Sun, 2006).

Among these effects, cavitation is the most important, which can lead not only to the production of cavitation bubbles but also to the occurrence of microstreaming. The former can act as ice nuclei and increase ice-nucleation rate, while the latter can enhance the heat and mass transfer during the freezing process. Moreover, large ice crystals will fracture into smaller size crystals when subjected to the alternating acoustic stress. Resulting from these acoustic effects, power ultrasound has proved itself an effective tool to initiate the nucleation of ice crystals, control the size and shape of ice crystals, accelerate the rate of freezing, and improve the quality of frozen foods (Zheng and Sun, 2006).

The final main step for the safe consumption of frozen foods is the method of thawing. When the food is thawed from outside to inside, before its center is efficiently defrosted, the temperatures and conditions can cause microbial growth. Therefore, using refrigerated containers, microwave ovens, or water currents during the finalizing of the thawing to minimize the microbial growth is crucial (Augusto et al., 2018).

4.4 Ohmic heating

Ohmic heating is a thermal method that reduces energy input to food, thus reducing thermal damage and positively affecting quality of the food (Butz and Tauscher, 2002). Ohmic heating is a process in which electric current passes through the food product and due to the electrical resistance of the food, heating occurs as a result of the movement of ions. It is also called Joule heating, electric resistance heating, direct electric resistance heating, electroheating, and electroconductive heating (Barba et al., 2018; Lima, 2007). Ohmic heating can be used to produce heat within the product if materials contain sufficient water and electrolytes to allow the passage of electric current (Knirsch et al., 2010).

As opposed to conventional heating's time-consuming convection and conduction heat transfers, ohmic heating is volumetric and has reduced heat transfer time. The electrical conductivity of the food or food mixture is the most significant parameter in ohmic heating. For the solid–liquid mixtures, as an ideal situation, they are required to have equal electrical conductivities, thus heat at the same rate. Density and specific heat of the product also influences the temperature distribution. Besides, viscosity of the fluid is important at ohmic heating, as higher viscosity fluids cause faster ohmic heating than lower viscosity fluids.

It is a green technology because it is a very rapid process, therefore heat losses from the product are very small and environmental losses are minimized. Also, energy losses are significantly reduced due to direct application of electrical energy to the product (Lyng and McKenna, 2011).

Ohmic heating has various advantages like liquid-particle mixtures can be heated uniformly, ultra-high temperature processing temperatures can be reached quickly, product damage is less because there are no hot surfaces for heat transfer, and it has high energy conversion yields and low capital cost.

Ohmic heating can be used as a continuous in-line heating method of pumpable foods for cooking and sterilization of viscous liquids and mixtures containing particulate food products (Baysal and Icier, 2010). It can be used in several food processes, such as cooking, blanching, pasteurization, sterilization, fermentation, dehydration, evaporation, and extraction, specifically in highly viscous or particulate foods. It can be applied to liquids, solids, and liquid–solid mixture foods like liquid egg, whole fruits such as strawberry, juices, sauces, stews, meats, seafood, pasta, and soups (Knirsch et al., 2010; Lima, 2007; Baysal and Icier, 2010).

Naturally, microbial inactivation in ohmic heating is thermal but due to the existence of electric field, slight nonthermal cell wall damage can be observed. As a main outcome of this effect, the *D* value observed for the microbial inactivation under ohmic heating is reduced when compared with traditional heating methods (Knirsch et al., 2010).

4.5 Microwaves

Microwave heating is being used in household and industrial food preparation and processing. It is preferred because of its volumetric origin, fast temperature increase, controllable heat deposition, and simple sanitation conditions.

The process is rapid and the come-up time required to reach the desired temperature is minimum. Therefore, microwave heating is preferred for pasteurization and sterilization. Also, the High Temperature Short Time (HTST) process is a conventional method and is not convenient for solid foods due to slow heat conduction, which overheats the surface of the solid. Microwave heating can accomplish this slow thermal diffusion of conventional heating (Ahmed and Ramaswamy, 2007).

Microwaves are electromagnetic waves that have frequencies between 300 MHz and 300 GHz and wavelengths from 1 to 0.001 m. Microwave heating influences the polar molecules of the material. In this way, electromagnetic field energy transforms into thermal energy. In conventional heating, heat diffuses in from the surface of the material and heat transfer takes place first by convention and then conduction, which requires a longer time. Whereas, microwave heating generates volumetric heat, which means that materials can absorb microwave energy internally and convert it into heat (Vadivambal and Jayas, 2010).

This volumetric heat can significantly decrease the total heating time and so the high temperatures needed for commercial sterilization to destroy the microorganisms are enhanced and thermal degradation of the desired components is reduced (Ahmed and Ramaswamy, 2007).

For industrial, scientific, and medical use, only limited frequencies (915 or 2450 MHz) are authorized to prevent interference with radio frequencies used for telecommunication (Barba et al., 2018; Ahmed and Ramaswamy, 2007).

Microwave heating in foods emerges when the electrical energy of an electromagnetic field in a microwave cavity is coupled with food and due to subsequent distribution in the food product. This causes a rapid increase in temperature within the product. Molecular interaction with the electromagnetic field provides microwave energy at the molecular level. This situation occurs especially through the dipole shift of polar solvents and molecular friction resulting from the conductive migration of dissolved ions. Therefore, dipole rotation and ionic polarization are the main mechanisms. Because of this internal molecular friction, heat is produced quickly (Ahmed and Ramaswamy, 2007).

Processes in which microwave energy can be used are pasteurization, sterilization, drying, cooking, baking, thawing, tempering, blanching, reheating, and even combined with freezing (Barba et al., 2018; Osaili, 2012). Microwave has several more advantages. It diffuses inside the food materials with reduced processing time and energy. Because the heat transfer is rapid, the flavors, sensory qualities, and color of the foods are well preserved as well as the nutritional and vitamin content. It has higher heating efficiency and suitable sanitation conditions. The system maintenance is cost-effective, suitable for heat-sensitive fluids, quiet, and without any gas outlet.

The disadvantages of microwave heating are usually related to nonuniform heating, which may produce hot and cold spots within the same food item. Main problems are low quality end product, overheating, and inability to ensure the microbial safety. But the major inconvenience is the presence of hot spots in various regions resulting from product geometry (Vadivambal and Jayas, 2010).

The industrial applications of microwave can be listed as follows: tempering of fish, meat, and poultry; precooking of bacon; sausage cooking; baking; drying; blanching of vegetables; concentration; puffing; and foaming.

4.6 Radio frequency

The RF, which is at higher frequencies than microwaves, occupies a region between 1 and 300 MHz in the spectrum of electromagnetic fields although the main frequencies used for industrial heating lie in the range 10–50 MHz. Like microwaves, only selected frequencies (namely 13.56 ± 0.00678, 27.12 ± 0.16272, and 40.68 ± 0.02034 MHz) are permitted for industrial, scientific, and medical applications (Marra et al., 2009). RF heating is also called high frequency dielectric heating.

RF heating enables associated rapid and uniform heat distribution, large penetration depth, and lower energy consumption, which makes this technology promising for food applications. RF heating can be used for drying, baking, and thawing of frozen meat and in meat processing. However, its use in continuous pasteurization and sterilization of foods is rather limited. During RF heating, applied alternating electric field causes the molecules and ions to oscillate, which results in molecular friction. Thus, heat is generated inside the product. When all the other conditions are kept constant, RF heating is affected mainly by the dielectric properties of the product (Piyasena et al., 2003).

RF heating of foods is clearly affected by dielectric properties. On the other hand, dielectric properties are influenced by several factors like frequency level, temperature, and properties of food, such as viscosity, water content, and chemical composition. Therefore, these properties should be considered when creating the heating system.

In contrast to conventional systems having large temperature gradients because of the transfer of heat from hot medium to cooler product, RF heating transfers the electromagnetic energy directly into the product and starts the frictional interaction between molecules causing volumetric heating (i.e., heat is generated within the product) (Piyasena et al., 2003).

Application areas of RF heating are cooking processed meat, heating bread and dehydrating vegetables, thawing of frozen products, postbaking (final drying) of cookies and crackers, sausage emulsion pasteurization, sterilization, and continuous flow aseptic processing and packaging systems (Piyasena et al., 2003). RF can be used to reduce microbial contamination and improve food safety and quality. There are studies on the reduction of microbial contamination of fresh carrots, meat, apple juice, milk, apple cider, and pork meat products by RF (Osaili, 2012).

As seen from the literature, RF heating has also been utilized for the pasteurization of dairy and pasta products, for the sterilization of low-acid foods, microbial inactivation of salmon caviar, ground beef pathogens, and other food materials (Barba et al., 2018).

4.7 Inhibition of oxidation in foods

Oxygen affects preserved foods and beverages in different ways such as rancidity of unsaturated fats, browning and loss of vitamin C (ascorbic acid) of fruits and vegetables, deterioration of bakery products, discoloration of fresh meat, and deterioration of flavor of beverages. The quality loss or spoilage of food products caused by oxygen strongly influences the consumer acceptance of products, which in turn causes significant amount of food loss or waste, which means the loss of economic value. Therefore, it is essential to understand and inhibit oxidation in food systems.

A free radical chain reaction that is responsible for the mechanism of lipid oxidation can be separated into three phases: initiation, propagation, and termination.

Initiation phase starts with the formation of free radicals by hydrogen abstraction reactions through metal ions, light, radiation, or other agents. If there is oxygen in the environment, it reacts with these highly reactive carbon-centered free radicals generating hydroperoxyl radicals.

In the propagation phase these hydroperoxyl radicals can abstract a hydrogen from a lipid, generating peroxy radical being very reactive. Peroxy radical abstracts a hydrogen from another polyunsaturated fatty acid molecule and forms hydroperoxide and alkyl free radical. Hydroperoxide can become the source of additional reactive radicals under the action of metal ions and causes the formation of free peroxy and alkoxy radical. This chain reaction sequence can be repeated many times. Hydroperoxides also undergo a variety of decomposition reactions generating off, rancid, stale flavor, or aroma-forming compounds.

In the termination phase free radicals in the medium interact with each other and form some nonradical compounds. The oxidation process is interrupted as substrates become depleted and radical recombination reactions begin to dominate. Nonlipid food materials can also undergo oxidative changes such as pigment degradation, oxidative vitamin loss, and protein oxidation (Pokorny, 2007; Berdahl et al., 2010).

4.7.1 Types of antioxidants

Antioxidants are the substances that can protect materials against autoxidation by delaying the start or slowing the rate of oxidation. There are many compounds both natural and synthetic that possess antioxidant activity. Generally, lipid-soluble antioxidants such as monohydric or polyhydric phenols with different ring structures are used in food systems. Antioxidants have mainly six groups, as seen in Fig. 4.4.

Maximum performance is achieved when the primary antioxidants are used in combination with different metal chelating agents or other phenolic antioxidants; this is the synergistic effect (Nawar, 1996; Berdahl et al., 2010). Major types of the inhibitors can be listed as singlet oxygen quenchers, chelating agent, synergists, hydroperoxide deactivators, and antioxidants (Fig. 4.5). Singlet oxygen quenchers such as carotenes convert singlet oxygen into triplet oxygen, chelating agents such as citric acid bind heavy metals into inactive complexes, synergists such as ascorbyl palmitate regenerate antioxidants, hydroperoxide deactivators such as cystein react with hydroperoxides, and antioxidants such as tocopherols react with free radicals (Pokorny, 2007).

Synthetic antioxidants such as butylated hydroxyanisole (BHA), butylated hydroxytoluene (BHT), propyl gallate (PG), and tertbutylhydroquinone (TBHQ) are the ones that are widely used in foods. However, there are some studies indicating that high concentrations of certain synthetic antioxidants such as BHA and BHT may cause weak carcinogenic effects in some animals. Even the use of TBHQ is restricted in Japan. Fortunately, natural antioxidants such as tocopherols, ascorbic acid, erythrobic acid, or their salts and derivatives such as ascorbyl palmitate, as well as extracts of rosemary and sage, have found prevalent applications in the food industry (Shahidi and Zhong, 2010).

The fat-containing foods can be stored for only a limited period of time during slow oxidation where the food decomposition has not started yet. In the beginning of storage the oxidation rate is very slow, therefore this stage is called the induction period, where the shelf life of the food product can be extended by addition of antioxidants. However, antioxidants are not able to eliminate the oxidation reactions

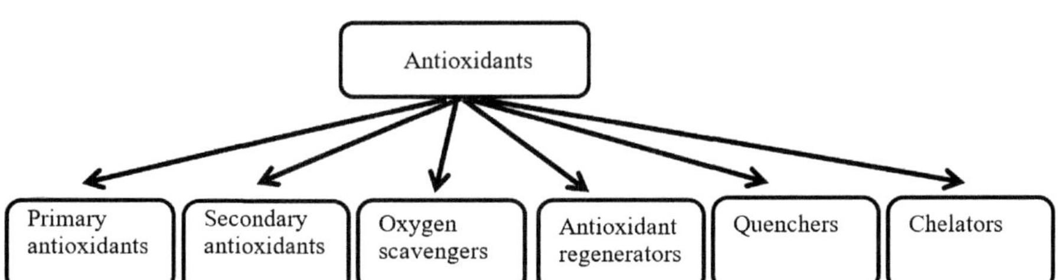

Figure 4.4 Types of antioxidants.

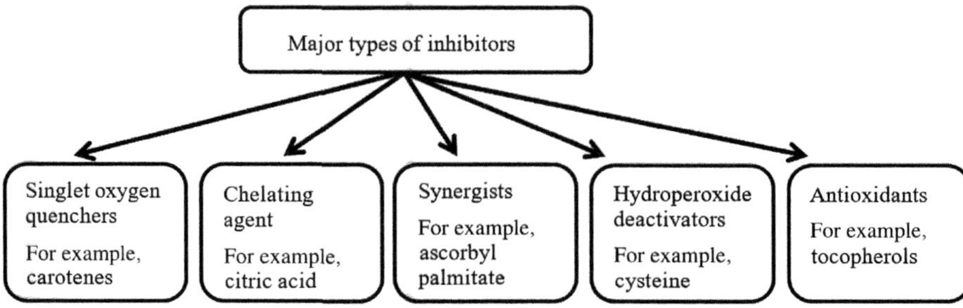

Figure 4.5 Major types of inhibitors.

although they are active during extended storage time. Some of the specific effects of oxidation-induced degradation in foods can be listed as follows:

1. Rancid taste and aroma formation in fats and fat-containing foods
2. Discoloring in pigments
3. Formation of toxic oxidation products
4. Taste and aroma loss and disorders in product
5. Changes in texture
6. Decrease in nutritional value due to destruction of vitamins (A, D, and E) and essential fatty acids (especially linoleic acid)

Reagents that cause or accelerate oxidation are primarily oxygen and also metal ions such as iron and copper, light, temperature, some pigments, and the rate of unsaturation.

Oxidation of food products is inhibited in four ways:

- Minimizing the influence of the physical factors during processing and storage such as air, light, and high temperature by modern packaging methods.
- Inhibiting the autoxidation of lipids, which is initiated by free radicals. Autoxidation process can be hindered by antioxidants, which are chain-breaking inhibitors, or by preventive inhibitors.
- Inhibiting the photosensitized oxidation. Physical quenchers such as tocopherol and carotenoids prevent single electron transfer reaction of the primary excited molecule.
- Enzymes such as lipoxygenase can be inhibited by flavonoids, phenolic acids, and gallates and also by heating. As heating also causes nonenzymatic oxidation it may cause oxidation to increase.

Antioxidants are the most effective substances in the food industry that inhibit the degradation and rancidity of food for a certain time by delaying the effect of atmospheric oxygen under normal temperature conditions during production, storage, transport, and marketing of various foodstuffs. They do not increase the quality of the food and do not change the flavor and smell. The desired quality can only be achieved by providing good raw material, correct production technique, and proper packaging and storage conditions. For proper and effective use of antioxidants, it is necessary to know the mechanism of oxidation, the functions of the antioxidant used, and to add antioxidant to the food before the oxidation starts (Çakmakçı and Gökalp, 1992; Yanishlieva-Maslarova, 2001; Cichello, 2015; Pokorny, 2007).

4.7.2 Use of antioxidants in food

Various antioxidants perform at different effectiveness when used in oils and fat-containing foods owing to the differences in their molecular structure. Factors that affect the potential application of antioxidant in a food system can be summarized as convenience of association with food, performance characteristics, pH sensitivity, discoloring or off flavor tendency, cost, and availability. Moreover, the presence of prooxidants and antioxidants present in the food itself or formed during processing has to be considered while predicting the functionality of the added antioxidant in food. Thus, selecting the appropriate antioxidant or combination of antioxidants for a food product is very complicated. Also, the hydrophilic–lipophilic properties of antioxidants influence their effectiveness in food applications. Vegetable origin food products are rarely stabilized by antioxidant addition compared with animal origin food products, which might be due to the presence of natural antioxidants such as tocopherols, carotenoids, or flavonoids in vegetable origin foods. Oxidation is catalyzed by a group of enzymes and lipoxygenases in vegetable origin foods.

Synthetic antioxidants are generally effective in fish oils, vegetable oils, animal fats and oils, and low-fat snack foods. However, in the last decade or two natural antioxidants have received increasing interest due to consumer preferences (Nawar, 1996; Yanishlieva-Maslarova, 2001).

4.7.2.1 Protection of fats and oils

The stability of animal fats such as milk fat or beef fat against oxidation is low despite the relatively low degree of unsaturation due to the presence of natural antioxidants in very low quantities. On the other hand both synthetic and natural antioxidants are very active in the stabilization of animal fats. Mixtures of antioxidants and synergists are used for stabilization. Lipid-soluble antioxidants give good results, but polar antioxidants can also be used. BHA is less effective in animal fats and shortenings when used alone than BHT or gallates, but its effectiveness increases with added synergists.

It is very difficult to stabilize vegetable oils due to unsaturated fatty acids; however the presence of natural antioxidants, mainly tocopherols in edible oils, is their advantage. Tocopherols as natural antioxidants exert their maximum antioxidant activity at relatively low levels almost equal to their concentration in vegetable oils. The addition of phenolic antioxidants to vegetable oils generally has limited efficacy, but the addition of synergists is beneficial. Useful inhibitors in vegetable oils are mainly ascorbyl palmitate, phospholipids, or organic polyvalent acids. BHA and BHT are more effective in animal fats than they are in vegetable oils and they have no significant effect on the stability of margarine. The efficacy of the antioxidants of TBHQ is equal to BHA but both have greater efficacy than BHT. Among the synthetic antioxidants PG has also been widely used in fats and oils, meat products, confectionery, nuts, milk products, fish products, margarine, and baked goods at levels between 0.001% and 0.04%. Before the deodorization of edible oils citric acid is often added and its decomposition products are also efficient as synergists of tocopherols (Pokorny, 2007; Berdahl et al., 2010; Pokorny, 2003, Pokorny and Trojakova, 2001).

4.7.2.2 Protection of nuts and oil seeds

Protection of peanuts that are roasted is difficult except for the application of spices on their surfaces. However, it is better to store them in a vacuum or in an inert atmosphere. After the roasting process in hot oil, about 2%–3% oil remains on the surface of the peanut. Addition of antioxidants into frying oils such as rice bran oil containing natural antioxidants improves the shelf life of nuts roasted in soybean or rapeseed oils. Moreover, a mixture of salt and TBHQ can be used for the stabilization. For the crushed or ground seeds used to produce a paste, like peanut butter, addition of antioxidants such as Tocomix D (a mixture of a- and d-tocopherols, citrate esters of monoacylglycerols) and Embanox 10 (a mixture of BHT and BHA) may be useful to improve stability during storage. Also, rosemary oleoresins were actively used for the stabilization of nuts (Pokorny and Trojakova, 2001).

4.7.2.3 Protection of cereal products

Cereal products such as peeled rice, white flour, or grits do not need to be stabilized. But to increase shelf life of whole grain flours enzymes have to be inactivated. Antioxidants such as rice bran, aqueous extracts from other whole grains or brans, tea extracts, and fruit extracts may be added to breakfast cereals with good results. Also, natural amino acids methionine and cysteine, phospholipids, and uric acid were also active as antioxidants in breakfast cereals. Antioxidants are mostly added to the extruded products with flour and other additives to the extruder barrel for homogeneous distribution. Extruded snack products' natural coloration is stabilized generally by an oil-soluble liquid rosemary extract (4942 Rosmanox) and its mixture with tocopherols (4993 Rosmanox E). BHA and TBHQ are also used in some of the snacks containing cottonseed oil and corn flour.

Some cereal products such as cookies, crackers, and cakes contain added fat. They should be stabilized by antioxidants to extend the shelf life. Sugar-snap cookies may be stabilized by BHA, but natural antioxidants are preferable by the consumers. Ferulic acid and sodium phytate, casein, whey proteins, or Maillard reaction products may be used instead of BHA in cookies. Caffeic acid and roasted coffee bean powder or extract being more active than tocopherol and chlorogenic acid, which are the components of coffee bean, have been added to cookies. Synergists like ascorbic and erythorbic acids, citric acid, and its isopropyl ester are also used with tocopherols. On the other hand, flavoring spices such as lemongrass extracts, clove leaves, black pepper leaves, and turmeric also increase the shelf life of cakes. Addition of spearmint, peppermint, and basil or their diethyl ether extracts gives better antioxidant effect than BHA at specific concentrations in cracker biscuits (Pokorny and Trojakova, 2001; Shahidi and Chandrasekara, 2015).

4.7.2.4 Protection of fruits and vegetables

As the lipid content in fruits and vegetables is very low the subject is not the lipid oxidation. The shelf life of fruits and vegetables is limited by factors like enzymatic browning, which affects the sensory properties. Therefore, to prevent oxidation of polyphenols, which are formed during enzymatic browning, antioxidants are added to fruits and mushrooms. Enzymes responsible for the oxidation of polyunsaturated fatty acids such as lipoxygenases are deactivated by blanching and then natural antioxidants, generally flavonoids, are added to protect lipid oxidation to achieve the best stabilization. Moreover, essential oils present in fruits show antioxidant activity, but they are also oxidized and have to be protected by similar antioxidants. For example, orange juice is stabilized with the combination of ascorbic acid, phenolic acids, and pasteurization. For the stability of mashed potatoes, ascorbyl palmitate or a mixture of rosemary, thyme, and marjoram are used more efficiently than α-tocopherol or TBHQ (Pokorny and Trojakova, 2001; Shahidi and Chandrasekara, 2015).

4.7.2.5 Protection of meat products

The application of antioxidants for animal products is very useful due to the lack of natural antioxidants in their structure. Only small amounts of antioxidants are efficient because the oxidation sensitive polyunsaturated fatty acids are relatively low in fats of land mammals.

There is an oxidative stability difference between animal species and muscle types within a species due to some endogenous factors that control the oxidation such as the presence of active antioxidants, oxygen deactivating enzymes, and on the other hand prooxidants like iron and ascorbic acid. Stability of the products is dependent on the balance between these factors. For example, the difference between the stability of meat species is that the most stable one is beef, followed by pork, chicken, turkey, and finally fish. Additionally, within a species such as poultry, the white meat is more durable against oxidation than the dark meat.

Ground rosemary leaves or rosemary oleoresin are widely used natural antioxidants in meat products. As an example, rosemary antioxidant was found to be effective for the stability of cooked minced pork or frozen pork sausage during storage. Also, rosemary oleoresin was efficiently used in reconstituted raw or cooked pork steaks or in reconstituted chicken nuggets as tripolyphosphate was used as a metal chelating agent in both of the meat products. Besides meat lipids were efficiently stabilized by green tea catechins. Stability of beef patties stored at 4°C is provided by tocopherols at the defined concentration.

Since ancient times, smoke has been traditionally used for the preservation of meat products. Ash and beech wood, which are phenolic-rich smoke sources, may extend the stability of lard or pork meat during storage, which is important in smoked meat production. During the curing process sodium nitrite addition prevents pork meat oxidation. Treatment of nitrite during cold storage increases the antioxidant activity of pork and beef. Maillard products in sausages and soy sauce in pork patties also protect them against oxidative degradation. Widely used antioxidants in meat products are either lipid soluble (tocopherols and carotenoids) or water-soluble (ascorbic acid, dipeptides, and plant phenolics or polyphenolics; raw meat will also contain antioxidant enzymes). Fish oils contain fatty acids with four to six double bonds, therefore they are susceptible to oxidation and it is very difficult to stabilize them against rancidification. The most widely used natural antioxidants for fish product preservation are tocopherols (Pokorny, 2003; Cuppett, 2001).

4.7.2.6 Protection of packaged foods

For distribution of some foods, packaging is needed. Therefore, the packaging material is very important for the preservation of these foods. Antioxidants may be added to the package to inhibit the oxygen diffusion, if the material is permeable to oxygen. The added antioxidants may migrate to the food, especially high-fat foods, from the package. Thus, the antioxidants that can be used in packaging should also be those that are allowed in foods, and their usage amounts in the material should be such that their quantities in foods do not exceed the legal limits. Even if the food manufacturer does not want to preserve the food in this way, packaging materials can be protected against oxidation by adding antioxidants already in the plant during the production.

On the other hand, antioxidants such as β-carotene, BHA, tocopherols, and ascorbic acid that are used in foods to prevent the oxygen degradation of food molecules such as lipids cannot absorb the oxygen in the void space of the packaging or between food particles. Oxygen interceptors are compounds that prevent oxygen from reaching the food product and they can be used both in the food product and package materials against autooxidation. Oxygen absorbers (scavengers) are materials that remove the oxygen from void space around the food particles and are generally used within a satchel that is added to modified atmosphere packaging. Their working mechanism is that they react with oxygen and moisture through the oxidation of a salt such as iron carbonate. Oxygen absorbers are applied to a wide variety of food types for preservation including bread and biscuits, fruit and vegetables, nut products, fish and seafood products, and meat products (Pokorny, 2007; Cichello 2015).

Water content of the foods has a significant effect on the efficiency of antioxidants in foods. Dry foods such as dried soups, dried milk, dried meats, etc. are susceptible to oxidation, because air oxygen can freely reach to the lipid film on nonlipidic particles through the tiny channels that occurred after the removal of water. Thus, autoxidation reaction initiation rate is very high and the antioxidants can be decomposed during processing and storage, which in turn makes the food stabilization less effective. Water-containing foods are protected from the oxygen by carbohydrates or a layer of hydrated proteins due to the more stable lipid fraction. Nonpolar antioxidants are more effective than the polar antioxidants because polar antioxidants may pass to the aqueous phase causing an activity loss. Other food components like proteins also have protective action and they may act as synergists of the inhibitors, which improves the antioxidant effect. Protein amine groups contribute to the protection of foods against oxidation by reacting with lipid peroxides and thus decreasing the level of free radicals. Chelating agents and heavy metals like heme derivatives in animal products are often present in food products as natural components, as well. Therefore, it is necessary to optimize the mixture of inhibitors and test the antioxidants to stabilize any kind of food product due to the complex structure of foods. (Pokorny, 2007).

4.8 Hurdle concept

Nowadays there is an increasing interest in food products with fewer chemical additives and physical injuries, which creates new opportunities for the hurdle technology concept of food preservation. Sensory and nutritive quality as well as the microbial safety and stability of the foodstuffs are based on the application of combined preservation methods, called hurdles. Widely used important hurdle factors in food preservation are water activity (a_w), temperature (high or low), redox potential (E_h), acidity (pH), competitive microorganisms (e.g., lactic acid bacteria), and preservatives (e.g., nitrite, sorbate, sulfite, antioxidants).

The hurdle concept is mainly based on the fact that many inhibitory factors (hurdles) can be effective when combined, although they are not able to prevent microbial growth individually. Unfortunately, the use of severe preservation methods may cause undesirable changes in food quality such as loss of nutrients, loss of flavor and aroma compounds, changes in texture, nonenzymatic browning, and protein denaturation, which may be limiting factor. Therefore, when two or more preservation methods are combined, due to their increased effect that is being summed, intensity of each method may be reduced. If a food is preserved only by high salt or low acidity it may be organoleptically unacceptable by the consumers, but when low salt is combined with low acidity then the consumer acceptability level will be higher.

To give an example, fruit jams and jellies would be appropriate that are produced from the similar concentrations of fruit and sugar to about 65–72 Brix. The main reason for their conservation is the low water activity of the product; however, preservation is achieved by a sum of methods such as low a_w, low pH for the formation of pectin gel, and thermal processing during evaporation for the concentration. Another good example of the hurdle concept is frozen concentrated orange juice preserved by the sum of thermal processing (pasteurization), water activity reduction by concentration, and freezing. Among meat products, sausage is an example of pasteurized product that still needs to be stored refrigerated, although it contains antimicrobial agents like nitrates and nitrites. Additionally, it will have longer shelf life when packed under vacuum (lower E_h). Over the last 20 years, hurdle technology has become more widespread and wisely applied as important conservation factors (e.g., temperature, pH, a_w, E_h, competitive microorganisms, preservatives) and their interactions have become better known (Augusto et al., 2018; Hamad, 2012; Leistner, 2000).

4.9 Conclusion

To preserve foods the application of heat is a very important process. Existing commercial thermal processing technology needs to be developed and improved in terms of final product safety and quality with the optimization of process efficiency. So far, the innovative electroheating technologies such as RF heating, microwave heating, and ohmic heating have substituted for the widely used commercial preservation processes. These technologies have been used to inactivate microorganisms in different types of food. In terms of oxidation inhibition in food products, there is an increasing focus on the determination of new, effective, and natural antioxidants. Another progress in this field is to reduce the concentration of added antioxidants to foods, which can be provided with the combination of phenolic substances, which are known as primary antioxidant with synergists (Generally Recognized as Safe status). Another way of removing oxygen is the addition of oxygen scavengers such as a combination of D-glucose and glucose oxidase to the packaging material. Finally there is an increasing tendency to more natural preservation techniques with the addition of fewer additives including antioxidants.

References

Ahmed, J., Ramaswamy, H.S., 2007. Microwave pasteurization and sterilization of foods. In: Rahman, M.S. (Ed.), Handbook of Food Preservation, second ed. CRC Press, Boca Raton, London, New York, pp. 691–711.

Augusto, P.E.D., Soares, B.M.C., Castanha, N., 2018. Conventional technologies of food preservation. In: Barba, F.J., Sant'Ana, A.S., Orlien, V., Koubaa, M. (Eds.), Innovative Technologies for Food Preservation. Academic Press, London, UK, pp. 3–23.

Awuah, G.B., Ramaswamy, H.S., Economides, A., 2007. Thermal processing and quality: principles and overview. Chem. Eng. Process. 46, 584–602.

Barba, F.J., Ahrne, L., Xanthakis, E., Landerslev, M.G., Orlien, V., 2018. Innovative technologies for food preservation. In: Barba, F.J., Sant'Ana, A.S., Orlien, V., Koubaa, M. (Eds.), Innovative Technologies for Food Preservation. Academic Press, London, UK, pp. 25–51.

Baysal, A.H., Icier, F., 2010. Inactivation kinetics of *Alicyclobacillus acidoterrestris* spores in orange juice by ohmic heating: effects of voltage gradient and temperature on inactivation. J. Food Prot. 73 (2), 299–304.

Berdahl, D.R., Nahas, R.I., Barren, J.P., 2010. Synthetic and natural antioxidant additives in food stabilization: current applications and future research. In: Decker, E.A., Elias, R.J., McClements, D.J. (Eds.), Oxidation in Foods and Beverages and Antioxidant Applications. Woodhead Publishing Limited, Cambridge, pp. 272–313.

Butz, P., Tauscher, B., 2002. Emerging technologies: chemical aspects. Food Res. Int. 35, 279–284.

Çakmakçı, S., Gökalp, H.Y., 1992. Gıdalarda Kısaca Oksidasyon; Antioksidantlar ve Gıda Sanayinde Kullanımları. Atatürk Üniversitesi Ziraat Fakültesi Dergisi 23 (2), 174–192.

Chemat, F., Rombaut, N., Meullemiestre, A., Turk, M., Perino, S., Fabiano-Tixier, A.-S., et al., 2017. Review of green food processing techniques. Preservation, transformation, and extraction. Innovative Food Sci. Emerging Technol. 41, 357–377.

Cheng, X., Zhang, M., Xu, B., Adhikari, B., Sun, J., 2015. The principles of ultrasound and its application in freezing related processes of food materials: a review. Ultrason. Sonochem. 27, 576–585.

Cheng, L., Sun, D.-W., Zhu, Z., Zhang, L., 2017. Emerging techniques for assisting and accelerating food freezing processes: a review of recent research progresses. Crit. Rev. Food Sci. Nutr. 57 (4), 769–781.

Cichello, S.A., 2015. Oxygen absorbers in food preservation: a review. J. Food Sci. Technol. 52 (4), 1889–1895.

Cuppett, S.L., 2001. The use of natural antioxidants in food products of animal origin. In: Pokorny, J., Yanishlieva, N., Gordon, M. (Eds.), Antioxidants in Food-Practical Applications. CRC Press, Boca Raton, pp. 293–318.

FAO, 2011. Global food losses and food waste – extent, causes and prevention. <http://www.fao.org/docrep/014/mb060e/mb060e00.pdf>.

FAO, 2015. Global initiative on food loss and waste reduction. <http://www.fao.org/3/a-i4068e.pdf>.

Hamad, S.H., 2012. Factors affecting the growth of microorganisms in food. In: Bhat, R., Alias, A.K., Paliyath, G. (Eds.), Progress in Food Preservation. John Wiley & Sons, Ltd, West Sussex, UK, pp. 405–426.

Karel, M., Lund, D.B., 2003a. Heat processing. In: Karel, M., Lund, D.B. (Eds.), Physical Principles of Food Preservation, second ed. Marcel Dekker, Inc, New York, USA, pp. 170–236.

Karel, M., Lund, D.B., 2003b. Storage at chilling temperatures. In: Karel, M., Lund, D.B. (Eds.), Physical Principles of Food Preservation, second ed. Marcel Dekker, Inc, New York, USA, pp. 237–275.

Kiani, H., Sun, D.-W., 2011. Water crystallization and its importance to freezing of foods: a review. Trends Food Sci. Technol. 22, 407–426.

Knirsch, M.C., Alves dos Santos, C., Vicente, A.A.M.O.S., Penna, V.T.C., 2010. Ohmic heating-a review. Trends Food Sci. Technol. 21, 436–441.

Leistner, L., 2000. Basic aspects of food preservation by hurdle technology. Int. J. Food Microbiol. 55 (2000), 181–186.

Li, B., Sun, D.-W., 2002. Novel methods for rapid freezing and thawing of foods-a review. J. Food Eng. 54, 175–182.

Lima, M., 2007. Food preservation aspects of ohmic heating. In: Rahman, M.S. (Ed.), Handbook of Food Preservation, second ed. CRC Press, Boca Raton, London, New York, pp. 742–750.

Lyng, J.G., McKenna, B.M., 2011. Ohmic heating of foods. In: Proctor, A. (Ed.), Alternatives to Conventional Food Processing. RSC Publishing, Cambridge, UK, pp. 307–340.

Marra, F., Zhang, L., Lyng, J.G., 2009. Radio frequency treatment of foods: review of recent advances. J. Food Eng. 91, 497–508.

Martindale, W., Schiebel, W., 2017. The impact of food preservation on food waste. Br. Food J. 119 (12), 2510–2518.

Nawar, W.W., 1996. Lipids. In: Fennema, O.R. (Ed.), Food Chemistry. Marcel Dekker Inc, New York, USA, pp. 225–319.

Osaili, T.M., 2012. Developments in the thermal processing of food. In: Bhat, R., Alias, A. K., Paliyath, G. (Eds.), Progress in Food Preservation. John Wiley & Sons Ltd, West Sussex, UK, pp. 211–230.

Piyasena, P., Dussault, C., Koutchma, T., Ramaswamy, H.S., Awuah, G.B., 2003. Radio frequency heating of foods: principles, applications and related properties—a review. Crit. Rev. Food Sci. Nutr. 43 (6), 587–606.

Pokorny, J., 2003. Natural antioxidants. In: Zeuthen, P., Sùrensen, L.B. (Eds.), Food Preservation Techniques. CRC Press, Boca Raton, pp. 31–45.

Pokorny, J., 2007. Antioxidants in food preservation. In: Rahman, M.S. (Ed.), Handbook of Food Preservation. CRC Press, Boca Raton, pp. 260–281.

Pokorny, J., Trojakova, L., 2001. The use of natural antioxidants in food products of plant origin. In: Pokorny, J., Yanishlieva, N., Gordon, M. (Eds.), Antioxidants in Food-Practical Applications. CRC Press, Boca Raton, pp. 363–380.

Rahman, M.S., Velez-Ruiz, J.F., 2007. Food preservation by freezing. In: Rahman, M.S. (Ed.), Handbook of Food Preservation, second ed. CRC Press, Boca Raton, London, New York, pp. 635–665.

Ramesh, M.N., 2007a. Pasteurization and food preservation. In: Rahman, M.S. (Ed.), Handbook of Food Preservation, second ed. CRC Press, Boca Raton, London, New York, pp. 571–583.

Ramesh, M.N., 2007b. Canning and sterilization of foods. In: Rahman, M.S. (Ed.), Handbook of Food Preservation, second ed. CRC Press, Boca Raton, London, New York, pp. 585–623.

Shahidi, F., Zhong, Y., 2010. Novel antioxidants in food quality preservation and health promotion. Eur. J. Lipid Sci. Technol. 112, 930–940.

Shahidi, F., Chandrasekara, A., 2015. The use of antioxidants in the preservation of cereals and low-moisture foods. In: Shahidi, F. (Ed.), Handbook of Antioxidants for Food Preservation. Woodhead Publishing Limited, Cambridge, pp. 413–428.

Tavman, S., Yilmaz, T., 2018. Freezing of dairy products. In: Conto, F., Del Nobile, M.A., Faccia, M., Zambrini, A.V., Conte, A. (Eds.), Advances in Dairy Products. John Wiley & Sons Ltd, West Sussex, UK, pp. 266–273.

Vadivambal, R., Jayas, D.S., 2010. Non-uniform temperature distribution during microwave heating of food materials—a review. Food Bioprocess Technol. 3, 161–171.

Yanishlieva-Maslarova, N.V., 2001. Inhibiting oxidation. In: Pokorny, J., Yanishlieva, N., Gordon, M. (Eds.), Antioxidants in Food-Practical Applications. CRC Press, Boca Raton, pp. 35–83.

Zheng, L., Sun, D.-W., 2006. Innovative applications of power ultrasound during food freezing processes—a review. Trends Food Sci. Technol. 17, 16–23.

Further reading

Guizani, N., Mothershaw, A., 2007. Fermentation as a method for food preservation. In: Rahman, M.S. (Ed.), Handbook of Food Preservation, second ed. CRC Press, Boca Raton, London, New York, pp. 215–236.

Nonthermal food processing/ preservation technologies

5

Elisabete M.C. Alexandre[1,2], *Carlos A. Pinto*[1], *Sílvia A. Moreira*[1,2], *Manuela Pintado*[2] *and Jorge A. Saraiva*[1]
[1]QOPNA & LAQV-REQUIMTE, Department of Chemistry, University of Aveiro, Aveiro, Portugal, [2]Center for Biotechnology and Fine Chemistry – Associated Laboratory, School of Biotechnology, Catholic University of Portugal, Porto, Portugal

Chapter Outline

5.1 Introduction

In 2011, about one third of all the food produced worldwide was lost or wasted, corresponding to approximately to 1.3 billion ton of food waste per year (Tonini et al., 2018). In Europe a food waste was estimated of roughly 173 kg per capita, which represents about 170 Mt. of CO_2 and an economic loss of about 143 billion of euros per year. Nevertheless, food waste can be classified as unavoidable or avoidable. The avoidable fraction relates to food that is thrown out while still edible, representing not only a waste of food, but also a waste of resources such as energy, land-use demands, or other materials necessary to the production process (Quested and Johnson, 2009). The unavoidable portion represents the waste arising from food or drink preparation that is not, and has not been, edible under normal circumstances, such as bones, eggshells, fruit peels, among others (Quested and Johnson, 2009).

The food industry must deal with the food waste problem throughout the entire food chain, from the initial agricultural production till the final consumer purchase and consumption. Global sustainability, food security, and the protection of natural resources, environment, and human health can be achieved by the choice of the

Saving Food. DOI: https://doi.org/10.1016/B978-0-12-815357-4.00005-5

correct processing methodology on the different phases of the food chain (Saini et al., 2018).

A simple definition of food processing is related with the conversion of raw livestock and agricultural products (raw materials/ingredients) into a widely diverse range of products for human consumption. In other words, it consists in the transformation of raw or cooked ingredients by physical or chemical means, into other marketable food products that are more suitable or appealing to be consumed, always aiming to extend their shelf life or improve their quality (Compton et al., 2018). The main purpose of food processing is its preservation, but later it was discovered that organoleptic characteristics such as taste, appearance, flavor, color, aroma, texture and shape, or even nutritional value can be attractively improved by applying different processing methods. Today, these are effectively the main goals of the food industries; on one side they intend to extend foods shelf life by inhibiting microbiological and biochemical changes, and on the other side they yield improvements of the organoleptic characteristics preserving food's nutritional characteristics (Alexandre et al., 2013).

Consumers are increasingly aware of natural food benefits on human health, thus, the demand for high-quality food products, with fresh-like properties and high nutrient content, led to the necessary development of new food processing methods.

The traditional heat treatments are efficient in microbial inactivation, reducing product decay, and attaining safety targets. However, they have a significant impact on organoleptic quality, namely in texture, color, aroma, flavor, taste, and nutritive value, which are normally severely affected by temperature. Nonthermal and eco-friendly processing methodologies, such as high pressure processing (HPP), pulsed electric fields (PEF), and ultrasounds (US) have been studied by both industry and academia, in an attempt to meet the challenges of producing safe processed foods with a high-quality standard (Balasubramaniam et al., 2015). HPP and PEF are among the most commercial technologies being studied and some processed products can already be found on the market.

HPP is an increasingly implemented technology for food "cold" pasteurization that is also showing great promise for other food, pharmaceutical, and biotechnological applications. The use of high pressures involves the exposition of liquid or solid foods to pressures that can range from 100 to 1000 MPa, from a millisecond pulse to over 20 minutes. Depending on some factors such as food properties, initial food temperature, and applied pressure, the adiabatic temperature can increase 4°C–6°C for each pressure increase of 100 MPa. However, the products' temperature during holding time can be below 0°C or above 100°C depending on the requirements for the final intended product (Pereira et al., 2010; Alexandre et al., 2017a,b,c,d,e). Room or chilled temperatures could be especially useful for "cold" food pasteurization, assuring pathogenic and spoilage bacteria, yeasts, molds, viruses, and spores inactivation (Balasubramaniam et al., 2015). PEF technology involves the application of short duration pulses (from several nanoseconds to several milliseconds), of high electric-field strengths (from 100 V/cm to 80 kV/cm), and specific energy input in the range of 50–1000 kJ/kg (Fincan et al., 2002; Toepfl et al., 2007; Vorobiev et al., 2008; Koubaa et al., 2015). This technology is able to inactivate pathogenic microorganisms such as *Listeria monocytogenes*, *Escherichia coli*, and *Salmonella typhimurium*, without significant loss of the organoleptic and nutritional properties of food. US technology can be classified in two categories: high frequency or high intensity. The one usually applied in the food industry is the high-intensity US, also called power US, which operates at frequencies that range between 20 and 100 kHz. This technology is mainly used to improve processes that include oil extraction, microbial and enzyme inactivation, and starch–protein complex disintegration (Feng et al., 2011).

Nonthermal innovative food processing technologies have potential, through process intensification, to be environmentally sustainable by reducing energy and water consumption and at same time achieve food security and quality and extend shelf life of food products or expand the shelf-stable product spectrum. They play an important role for food industries contributing also to product innovation and providing more diversity of food industry products and more competitive and efficient processes (Knoerzer, 2016).

Food waste and byproducts generated during processing can be a great source of high added-value compounds, which, due to their different characteristics, can be applied as food additives and/or as nutraceuticals (Saini et al., 2018). Thus, the extraction of such interesting compounds can be another way to reduce food waste and promote food saving. Thus, the selection of an effective extraction process, depending on the previous food processing, can lead to the extraction and purification of bioactive compounds and should be seen as an ultimate goal in the industrial-scale processes. In agro-food waste management procedures, there are several extraction techniques commonly used, but the ideal should be not expensive, quantitative, fast, and nondestructive. Nevertheless, conventional extraction methods have many disadvantages, such as the health hazards of toxic organic solvents and the high volume needed, the long operation or reaction times, the high energy input, the low extraction selectivity, among others (Alexandre et al., 2017a,b,c,d,e). These limitations led to the need to develop other more environmentally friendly technologies. Therefore, green extraction methods such as HPP, PEF, and US have also been used as extraction assisting technologies to decrease food waste.

In this chapter, some of the main technological fundamentals of HPP, PEF, and US are reviewed. The impact of each technology on food composition and microorganisms, their applications and technological advantages and limitations, and their use as extraction methodology are discussed, too.

5.2 Quality indicators for processed food

Food processing aims to guarantee not only the preservation of organoleptic properties of fresh products but also food safety during food consumption.

Over the last 10 years, food quality was more intensively discussed by the public, governments, and the food industry. The increasingly complex and extended food supply chain significantly increases the risk of food presenting lower quality (Liu, 2018).

Food quality includes chemical, physical, and sensory characteristics of food products. These properties may be assessed by sensory inspection at purchase time or require quantification through instrumental measurements. For example, consumers through vision, touch, or smell can easily evaluate external attributes such as overall appearance, size, shape, aroma, color, state of maturation, and defects. But internal properties or hidden characteristics such as taste, flavor, texture, nutritive value, and wholesomeness cannot be assessed by consumer before purchase and some of that only can be evaluated through some specific instrumental measurements (Anonymous, 2002, Rico et al., 2007, Alexandre et al., 2012). Quality is also associated with food components (soluble solids content, starch content, carotenoids, sugar, ascorbic acid, total flavonoids, total phenolic, total tannins, total anthocyanins, antioxidant activity, among others) that have a positive and functional impact on human health. Fruits and vegetables are important sources of vitamins, fibers, minerals, and antioxidants that play an essential role in the human diet preventing numerous diseases (Abasi et al., 2018). In fruit, the external indicators of quality are the shape, size, skin color, and general appearance; nevertheless other internal characteristics are also important, such as soluble solids content, titratable acidity, ratio of soluble solids to titratable acidity, pH, starch and sugar content, carotenoids, ascorbic acid, total flavonoids, total phenolic, and antioxidant activity. For vegetables, nutritional composition, antioxidant activity, along with enzymatic activity and texture are the most important quality properties. Steak, roast, cubed, minced properties, color, fat, protein content, and marbling were important quality properties for meat products (Abasi et al., 2018). Fish is known to be an important supplier of micronutrients such as vitamins A and D, minerals such as calcium, phosphorus, magnesium, iron, zinc, selenium, fluorine and iodine, and essential fatty acids (Arino et al., 2005). However, one of the most important quality indicators common to all food products is food safety.

Food safety is related with the guarantee that food will not cause harm to the consumer when it is prepared and eaten according to its intended use (Alexandre et al., 2013). Microbial counts of Gram-positive (such as *Bacillus cereus*, *Staphylococcus aureus*, *Methicillin-resistant S. aureus*, and *L. monocytogenes*) or Gram-negative (such as *Salmonella enteritidis*, *Pseudomonas aeruginosa*, and *E. coli*) food contaminant/pathogenic bacteria or even counts of total bacteria, coliform bacteria, or molds and yeasts are frequently considered food quality features that can be used as indicators of processed food safety (Alexandre et al., 2013). Some examples of unsafe food that commonly provoke food-borne diseases include uncooked foods of animal origin, fruits and vegetables contaminated with feces, raw shellfish and industrial pollution. Raw products often contain a great diversity of microbial flora that can be involved in food-borne outbreaks and may cause serious illnesses. The pathogens/commodities most frequently responsible in the 21st century for outbreak-related illnesses were norovirus/leafy vegetables, *Clostridium perfringens*/poultry, *Salmonella*/vine-stalk vegetables, and *C. perfringens*/beef (Fung et al., 2018). The symptoms of food contamination frequently include headache, muscle pain, nausea, fatigue, chills or fever, stomach or abdominal pain, vomiting, and diarrhea. Children and elderly are particularly vulnerable groups and food poisoning may be even a fatal illness for those due to some specific pathogenic bacteria (Alli, 2004; Behling et al., 2010; Waite et al., 2010; Alexandre et al., 2012). Other beneficial gut microbiota such as *Lactobacillus* and *Bifidobacterium* strains also can be used as quality markers, but in a different way since these microorganisms are beneficial and do not compromise the consumer's health.

5.3 Food contamination sources

There are many possible scenarios where food contamination may happen. Food products can become contaminated at any point from farm to table, as well as during production, processing, shipping, and distribution.

Food contamination can be classified as biological, chemical, physical, or of cross-contamination nature. Briefly, biological contamination is when food is contaminated with infectious bacteria (such as *Salmonella* spp. and *L. monocytogenes*) or toxin-producing organisms (such as *Clostridium botulinum*) and viruses (such as norovirus), which are a common cause of food poisoning and food spoilage. Chemical contamination happens when a food product is exposed to chemicals (such as food additives, heavy metals, dioxins, radionuclides, veterinary drug residues and pesticides residues, as well as contaminants from processing and packing or other environmental contaminants) that can lead to chemical food poisoning. Physical contamination occurs when foreign objects (such as metal filings, glass, jewelry, stones, or bone chips) contaminate foods and sometimes these objects can also be the vehicle for biological contamination (cross-contamination) if they harbor dangerous bacteria. Cross-contamination occurs when bacteria or pathogens are transported from one object to another that will contaminate food products (Mitchell et al., 2014).

Food contamination during food processing has been extensively reviewed by Nerín et al. (2016). Some of the most important contamination sources during processing are related with:

1. *external raw food contamination due to environmental contamination* (example of pesticides, fertilizers, toxic heavy metals, antibiotic residues, etc.);
2. *transport of raw materials to the factory where they will be processed* (caused by vehicle exhaust from petrol and diesel or cross contamination in the vehicle used for food transportation);
3. *food conditioning, which involves the storage of raw materials, preheating, disinfection, cleaning, and sterilization steps* (some common surfactants used to clean are quaternary ammonium compounds and nonionic surfactants);
4. *heating, which includes boiling, cooking, baking, frying, or combining with other ingredients at high temperature in an oven or in a reactor* (certain toxics compounds such as acrylamide, nitrosamines chloropropanols, furanes, or Polycyclic Aromatic Hydrocarbons (PAHs) can be formed in foods processing);

5. *food packaging* (different additives such as antioxidants, stabilizers, slipping agents, or plasticizers are commonly added to the package polymers to improve material properties that can end up transferred to food in a phenomenon called migration);
6. *transport, storage, and food distribution* (packaging material properties can be affected, low temperature and humidity) (Nerín et al., 2016).

It is extremely probable that biological, chemical, physical, or cross-contamination happen industrially. Today, the methods used by industry to assure product quality and safety are essentially thermal (such as blanching, pasteurization, and sterilization) or simply water washes with chemical sanitizer solutions (such as chlorine, hydrogen peroxide, and acid solutions). Although these methods guarantee the products' safety, many times they are inefficient to preserve fresh food quality such as the indicators of texture, color, nutritional quality; also, sometimes off-flavors can be formed, and they are expensive due to energy consumption and time-consuming methods. For all of these reasons it is extremely important to develop nonthermal preservation methods that can maintain or improve fresh food quality while assuring food safety for consumers at same time.

5.4 Nonthermal emerging processing technologies

In the last few decades, new nonthermal food processing methods have been faced with a challenge to accelerate the shift towards sustainable development and production. The identification of solutions that can enhance productivity and sustainability along the supply chain while helping the sector cope with climate change issues is urgent (Cavaliere et al., 2018). In fact, food waste gives rise to a heavy environmental burden. The Food and Agricultural Organization of the United Nations states that around 30% of global food production ends up as food waste and its effect on climate change is catastrophic since 95% of waste ends up at landfill sites, where it is converted into methane and other greenhouse gases (McCarthy et al., 2017). Researchers highlighted the negative externalities linked to food waste and showed that emergent nonthermal technologies are able to reduce food wastes through the food shelf-life extension, which is considered to be one of the most sustainability-driving food innovations contributing to sustainable development (Cavaliere et al., 2018). The transition towards these new methodologies of food production and consumption will depend on the sector's capacity to introduce innovative and emergent approaches and strategies at any level of the supply chain (Cavaliere et al., 2018). However, emergent technologies such as HPP, PEF, and US have been studied by both industry and academia, in an attempt to minimize the disadvantages of conventional thermal methods. Recently, these technologies have also been studied as extraction methods to recover bioactive compounds, which present important biologic activities, from food byproducts and residues, contributing to food waste valorization and also reducing its impact on environment (Alexandre et al., 2017a,b,c,d,e). HPP, PEF, and US, which will be discussed in this section, are among the most studied nonthermal alternative processing and extraction methods.

5.4.1 High pressure processing technology

5.4.1.1 Technological fundamentals

The great and increasing interest of the food industry in HPP, as a "cold" pasteurization method, is mainly due to the low effect of this technology on functional and nutritional properties of food products and their capacity to inactivate/destroy microbial loads of foods (Barba et al., 2012; Barba et al., 2015,a,b). The high pressure effect is based on two fundamental physical principles: *Pascal's isostatic principle* and *Le Chatelier's principle. Pascal's isostatic principle* claims that pressure applied to a sample is transmitted uniformly and instantaneously by the entire food sample whether in direct contact or in a flexible container, regardless of its shape, volume, size, or geometry, unlike thermal treatment, which has slower heating points. Therefore, HPP technology is easier to implement industrially than thermal treatments (Neetoo et al., 2012; Misra et al., 2017). *Le Chatelier's principle* states that any change made in an equilibrium system (chemical reaction, phase transition or modifications of molecular configurations) accompanied by a volume decrease is favored by a high pressure increase, whereas reactions that involve a volume increase will be inhibited (Neetoo et al., 2012). Thus, the pressure has a huge influence on the biochemical reaction rates that occur in foods since most of these reactions are often involved in volume changes. Additionally, the enzymatic reactions may be affected by the pressure compromising the energy production (Welti-Chanes et al., 2005; Venugopal, 2006).

Food products are typically vacuum-packaged and placed inside a pressure basket, being after loaded into the pressure vessel and then HPP occurs in three distinct phases. The first one, *pressure boost*, happens during a short time and is related with the increase of pressure until that the desired treatment pressure is achieved; the second one, *pressure maintaining*, consists of pressure maintenance for the desired period; and the third one, *pressure relief*, is related with the pressure relief that happens from the target level to atmospheric pressure in a couple of seconds (Huang et al., 2013). High pressure can be generated by three different ways:

1. *by direct compression*, which requires dynamic pressure seals between piston and vessel surface;
2. *by indirect compression*, which requires static pressure seals; and
3. *by heating the pressure medium*, which is usually water combined with mineral or vegetable oil for lubrication with anticorrosive purposes (Ohlsson, 2002; Welti-Chanes et al., 2005).

In fact, the most common transmitting fluids are water, food grade glycol/water solutions, silicone oil, sodium benzoate solutions, ethanol solutions, and castor oil (Balasubramaniam et al., 2008). During HPP, food products are submitted to pressures that may range from 100 to 800 MPa in batch or semicontinuous processes, from a millisecond pulse to over 20 minutes, at temperatures that can be from below 0°C to above 100°C (Muntean et al., 2016). The processing temperature can be controlled by cooling jackets, heat exchangers, or by recirculation of the cooling/heating medium. The food temperature always increases through the adiabatic heating, around 3°C per 100 MPa at 25°C [but it can be significantly higher (8°C–9°C/100 MPa) for more compressible food ingredients such as fats], but if no heat is exchanged from the pressure vessel walls during the holding time (second phase), the temperature decreases to the original temperature on the decompression cycle (Guan et al., 2005; Pereira et al., 2010; Muntean et al., 2016). The packaging material of food products must be flexible since foods decrease in volume under pressure and regain volume during decompression (Lou et al., 2015) and currently, industrial HPP of solid foods can only be done in batch mode (Fig. 5.1) (Considine et al., 2008).

However, HPP and PEF are two of the most commercial technologies and some processed products can be even found on the market. Indeed, there is an increasingly high number of high pressure equipment operating worldwide (Fig. 5.2), the

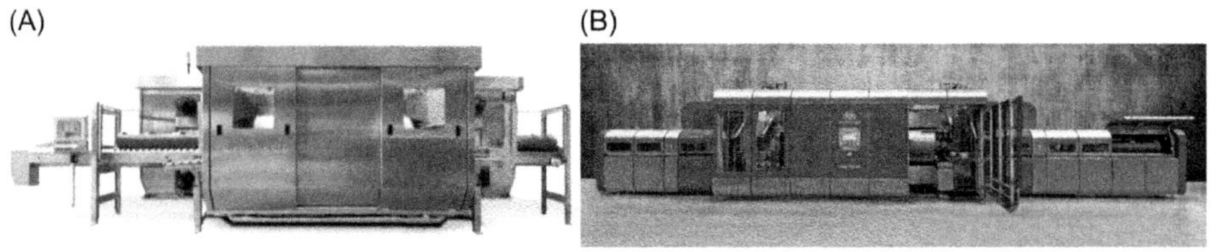

Figure 5.1 Industrial scale equipment of high pressure processing (in batch mode).
Source: Courtesy Hiperbaric S.A (A) Model 135, www.hiperbaric.com and UHDE (B) Model 350–60, www.thyssenkrupp-industrial-solutions.com/en/high-pressure-processing.

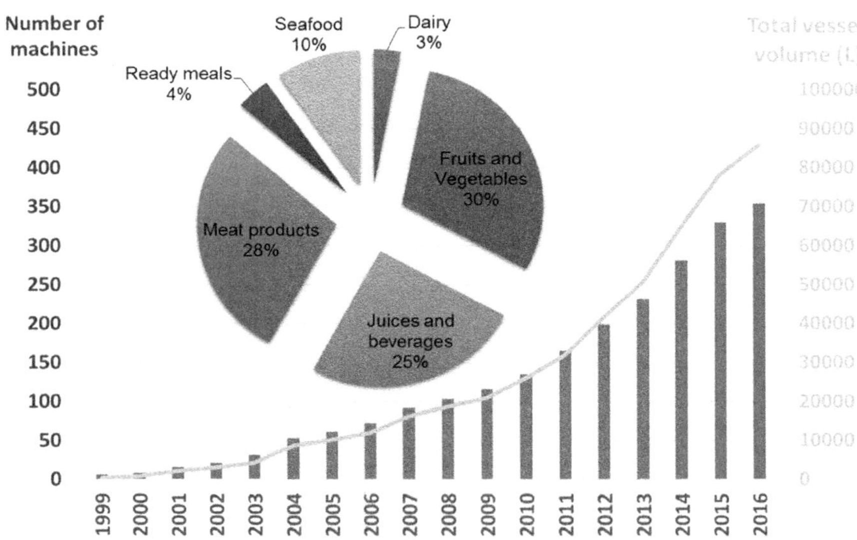

Figure 5.2 Evolution of the total number of high pressure industrial machines operating worldwide and global high pressure processing submarket share in 2015.
Source: Courtesy Hiperbaric S.A.

major suppliers being Avure Technologies (USA), Hiperbaric (Spain), and UHDE High Pressure (Germany) (Barba et al., 2016). Furthermore, Visiongain (2017) estimates that the HPP food equipment market will exceed 13 billion dollars in 2017 and predicts a strong revenue growth of HPP market through to 2027, primarily due to health-driven consumers.

5.4.1.2 Effect on food composition, microorganisms, and applications

The effect of HPP on molecules with a low molecular weight is minimal. Small molecules such as vitamins, alkaloids, saponins, flavonoids, peptides, fatty acids, and saccharides remain relatively undamaged compared with thermal processing since the covalent bonds have very low compressibility at high pressure. In this way the nutritional value and the quality of food is preserved during the HPP. High pressure only affects noncovalent bonds such as hydrogen, ionic, and hydrophobic bonds acting selectively. Therefore, only the native structure and functionality of macromolecules such as proteins, enzymes, lipids, polysaccharides, and nucleic acids may be disrupted by HPP (Jun et al., 2011; Huang et al., 2013; Jun, 2013). The primary structure of proteins is unaffected by pressure due to covalent bonds while the secondary, tertiary, or quaternary structure are changed, due to the disruption of ionic bonds, hydrogen bonds, and hydrophobic and electrostatic interactions responsible for the protein structure (Rendueles et al., 2011). Thus, some compounds are irreversibly changed during HPP such as the gelatinization of carbohydrates that can be achieved through pressure increases rather than through temperature increases and proteins can be denatured at high temperature (Muntean et al., 2016).

The impact of high pressure in the microbiological food loads is similar to that previously discussed for chemical components of food. The microbial inactivation results in a multiplicity of injuries accumulated in different components of the cell including cell membranes, nucleoids, ribosomes, proteins, and enzymes (Considine et al., 2008; Daryaei et al., 2012). The cell membrane of the microorganisms is composed by a phospholipid bilayer with proteins and lipids, which are affected by high pressure. In fact, the proteins are essential to many functions of the bacterial cell and their denaturation by pressure will compromise the microorganism survival. Lipids also change their conformation and packing, altering the membrane fluidity. Thus, protein–lipid interactions are weakened and the membrane becomes less permeable reducing the transmembrane transport (Balasubramaniam et al., 2015). Below 300 MPa, most of these reactions can be reversible but above 300 MPa, irreversible cell damages can be achieved breaking the cell membrane integrity and the flow of internal substances, leading to bacterial death (Huang et al., 2014). HPP also changes ribosome configurations interfering with normal protein biosynthesis and inhibiting the protein repair system. Moreover, HPP also affects the normal DNA replication and gene transcription due to condensation of the genetic materials leading to degradation of the chromosomal DNA (Huang et al., 2014). Yeasts and molds are relatively sensitive to HPP being inactivated within a few minutes by 300–400 MPa at room temperature.

However, spore-forming microorganisms are highly resistant when in spore form, being required a combination of high pressure (exceeding 1000 MPa) and heat (above 80°C) to attain a significant log reduction of spores in food products. Viruses show a wide range of sensitivity in response to HPP (Muntean et al., 2016).

The efficacy of HPP is influenced by several factors such as the *treatment time* (time to achieve the desired pressure, holding time, and decompression time); *temperature* (initial temperature, processing temperature, and distribution temperature in the pressure vessel due to the adiabatic heating), *pressure* level used and the nature of pressure (batch or continuous), *food matrix characteristics* (such as pH, composition, and water activity), *packing material* and type of *microorganisms* present in food products (Venugopal, 2006; Alexandre et al., 2017a,b,c,d,e). The effect of high pressure on some of the most important food pathogens was reviewed by Alexandre et al. (2017a,b,c,d,e) and in general, the increase of the pressure's treatment time and/or the increase of the pressure level cause an increase of the microbial inactivation. For example, *E. coli* was analyzed in carrot, strawberry/blueberry, apricot, orange, and cherry juices as well as in raw milk cheese and dry fermented salami and microbial log reductions obtained by HPP were between 0.5 and 8.0 (Rodríguez et al., 2005; Van Opstal et al., 2005; Bayındırlı et al., 2006; Porto-Fett et al., 2010; Tadapaneni et al., 2014); *S. enteritidis* was analyzed in nuts and apricot, orange, and cherry juice showing log reductions between 1.0 and 7.3 log cycles due to HPP (Bayındırlı et al., 2006; Prakash, 2013); *S. typhimurium* was studied in strawberry/blueberry beverage and dry fermented salami presenting log reductions between 1.9 and 6.0 log cycles (Porto-Fett et al., 2010; Tadapaneni et al., 2014); *L. monocytogenes* was analyzed in strawberry/blueberry beverage, dry fermented pork sausage, dry fermented salami, and raw cow's milk cheese showing microbial reductions between 0.60 and 6.34 log cycles after HPP (Rodríguez et al., 2005; Jofré et al., 2009; Porto-Fett et al., 2010; Tadapaneni et al., 2014); and *S. aureus* was studied in dry fermented pork sausage, raw cheese, and apricot, orange, and cherry juice obtaining log reductions between 0.30 and 5.7 after HPP (Rodríguez et al., 2005; Bayındırlı et al., 2006; Jofré et al., 2009), depending on the food product and HPP conditions. The effect of HPP on fish meat quality was reviewed by Oliveira et al. (2017), while Hygreeva et al. (2016) reviewed the quality of processed meat products through HPP and Oey et al. (2008) reviewed the effect of HPP on color, texture, and flavor of fruit- and vegetable-based food products. In general, HPP had low impact on quality parameters of food products and some researchers, such as Fernandez et al. (2018), are optimizing the processing conditions to preserve quality attributes of a mixed fruit and vegetable smoothie, keeping in mind food quality and safety requirements.

HPP could be used for some commercial applications such as the pasteurization of meats and vegetables; pasteurization and sterilization of fruits, sauces, yogurts, and salad dressings; and decontamination of high risk and high value heat sensitive ingredients including flavorings and vitamins. In fact, there are already some HPP food products available on the market such as vegetables and fruit juices, salsas, dressings, meats, ready-to-eat meats and poultry, seafood, shellfish, and fish products (Muntean et al., 2016).

5.4.1.3 Advantages and limitations

As mentioned before, the major advantage of the HPP consists of the preservation of fresh food characteristics, namely sensorial and nutritional properties, extending shelf-life and improving food quality and safety, due to microbial and enzymatic inactivation (Rendueles et al., 2011). In other words, high pressure inactivates vegetative bacteria and spores (at higher temperatures) preserving nutrients, colors, and flavors. The impact of high pressure on food products is homogeneous, uniform, and instantaneous and there is no evidence of toxicity. Moreover, HPP reduces processing times and involves fewer energy requirements than thermal treatments; only a small amount of energy is necessary to achieve the desired processing pressure, and it is environmentally friendly with few effluents generated (it only needs water, which is usually recycled). HPP reduces or eliminates the need for chemical preservatives and their applicability to packaged foods may avoid microbial contaminations. Besides, food products have a positive consumer appeal (Alexandre et al., 2013; Muntean et al., 2016).

The biggest limitation for the industrial implementation of this technology is related with the high capital cost, that is, the equipment is still expensive and inevitably requires high investment. A set of HPP equipment costs between 0.5 and 2.5 million euros depending on the capacity and operating parameter range of the equipment (Galanakis, 2013; Huang et al., 2017). Intermittent operation and small workload (batch processing) are other limitations that increase the cost of production. Moreover, HPP allows inactivation of vegetative microorganisms but is insufficient to substantially destroy spores at room temperature (Balasubramaniam et al., 2015) and may have a reversible or irreversible and partial or complete unfolding effect on the enzyme structure being usually less efficient in enzymatic inactivation than heat treatments (Gong et al., 2015; Aghajanzadeh et al., 2018). Moreover, HPP products need transportation and storage under refrigeration; this process is not applicable to food products with low water content (foods should have 40% free water for antimicrobial effect) and there are limited packing options (the packaging material must be compressible to be suitable for HPP) (Muntean et al., 2016; Huang et al., 2017).

As extraction method, high pressure also is considered environmentally friendly since it complies with standards set by the EPA (2015), which include the reduction of synthetic and organic solvents used, and the reduction of operational time and energetic consumption, allowing higher yields and final extracts with a higher quality (Azmir et al., 2013).

5.4.1.4 High pressure processing as extraction method

As well as for HPP, extraction processes assisted by high pressure also occur in three distinct phases: pressure boost, pressure maintaining, and pressure relief. In the first one, the plant tissues and solvent are submitted to increasing pressures and the pressure differential generated between the interior and the exterior of the cell lead to cell deformation and wall damage. If the pressure applied is not enough to exceed the deformation limit of the membranes, the extraction solvent can permeate into the cell through some wall channels and the cells are quickly filled with solvent. Nevertheless, if the pressure applied is enough to exceed the deformation limit of the cell, the pressure will cause damage in the membrane structures and exterior wall, leading to the formation of cracks (Huang et al., 2013; Alexandre et al., 2017a,b,c,d,e). If in the first stage the pressure applied was not enough to exceed the deformation limit of the membranes, this may happen during the second stage and the solvent will permeate quickly into the cells, allowing the bioactive compounds' dissolution (Huang et al., 2013; Alexandre et al., 2017a,b,c,d,e). In the third stage the high pressure acts selectively, and may disrupt large molecule structure or even the cell structure (e.g., enzymes, proteins, and lipids) leaving small molecules (e.g., vitamins, pigments, fragrance ingredients, alkaloids, saponins, or flavonoids) unaffected, since only noncovalent bonds (such as hydrogen bonds, ionic bonds, and hydrophobic interactions) will be affected (Jun et al., 2011; Huang et al., 2013; Jun, 2013). In conclusion, high pressure can reduce the resistance to mass transfer during the extraction of important bioactive compounds from the solvent, mainly due to their effect in the tissues, cellular wall, membrane, and organelles (Shouqin et al., 2007; Huang et al., 2013).

The different parameters to take into consideration for HPP extraction are, in order of importance, extraction temperature, pressure level, solvent and its concentration, ratio of solvent to raw material, and holding pressure time (Chen et al., 2009). The most important parameter is the extraction temperature since the efficiency of thermosensitive compounds' extraction depends on it (Prasad et al., 2009). The solvent choice, and its concentration, are also closely related to the components to extract, and besides being nontoxic, it should be easy to remove from the final extract. The pressure holding time is the period of time needed to achieve the equilibrium between inside and outside of the cells, in terms of solvent and pressure (Xi et al., 2009).

High pressure has been used as an extraction method to recover several bioactive compounds from herbal materials and fruit residues. For example, total phenolic compounds and flavonoids have been extracted from Korean barberry and deodeok (Qadir et al., 2009; He et al., 2010; Lee et al., 2010; He et al., 2011) and from papaya seeds (Briones-Labarca et al., 2015) and citrus peels (Casquete et al., 2014; M'hiri et al., 2014; Casquete et al., 2015). Lycopene and carotenoids were extracted from tomato wastes (Jun, 2006; Xi, 2006; Strati et al., 2015), pectin from orange (Guo et al., 2012) and honey pomelo peels (Guo et al., 2014), mangiferin and lupeol from mango peels (Ruiz-Montanez et al., 2014), ginsenosides from ginseng (Shouquin et al., 2006; Shin et al., 2010; Lee et al., 2011), salidroside from rhodiola (Zhang et al., 2007; Bi et al., 2009), catechins and caffeine from green tea (Xi, 2009; Xi et al., 2010), deoxyschisandrin and y-schisandrin from Magnolia berry (Liu et al., 2009), and podophyllotoxin and 4'-demethylpodophyllotoxin from hance (Zhu et al., 2012). In general, high pressure extraction increases the extraction yields when compared with traditional thermal methods. The optimum conditions were extensively reviewed and discussed by Alexandre et al. (2017a,b,c,d,e) but will be mainly dependent on the target compound to be extracted and of course the plant material used for the extractions.

5.4.2 Ultrasounds

5.4.2.1 Technical fundamentals

Sounds can be defined as the continuous propagation of a mechanic compression or wave that cause particles vibration longitudinally propagated throughout a medium, according to the number of events/repetitions in a certain period of time (frequency). Sounds, according to the frequency range, can be categorized as infrasounds (frequencies up to 20 Hz), acoustic sounds (up to 20 kHz, audible to the human ear), and US (above 20 kHz and up to 10 MHz) (Leong, 2016). US are widely used in the medical field, namely in imagology (such as to visualize the fetus in utero, among other applications).

According to Tiwari (2015), US can be classified according to the intensity of the sonication as low-intensity sonication (<1 W/cm^2, used in several processes of quality and control and physical state of matter) or high-intensity sonication (10–1000 W/cm^2, usually used for extraction and food processing).

The use of US as a nonthermal alternative to the conventional thermal pasteurization procedures is raising quite a bit of interest (meeting consumer's demands for fresh, better tasting, healthier, and minimally processed products), not only for that purpose, but also as an assistant for other processes, such as freezing, cutting, drying, tempering, and sterilization (Chemat et al., 2011).

An US apparatus can simply consist of a water bath, or in more complex apparatus, with an US probe, as shown in Fig. 5.3, along with the propagation media (liquid/fluid food product or a solvent mixed to the interesting matrix). From an industrial point of view, US can be used in batch, semibatch, or continuous operation mode, allowing to obtain high yields (Tiwari, 2015). Nevertheless, as it is not an "in package" processing technology, such as HPP or PEF (in certain conditions),

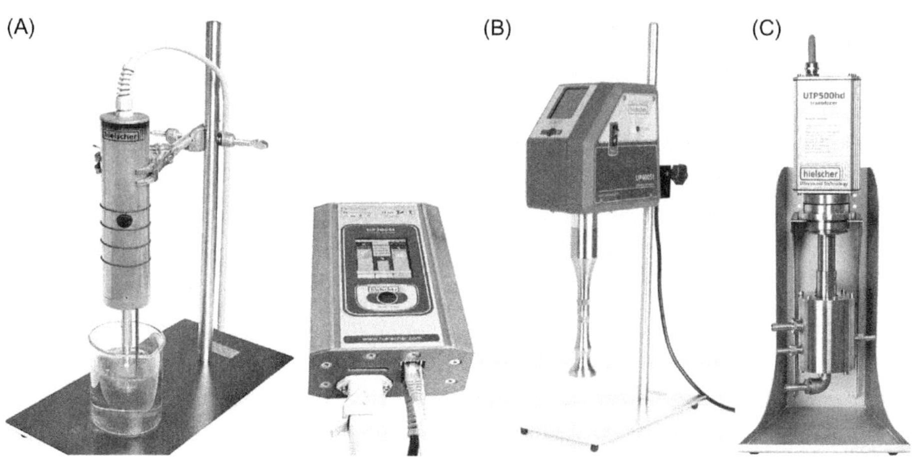

Figure 5.3 Hielscher system (with probe) with different potencies: (A) UP200St—200W, (B) UP400St—400W, and (C) UIP500hdT—500W, available for various applications. Source: Courtesy Hielscher Ultrasonics GmbH 2019 (www.hielscher.com).

it requires aseptic packaging to ensure that contaminants are not reintroduced in the processed product.

5.4.2.2 Effect on food composition, microorganisms, and applications

Piyasena et al. (2003) described the main factors that affect the resistance of microorganisms to ultrasonic treatment as (1) amplitude of the US waves, (2) treatment time, (3) food product properties (bulkier food products may not propagate the waves uniformly, especially nonhomogeneous foods), (4) food composition, and (5) operation temperature. The size, shape, and type of microorganism (Gram-negative vs Gram-positive) are also to be considered in inactivation studies by US, as Gram-positive bacteria are known to be more resistant than Gram-negative bacteria due to their thick peptidoglycan layer, while larger cells are more sensitive to US than the smaller ones, since the former have a larger contact area and "target points" to the US (Chemat et al., 2011).

Microbial inactivation by US is achieved by cavitation, which consists of the creation of bubbles by the ultrasonic waves on the cell membrane, causing local changes in pressure and temperature, leading to its breakdown/rupture, with consequent release of organelles and DNA damage by the production of free radicals (Ercan and Soysal, 2013). Similarly to other nonthermal processing technologies, US can only inactivate, generally, vegetative microorganisms, remaining the endospores, ascospores, and some enzymes not inactivated by US. Ultrasonic pasteurization is more common for liquid foods, such as juices, with application being the main review approach in this section. Gómez-López et al. (2010) sonicated raw orange juice at 20 kHz and 500 W for 8 minutes at 10°C and were able to reduce the total aerobic mesophile count in 1.38 log units, while yeast and molds counts were reduced about 0.56 log units, leading to a shelf-life extension up to 10 days under refrigeration (4°C), compared with the unprocessed juice (spoiled by the sixth day). The authors also reported a reduction on the ascorbic acid content and color degradation in sonicated juice. In another study, Yuan et al. (2009) reported a reduction of about 60% of *Alicyclobacillus acidoterrestris* (a heat-resistant bacterium that is prevalent in acidic fruit juices) in apple juice after US treatment at 24 kHz, 300 W, and 30 minutes, achieving higher inactivation rates (up to 80%) when the treatment time was increased to 60 minutes. The total sugar content, transmittance, and color value decreased with treatment intensity and time increase, while the titratable acidity increased with treatment intensity and time increase. Nevertheless, according to the authors, these alterations did not significantly change the appearance of the apple juice.

Regarding enzymatic inactivation, US, especially when combined with high temperatures, can inactivate several enzymes that compromise the organoleptic characteristics of food during storage. The mechanism for enzymatic inactivation by US starts with the cavitation phenomenon, which leads to the creation of bubbles that collapse as consequence of the increased pressure and temperature. Furthermore, the shock waves also create strong shear and streaming, which may cause changes in the secondary and tertiary structures of proteins, leading to the destabilization of hydrogen bonds and van der Waals interactions, leading to enzymatic inactivation. Free radicals (which form as a consequence of cavitation) may interact with proteins' amino acids (sonochemical reaction), which are related with the stability of the enzyme, substrate binding site, or catalytic function, thus compromising its activity (Raymond et al., 2011).

5.4.2.3 Advantages and limitations

According to Chemat et al (2011), the main advantages of using US in food processing are:

1. the effectiveness of mixing and micromixing products,
2. faster mass and energy transferences,
3. reduced temperature increments during processing,
4. selective components extraction,
5. reduced equipment size,
6. lower costs (when compared with other nonthermal processing technologies, such as HPP or PEF),
7. they are an easy to use technology,
8. they are considered environmentally friendly (due to the reduced energetic costs associated), among others.

Similarly to PEF, US can be operated in continuous operation mode, allowing higher yields from an industrial integration point of view. Nevertheless, some drawbacks are to be pinpointed, such as the fact that US are highly dependent on the sample matrix and on the presence of a disperse phase, which may reduce the effectiveness of the method by wave attenuation, with optimization necessary for each case (Alexandre et al., 2017a,b,c,d,e).

5.4.2.4 Ultrasounds as extraction technique

As aforementioned, the need for replacing hazardous organic solvents for extraction processes lead to the search for novel extraction techniques that reduce the volume of solvents, to meet consumers' demands for chemical-free food products, as well to improve the environmental sustainability of food related processes and waste management.

The success of ultrasounds as an extraction technique relies on four essential mechanisms, which are, according to Tiwari (2015):

1. the improved mass transference by acoustic streaming and turbulent mixing,
2. the damages in the solvent-matrix interfaces (matrix surface in contact with the solvent) caused by shock waves and micro jets,
3. the collisions of interparticles at high velocities, and
4. the disintegration of the matrix to maximize the surface contact area.

Ultrasounds, when in contact with a sample matrix, produce sound waves, which are then are able to generate a phenomenon called cavitation, which is characterized by the production, growth, and collapse of a bubble inside the cell (Adetunji et al., 2017). The bubble's growth bubbles are related with the expansion/compression cycles that affect the cellular structure by pulling/pushing the molecular content apart/together, respectively. The consequent disruption of the bubbles inside the cell wall lead to the damage of its structure, thus facilitating the mass transfer from the cellular particles into the solvent (Ebringerová and Hromádková, 2010).

At the final stages of the extraction, the compounds diffuse from the cell into the solvent. The factors that most affect ultrasound-assisted extractions are related to the sample matrix, due to properties such as its hardness, structure, composition, moisture content, and particle size. Also, the frequency, power, pressure, temperature, time of sonication, and the chosen solvent can significantly change the final extraction yields since all these factors can greatly affect the ultrasound intensity by sound waves attenuation (Alexandre et al., 2017a,b,c,d,e).

Several studies suggested the use of ultrasounds combined with green solvents as water, ethanol, and sunflower oil to enhance the extraction of antioxidant compounds (mainly phenolic compounds), anthocyanins, carotenoids, and polysaccharides, while decreasing the use of organic solvents (Alexandre et al., 2017a,b,c,d,e). The use of mildly nonpolar solvents, as ethanol and acetone, may allow to obtain higher extraction yields of lipid fractions from biomass, in attempts to replace the traditional hazardous solvents used for lipids extraction, such as *n*-hexane or others.

5.4.3 Pulse electric fields

5.4.3.1 Technological fundamentals

PEF is described as a nonthermal food pasteurization technique that relies on the application of short duration pulses (from several nanoseconds up to 1 ms) of moderate to high electric field strengths (up to 80 kV/cm) and low energy (up to 10 kJ/kg) to products placed between two electrodes in a chamber, at mild-high temperatures (Bobinaitė et al., 2014) to inactivate both spoilage and pathogenic vegetative microorganisms, at room temperature (Sánchez-Vega et al., 2015), especially in acidic food products (Shahbaz et al., 2018). The PEF equipment can be described, in a simplistic way, as a pulse power supply and a treatment chamber. It can be operated discontinuously and continuously; the latter is more interesting for the food industry, as it allows to treat higher amounts of product at the same time.

The most commonly used continuous units are formed by a pump that inlets the product into the treatment chamber, wherein the electrodes discharge the current on the food product, which is then pulled out (Fig. 5.4). The treated product is then aseptically packed to avoid recontamination. The industrial application of PEF is often coupled with preheating systems ensuring higher microbial inactivation rates when compared with the conventional pasteurization processes (and to nonthermal PEF). These methodologies allow to reduce the impact of thermal processing on the organoleptic attributes of the food products. After the PEF treatment chamber, a cooling system is often used to quickly cool the processed product, as a result of the temperature increase due to the high conductivity of the product itself or the

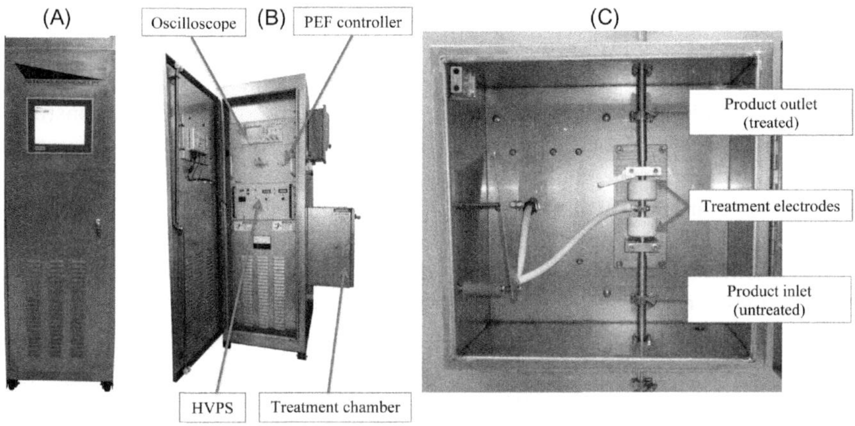

Figure 5.4 Laboratorial/pilot scale pulsed electric field (PEF) equipment, showing the control panel (A), the oscilloscope, the PEF controller and the high voltage power supply (HVPS) (B), and the treatment chamber (C).
Source: Courtesy Diversified Technologies (http://www.divtecs.com/).

high inlet temperatures (Loeffler, 2010). As no heat is applied (even though the temperature increases during the treatment are widely dependent on the product conductivity), the resulting food products present fresh-like attributes, thus answering the consumers' demands for minimally processed food products (Li et al., 2016).

The first experiments concerning the application of PEF on food products were carried out in the 1950s, although it wasn't until 1995 that the first continuously operated equipment was launched, resulting from a consortium between food processors, universities, and equipment manufacturers in Europe and the United States. Nevertheless, it was only in 2006 that the first PEF system was installed for operation in the juice industry, in the United States (Sitzmann et al., 2016).

5.4.3.2 Effect on food composition, microorganisms, and applications

The efficacy of the application of strong electric fields for food safety is related to the formation of pores on microorganism's membranes (a phenomenon called electroporation), which leads to the loss of cellular integrity and viability. The feasibility of PEF on microbial inactivation is well reported in the literature, namely in acidic fruit juices, milk, and liquid eggs. According to Barba et al. (2015,a,b), several factors affect the effectiveness of PEF as a microbial inactivation strategy, namely the field strength, treatment temperature and time, pulse width, pulse frequency, pulse shape (monopolar, bipolar, square, exponential, etc.), electrode distance (treatment gap of the chamber), polarity, and applied energy. This emergent nonthermal technology can be used to improve the quality and safety of dairy products, such as milk, whey, infant formulas, liquid eggs, and fruit juices, to enhance their shelf-stability.

Sharma et al. (2014) studied the effect of PEF treatment (23–28 kV, 17–101 μs, 4.2 mL/s) in whole milk and reported that, due to the product specific constitution, no significant microbial load reductions were found in milk treated at low temperatures (4°C), even when using higher electric strengths and treatment times. Nevertheless, considerable microbial load reductions were found when the milk was preheated before the PEF treatment, as the PEF was more effective after higher preheating temperatures. The same authors also reported that the matrix composition had a major role on the microbial inactivation after PEF, as more pronounced microbial load reductions were found in skimmed milk than in whole milk. This could be associated with whole milk's richness in fat and proteins, which confer a protective effect against bacterial membrane electroporation during PEF treatments. A known protection mechanism is the thick peptidoglycan layer present in Gram-positive bacteria, which are known to be more resistant to electric field pulses than Gram-negative, since that specific layer helps to protect bacteria from electroporation.

In another study, the effect of PEF treatment on raw milk for cheese production was accessed, with the outcomes revealing that treated milk (20 and 30 kV/cm, 2 μs, 120 pulses, 20°C) presented higher curd firmness than thermally pasteurized milk (63°C, 30 minutes), with this parameter decreasing with higher treatment intensities. The same study reported that the rennet coagulation time was lower after PEF treatment, which increased as the intensity also increased (Yu et al., 2009). These results led to the formation of an improved, stronger gel, probably due to the partial protein denaturation that occurs at higher intensity treatments, as stated by Yu et al. (2009). Denaturation is known to be due to protein molecule polarization, dissociation of noncovalent bonds from the quaternary structure of the protein, exposure of entrapped hydrophobic amino acids, and/or sulfhydryl groups exposure as a result of protein conformational changes. During PEF treatment, the milk proteins can change their total charge leading to the establishment/modification of ionic interaction between proteins. Milk functional aspects such as coagulation, emulsification capacity, and foaming can be changed after PEF treatment due to the modification of protein network (Floury et al., 2006; Perez et al., 2004).

The most popular application of PEF in the food industry is in acidic fruit juices treatment, namely on strawberry, apple, orange, and pear juices (among others), since their low pH acts as a hurdle that hinders endospore germination and outgrowth, thus expanding the juice shelf-life, which is difficult in low acidic food products. Timmermans et al. (2014) studied the inactivation of *Salmonella panama*, *E. coli*, *L. monocytogenes*, and *Saccharomyces cerevisiae* inoculated in fruit juices with different pH values, namely apple (3.5), orange (3.7), and watermelon juices (5.3), using a continuous-flow PEF system (frequencies of 120–964 Hz, 2 μs monopolar pulses, 20 kV/cm at a flow of 14 mL/min). The results showed that, for the same treatment conditions, *S. cerevisiae* was more sensible than the other microorganisms under study, proving that bacteria need more energy expense to be inactivated than yeasts. At temperatures above 35°C, a combined effect of temperature and electric field pulses on the microbial inactivation was observed.

When it comes to the quality parameters, the use of nonthermal approaches to preserve fruit juices can be very promising regarding the preservation of thermolabile compounds, such as vitamin C, which are lost during thermal pasteurization procedures. Nonetheless, general physicochemical parameters (such as pH, Brix degree, conductivity, and hydroxyethyl furfural) are not affected by PEF processing. Otherwise, viscosity is known to increase in PEF treated juices, as well as color degradation (Koubaa et al., 2017).

Concerning the application of nonthermal PEF for microbial inactivation in alcoholic beverages, such as beer and wine, Puértolas et al. (2009) reported a 3 log units reduction (after a treatment at 22–31 kV/cm, 1 Hz, up to 100 pulses, at 24°C) of *Dekkera anomala*, *Dekkera bruxellensis*, *Saccharomyces bayanus*, *Lactobacillus plantarum*, and *Lactobacillus hilgardii*. These microorganisms are responsible for off-flavors production by using the residual sugars of wine to produce D-lactic and acetic acids, thus changing the organoleptic characteristics of wine. As sulfur dioxide (SO_2) is a chemical recognized as harmful and usually used for microbial inhibition in wines, those authors also stated that nonthermal PEF could be an interesting technology for the production of additive-free wines.

Puértolas et al. (2009) also reported that, after a PEF treatment (35 kV/cm, 2296 μs, 2122 kJ/L, 14.7°C), *Bacillus subtilis* populations inoculated in beer were reduced in 8.4 log units, while a treatment performed at 45 kV/cm, 804 μs, 1169 kJ/L at 4°C reduced the population of *L. plantarum* in about 7.0 log units, as well of the pathogenic *Salmonella choleraesuis* in about 5.7 log units, when compared with thermal pasteurization.

Pasteurized liquid eggs are quite popular and useful, not only for the food industry, but also for hotels, restaurants, and cafes (HoReCa channel services) that have to ensure the quality and safety of their egg-based products. The conventional thermal pasteurization procedure of whole egg products (61.1°C for 3.5 minutes) is reported to be the minimal temperature/time binomial to ensure the inactivation of pathogenic microorganisms (Puértolas et al., 2009), as lowering the processing temperature can result in nonpasteurized whole eggs, while higher temperatures result in overheating, leading to egg coagulation and the formation of films (Li-Chan et al., 1995). The possibility of applying PEF on liquid eggs as a nonthermal alternative (despite its high conductivity) is gaining particular interest. Puértolas et al. (2009) reported 3 log units' reduction of *B. cereus* reduction in liquid whole egg after a PEF treatment of 40 kV/cm, 2.5 μs pulse width, 155 pulses for 360 μs, at 20°C, while a 1.3 log reduction was found for *E. coli* O157:H7 after exposure to electric fields of 15 kV/cm, up to 500 pulses at 1 Hz, at 0°C (Amiali et al., 2004). *Salmonella* sp., one of the most prevalent pathogenic bacteria in eggs and responsible for food poisoning outbreaks (such as gastroenteritis), was reduced by about 3 log units after a PEF treatment of 40 kV/cm, 300 Hz, 5 pulses for 15 μs, below 35°C (Monfort et al., 2010), showing the potential of this technology to pasteurize liquid whole eggs, without compromising its quality attributes.

5.4.3.3 Advantages and limitations

As previously discussed, the use of PEF as a nonthermal food processing technique has advantages over the conventional thermal processes, since it avoids the destruction of food products' thermolabile compounds, such as proteins, vitamins, and enzymes, providing safe and shelf-stable food products with natural-like sensorial attributes (Vega-Mercado et al., 2007).

Nonetheless, some drawbacks to PEF processing should be pointed out, for example, the temperature rise during the pulse application, as a result of the product conductivity (higher conductivities result in higher temperature rise during processing), thus decreasing the effect of nonthermal processing, leading to loss of sensorial attributes, similarly to the conventional thermal processes (Toepfl et al., 2006).

The transition between laboratorial/pilot scales to industrial units is still a hard task, due to the multiplicity of variables influencing the feasibility of PEF nonthermal pasteurization, such as the product flow, electrodes distance, capillary diameter, and product viscosity and conductivity (Shahbaz et al., 2018).

Industrial PEF units also require aseptic filling at the end of the processing line, which is expensive and can introduce contamination into the processed food products if it is not properly carried out and maintained (Toepfl et al., 2006).

5.4.3.4 Pulse electric fields as extraction method

Electric fields are also being widely explored as an alternative to the conventional extraction methods, to obtain interesting bioactive compounds, that are, most of the time, degraded by high temperature. The extraction of selective molecules by PEF can be applied by diffusion in solvent (solvent extraction) or by application of pressing procedures (expression) (Vorobiev and Lebovka, 2016). The feasibility of PEF as an extraction technique relies on cell membranes' electroporation that allows the solvents to penetrate the cells and solubilize the interest compounds (according to their polarity) (Saldaña et al., 2017; Yan et al., 2017). The electroporation phenomenon occurs when the cells are exposed to an electric field and the consequence charge accumulation on cell membranes leads to an increase of the transmembrane potential and the consequential formation of pores. The cell membrane permeability can increase drastically and probably result in cell breakdown, which is often a key processing step in food and bioengineering operations. The extent of electroporation is dependent on several parameters, such as treatment time, type of pulse waveform, number of pulses, the plant material components, and electric-field intensity. The duration and number of pulses should be limited to avoid significant temperature elevations that are usually lower than 3°C–5°C (Donsì et al., 2010).

Additionally, PEF can also enhance osmotic dehydration, allowing to obtain higher extraction yields of value-added compounds.

The main advantages of PEF as an emergent extraction methodology compared with the conventional thermal extraction procedures are:

1. the improved extraction yields,
2. enhanced mass transfer rates,

3. short extraction times,
4. milder extraction parameters (lower extraction temperatures and lower volume of solvents),
5. lower degradation of thermolabile compounds (proteins, vitamins, flavors, aromas, etc.),
6. lower environmental impact (reduction of the energy costs), and
7. easier extract purification (reduced grinding) (Barba et al., 2015,a,b).

Several studies report the possibility of using PEF as an improvement for fruit maceration to electroporate the vegetal cells to extract higher amounts of high added-value compounds such as carbohydrates, polyphenols, tannins, carotenoids, etc., using green-solvents such as water and ethanol. For example, Brianceau et al. (2015) reported an increase of phenolic compounds extraction from fermented grape pomace using PEF assisted extraction, obtaining higher yields when compared with control samples. Additionally, the authors demonstrated the selective nature of PEF for extracting anthocyanins (with higher ratios of anthocyanins:flavan-3-ols after PEF assisted extraction), using ethanol as solvent. During PEF extraction the cell network acts as a barrier for the passage of some undesirable compounds, whereas the cell wall remains intact, acting as a "filter" for a selective extraction process, allowing to obtain final extracts with high purity and high extraction yields.

The electric pulse intensity (among other factors, such as temperature, number of pulses, etc.) ultimately determine the efficiency of the extraction process, as observed by Parniakov et al. (2016), who combined aqueous extraction at several pH values and temperatures to PEF and high voltage electric discharges in mango peels. The authors obtained final extracts rich in proteins, carbohydrates, and antioxidant compounds (especially phenolic compounds). The combination of PEF to high voltage electric discharges may be an interesting approach to increase the extraction yields, yet it triggers the production of contaminants (as a result of electrolysis of the extracted compounds, free reactive radicals, etc.), which are inconvenient on value-added extracts (Parniakov et al., 2015).

5.5 Final remarks

Nonthermal emerging processing technologies have a great potential to produce safe food products of high quality. These processing methodologies can minimize the adverse effect of heat verified on conventional methods and there is an increased interest in their application to meet the increasing consumer demand for nutritious foods, in terms of bioactive compound retention and sensorial characteristics and at the same time ensuring food safety standards. These technologies are efficient, rapid, and reliable as an alternative for improving the food quality. Besides, they also have the potential to develop new food products having distinctive functionality. The possible combination of different technologies could be also improve food products quality and, at this time, could be a good starting point for interested researchers to develop their knowledge in this field.

HPP and PEF have already been implemented at industrial scales and some products are even available on the market. However, extensive work remains unchartered to simplify the processes. Thus, an active and strict collaboration between scientists and engineers from different disciplines such as electronic, mechanical, information technology, food technology, among others, is urgently required to reduce the cost of equipment, establish possible processing modes (batch, continuous, semicontinuous), as well as increase efficiency of labor and utilities.

Acknowledgments

This work was supported by National Funds from FCT−Fundação para a Ciência e a Tecnologia through project UID/Multi/50016/2013 and by FCT/MEC by the financial support to the QOPNA research Unit (FCT UID/QUI/00062/2019), through national funds and where applicable cofinanced by the FEDER, within the PT2020 Partnership Agreement. Carlos A. Pinto and Sílvia A. Moreira are grateful for the financial support of this work from "Fundação para a Ciência e Tecnologia—FCT" through the Doctoral Grants SFRH/BD/137036/2018 and SFRH/BD/110430/2015 and Elisabete M.C. Alexandre for the Postdoctoral Grant SFRH/BPD/95795/2013.

References

Abasi, S., Minaei, S., Jamshidi, B., Fathi, D., 2018. Dedicated non-destructive devices for food quality measurement: a review. Trends Food Sci. Technol. 78, 197−205.

Adetunji, L.R., Adekunle, A., Orsat, V., Raghavan, V., 2017. Advances in the pectin production process using novel extraction techniques: a review. Food Hydrocolloids 62, 239−250.

Aghajanzadeh, S., Ziaiifar, A.M., 2018. A review of pectin methylesterase inactivation in citrus juice during pasteurization. Trends Food Sci. Technol. 71, 1−12.

Alexandre, E.M.C., Brandão, T.R.S., Silva, C.L.M., 2012. Emerging technologies to improve the safety and quality of fruits and vegetables. Novel Technologies in Food Science. Springer Science + Business Media, LLC, pp. 261−297.

Alexandre, E.M.C., Brandão, T.R.S., Silva, C.L.M., 2013. Novel thermal and non-thermal food processing technologies. Processed Foods: Quality, Safety Characteristics and Health Implications. Nova Science Publishers, Inc, pp. 1−34.

Alexandre, E.M.C., Brandão, T.R.S., Silva, C.L.M., 2017a. Emerging Sanitation Techniques for Fresh-Cuts. Fresh-Cut Fruits and Vegetables: Technology, Physiology and Safety. CRC Press, Taylor and Francis.

Alexandre, E.M.C., Castro, L.M.G., Moreira, S.A., Pintado, M., Saraiva, J.A., 2017b. Comparison of emerging technologies to extract high-added value compounds from fruit residues: pressure- and electro-based technologies. Food Eng. Rev. 1−23.

Alexandre, E.M.C., Inácio, R.S., Ribeiro, C., Castro, S.M., Teixeira, P., Pintado, M., et al., 2017c. Effect of Commercial Emerging Non-Thermal Technologies on Food Products: Microbiological Aspects. In Food Safety and Protection. CRC Press, Taylor and Francis.

Alexandre, E.M.C., Moreira, S.A., Castro, L.M.G., Pintado, M., Saraiva, J.A., 2017d. Emerging technologies to extract high added value compounds from fruit residues: sub/supercritical, ultrasound-, and enzyme-assisted extractions. Food Rev. Int. 34 (6), 1–32.

Alexandre, E.M.C., Moreira, S.A., Pintado, M., Saraiva, J.A., 2017e. Emergent extraction technologies to valorize fruit and vegetable residues, Agricultural Research Updates, vol. 17. Nova Science Publishers, Inc, pp. 37–79.

Alli, I., 2004. Food Quality Assurance: Principles and Practices. CRC Press LLC, USA.

Amiali, M., Ngadi, M.O., Raghavan, V.G.S., Smith, J.P., 2004. Inactivation of *Escherichia coli* O157:H7 in liquid dialyzed egg using pulsed electric fields. Food Bioprod. Process. 82 (2 C), 151–156.

Anonymous, 2002. Food safety and quality assurance issues. Improving the Safety and Quality of Fresh Fruits and Vegetables: A Training Manual for Trainers. University of Maryland, Food and Drug Administration, pp. V1–V27.

Arino, A., Beltran, J., Herrera, A., Roncales, P., 2005. Fish. Encyclopedia of human nutrition. In: Caballero, B., Allen, L., Prentice, A. (Eds.), The Boulevard, second ed. Elsevier Academic Press Ltd., Langford Lane, Kidlington, Oxford, UK, pp. 247–256.

Azmir, J., Zaidul, I.S.M., Rahman, M.M., Sharif, K.M., Mohamed, A., Sahena, F., et al., 2013. Techniques for extraction of bioactive compounds from plant materials: a review. J. Food Eng. 117 (4), 426–436.

Balasubramaniam, V.M., Farkas, D., 2008. High-pressure food processing. Food Sci. Technol. Int. 14 (5), 413–418.

Balasubramaniam, V.M., Martínez-Monteagudo, S.I., Gupta, R., 2015. Principles and application of high pressure-based technologies in the food industry. Annu. Rev. Food Sci. Technol. 6 (1), 435–462.

Barba, F.J., Esteve, M.J., Frígola, A., 2012. High pressure treatment effect on physicochemical and nutritional properties of fluid foods during storage: a review. Compr. Rev. Food Sci. Food Saf. 11 (3), 307–322.

Barba, F.J., Parniakov, O., Pereira, A., Wiktor, A., Grimi, N., Boussetta, N., et al., 2015a. Current applications and new opportunities for the use of pulsed electric fields in food science and industry. Food Res. Int. 77, 773–798.

Barba, F.J., Terefe, N.S., Buckow, R., Knorr, D., Orlien, V., 2015b. New opportunities and perspectives of high pressure treatment to improve health and safety attributes of foods. a review. Food Res. Int. 77, 725–742.

Barba, F.J., Orlien, V., Mota, M.J., Lopes, R.P., Pereira, S.A., Saraiva, J.A., 2016. In: Galanakis, C.M. (Ed.), Implementation of Emerging Technologies. Innovation Strategies in the Food Industry. Academic Press Inc., London, pp. 117–148.

Bayındırlı, A., Alpas, H., Bozoğlu, F., Hızal, M., 2006. Efficiency of high pressure treatment on inactivation of pathogenic microorganisms and enzymes in apple, orange, apricot and sour cherry juices. Food Control 17 (1), 52–58.

Behling, R.G., Eifert, J., Erickson, M.C., Gurtler, J.B., Kornacki, J.L., Line, E., et al., 2010. Selected pathogens of concern to industrial food processors: infectious, toxigenic, toxico-infectious, selected emerging pathogenic bacteria. Principles of Microbiological Troubleshooting in the Industrial Food Processing Environment. Springer Science + Business Media, LLC, pp. 5–62.

Bi, H.M., Zhang, S.Q., Liu, C.-J., Wang, C.Z., 2009. High hydrostatic pressure extraction of salidroside from *Rhodiola sachalinensis*. J. Food Process Eng. 32 (1), 53–63.

Bobinaitė, R., Pataro, G., Lamanauskas, N., Šatkauskas, S., Viškelis, P., Ferrari, G., 2014. Application of pulsed electric field in the production of juice and extraction of bioactive compounds from blueberry fruits and their by-products. J. Food Sci. Technol. 52 (September), 5898–5905.

Brianceau, S., Turk, M., Vitrac, X., Vorobiev, E., 2015. Combined densification and pulsed electric field treatment for selective polyphenols recovery from fermented grape pomace. Innovative Food Sci. Emerging Technol. 29 (29), 2–8.

Briones-Labarca, V., Plaza-Morales, M., Giovagnoli-Vicuña, C., Jamett, F., 2015. High hydrostatic pressure and ultrasound extractions of antioxidant compounds, sulforaphane and fatty acids from Chilean papaya (*Vasconcellea pubescens*) seeds: effects of extraction conditions and methods. LWT-Food Sci. Technol. 60 (1), 525–534.

Casquete, R., Castro, S.M., Villalobos, M.C., Serradilla, M.J., Queirós, R.P., Saraiva, J.A., et al., 2014. High pressure extraction of phenolic compounds from citrus peels. High Pressure Res. 34 (4), 447–451.

Casquete, R., Castro, S.M., Martín, A., Ruiz-Moyano, S., Saraiva, J.A., Córdoba, M.G., et al., 2015. Evaluation of the effect of high pressure on total phenolic content, antioxidant and antimicrobial activity of citrus peels. Innovative Food Sci. Emerging Technol. 31, 37–44.

Cavaliere, A., Ventura, V., 2018. Mismatch between food sustainability and consumer acceptance toward innovation technologies among Millennial students: the case of shelf life extension. J. Clean. Prod. 175, 641–650.

Chemat, F., Zill-E-Huma, Khan, M.K., 2011. Applications of ultrasound in food technology: processing, preservation and extraction. Ultrason. Sonochem. 18 (4), 813–835.

Chen, R., Meng, F., Zhang, S., Liu, Z., 2009. Effects of ultrahigh pressure extraction conditions on yields and antioxidant activity of ginsenoside from ginseng. Sep. Purif. Technol. 66, 340–346.

Compton, M., Willis, S., Rezaie, B., Humes, K., 2018. Food processing industry energy and water consumption in the Pacific Northwest. Innovative Food Sci. Emerging Technol. 47, 371–383.

Considine, K.M., Kelly, A.L., Fitzgerald, G.F., Hill, C., Sleator, R.D., 2008. High-pressure processing-effects on microbial food safety and food quality. FEMS Microbiol. Lett. 281 (1), 1–9.

Daryaei, H., Balasubramaniam, V.M., 2012. Microbial decontamination of food by high pressure processing. Microbial Decontamination in Food Industry. Woodhead Publishing, Cambridge, pp. 370–406.

Donsì, F., Ferrari, G., Pataro, G., 2010. Applications of pulsed electric field treatments for the enhancement of mass transfer from vegetable tissue. Food Eng. Rev. 2, 109–130.

Ebringerová, A., Hromádková, Z., 2010. An overview on the application of ultrasound in extraction, separation and purification of plant polysaccharides. Central Eur. J. Chem. 8 (2), 243–257.

EPA, 2015. Enviromental Protection Agency, USA. <http://www.epa.gov/greenchemistry/pubs/about_gc.html> (accessed July, 2018.).

Ercan, S.Ş., Soysal, Ç., 2013. Use of ultrasound in food preservation. Nat. Sci. 05 (8), 5–13.

Feng, H., Yang, W., 2011. Ultrasonic processing. Nonthermal Processing Technologies for Food, in Nonthermal Processing Technologies for Food. Blackwell Publishing Ltd, pp. 135–154.

Fernandez, M.V., Denoya, G.I., Agüero, M.V., Jagus, R.J., Vaudagna, S.R., 2018. Optimization of high pressure processing parameters to preserve quality attributes of a mixed fruit and vegetable smoothie. Innovative Food Sci. Emerging Technol. 47, 170–179.

Fincan, M., Dejmek, P., 2002. In situ visualization of the effect of a pulsed electric field on plant tissue. J. Food Eng. 55 (3), 223–230.

Floury, J., Grosset, N., Leconte, N., Pasco, M., Madec, M.N., Jeantet, R., 2006. Continuous raw skim milk processing by pulsed electric field at nonlethal temperature: effect on microbial inactivation and functional properties. Lait 86 (1), 43–57.

Fung, F., Wang, H.S., Menon, S., 2018. Food safety in the 21st century. Bull. W. H. O. 77, 347–351.

Galanakis, C.M., 2013. Emerging technologies for the production of nutraceuticals from agricultural by-products: a viewpoint of opportunities and challenges. Food Bioprod. Process. 91 (4), 575–579.

Gómez-López, V.M., Orsolani, L., Martínez-Yépez, A., Tapia, M.S., 2010. Microbiological and sensory quality of sonicated calcium-added orange juice. LWT – Food Sci. Technol. 43 (5), 808–813.

Gong, Y., Yu, J.Y., Qian, P., Meng, J., Zhang, X.J., Lu, R.R., 2015. Comparative study of the microbial stability and quality of carrot juice treated by high-pressure processing combined with mild temperature and conventional heat treatment. J. Food Process Eng. 38 (4), 395–404.

Guan, D., Hoover, D.G., 2005. Emerging Decontamination Techniques for Meat. Improving the Safety of Fresh Meat. In Improving the Safety of Fresh Meat. Woodhead Publishing Limited, pp. 388–417.

Guo, X., Han, D., Xi, H., Rao, L., Liao, X., Hu, X., et al., 2012. Extraction of pectin from navel orange peel assisted by ultra-high pressure, microwave or traditional heating: a comparison. Carbohydr. Polym. 88 (2), 441–448.

Guo, X., Zhao, W., Pang, X., Liao, X., Hu, X., Wu, J., 2014. Emulsion stabilizing properties of pectins extracted by high hydrostatic pressure, high-speed shearing homogenization and traditional thermal methods: a comparative study. Food Hydrocolloid 35, 217–225.

He, X., Kim, S.S., Park, S.J., Seong, D.H., Yoon, W.B., Lee, H.Y., et al., 2010. Combined effects of probiotic fermentation and high-pressure extraction on the antioxidant, antimicrobial, and antimutagenic activities of deodeok (*Codonopsis lanceolata*). J. Agric. Food Chem. 58 (3), 1719–1725.

He, X., Yoon, W.B., Park, S.J., Park, D.S., Ahn, J., 2011. Effects of pressure level and processing time on the extraction of total phenols, flavonoids, and phenolic acids from Deodeok (*Codonopsis lanceolata*). Food Sci. Biotechnol. 20 (2), 499–505.

Huang, H.W., Hsu, C.P., Yang, B.B., Wang, C.Y., 2013. Advances in the extraction of natural ingredients by high pressure extraction technology. Trends Food Sci. Technol. 33 (1), 54–62.

Huang, H.W., Lung, H.M., Yang, B.B., Wang, C.Y., 2014. Responses of microorganisms to high hydrostatic pressure processing. Food Control 40, 250–259.

Huang, H.W., Wu, S.J., Lu, J.K., Shyu, Y.T., Wang, C.Y., 2017. Current status and future trends of high-pressure processing in food industry. Food Control 72, 1–8.

Hygreeva, D., Pandey, M.C., 2016. Novel approaches in improving the quality and safety aspects of processed meat products through high pressure processing technology – a review. Trends Food Sci. Technol. 54, 175–185.

Jofré, A., Aymerich, T., Garriga, M., 2009. Improvement of the food safety of low acid fermented sausages by enterocins A and B and high pressure. Food Control 20 (2), 179–184.

Jun, X., 2006. Application of high hydrostatic pressure processing of food to extracting lycopene from tomato paste waste. High Pressure Res. 26 (1), 33–41.

Jun, X., 2013. High-pressure processing as emergent technology for the extraction of bioactive ingredients from plant materials. Crit. Rev. Food Sci. Nutr. 53 (8), 837–852.

Jun, X., Deji, S., Ye, L., Rui, Z., 2011. Micromechanism of ultrahigh pressure extraction of active ingredients from green tea leaves. Food Control 22 (8), 1473–1476.

Knoerzer, K., 2016. Nonthermal and innovative food processing technologies. Reference Module in Food Science. Elsevier.

Koubaa, M., Barba, F.J., Bursać Kovačević, D., Putnik, P., Santos, M.D., Queirós, R.P., et al., 2017. Pulsed electric field processing of fruit juices. In Fruit Juices: Extraction, Composition, Quality and Analysis. Elsevier, pp. 437–449.

Koubaa, M., Roselló-Soto, E., Šic Žlabur, J., Režek Jambrak, A., Brnčić, M., Grimi, N., et al., 2015. Current and new insights in the sustainable and green recovery of nutritionally valuable compounds from *Stevia rebaudiana* Bertoni. J. Agric. Food Chem. 63 (31), 6835–6846.

Lee, H.S., Lee, H.J., Yu, H.J., Ju do, W., Kim, Y., Kim, et al., 2011. A comparison between high hydrostatic pressure extraction and heat extraction of ginsenosides from ginseng (*Panax ginseng* CA Meyer). J. Sci. Food Agric. 91 (8), 1466–1473.

Lee, H.Y., He, X., Ahn, J., 2010. Enhancement of antimicrobial and antimutagenic activities of Korean barberry (*Berberis koreana* Palib.) by the combined process of high-pressure extraction with probiotic fermentation. J. Sci. Food Agric. 90 (14), 2399–2404.

Leong, T., 2016. Reference Module in Food Science. Elsevier.

Li, X., Farid, M., 2016. A review on recent development in non-conventional food sterilization technologies. J. Food Eng. 182, 33–45.

Li-Chan, E.C.Y., Powrie, W.D., Nakai, S., 1995. The chemistry of eggs and egg products. Egg Sci. Technol. 4, 105–177.

Liu, C., Zhang, S., Wu, H., 2009. Non-thermal extraction of effective ingredients from *Schisandra chinensis* Baill and the antioxidant activity of its extract. Nat. Prod. Res. 23 (15), 1390–1401.

Liu, G., 2018. The impact of supply chain relationship on food quality. Procedia Comput. Sci. 131, 860–865.

Loeffler, M.J., 2010. Generation and Application of High Intensity Pulsed Electric Fields in Pulsed Electric Fields Technology for the Food Industry: Fundamentals and Applications. Springer, pp. 27–71.

Lou, F., Neetoo, H., Chen, H., Li, J., 2015. High hydrostatic pressure processing: a promising nonthermal technology to inactivate viruses in high-risk foods. Annu. Rev. Food Sci. Technol. 6, 389–409.

M'hiri, N., Ioannou, I., Ghoul, M., Boudhrioua, N.M., 2014. Extraction methods of citrus peel phenolic compounds. Food Rev. Int. 30 (4), 265–290.

McCarthy, B., Liu, H.B., 2017. Food waste and the 'green' consumer. Aust. Mark. J. 25, 126–132.

Misra, N.N., Koubaa, M., Roohinejad, S., Juliano, P., Alpas, H., Inácio, R., et al., 2017. Landmarks in the historical development of twenty first century food processing technologies. Food Res. Int. 97, 318–339.

Mitchell, R., Mitchell, B., 2014. Public health measures: management of food safety in food service sector, Encyclopedia of Food Safety, vol. 1. Elsevier, Inc, pp. 133–139.

Monfort, S., Gayán, E., Saldaña, G., Puértolas, E., Condón, S., Raso, J., et al., 2010. Inactivation of *Salmonella typhimurium* and *Staphylococcus aureus* by pulsed electric fields in liquid whole egg. Innovative Food Sci. Emerging Technol. 11 (2), 306–313.

Muntean, M.V., Marian, O., Barbieru, V., Cătunescu, G.M., Ranta, O., Drocas, I., et al., 2016. High pressure processing in food industry – characteristics and applications. Agric. Agric. Sci. Procedia 10, 377–383.

Neetoo, H., Chen, H., 2012. Application of high hydrostatic pressure technology for processing and preservation of foods. Progress in Food Preservation. Wiley-Blackwell, pp. 247–276.

Nerín, C., Aznar, M., Carrizo, D., 2016. Food contamination during food process. Trends Food Sci. Technol. 48, 63–68.

Oey, I., Lille, M., Van Loey, A., Hendrickx, M., 2008. Effect of high-pressure processing on colour, texture and flavour of fruit- and vegetable-based food products: a review. Trends Food Sci. Technol. 19 (6), 320–328.

Ohlsson, T., 2002. Minimal processing of foods with thermal methods. Minimal Processing Technologies in the Food Industry. Woodhead Publishing Limited, pp. 4–33.

Oliveira, F.A.D., Neto, O.C., Santos, L.M.R.D., Ferreira, E.H.R., Rosenthal, A., 2017. Effect of high pressure on fish meat quality – a review. Trends Food Sci. Technol. 66, 1–19.

Parniakov, O., Roselló-Soto, E., Barba, F.J., Grimi, N., Lebovka, N., Vorobiev, E., 2015. New approaches for the effective valorization of papaya seeds: extraction of proteins, phenolic compounds, carbohydrates, and isothiocyanates assisted by pulsed electric energy. Food Res. Int. 77, 711–717.

Parniakov, O., Barba, F.J., Grimi, N., Lebovka, N., Vorobiev, E., 2016. Extraction assisted by pulsed electric energy as a potential tool for green and sustainable recovery of nutritionally valuable compounds from mango peels. Food Chem. 192, 842–848.

Pereira, R., Vicente, A., 2010. Environmental impact of novel thermal and non-thermal technologies in food processing. Food Res. Int. 43 (7), 1936–1943.

Perez, O.E., Pilosof, A.M.R., 2004. Pulsed electric fields effects on the molecular structure and gelation of β-lactoglobulin concentrate and egg white. Food Res. Int. 37 (1), 102–110.

Piyasena, P., Mohareb, E., McKellar, R., 2003. Inactivation of microbes using ultrasound: a review. Int. J. Food Microbiol. 87 (3), 207–216.

Porto-Fett, A.C.S., Call, J.E., Shoyer, B.E., Hill, D.E., Pshebniski, C., Cocoma, G.J., et al., 2010. Evaluation of fermentation, drying, and/or high pressure processing on viability of *Listeria monocytogenes*, *Escherichia coli* O157:H7, *Salmonella* spp., and *Trichinella spiralis* in raw pork and Genoa salami. Int. J. Food Microbiol. 140 (1), 61–75.

Prakash, A., 2013. Non-thermal processing technologies to improve the safety of nuts. In: Harris, L.J. (Ed.), Improving the Safety and Quality of Nuts. Woodhead Publishing Series in Food Science, Technology and Nutrition, pp. 37–40.

Prasad, N.K., Yang, B., Zhao, M., Wang, B.S., Chen, F., Jiang, Y., 2009. Effects of high-pressure treatment on the extraction yield, phenolic content and antioxidant activity of litchi (*Litchi chinensis* Sonn.) fruit pericarp. Int. J. Food Sci. Technol. 44, 960–966.

Puértolas, E., López, N., Condón, S., Raso, J., Álvarez, I., 2009. Pulsed electric fields inactivation of wine spoilage yeast and bacteria. Int. J. Food Microbiol. 130 (1), 49–55.

Qadir, S.A., Kwon, M.C., Han, J.G., Ha, J.H., Chung, H.S., Ahn, J., et al., 2009. Effect of different extraction protocols on anticancer and antioxidant activities of *Berberis koreana* bark extracts. J. Biosci. Bioeng. 107 (3), 331–338.

Quested, T., Johnson, H., 2009. Household Food and Drink Waste in the UK, October. <http://www.wrap.org.uk/sites/files/wrap/Household_food_and_drink_waste_in_the_UK_-_report.pdf> (accessed July 2018.).

Raymond, M., Gamage, M., Terefe, N.S., Knoerzer, K., 2011. Ultrasound in enzyme activation and inactivation. Ultrasound Technologies for Food and Bioprocessing, Food Engineering Series. Springer, pp. 369–404.

Rendueles, E., Omer, M.K., Alvseike, O., Alonso-Calleja, C., Capita, R., Prieto, M., 2011. Microbiological food safety assessment of high hydrostatic pressure processing: a review. LWT – Food Sci. Technol. 44 (5), 1251–1260.

Rico, D., Martín-Diana, A.B., Barat, J.M., Barry-Ryan, C., 2007. Extending and measuring the quality of fresh-cut fruit and vegetables: a review. Trends Food Sci. Technol. 18 (7), 373–386.

Rodríguez, E., Arqués, J.L., Nuñez, M., Gaya, P., Medina, M., 2005. Combined effect of high-pressure treatments and bacteriocin-producing lactic acid bacteria on inactivation of *Escherichia coli* O157: H7 in raw-milk cheese. Appl. Environ. Microbiol. 71 (7), 3399–3404.

Ruiz-Montanez, G., Ragazzo-Sanchez, J.A., Calderon-Santoyo, M., Velazquez-de la Cruz, G., Ramirez de Leon, J.A., Navarro-Ocana, A., 2014. Evaluation of extraction methods for preparative scale obtention of mangiferin and lupeol from mango peels (*Mangifera indica* L.). Food Chem. 159, 267–272.

Saini, R.K., Moon, S.H., Keum, Y.S., 2018. An updated review on use of tomato pomace and crustacean processing waste to recover commercially vital carotenoids. Food Res. Int. 108, 516–529.

Saldaña, G., Cebrián, G., Abenoza, M., Sánchez-Gimeno, C., Álvarez, I., Raso, J., 2017. Assessing the efficacy of PEF treatments for improving polyphenol extraction during red wine vinifications. Innovative Food Sci. Emerging Technol. 39, 179–187.

Sánchez-Vega, R., Elez-Martínez, P., Martín-Belloso, O., 2015. Influence of high-intensity pulsed electric field processing parameters on antioxidant compounds of broccoli juice. Innovative Food Sci. Emerging Technol. 29, 70–77.

Shahbaz, H.M., Kim, J.U., Kim, S.-H., Park, J., 2018. Advances in nonthermal processing technologies for enhanced microbiological safety and quality of fresh fruit and juice products. Food Processing for Increased Quality and Consumption. Elsevier, pp. 179–217.

Sharma, P., Bremer, P., Oey, I., Everett, D.W., 2014. Bacterial inactivation in whole milk using pulsed electric field processing. Int. Dairy J. 35 (1), 49–56.

Shin, J.S., Ahn, S.C., Choi, S.W., Lee, D.U., Kim, B.Y., Baik, M.Y., 2010. Ultra-high pressure extraction (UHPE) of ginsenosides from Korean Panax ginseng powder. Food Sci. Biotechnol. 19 (3), 743–748.

Shouquin, Z., Ruizhan, C., Hua, W., Changzheng, W., 2006. Ginsenoside extraction from *Panax quinquefolium* L. (American ginseng) root by using ultrahigh pressure. J. Pharm. Biomed. Anal. 41 (1), 57–63.

Shouqin, Z., Ruizhan, C., Changzheng, W., 2007. Experiment study on ultrahigh pressure extraction of ginsenosides. J. Food Eng. 79 (1), 1–5.

Sitzmann, W., Vorobiev, E., Lebovka, N., 2016. Applications of electricity and specifically pulsed electric fields in food processing: historical backgrounds. Innovative Food Sci. Emerging Technol. 37, 302–311.

Strati, I.F., Gogou, E., Oreopoulou, V., 2015. Enzyme and high pressure assisted extraction of carotenoids from tomato waste. Food Bioprod. Process. 94, 668–674.

Tadapaneni, R.K., Daryaei, H., Krishnamurthy, K., Edirisinghe, I., Burton-Freeman, B.M., 2014. High-pressure processing of berry and other fruit products: implications for bioactive compounds and food safety. J. Agric. Food Chem. 62 (18), 3877–3885.

Timmermans, R.A.H., Nierop Groot, M.N., Nederhoff, A.L., van Boekel, M.A.J.S., Matser, A.M., Mastwijk, H.C., 2014. Pulsed electric field processing of different fruit juices: impact of pH and temperature on inactivation of spoilage and pathogenic microorganisms. Int. J. Food Microbiol. 173, 105–111.

Tiwari, B.K., 2015. Trends in analytical chemistry ultrasound : a clean, green extraction technology. Trends Anal. Chem. 71, 100–109.

Toepfl, S., Heinz, V., Knorr, D., 2006. Application of pulsed electric fields in liquid processing. In: Paper Presented at the EFFOST Conference Processing Developments for Liquids, EFFoST, Kolonia, 2006.

Toepfl, S., Heinz, V., Knorr, D., 2007. High intensity pulsed electric fields applied for food preservation. Chem. Eng. Process.: Process Intensif. 46 (6), 537–546.

Tonini, D., Albizzati, P.F., Astrup, T.F., 2018. Environmental impacts of food waste: learnings and challenges from a case study on UK. Waste Manage. 76, 744–766.

Van Opstal, I., Vanmuysen, S.C.M., Wuytack, E.Y., Masschalck, B., Michiels, C.W., 2005. Inactivation of *Escherichia coli* by high hydrostatic pressure at different temperatures in buffer and carrot juice. Int. J. Food Microbiol. 98 (2), 179–191.

Vega-mercado, H., Gongora-nieto, M.M., Barbosa-Cánovas, G.V., Swanson, B.G., 2007. Pulsed electric fields in food preservation, Handbook of Food Preservation, second ed. CRC Press, Taylor and Francis, pp. 783–813.

Venugopal, V., 2006. High Pressure Processing. Seafood Processing - Adding Value Through Quick Freezing, Retortable Packaging, and Cook-Chilling. Taylor and Francis Group, LLC, pp. 319–340.

Visiongain, 2017. Food High Pressure Processing (HPP) Technologies Market 2017–2027: Top Companies Providing Pascalization, Bridgmanization Equipment and Tolling Services for Meat and Poultry, Fruit and Vegetable, Seafood and Fish, Juices and Beverages, Dairy, Sauces and Dips. <https://www.visiongain.com/Report/1880/Food-High-Pressure-Processing-(HPP)-Technologies-Market-2017-2027> (accessed 11.11.17.).

Vorobiev, E., Lebovka, N., 2008. Electrotechnologies for Extraction From Food Plants and Biomaterials. Springer.

Vorobiev, E., Lebovka, N., 2016. Selective extraction of molecules from biomaterials by pulsed electric field treatment, Handbook of Electroporation, vol. 8. Springer International Publishing, Cham, pp. 1–16.

Waite, J.G., Yousef, A.E., 2010. Overview of food safety. Processing Effects on Safety and Quality of Foods. CRC Press, Taylor and Francis Group, pp. 11–66.

Welti-Chanes, J., López-Malo, A., Palou, E., Bermúdez, D., Guerrero-Beltrán, J.A., Barbosa-Cánovas, G.V., 2005. Fundamentals and applications of high pressure processing to foods. Novel Food Processing Technologies. Marcel Dekker, pp. 157-152.

Xi, J., 2006. Effect of high pressure processing on the extraction of lycopene in tomato paste waste. Chem. Eng. Technol. 29 (6), 736–739.

Xi, J., 2009. Caffeine extraction from green tea leaves assisted by high pressure processing. J. Food Eng. 94 (1), 105–109.

Xi, J., Shen, D., Zhao, S., Lu, B., Li, Y., Zhang, R., 2009. Characterization of polyphenols from green tea leaves using a high hydrostatic pressure extraction. Int. J. Pharm. 382 (1-2), 139–143.

Xi, J., Zhao, S., Lu, B., Zhang, R., Li, Y., Shen, D., et al., 2010. Separation of major catechins from green tea by ultrahigh pressure extraction. Int. J. Pharm. 386 (1–2), 229–231.

Yan, L.G., He, L., Xi, J., 2017. High intensity pulsed electric field as an innovative technique for extraction of bioactive compounds - a review. Crit. Rev. Food Sci. Nutr. 57 (13), 2877–2888.

Yu, L.J., Ngadi, M., Raghavan, G.S.V., 2009. Effect of temperature and pulsed electric field treatment on rennet coagulation properties of milk. J. Food Eng. 95 (1), 115–118.

Yuan, Y., Hu, Y., Yue, T., Chen, T., Lo, Y.M., 2009. Effect of ultrasonic treatments on thermoacidophilic *Alicyclobacillus acidoterrestris* in apple juice. J. Food Process. Preserv. 33 (3), 370–383.

Zhang, S.Q., Bi, H., Liu, C., 2007. Extraction of bio-active components from *Rhodiola sachalinensis* under ultrahigh hydrostatic pressure. Sep. Purif. Technol. 57 (2), 277–282.

Zhu, Q., Liu, F., Xu, M., Lin, X., Wang, X., 2012. Ultrahigh pressure extraction of lignan compounds from *Dysosma versipellis* and purification by high-speed counter-current chromatography. J. Chromatogr. B 905, 145–149.

Innovative packaging that saves food

6

Vilásia Guimarães Martins, Viviane Patrícia Romani, Paola Chaves Martins and Gabriel da Silva Filipini
Federal University of Rio Grande, School of Chemistry and Food, Rio Grande, Brazil

Chapter Outline

6.1 Introduction

About 1.3 billion tons of food are wasted and lost in the world every year; this statistic represents approximately one-third of the food produced for consumption (FAO, 2015). Most food losses are related to microbiological growth and oxidative processes that affect food during the many stages of production (Prakash et al., 2015). Thus, several technologies have been studied to reduce food losses, among them the development of packaging capable of reducing or delaying deterioration

Saving Food. DOI: https://doi.org/10.1016/B978-0-12-815357-4.00006-7

processes (Otoni et al., 2016). To conserve foods while reducing losses, packaging has significant functions due to its ability to keep the product safe from external damage. If no packaging was used, food waste would be higher than it actually is. Despite the basic functions of packaging regarding food protection, it might be used as an active keeper of product quality (Robertson, 2013). This means an increase in shelf life through prevention of antimicrobial growth and oxidative processes, which are the main causes of food degradation (López-De-Dicastillo et al., 2012). The use of new technologies in the development of packaging is an essential strategy to decrease the problem of food waste.

Among the new trends in designing materials for food packaging, the development of sustainable materials from agriculture suggests better use of these resources while decreasing environmental pollution (Benbettaïeb et al., 2016a). Intelligent systems capable of providing information about the events inside and/or outside the packaging environment (Majid et al., 2016), and active materials that have a positive interaction with the food product, reducing and slowing down its degradation, are new types of packaging that are being developed, mainly with natural compounds (Ahmed et al., 2017).

The latter have been studied as additives in packaging formulations due to their ability to confer functional properties on the packaging, making it active (Dainelli et al., 2008). Active packaging (AP) can be characterized as a system that interacts between food and packaging either by direct contact or by migration of compounds to the headspace (Adilah et al., 2018). The use of AP aims to increase shelf life, maintain product quality (Majid et al., 2016), reduce food losses, and reduce foodborne diseases (Krepker et al., 2017). Although it has many benefits, active and biodegradable packaging still needs to be improved, and the use of natural additives can contribute to improvement and/or maintenance of packaging properties. In addition, other strategies such as polymer modification technologies (Majeed et al., 2017), blend formulation (Sun and Xiong, 2014), plasma application (Oh et al., 2016), and others can be applied to improve packaging.

This chapter is focused on existing and innovative packaging solutions and barriers to minimizing food waste. Thus, AP will be further discussed because of its role in extending the shelf life of food, minimizing its degradation and, as a consequence, reducing food waste. Among AP technologies, special attention is given to films because of the growing number of studies and interest in this field.

6.2 Innovations in food packaging

6.2.1 Sustainable packaging

Packaging performs important tasks to maintain the quality of food products, from processing and manufacturing, through handling and storage, until transportation to the final consumer. The four primary functions of packaging are containment, protection, convenience, and communication. All these functions must be considered in the package development process. Without packaging materials, handling of food would be messy and inefficient as well as costly, and marketing would not be possible. In developed countries, the packaging sector represents around 2% of the gross national product, and half of it is used for food packaging (Han, 2005; Robertson, 2013). To maintain the quality and safety of food and extend its shelf life, packaging must prevent unfavorable conditions such as the presence of spoilage microorganisms, chemical contaminants, oxygen, moisture, light, and external force. Thus, the packaging material is required to offer physical protection and proper physicochemical conditions for products, for example, hindering gain or loss of moisture, preventing microbial contamination and acting as a barrier against permeation of water vapor, oxygen, carbon dioxide, and other volatile compounds, in addition to the basic properties of packaging (Brown et al., 2011; Rhim et al., 2013; Singh and Singh, 2005).

Petroleum-based polymers are the materials most used for food packaging due to their low cost, convenience, processability, and excellent physicochemical properties. Packaging represents more than 40% of plastics, and half of them are for food packaging, including films, sheets, bottles, cups, trays, and others. Polyethylene (PE), polypropylene, and polyethylene terephthalate are polymers widely used in the packaging sector, but polyvinyl chloride and polystyrene (PS) are the materials most used for food packaging because of their excellent mechanical and water barrier properties. The large-scale use of plastics has led to environmental concerns due to their nonbiodegradable nature, besides the nonrenewable nature of petroleum. Thus, seeking for renewable and biodegradable raw materials as substitutes for conventional plastics has attracted attention (Etxabide et al., 2016; Ferreira et al., 2016; Rhim et al., 2013).

Innovative packaging solutions have been made using renewable biological resources, generally called biopolymers. These biomacromolecules can be obtained from the metabolism of microorganisms (e.g., polyhydroxy butyrate), chemical synthesis of bioderived monomers (e.g., polylactate), plant biomass (e.g., starch and cellulose), and byproducts from food industries (e.g., gelatin, whey protein and chitosan) (Benbettaïeb et al., 2016a; Rhim et al., 2013).

Different polymers obtained from agricultural origin are available for the development of biodegradable materials. Polysaccharides, proteins, and lipids are raw materials used to prepare agropolymers through different techniques:

- they can be extracted and purified for use;
- they can be used as fermentation substrates to produce microbial polymers;
- and they can be used as fermentation substrates in the production of monomers or oligomers that will be polymerized by conventional chemical processes (Guilbert and Gontard, 2005).

The use of extracted and purified polysaccharides and proteins as raw materials for the development of film packaging has been investigated extensively. Proteins are polymers composed of peptide bonds and are widely used to produce films. In the process of forming films, it is necessary to denature proteins to form more extended structures that are essential for film formation. After denaturation, protein chains tend to associate, resulting in stronger films, but they will be less flexible and permeable to gases, vapors, and liquids (Wittaya, 2012). Many protein matrices are used for the preparation of food packaging, such as sunflower seed meal (Song et al., 2013), triticale (Aguirre et al., 2013), soy (González and Igarzabal, 2013; Ortiz et al., 2013; Zhao et al., 2016), fish protein isolate (Arfat et al., 2014; Romani et al., 2018a), skate skin gelatin (Lee et al., 2016a), whey (Akcan et al., 2017; Beristain-Bauza et al., 2017; Boyaci et al., 2016), fish gelatin (Adilah et al., 2018; Dazaa et al., 2016; Shakila et al., 2012), gelatin (Bodini et al., 2013), sodium and calcium caseinates (Arrieta et al., 2013), chitosan (Caro et al., 2016; Rubilar et al., 2013; Siripatrawan and Noipha, 2012; Siripatrawan and Vitchayakitti, 2016), quinoa (Caro et al., 2016), gliadin (Balaguer et al., 2013), shrimp muscle proteins (Gómez-Estaca et al., 2014), egg-white protein (Kavas and Kavas, 2016), fish skin gelatin (Tongnuanchan et al., 2013), sunflower protein concentrate (Salgado et al., 2013), red pepper seed meal protein (Lee et al., 2016b), and zein (Ünalan et al., 2013), among others.

Carbohydrates are characterized as macromolecules distributed in nature. Several polysaccharides are attractive for use in film-forming processes due to their biodegradability, wide availability, and ability to form thin films with good mechanical and barrier properties against gas, oil, and lipids. In the literature, different studies can be found using carbohydrate rich matrices as rice starch (Borges et al., 2015; Romani, et al., 2017; Woggum et al., 2015), quinoa starch (Pagno et al., 2015), carrageenan (Shojaee-Aliabadi et al., 2013; Soni et al., 2016), cassava starch (Chiumarelli and Hubinger, 2014; Muller et al., 2017; Piñeros-Hernandez et al., 2017; Reis et al., 2015; Rodrigues et al., 2014; Souza et al., 2013), tamarind starch (Meenatchisundaram et al., 2016), gum cordia (Haq et al., 2016), corn starch (Ghasemlou et al., 2013; Meira et al., 2016; Moreno et al., 2014), soluble soybean polysaccharide (Salarbashi et al., 2013), potato starch (Liu et al., 2017; Nisa et al., 2015), tara gum (Antoniou et al., 2014), pea starch (Saberi et al., 2017), methylcellulose (Yu et al., 2014), wheat starch (Bonilla et al., 2013), quince seed mucilage (Jouki et al., 2014), agar (Arancibia et al., 2014; Arfat et al., 2017), and banana powder (Orsuwan et al., 2016), among others.

Biodegradable polymers extracted from food processing waste have also been investigated since this process may add value to these byproducts. An example of a polymer recovered from byproducts is whey protein, the liquid fraction obtained during the processing of cheese. Many authors have used this product for the formulation of films for food packaging (Akcan et al., 2017; Azevedo et al., 2017; Beristain-Bauza et al., 2017; Boyaci et al., 2016; Kokoszka et al., 2010; Schmid et al., 2018; Zinoviadou et al., 2009).

In the literature, there are many studies about the production of films from biodegradable polymers, as mentioned previously. However, more research is necessary about materials and combinations of materials and additives capable of replacing the conventional polymers used in the manufacture of packaging. The use of biodegradable polymers is still restricted because the polymers form films with some poor properties, such as mechanical properties, barrier properties, solubility, and water resistance. The addition of active compounds to films may be a suitable alternative to this problem. In addition, the active compounds can be used as agents with the ability to provide different bioactivities to the films, extending the shelf life of food products; also, it is possible to use other technologies such as chemical, physical, and enzymatic modifications to improve the properties of the films.

6.2.2 Intelligent packaging

Intelligent packaging (IP) is a system responsible for communication of events in the package environment. It has the capacity to detect, sense, record, trace, communicate, and apply science logic, to facilitate consumer decisions, providing information and warning about problems that could happen (Poças et al., 2008). This type of technology is not designed to extend the shelf life of food products, as AP is, but to monitor the conditions of the packaged food. The European Commission (2004) defines IP as "materials and articles that monitor the condition of packaged food or the environment surrounding the food." According to Ghaani et al. (2016), IP aims to provide information for manufacturers, retailers, and consumers about food quality. These systems might provide information about the freshness of the product and if the shelf life has expired, because sometimes foods are not properly stored and distributed, which compromises their shelf life even before the expiration date. As well as information about events inside the packaging, IP may monitor events outside the packaging environment (Majid et al., 2016). Thus, its functionality is also important to measure effectiveness of the cold chain, for example. In this way, as mentioned by Dainelli et al. (2008), this technology provides detailed knowledge throughout the supply chain by identifying critical points by the use of labels which can be attached to, incorporated into, or printed on the packaging material.

To obtain intelligent functionality, indicators, sensors, and devices, as well as their combination, can be used to communicate information about the product. Indicators provide information through visual changes about events such as temperature and pH variations in a food product and/or its environment, while biosensors have the capacity to detect, record, and transmit precise information regarding biological reactions inside the packaging (Realini and Marcos, 2014). Among IP technologies, some examples are: time/temperature indicators, gas indicators, pH indicators, biosensors, and radiofrequency tags (Poças et al., 2008).

In addition to the mentioned functions of IP, it can be used to check the effectiveness of AP (Kerry et al., 2006). When working synergistically, intelligent and AP yield the definition of smart packaging, which is a concept that combines the benefits of both technologies (Vanderroost et al., 2014).

Other systems of IP, such as self-heating and self-cooling systems used for temperature control, are important. Self-heating of packaging occurs as a result of exothermic reactions using substances such as calcium or magnesium oxide or water, while self-cooling packaging induces an evaporative cooling effect by evaporation of external compounds such as water removing heat and being absorbed on the surface (Brody et al., 2008).

6.2.3 Active packaging

AP is designed as a system in which the packaging, the product, and the environment within the packaging have a positive interaction, increasing shelf life and/or ensuring microbial safety while maintaining the quality of the food (Ahmed et al., 2017; Fang et al., 2017). The term "active packaging" in the United States describes packaging that prevents contamination or degradation of food through the barrier to the external environment and interaction with the atmosphere within the package (Ettinger, 2002). According to the European Union Guidance to the Commission Regulation No. 450/2009 (EU, 2009), AP is a type of packaging that presents extra functions in addition to the traditional protection barrier against the external environment. This type of packaging might absorb chemicals released from the food and/or release compounds (e.g., preservatives) to the food and the surrounding environment.

The capacity of AP to extend the shelf life of food products occurs through the regulation of various aspects responsible for food degradation, such as physiological (e.g., respiration of fresh fruit and vegetables), chemical (e.g., lipid oxidation) and physical processes (e.g., dehydration), and microbiological aspects. Several types of additive may be used to scavenge or absorb oxygen, carbon dioxide, ethylene, moisture, and/or odor and flavors, to release/emit oxygen, carbon dioxide, moisture and preservatives, and to control temperature (Hosseinnejad, 2014).

Among AP technologies, materials that release active substances to preserve food are particularly important. Many forms of this special type of packaging involve the use of films of polymeric materials that act as carriers for different active compounds (Kuorwel et al., 2015). One issue in food preservation is the use of synthetic additives, which are associated with different adverse effects on human health. Thus, natural substances such as plant extracts and essential oils (EOs) are important due to their benefits for food preservation as well as in human health, in addition to being natural products (Ribeiro-Santos et al., 2017). Besides that, enzymes such as lysozyme and lactoferrin from animal sources, bacteriocins from microbial sources, and biopolymers such as chitosan also demonstrate properties important for food preservation (Aloui and Khwaldia, 2016).

In this context of active substances, other important raw materials that have valuable components such as proteins and phenolic compounds in their composition are waste and byproducts from the food industry. The reuse of these raw materials could lessen environmental pollution while conferring functional properties for packaging due to their film-forming ability and active properties, such as antioxidant and antimicrobial activity. In summary, AP plays a vital role in food safety and security, guaranteeing food with adequate conditions for consumption and minimizing food waste.

6.3 Application of active packaging in foods

Use of AP is one of the most dynamic techniques related to food preservation. Positive interaction between the packaging, product, and environment promotes several advantages such as extension of the shelf life and an increase in food safety (Fang et al., 2017; Salgado et al., 2015). For these interactions to occur, some factors, such as the kind of additive used, polymer characteristics, process conditions, and the type of food to be packed, must be observed (Majid et al., 2016).

Different compounds may be used as AP additives, including compounds such as enzymes, bacteriocin compounds, organic acids, protein hydrolysates, polyphenols, natural extracts, EOs, and others. There is a high demand for natural sources for use as active compounds due to the potential health risks posed by synthetic compounds. Many of these natural products can be obtained directly from the original products or from industrial waste derived from the processing of vegetables, fruits, wines, beers, and meats (Salgado et al., 2015). Consequently, this promotes the reuse of waste, reducing production costs and providing natural compounds with several industrial applications.

Natural compounds may have many functional properties as additives to biodegradable packaging. Depending on their characteristics and appearance, they can confer bioactive properties such as antioxidant (Bitencourt et al., 2014), antimicrobial (Kwon et al., 2017), antifungal (Aloui and Khwaldia, 2016), and antibrowning activity (Romani et al., 2018b). The antioxidant and antimicrobial activity attributed to these additives is, in general, related to the presence of chemical compounds characteristic of each group of additives. Natural compounds such as EOs and natural extracts, for example, present large amounts of phenolic compounds that are responsible for their bioactivity (Ganiari et al., 2017).

The antimicrobial activity of natural compounds is due to several factors, as mentioned above, but it is believed that the principal ones responsible for antimicrobial activity are phenolic compounds. These compounds have volatile characteristics; during application, they volatilize to the headspace of the packaging and produce an antimicrobial effect. The compounds interact with the cell membrane of the microorganism, causing structural changes and destabilization of the membrane (Muriel-Galet et al., 2015). In addition to affecting antimicrobial processes, phenolic compounds can act as antioxidants, inhibiting oxidative processes. The presence of phenolic compounds in the composition of natural additives is the major factor responsible for antioxidant performance since the phenolic compounds present hydrogen atoms to be donated during an oxidation process, thus preventing lipid oxidation (Ganiari et al., 2017). However, for their bioactivity to act effectively, some characteristics must be observed when the active compounds are applied, such as food composition (protein, lipid, moisture, etc.), the presence of a vacuum, the amount of oxygen and carbon dioxide, pH, salt concentration, and storage temperature, among others (Burt, 2004).

6.3.1 Essential oils

EOs are used as additives in active food packaging due to their characteristics of being safe for human health and having antioxidant and antimicrobial properties (Kavoosi et al., 2014). Most of the EOs used for the formulation of AP are recognized as Generally Recognized as Safe, making them capable of being used in food packaging without damaging the food and, consequently, the health of consumers (Atarés and Hiralt, 2016). Many types of EO have been studied, such as oregano (Kwon et al., 2017; Romani et al., 2017; Valencia-Sullca et al., 2018), clove (Salgado et al., 2013; Teixeira et al., 2014), cinnamon (Valencia-Sullca et al., 2018;

Zhang et al., 2015), garlic (Teixeira et al., 2014), thyme (Lee et al., 2016a), orange (Kavas and Kavas, 2016), and lemon (Peng and Li, 2014).

In the literature, many studies evaluating the bioactivity of EOs are found. Peng and Li (2014) developed chitosan films with the addition of three EOs (lemon, cinnamon, and thyme). The authors verified that the EOs had better performance in the inhibition of Gram-positive than Gram-negative bacteria and attributed this behavior to the microorganisms' characteristics. Gram-negative bacteria have a more resistant outer membrane, making the action of EOs more difficult. The authors verified synergy in the combination of oils and observed that the mixture of EOs did not increase the antimicrobial activity.

Salgado et al. (2013) produced films of sunflower protein with added clove EO. The authors evaluated the antioxidant potential of the films and verified that clove EO shows great antioxidant capacity. The authors used four methods to evaluate antioxidant power and concluded that clove EO acts in the oxidative process, eliminating free radicals and slowing their oxidation. This oxidative effect was attributed to the chemical characteristics of clove EO due to the presence of phenolic compounds in its structure. Jouki et al. (2014) studied the addition of oregano EO to quince seed mucilage films. The authors observed that an increase in the concentration of EO added to the films increased the antioxidant activity. This behavior was attributed to the higher concentration of phenolic compounds present in the higher amounts of oregano EO used in the films. The films evaluated by the authors showed an ability to reduce free radicals and inhibit the oxidative process.

Lee et al. (2016a) studied the development of gelatin films with added thyme EO, to be used as chicken tenderloin packaging. The effect on shelf life was verified through the antimicrobial activity provided to the products packaged with the films. The authors verified that the addition of thyme EO to films promoted a reduction in the growth of pathogenic bacteria such as *Listeria monocytogenes* and *Escherichia coli* O157:H7. Kavas and Kavas (2016) elaborated egg-white protein films with the addition of orange EO and applied them for preservation of Kashar cheese. The physicochemical properties of cheeses and antimicrobial activity were used to evaluate the process of food degradation. The authors verified that the cheeses packed with films containing the natural compounds showed a reduced microbial load during the period of 30 days of storage when compared with films without addition of the oil. This reduction in microbiological growth was attributed to the phenolic compounds present in the EO, which act on the microbiological membrane, inhibiting growth.

It is possible to observe that different EOs have different antimicrobial and antioxidant activity, which is expected because the activity is related to the compounds present in the oil; also, these substances will interact in a different way with the film matrix. This interaction could be good for the release of active compounds or could be adverse, depending on the situation.

6.3.2 Natural extracts

In addition to EOs, other natural extracts are widely used as natural additives for the preparation of AP. Different types of extract have been studied, such as green tea extract (Siripatrawan and Noipha, 2012), mango kernel extract (Adilah et al., 2018), curcuma ethanol extract (CEE) (Bitencourt et al., 2014), rosemary extract (Piñeros-Hernandez et al., 2017), yerba mate extract (Reis et al., 2015), longan seed (LS) extract (Sai-Ut et al., 2015), ethanol-propolis extract (Bodini et al., 2013), and grape seed extract (Rubilar et al., 2013), among others.

Bitencourt et al. (2014) produced gelatin films with added CEE. The extract concentration was varied from 0 to 200 g (CEE/100 g gelatin); the antioxidant activity of the films was improved as the concentration of CEE increased, due to the superior concentration of phenolic compounds present in the extract. Sai-Ut et al. (2015) elaborated gelatin films incorporating LS extract. The authors made a comparative study of the natural extract and the synthetic antioxidant butylated hydroxytoluene (BHT) used as a food additive. Antioxidant activity was verified by the 2,2-diphenyl-1-picrylhydrazyl (DPPH) method, and the authors concluded that both antioxidants used at the same concentration have the same antioxidant potential. However, when the antioxidant LS was evaluated at higher concentrations than the synthetic one (BHT), it showed higher antioxidant activity than BHT. In addition, the natural compound was more stable than the synthetic compound during the storage period. This study demonstrated the benefits that natural composites may confer on films and stored products, through slowing down the oxidative process. Saberi et al. (2017) developed pea starch–guar films with different natural extracts added such as green tea extract, macadamia extract, and blueberry ash extract. The authors investigated the antimicrobial activity of the films incorporating the extracts and verified a reduction in microbiological growth. The antimicrobial activity of natural extracts is related to the phenolic compounds and anthocyanins present in different extracts. The hydroxyls present in polymers interact with phenolic compounds, reducing their antimicrobial capacity; however, part of the activity is still maintained. The authors observed higher activity of the extracts for the inhibition of bacteria and attributed this behavior to the action of phenolic compounds in destabilizing microbiological membranes.

Several studies have evaluated the application of active films, but estimation of the shelf life of a product is a difficult task. Tests performed on food submitted to shelf-life studies assess, in general, oxidative processes, microbiological growth, water loss, colors, texture, and others. A large number of responses obtained in these tests, coupled with the variability of information in the literature about the quality of the products, makes it difficult to estimate an increase or reduction of shelf life. The data obtained provide, in general, a background of the behavior of active films in relation to a certain raw material. However, few studies are able to provide an estimation of the increase of shelf life in days. Qin et al. (2013) evaluated the addition of tea polyphenols to chitosan films for application in pork-meat patties. During the application process, the authors studied microbiological growth and physicochemical and sensorial properties. Through this, they were able to estimate that the shelf life of the pork-meat patties was increased by 6 days when using chitosan films with tea polyphenols. Meenatchisundaram et al. (2016) investigated the addition of spices (clove and cinnamon) to matrices of tamarind starch polysaccharides. The authors added the active compounds in the form of particles with reduced size and observed the application of these films in relation to the shelf life of white shrimp. They stored the white shrimp at two temperatures (4°C and 10°C) and evaluated the oxidative process, antimicrobial activity, and appearance. They concluded that with addition of the active compounds, the white shrimp stored at 4°C had a shelf life of 29 days, and the shrimp stored at 10°C had a shelf life of 21 days. The control sample showed a maximum shelf life of 10 days.

6.3.3 Modified atmosphere

The use of natural compounds added to film matrices has been a large target of studies, as mentioned previously. However, there are other types of AP capable of increasing the shelf life of products, such as the use of modified atmosphere. According to Kirtil et al. (2016), the process of modifying the atmosphere consists of changing the gases present in the packaging in which the food is stored. This process can be performed using different gases, mainly O_2, CO_2, and N_2. According to Lee (2016), CO_2 is widely used in this process since it behaves as an antimicrobial agent, improving the shelf life of foods and keeping them fresh. The antimicrobial activity of CO_2 is higher in chilled products because its solubility is increased in food at low temperatures, reducing microbiological growth. Regarding antioxidant activity, the use of N_2 and CO_2 helps to reduce O_2 levels, consequently decreasing the oxidative processes in foods prone to these alterations.

Janjarasskul et al. (2016) studied the shelf life of sponge cake packaged in two different commercial polymers and in the presence of different concentrations of an O_2 scavenger and an ethanol emitter. The authors evaluated parameters such as microbiological growth, lipid oxidation, water activity, color changes, texture, and sensorial acceptance. Thus, they found that by using the O_2 scavenger it was possible to maintain lower O_2 quantities during the storage period, and the ethanol emitter was responsible for reducing microbiological growth throughout the storage process. For microbiological growth, the authors mention that at the end of the storage period, synergistic effects of both modification agents were observed, since they were able to suppress the proliferation of aerobic and anaerobic microorganisms. As for the oxidative process, the authors report that the treatments with higher concentrations of O_2 scavenger were able to delay lipid oxidation of the product. Finally, they concluded that the shelf life of the sponge cake was extended by 42 days through the use of an atmosphere modified by O_2 scavengers and ethanol emitters.

Nugraha et al. (2015) studied the use of CO_2 and C_2H_4 absorbers during the storage of pears in commercial packaging. The authors evaluated internal browning of the product over 7 months at 1°C. The absorbers used were able to control the emission of gases generated during storage, while maintaining the characteristics of the product as compared with the standard. Wang et al. (2015) developed agar-based films with sodium carbonate and sodium glycinate added as CO_2 absorbers. The films were applied as absorbers in indirect contact with chilled mushrooms at 10°C for 5 days. The authors compared the treatments with each other and with a commercial film, checking microbiological growth, texture, weight loss, and flavor. After 5 days of storage, the mushrooms stored with the sodium carbonate film presented a microbiological count of 1.83 log CFU/g, while the mushrooms stored with the commercial film had microbiological growth of 2.73 log CFU/g. Therefore, this behavior indicates that the use of CO_2 absorbers supports the storage process, reducing microbial growth.

In general, the use of natural compounds combined with sustainable polymers provides eco-friendly packaging for the food industry. Addition of active compounds to the packaging process reduces the use of chemical additives in food and extends the shelf life of the products. These actions result in food preservation and consequently reduce the waste of packaged food. Besides conferring bioactivity to films, the addition of natural compounds may act to improve the properties of composite films. The structure of the natural compounds combined with the polymer matrix can modify the film properties, making them more resistant, less soluble, and less permeable.

6.4 Food packaging properties

The main objective of food packaging is to preserve the quality and safety of the food, providing adequate conditions of transport and storage until it reaches the consumer and avoiding losses in the production chain. There is a large diversity of packaging in the market, and it can vary depending on its design, specific material properties, functionality, and storage conditions, among other properties, as demonstrated in Fig. 6.1.

Packaging feasibility occurs when it is possible to justify the cost in relation to protecting the food. In this context, for each type of product, specific properties are desired to guarantee the safety of the food. In the packaging sector, the properties of polymers are extremely important because they affect the possibilities of application. In this context, the main properties are mechanical, thermal, optical, microstructural, barrier, solubility, and biodegradability (Arfat et al., 2017).

In the last decade, a great increase in environmental and social problems caused by the inappropriate use and large-scale disposal of synthetic packaging has been observed. This encourages research about types of packaging that are less harmful, such as biodegradable packaging from natural polymers (Dilkes-Hoffman et al., 2018). Although the use of natural polymers helps to reduce pollution of the environment, biodegradable packaging generally has poorer mechanical and barrier properties, which limit its application.

In general, films produced from natural polymers have less desirable properties than synthetic polymers; as a result, studies have been carried out to improve the properties of biodegradable films, to make them competitive and applicable in food packaging.

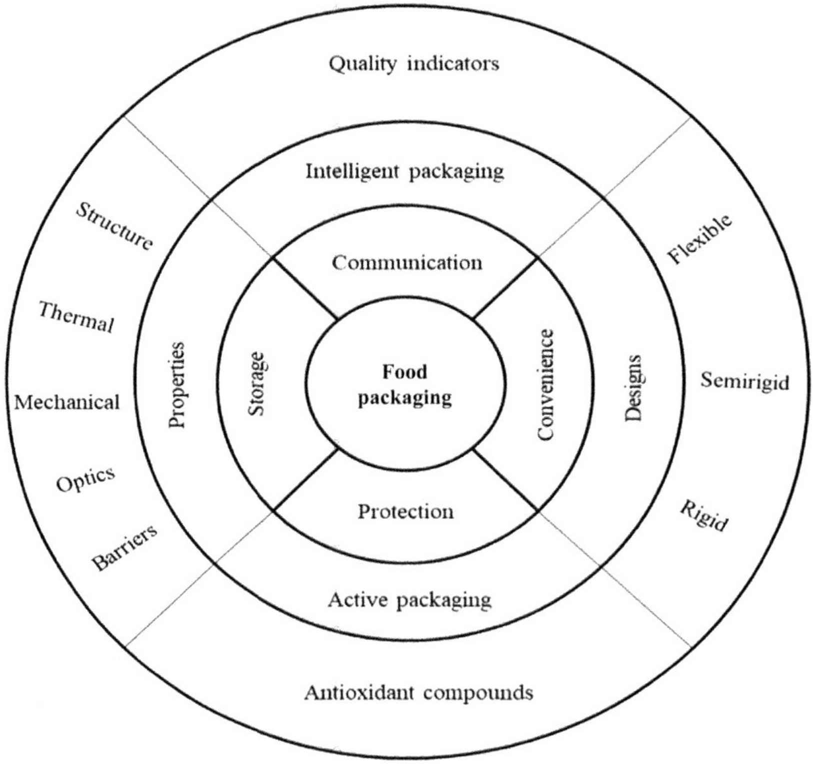

Figure 6.1 Scheme of some packaging functions and properties.

6.4.1 Mechanical properties

The mechanical properties of packaging films are associated with the behavior of the material against forces applied on their surface. Evaluation of mechanical parameters is extremely important in food packaging since through knowledge of them it is possible to carry out storage and distribution processes without loss, allowing adequate protection for each type of food. In general, requirements for the mechanical properties of packaging are good tensile strength (TS), adequate elongation, flexibility and resistance to drilling, to promote a physical barrier against external agents.

Elongation capacity refers to the increase in film length after application of opposing forces on its surface, being measured as a percentage. TS is given by the maximum deformation of the film to the breaking point, taking into account the cross-sectional area of the film (Ghasemlou et al., 2013), and perforation resistance is given by the maximum film deformation values up to the point of rupture, taking into account the maximum force exerted on a known area of the film surface (Song et al., 2000), as shown in Fig. 6.2.

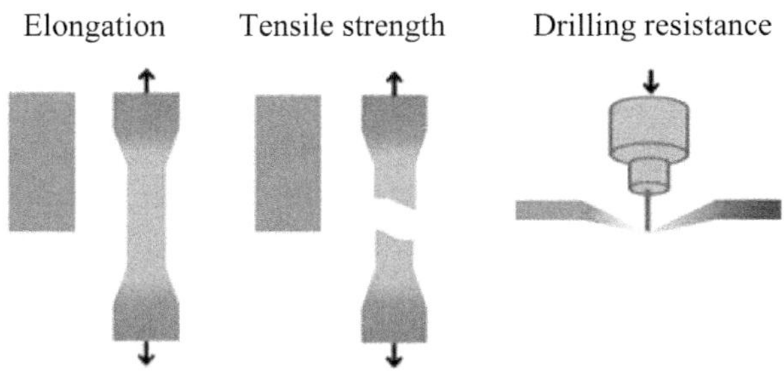

Figure 6.2 Scheme of some mechanical properties.

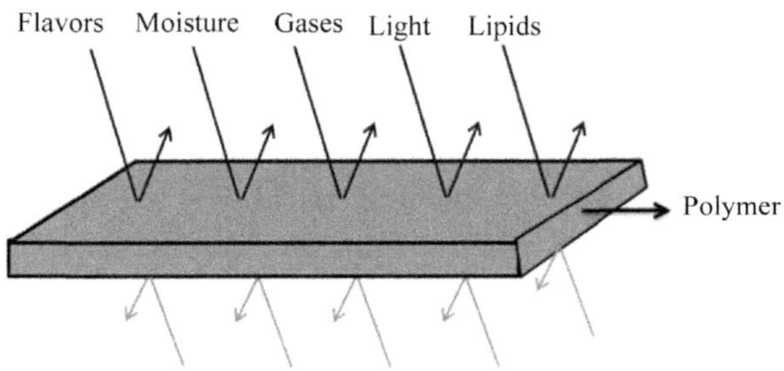

Figure 6.3 Scheme of packaging barriers.

These attributes may vary depending on the origin of the polymer, method of manufacture, thickness, and exposure to agents that promote changes in polymer structure, such as heat, moisture, light, and others. In addition to these factors, incorporation of active and antimicrobial compounds in materials often has the capacity to influence the mechanical properties. It was observed in a study with cassava starch composite films that addition of cinnamon EO resulted in a reduction of TS and an increase in elongation, indicating a loss of macromolecular mobility (Souza et al., 2013).

In the production of gelatin-based films with added CEE, improvements were observed in the mechanical properties inherent to interactions between phenolic compounds and peptides, which form covalent crosslinks that produce more cohesive and flexible matrices. It was observed that TS and elongation at break (EB) increased significantly at concentrations above 50 g of extract/100 g of gelatin (Bitencourt et al., 2014).

6.4.2 Barrier properties

The barrier property of packaging is characterized as the ability of the polymer to prevent exchanges with the external medium, resisting processes such as absorption or moisture loss, gas exchange, and lipid and light permeation, as illustrated in Fig. 6.3.

Water vapor permeability (WVP) is one of the most important barrier properties in food packaging. Through this parameter, it is possible to obtain relevant data about the resistance of a material to humidity and its storage needs, and to have some idea about application of the polymer for different products (Jarvis et al., 2017). The addition of compounds to the polymer can alter the structure of films and packaging, promoting better properties against moisture and providing greater protection to the stored product. In contrast, the addition of some compounds such as emulsifiers may increase WVP (Souza et al., 2013).

Gas barriers are also important parameters to be evaluated in the production of packaging, according to the needs of each stored product; they are very important for food preservation, since oxygen is a key factor in oxidation, initiating many deterioration reactions. Several factors can influence the permeability of films to gases, such as the chemical structure, cohesive energy density, free volume between molecules, crystallinity, orientation of the polymer chains, tacticity, and crosslinking (Sothornvit and Pitak, 2007).

In evaluating the addition of gold nanoparticles to quinoa films, researchers observed that their incorporation in the polymer caused a reduction in oxygen permeability, since the addition causes a delay in oxygen transport due to an increase in the tortuosity of the oxygen pathway (Pagno et al., 2015). On the other hand, some compounds have the capacity to increase gas permeability, such as demonstrated in studies regarding the production of chitosan films with carvacrol and grape seed extract, in which the oxygen barriers were reduced once the crystallinity of the polymer was affected by addition of the compound (Rubilar et al., 2013).

6.4.3 Optical properties

The visual aspect of packaging is related to its color and transparency; they are important attributes that influence the acceptability of the product by the consumer and have an unpredictable influence in relation to the protection of food against light permeability. In evaluation of packaging color, parameters related to the influence of polymer compounds on luminosity, color, and opacity are generally analyzed. In general, polymers with greater transparency and less difference in color are more desirable in the market because they provide the consumer a realistic view of the product to be purchased. On the other hand, some food products that are sensitive to light require more opaque packaging that offers protection against light, which can cause oxidation reactions. In a study with chia mucilage films, it was observed that the incorporation of oregano EO resulted in a change in the films color; when a higher concentration of oil was added, greater color variation occurred (Jouki et al., 2014).

6.4.4 Solubility in water

The solubility of packaging is characterized as the resistance or tolerance of the polymer to water or moisture. This property is totally related to the chemical structure and hydrophobicity of the surface of the material (Wang et al., 2017a).

Solubility is an important parameter for the characterization of packaging, taking into account that lower solubility leads to greater stability and less interaction with humidity.

In most cases, natural polymers present high solubility; in this context, some strategies such as the addition of compounds or a combination of different polymers may be used to control film solubility. In studies aimed at reducing the solubility of collagen films, the use of 50% starch in the formulation of the filmogenic solution resulted in a significant reduction of solubility (37.8%–26.7%), due to the interaction between starch granules and collagen forming highly cross-linked systems preventing penetration of water molecules into the films (Wang et al., 2017a).

Some films, such as those produced with the use of glucomannan in which they dissolve very quickly, have a high degree of solubility, which limits their application as packaging. To reduce this effect, one study used zein (30%) in films, promoting a reduction in solubility and allowing them to be immersed in water for 5 hours without being completely solubilized (Wang et al., 2017b). On the other hand, in some cases, high solubility may be appreciated such as in use for soluble sachets or soluble and edible films (Lai et al., 2018).

6.4.5 Thermal properties

Through thermal characterization, it is possible to describe the behavior of a material in different ranges of temperature; it can also indicate parameters for the production of packaging, taking into consideration production temperature as well as glass transition temperature, melt temperature, and crystallization of materials.

The techniques most used for thermal characterization of polymers are differential thermal analysis, thermogravimetric analysis (TGA), and differential scanning calorimetry. These techniques are essential for obtaining parameters such as loss of mass, melting point (T_m), glass transition temperature (T_g), boiling point (T_b), crystallization temperature (T_c), enthalpy of crystallization (ΔH), and enthalpy of melting (ΔH_m) (Erdohan et al., 2013).

In the production of biodegradable packaging, incorporation of active compounds may add functionalities and modify the inherent thermal properties of the polymer. For example, the use of extracts of BHT and green tea in potato starch films increases T_g and ΔH, possibly by the generation of hydrogen bonds between the starch and the active compound, strengthening the network of films and limiting molecular mobility (Nisa et al., 2015).

Evaluating the effect of xanthan and locust bean gum synergistic interaction on the characteristics of biodegradable edible films, TGA showed that composite films presented different thermal properties. This behavior was attributed to good miscibility, which decreased the free volume, resulting in lower molecular mobility of the polymer matrix as a result of interaction between the polymers, making it more resistant to loss of mass at high temperatures than film from a single polymer (Kurt et al., 2017).

6.4.6 Microstructural properties

The microstructural properties of polymers are crucial for good performance in the production of packaging. Through evaluation of these attributes, it is possible to observe the surface and sectional area as well as to verify homogeneity, roughness, crystallinity, and interactions among multilayers. It is also possible to verify the presence of faults, microholes, and ruptures that can lead to the loss of mechanical and barrier properties. More than this, it is even possible to identify conformations and molecular structures added to the films.

In the production of cassava starch antimicrobial films with added cinnamon EO and sucrose ester (emulsifier), it was observed that in the formulations with added emulsifier, the films presented a microstructure with a smooth, uniform, and regular surface. On the other hand, the absence of emulsifier caused a discontinuous structure, with lipid droplets dispersed in the polymer chain (Souza et al., 2013).

Using scanning electron microscopy in the evaluation of active films with sodium and calcium caseinate with carvacrol, homogeneity without ruptures, faults, or perforations was observed. On the other hand, when optical micrographs were evaluated, the presence of droplets on the surface was observed, due to the hydrophobicity of carvacrol, which forms an emulsion in the aqueous caseinate–glycerol solution (Arrieta et al., 2013). It is necessary to search for a balance when adding compounds, to avoid significant losses in the microstructural properties of the films.

6.4.7 Biodegradability

Biodegradability consists of the deterioration of a material in direct contact with nature, due to mechanical fragmentation or chemical modifications through the action of microorganisms and enzymes; finally, the polymer is converted to carbon dioxide, water, inorganic compounds, methane, and biomass. This process is influenced by external factors such as microbial load, temperature, light, humidity, and the pH of the soil (Calmon-Decriaud et al., 1998).

According to the European standard EN 13432, packaging must decompose by at least 90% by biological action in a period of 6 months to be considered biodegradable. In recent years, the search to produce biodegradable packaging has been intensified as a way of mitigating the pollution caused by the use of synthetic polymers of fossil origin.

Cassava starch films incorporated with lycopene nanocapsules present biodegradability of 36% in 15 days (Assis et al., 2017). Evaluating the biodegradability of starch films with added biomass and microalgae extract, the authors observed that films with biomass added presented 69.9% degradation and those with extract were degraded by 55.1% after 15 days in soil, indicating that the addition of biomass potentiates the decomposition process (Carissimi et al., 2018).

In the literature, several studies can be found about the production of films that are biodegradable as a result of the materials used. In contrast, evaluation of the materials degradation is often not made; thus this information is flawed, and the biodegradability of the material cannot be proven, since the weather conditions and degradation of these materials are not shown.

6.5 Strategies to improve the properties of films

Biopolymer films have been the focus of research and development of the new generation of environmentally friendly materials. However, some drawbacks still limit their commercial use, such as inferior mechanical and barrier performance compared with synthetic films [e.g., low-density polyethylene (LDPE), high-density polyethylene (HDPE), and cellophane] (Wihodo and Moraru, 2013). In general, protein and polysaccharide films have higher WVP than synthetic films because of the hydrophilic nature of these macromolecules. Regarding mechanical properties, films from biomacromolecules are usually weaker and have less elongation (Hernandez-Izquierdo and Krochta, 2008).

The main function of packaging is the protection of food from the surrounding environment. Lipid oxidation and microbial activity are important causes of quality loss in foods, and water passing through packaging has a vital role in the products since water is directly linked to these mechanisms of deterioration (Robertson, 2013). That is why packaging, in general, requires WVP to be as low as possible, and adequate mechanical strength and extensibility is required (Guo et al., 2014). Mechanical properties have different implications in films for food packaging depending on the product. Film wrap, for example, requires, in general, higher flexibility but TS also has a significant role in protecting the product from mechanical damage from processing until distribution to the consumer (Kaewprachu et al., 2017).

Different approaches have been tested to improve the physical properties of biopolymeric films. Among the strategies used to improve the properties of films are modifications by chemical, enzymatic, and physical methods and combination of the polymers with hydrophobic components, focused mainly on the improvement of mechanical and barrier properties (Bourtoom, 2009). Table 6.1 presents examples of studies focused on the improvement of biopolymeric films through different methods.

6.5.1 Chemical methods

Chemical treatments using acid, alkali, or crosslinking agents have been employed extensively to improve film properties. These methods are expected to decrease the permeability and improve the TS of films. Among the chemical agents used for modifications of biomacromolecules are glyoxal, glyceraldehyde, glutaraldehyde (GA), formaldehyde, boric acid, sodium trimetaphosphate, hydroxypropylation–acetylation agents, hydrogen peroxide, persulfate, organic acids, hydrochloric acid, and sulfuric acid, among others (Benbettaïeb et al., 2016a; Shah et al., 2016).

As presented in Table 6.1; López-De-Dicastillo et al. (2016) observed an improvement in water barrier and mechanical properties when using different concentrations of GA in methylcellulose films incorporated with plant extracts. The reduction in water sorption and swelling of films was a result of new bonds formed from the crosslinking. Incorporation of an active extract from maqui berry also

Table 6.1 Examples of researches focused in the improvement of biopolymeric films through different methods

Method	Film material	Agent or mechanism		Mechanical properties		Barrier properties	References
				TS (MPa)	*E* (%)	WVP	
Chemical	Methylcellulose	Glutaraldehyde	0%	5.1	31.3	$\sim 85 \times 10^{16}$ kg m/m^2 s Pa	López-De-Dicastillo et al. (2016)
			2.5%	10.3	69.1	$\sim 80 \times 10^{16}$ kg m/m^2 s Pa	
			5%	8.0	53.1	$\sim 70 \times 10^{16}$ kg m/m^2 s Pa	
			7.5%	6.3	55.5	$< 60 \times 10^{16}$ kg m/m^2 s Pa	
	Potato starch	Sodium hypochlorite	Native (3% starch)	3.53	85.2	~ 9.3 g mm/m^2 day kPa	Zavareze et al. (2012)
			Native (4% starch)	3.61	62.0	~ 9.3 g mm/m^2 day kPa	
			Native (5% starch)	4.87	58.3	~ 9.3 g mm/m^2 day kPa	
			Oxidized (3% starch)	6.07	84.9	~ 5.8 g mm/m^2 day kPa	
			Oxidized (4% starch)	6.39	45.3	~ 6.7 g mm/m^2 day kPa	
			Oxidized (5% starch)	9.12	38.8	~ 7.5 g mm/m^2 day kPa	
Enzymatic	Starch/gelatin	Transglutaminase	0 mg	2.60	72.19	2.5×10^{-4} g mm/h m^2 kPa	AL-Hassan and Norziah (2017)
			1 mg	3.81	61.93	3.0×10^{-4} g mm/h m^2 kPa	
			5 mg	3.99	57.42	3.0×10^{-4} g mm/h m^2 kPa	
			10 mg	5.89	47.75	3.0×10^{-4} g mm/h m^2 kPa	
	Fish myofibrillar protein	Transglutaminase	0%	7.16	167.49	2.38×10^{-9} g/s m Pa	Kaewprachu et al. (2017)
			1%	8.73	143.66	2.19×10^{-9} g/s m Pa	
			2%	9.68	132.76	2.11×10^{-9} g/s m Pa	
			3%	11.08	107.15	2.09×10^{-9} g/s m Pa	
			4%	13.10	85.61	2.02×10^{-9} g/s m Pa	

Physical	Soybean	Plasma treatment	Untreated	3.2	26.1	4.1 g mm/h m^2 kPa	Oh et al. (2016)
			Oxygen	3.3	30.0	3.8 g mm/h m^2 kPa	
			Nitrogen	3.4	29.1	4.0 g mm/h m^2.kPa	
			Air	3.3	27.9	4.0 g mm/h m^2 kPa	
			Helium	3.3	29.7	4.1 g mm/h m^2 kPa	
			Argon	3.2	29.8	3.7 g mm/h m^2 kPa	
	Whey protein	Ultraviolet radiation	Untreated	1.71	15.05	1.02 g mm/h m^2 kPa	Díaz et al. (2016)
			Heat treated	2.98	18.78	1.14 g mm/h m^2 kPa	
			Solution—0.12 J/cm^2	1.57	13.78	1.06 g mm/h m^2 kPa	
			Solution—0.4 J/cm^2	1.86	10.17	1.06 g mm/h m^2 kPa	
			Solution—12.0 J/cm^2	3.11	10.06	0.97 g mm/h m^2 kPa	
			Film—0.12 J/cm^2	1.49	12.74	1.09 g mm/h m^2 kPa	
			Film—0.4 J/cm^2	1.93	11.37	1.12 g mm/h m^2 kPa	
			Film—12.0 J/cm^2	2.04	11.55	1.03 g mm/h m^2 kPa	
Blending with other materials	Corn starch	Chitosan	0%	~3	~55	7.89×10^{-10} g/m s Pa	Ren et al. (2017)
			21%	~4.5	~87	$\sim 1.2 \times 10^{-10}$ g/m s Pa	
			41%	~5	~115	$\sim 1.5 \times 10^{-10}$ g/m s Pa	
			61%	~6.5	~125	$\sim 1.8 \times 10^{-10}$ g/m s Pa	
			81%	~2.5	~105	$\sim 3 \times 10^{-10}$ g/m s Pa	
	Chitosan	Flax cellulose nanocrystals	0%	5.4	37.16	7.69×10^{13} g cm/cm^2 s Pa	Mujtaba et al. (2017)
			5%	5.86	45.29	9.33×10^{13} g cm/cm^2 s Pa	
			10%	5.99	36.32	10.67×10^{13} g cm/cm^2 s Pa	
			20%	6.67	35.34	12.15×10^{13} g cm/cm^2 s Pa	
			30%	6.28	37.49	14.12×10^{13} g cm/cm^2 s Pa	

TS, tensile strength; *E*, elongation at break; *WVP*, water vapor permeability.

contributed to a decrease in the WVP of the films. According to these authors, the extract probably induces changes in the hydrophilic nature of the matrix, forming internal hydrogen bonds that limit affinity with water molecules or induce formation of hydrophobic domains where transport of water molecules is inhibited, thus increasing the tortuosity factor for mass transfer. For the mechanical properties, TS increased due to the formation of a more stable network, while EB decreased, indicating the establishment of new linkages between the polymer chains and GA. Besides that, the addition of GA produced a yellowish color. Thermal stability was also influenced by GA; the crosslinked films showed an additional degradation stage, corresponding to degradation of the new linkage. Despite the improved performance of films caused by GA crosslinking, its use is limited in materials for food packaging due to concerns of toxicity in humans.

Dialdehyde polysaccharides are ideal crosslinking agents whose aldehyde groups can crosslink with the 3-amino groups of lysine or hydroxylysine side groups of protein by formation of Schiff bases. Guo et al. (2014) oxidized xanthan gum with different aldehyde content successfully prepared by periodate oxidization and used it as a crosslinking agent for edible gelatin films. These authors observed a decrease in WVP with an increase in the oxidization level of xanthan, which might be associated with the compact and dense network created by crosslinking leading to a more tortuous path to the water molecules. An increase in TS and decrease in elongation were also observed on incorporation of xanthan gum due to the electrostatic forces, hydrogen bonding, and van der Waals forces in the gelatin film matrix. These results suggest crosslinking between gelatin and xanthan gum. According to Mu et al. (2012), higher TS values are, in general, associated with lower EB values, which result from a harder structure of the films.

Zavareze et al. (2012) investigated the effects of sodium hypochlorite oxidation of potato starch on the physicochemical and textural properties of starch, in addition to the WVP and mechanical properties of the potato starch films produced. They observed that films produced from oxidized potato starch had decreased solubility, elongation, and WVP values in addition to increased TS compared with native starch films. Changes in mechanical properties were attributed to the presence of carbonyl and carboxyl groups in the oxidized starch, which may induce hydrogen bonds between the OH^- groups of the amylose and amylopectin molecules; these linkages provide more structural integrity in the polymeric matrix, increasing TS and affecting flexibility. The decrease in WVP of films as a result of starch oxidation was explained by the reduction in moisture adsorption due to the replacement of hydrophilic hydroxyl groups by more hydrophobic aldehyde groups.

Other studies have reported changes in the physicochemical properties of chemically treated films developed with biomacromolecules. Woggum et al. (2015) obtained increased film solubility, EB, and transparency using hydroxypropylation of rice starch. Seligra et al. (2016) observed a reduction in WVP when using citric acid in starch. The use of hydrophilic inorganic salts resulted in increased water absorption, plasticization, and EB, and decreased crystallinity and TS in the study of Jiang et al. (2016).

6.5.2 Enzymatic methods

Another approach for crosslinking of polymer chains to improve film properties is the use of enzymes (Bourtoom, 2009), as demonstrated in Table 6.1. Enzymatic processes are important because chemical strategies might provide toxicity inappropriate for food applications. Most of these treatments are performed by the use of transglutaminases (TGs), which are food-grade enzymes that use the acyltransferase mechanism, linking the γ-carboxamide (acyl donor) of a glutamine residue to the γ-amine (acyl acceptor) of lysine residues along protein chains (Mahmoud and Savello, 1992). Besides the use of TGs, proteins can also be crosslinked with horseradish peroxidase (EC 1.11.1.7), an enzyme that catalyzes the oxidation of tyrosine residues to form di-, tri-, and tert-tyrosine, which are responsible for promoting protein crosslinking (Stuchell and Krochta, 1994).

AL-Hassan and Norziah (2017) incorporated 1, 5, and 10 mg of TG with activity of 100 U/g powder in sago starch/fish gelatin and observed a significant increase in TS with all concentrations of TG in comparison to the control film, while a reduction in EB was observed. According to the authors, TG induces crosslinking through covalent isopeptide bonds; however, the sorption sites responsible for the films hydrophilicity were not influenced. This might be due to the presence of a hydrophilic plasticizer. The degree of crosslinking of sago starch/fish gelatin films was superior with a higher concentration of TG (1 mg−30.9%; 5 mg−47.2%; 10 mg−53.0%), explaining the behavior observed in the mechanical properties of the films. In addition, the transmittance percentage of amide I and II bands, which indicates protein presence, decreased with higher enzyme concentrations.

TG was also used by Kaewprachu et al. (2017) to improve the mechanical and physical properties of fish myofibrillar protein films. These authors incorporated different concentrations of microbial TG (0%, 1%, 2%, 3%, and 4% w/w, based on protein content). As expected, this study showed an increase in compactness but a reduction in elasticity indicated by TS and EB due to the formation of intermolecular crosslinking. In contrast to AL-Hassan and Norziah (2017), Kaewprachu et al. (2017) observed slight decreases in WVP when TG content increased from 1% to 4%, suggesting a reduction in the free volume of the film matrix due to more crosslinking in the protein structure. This study also showed decreases in the moisture, solubility, and degree of swelling of films with the use of TG. Lightness decreased while yellowness increased, and thermal stability was improved. Fourier transform infrared spectroscopy studies confirmed the formation of crosslinks and conformational changes in the fish myofibrillar protein films.

6.5.3 Physical methods

Physical methods for the improvement of film properties involve exposure to ionizing radiation of the raw materials before film development, as well as of the films after their preparation. Ionizing irradiation is responsible for conformational changes, oxidation of amino acids, rupture of covalent bonds, and the formation of free radicals, in addition to recombination and polymerization (Wihodo and Moraru, 2013).

Cold plasma (CP) treatment is a technology used for surface functionalization, etching, polymer degradation, and crosslinking. It consists of ultraviolet photons, electrons, positive and negative ions, free radicals, and excited and nonexcited molecules and atoms, which break covalent bonds and initiate chemical reactions (Kim et al., 2014). Among the advantages of CP treatment are no use of hazardous solvents, uniformity of treatment, and no thermal damage to materials including heat-sensitive biopolymers (Morent et al., 2011).

Oh et al. (2016) investigated the effect of different CP-forming gases on the properties of defatted soybean meal films, and the effect of the films on the storage stability of smoked salmon. The optimal conditions for plasma generation were estimated to be 15 minutes and power of 400 W. Among the gases used (oxygen, nitrogen, air, helium, and argon), no effect on TS was observed, while all of them, except air, increased elongation (Table 6.1). According to the authors, this behavior might be induced by polymer degradation and functionalization during plasma treatment, in which branch scissions form low molecular weight molecules, increasing free volume in the film network and increasing flexibility. Regarding WVP, no significant differences were observed in plasma-treated compared with untreated films. In addition, according to the results, the functionalization reactions provided by CP were the reason for increases in elongation and ink adhesion, while etching effects were responsible for increases in roughness, ink adhesion, water contact angle, and biodegradability. When used in smoked salmon wrapping, CP-treated films were effective in slowing down lipid oxidation and maintaining hardness, suggesting a potentially improved O_2 barrier property.

UV light is another type of irradiation used to enhance film performance. This is nonionizing radiation absorbed by double bonds and aromatic rings of proteins, causing formation of free radicals in amino acids and leading to intermolecular covalent bonds responsible for changes in film properties (Wihodo and Moraru, 2013). The effects of UV radiation were investigated in whey protein film-forming solutions and preformed films at different doses (0.12, 4.0, and 12.0 J/cm^2) by Díaz et al. (2016). The UV treatment had a significant effect on most mechanical properties and solubility only when applied to the film-forming solution at the highest dose. This behavior was attributed to covalent bonds between aromatic amino acids and not to disulfide bonding. WVP and transmission were not affected by UV radiation. Besides that, UV light turned the films more yellow, green, and dark, and these effects were superior when applied to the film-forming solution. The authors commented that UV light treatment increased the concentration of sulfhydryl groups and induced formation of aggregates; however, no changes in the secondary structure of proteins were observed.

Other types of physical modification of biopolymers include the use of gamma irradiation (Shahabi-Ghahfarrokhi et al., 2015) and electron beams (Benbettaïeb et al., 2016b), among others. The effect of different gamma-ray dosages (3, 6, and 9 kGy) on the functional properties of kefiran biopolymer was investigated; the results demonstrated that they were dependent on the ratio of crosslinking between polymer chains and produced mono- and disaccharides by gamma irradiation. This treatment is an important method because it is a simple, cheap, and effective procedure with high penetration power (Shahabi-Ghahfarrokhi et al., 2015). In the study of Benbettaïeb et al. (2016b), plasticized fish gelatin film properties were affected by electron beam accelerator doses, which induce intermolecular crosslinking, increasing TS (improved by 30% for 60 kGy) and wettability (due to the increase in surface tension and its polar component), while weak effects on WVP were observed.

6.5.4 Blending with other materials

The combination of proteins and polysaccharides with different compounds is used extensively to enhance the physical properties of films. These combinations can be produced with hydrophobic materials (oils, fats, triglycerides, and waxes) (Benbettaïeb et al., 2016a; Rodrigues et al., 2014; Salgado et al., 2013), synthetic materials [PE, poly(vinyl alcohol), polyvinylidene chloride, and PS, among others] (Benbettaïeb et al., 2016a), and reinforcing agents (Arfat et al., 2017; Kadam et al., 2013; Ortega-Toro et al., 2014; Piyada et al., 2013) as well as by blending of different biomacromolecules (Benbettaïeb et al., 2016a; Guerrero et al., 2013; Romani et al., 2017; Sun and Xiong, 2014).

Ren et al. (2017) investigated the influence of chitosan concentration on the mechanical and barrier properties of corn starch/chitosan films because these are the most abundant natural polysaccharides and promising polysaccharides for food packaging. Concentrations of 21%, 41%, 61%, and 81% were studied, and positive effects on mechanical and barrier properties were observed. The incorporation of chitosan led to a significant increase in TS and elongation of films, which was attributed to a high degree of formation of intermolecular hydrogen bonds between polysaccharides and the interaction of plasticizer–polymer chains, which facilitates the sliding of chains. WVP decreased on incorporation of chitosan in corn starch films, which was attributed to reduced availability of hydrophilic groups, although chitosan is a hydrophilic polymer. The incorporation of chitosan also resulted in an increase in film solubility and total color differences (ΔE^*) and a decrease in crystallinity.

On the other hand, Mujtaba et al. (2017) incorporated cellulose nanocrystals (CNC) in chitosan films as reinforcing materials. CNC was obtained from flax fiber using acid hydrolysis and added to the films at concentrations of 5%, 10%, 20%, and 30%. The increase in the concentration of CNC in chitosan films resulted in a gradual increase in the mechanical properties due to interactions between the anionic CNC and cationic chitosan as well as stress transfer through the components interface. An increase in WVP was also observed with higher concentrations of CNC, which might be due to changes in crystallinity altering the adsorption percentage of water molecules, and the presence of hydroxyl and amine groups in chitosan films. The incorporation of CNC in chitosan films also significantly affected color, increasing ΔE^*. Homogeneity was confirmed by studying the morphology of the films, and the antimicrobial activity of films was improved on incorporation of flax CNC.

Since one of the main functions of packaging is to preserve the quality of packaged food, aiming to reduce food waste, protecting the product from the environment is crucial. That is why some specific properties, such as low sensitivity to water and good mechanical properties, are required for food packaging; these still remain an obstacle to using sustainable polymers instead of synthetic ones. Many studies have reported strategies to overcome these drawbacks, and despite the progress observed, further research is required because the improvement of some characteristics of the material is, in general, associated with a decrease in others.

References

Adilah, Z.A.M., Jamilah, B., Nur Hanani, Z.A., 2018. Functional and antioxidant properties of protein-based films incorporated with mango kernel extract for active packaging. Food Hydrocolloids 74, 207–218.

Aguirre, A., Borneo, R., Leon, A.E., 2013. Antimicrobial, mechanical and barrier properties of triticale protein films incorporated with oregano essential oil. Food Biosci. 1, 2–9.

Ahmed, I., Lin, H., Zou, L., Brody, A.L., Li, Z., Qazi, I.M., et al., 2017. A comprehensive review on the application of active packaging technologies to muscle foods. Food Control 82, 163–178.

Akcan, T., Estévez, M., Serdaroğlu, M., 2017. Antioxidant protection of cooked meatballs during frozen storage by whey protein edible films with phytochemicals from *Laurus nobilis L.* and *Salvia officinalis*. LWT—Food Sci. Technol. 77, 323–331.

AL-Hassan, A.A., Norziah, M.H., 2017. Effect of transglutaminase induced crosslinking on the properties of starch/gelatin films. Food Packag. Shelf Life 13, 15–19.

Aloui, H., Khwaldia, K., 2016. Natural antimicrobial edible coatings for microbial safety and food quality enhancement. Compr. Rev. Food Sci. Food Saf. 15, 1080–1103.

Antoniou, J., Liu, F., Majeed, H., Qazi, H.J., Zhong, F., 2014. Physicochemical and thermomechanical characterization of tara gum edible films: effect of polyols as plasticizers. Carbohydr. Polym. 111, 359–365.

Arancibia, M.Y., López-Caballero, M.E., Gómez-Guillén, M.C., Montero, P., 2014. Release of volatile compounds and biodegradability of active soy protein lignin blend films with added citronella essential oil. Food Control 44, 7–15.

Arfat, Y.A., Benjakul, S., Prodpran, T., Osako, K., 2014. Development and characterisation of blend films based on fish protein isolate and fish skin gelatin. Food Hydrocolloids 39, 58–67.

Arfat, Y.A., Ahmed, J., Jacob, H., 2017. Preparation and characterization of agar-based nanocomposite films reinforced with bimetallic (Ag-Cu) alloy nanoparticles. Carbohydr. Polym. 155, 382–390.

Arrieta, M.P., Peltzer, M.A., Garrigós, C., Jiménez, A., 2013. Structure and mechanical properties of sodium and calcium caseinate edible active films with carvacrol. J. Food Eng. 114, 486–494.

Assis, R.Q., Lopes, S.M., Costa, T.M.H., Flôres, S.H., Rios, A., de, O., 2017. Active biodegradable cassava starch films incorporated lycopene nanocapsules. Ind. Crops Prod. 109, 818–827.

Atarés, L., Hiralt, A., 2016. Essential oils as additives in biodegradable films and coatings for active food packaging. Trends Food Sci. Technol. 48, 51–62.

Azevedo, V.M., Borges, S.V., Marconcini, J.M., Yoshida, M.I., Neto, A.R.S., Pereira, T.C., et al., 2017. Effect of replacement of corn starch by whey protein isolate in biodegradable film blends obtained by extrusion. Carbohydr. Polym. 157, 971–980.

Balaguer, M.P., Lopez-Carballo, G., Catala, R., Gavara, R., Hernandez-Munoz, P., 2013. Antifungal properties of gliadin films incorporating cinnamaldehyde and application in active food packaging of bread and cheese spread foodstuffs. Int. J. Food Microbiol. 166, 369–377.

Benbettaïeb, N., Gay, J.P., Karbowiak, T., Debeaufort, F., 2016a. Tuning the functional properties of polysaccharide–protein bio-based edible films by chemical, enzymatic, and physical cross-linking. Compr. Rev. Food Sci. Food Saf. 15, 739–752.

Benbettaïeb, N., Karbowiak, T., Brachais, C.H., Debeaufort, F., 2016b. Impact of electron beam irradiation on fish gelatin film properties. Food Chem. 195, 11–18.

Beristain-Bauza, S., del, C., Mani-López, E., Palou, E., López-Malo, A., 2017. Antimicrobial activity of whey protein films supplemented with *Lactobacillus sakei* cell-free supernatant on fresh beef. Food Microbiol. 62, 207–211.

Bitencourt, C.M., Fávaro-Trindade, C.S., Sobral, P.J.A., Carvalho, R.A., 2014. Gelatin-based films additivated with curcuma ethanol extract: antioxidant activity and physical properties of films. Food Hydrocolloids 40, 145–152.

Bodini, R.B., Sobral, P.J.A., Favaro-Trindade, C.S., Carvalho, R.A., 2013. Properties of gelatin-based films with added ethanol-propolis extract. LWT—Food Sci. Technol. 51, 104–110.

Bonilla, J., Talón, E., Atarés, L., Vargas, M., Chiralt, A., 2013. Effect of the incorporation of antioxidants on physicochemical and antioxidant properties of wheat starch-chitosan films. J. Food Eng. 118, 271–278.

Borges, J.A., Romani, V.P., Cortez-Vega, W.R., Martins, V.G., 2015. Influence of different starch sources and plasticizers on properties of biodegradable films. Int. Food Res. J. 22, 2346–2351.

Bourtoom, T., 2009. Edible protein films: properties enhancement. Int. Food Res. J. 16, 1–9.

Boyaci, D., Korel, F., Yemenicioğlu, A., 2016. Development of activate-at-home-type edible antimicrobial films: an example pH-triggering mechanism formed for smoked salmon slices using lysozyme in whey protein films. Food Hydrocolloids 60, 170–178.

Brody, A.L., Bugusu, B., Han, J.H., Sand, C.K., McHugh, T.H., 2008. Innovative food packaging solutions. J. Food Sci. 73.

Brown, H., Williams, J., Kirwan, M., 2011. Packaged product quality and shelf life. Food and Beverage Packaging Technology. Wiley-Blackwell, Oxford, pp. 59–83.

Burt, S., 2004. Essential oils: their antibacterial properties and potential applications in foods —a review. Int. J. Food Microbiol. 94, 223–253.

Calmon-Decriaud, A., Bellon-Maurel, V., Silvestre, F., 1998. Standard methods for testing the aerobic biodegradation of polymeric materials. Review and perspectives. In: Bellon-Maurel, V., Calmon-Decriaud, A., Chandrasekhar, V., Hadjichristidis, N., Mays, L.W., Pispas, S., et al.,Blockcopolymers—Polyelectrolytes—Biodegradation. Advances in Polymer Science. Springer, Berlin.

Carissimi, M., Flôres, S.H., Rech, R., 2018. Effect of microalgae addition on active biodegradable starch film. Algal Res. 32, 201–209.

Caro, N., Medina, E., Díaz-Dosque, M., López, L., Abugoch, L., Tapia, C., 2016. Novel active packaging based on films of chitosan and chitosan/quinoa protein printed with chitosan-tripolyphosphate-thymol nanoparticles via thermal ink-jet printing. Food Hydrocolloids 52, 520–532.

Chiumarelli, M., Hubinger, M.D., 2014. Evaluation of edible films and coatings formulated with cassava starch, glycerol, carnauba wax and stearic acid. Food Hydrocolloids 38, 20–27.

Dainelli, D., Gontard, N., Spyropoulos, D., Zondervan-van den Beuken, E., Tobback, P., 2008. Active and intelligent food packaging: legal aspects and safety concerns. Trends Food Sci. Technol. 19, S103–S112.
Dazaa, L., Fujitaa, A., Granatob, D., 2016. Active packaging of fish gelatin films with *Morinda citrifolia* oil. Food Biosci. 16, 66–71.
Díaz, O., Candia, D., Cobos, Á., 2016. Effects of ultraviolet radiation on properties of films from whey protein concentrate treated before or after film formation. Food Hydrocolloids 55, 189–199.
Dilkes-Hoffman, L.S., Lane, J.L., Grant, T., Pratt, S., Lant, P.A., Laycock, B., 2018. Environmental impact of biodegradable food packaging when considering food waste. J. Clean. Prod. 180, 325–334.
Erdohan, Z.Ö., Çam, B., Turhan, K.N., 2013. Characterization of antimicrobial polylactic acid based films. J. Food Eng. 119, 308–315.
Ettinger, D.J., 2002. Active and intelligent packaging: a U.S. and EU perspective. From Packaginglaw.com. <http://www.packaginglaw.com/2558_.shtml> (accessed 15.12.17.).
Etxabide, A., Uranga, J., Guerrero, P., de la Caba, K., 2016. Development of active gelatin films by means of valorisation of food processing waste: a review. Food Hydrocolloids 68, 192–198.
EU. European Commission, 2004. Commission Regulation (EC) No 1935/2004 of 27 October 2004 on materials and articles intended to come into contact with food and repealing Directives 80/590/EEC and 89/109/EEC. Official Journal of the European Union.
EU. European Commission, 2009. Guidance to the commission regulation (EC) No 450/2009 of 29 May 2009 on active and intelligent materials and articles intended to come into contact with food. In Version 10. European commission health and consumers directorate-general directorate e-safety of the food chain. E6-innovation and sustainability.
Fang, Z., Zhao, Y., Warner, R.D., Johnson, S.K., 2017. Active and intelligent packaging in meat industry. Trends Food Sci. Technol. 61, 60–71.
FAO, 2015. Food loss and waste facts. <http://www.fao.org/resources/infographics/infographics-details/en/c/317265/> (accessed 10.01.18.).
Ferreira, A.R.V., Alves, V.D., Coelhoso, I.M., 2016. Polysaccharide-based membranes in food packaging applications. Membranes (Basel) 6, 1–17.
Ganiari, S., Choulitoudi, E., Oreopoulou, V., 2017. Edible and active films and coatings as carriers of natural antioxidants for lipid food. Trends Food Sci. Technol. 68, 70–82.
Ghaani, M., Cozzolino, C.A., Castelli, G., Farris, S., 2016. An overview of the intelligent packaging technologies in the food sector. Trends Food Sci. Technol. 51, 1–11.
Ghasemlou, M., Aliheidari, N., Fahmi, R., Shojaee-Aliabadi, S., Keshavarz, B., Cran, M.J., et al., 2013. Physical, mechanical and barrier properties of corn starch films incorporated with plant essential oils. Carbohydr. Polym. 98, 1117–1126.
Gómez-Estaca, J., Montero, P., Gómez-Guillén, M.C., 2014. Shrimp (*Litopenaeus vannamei*) muscle proteins as source to develop edible films. Food Hydrocolloids 41, 86–94.
González, A., Igarzabal, C.I.A., 2013. Soy protein—poly (lactic acid) bilayer films as biodegradable material for active food packaging. Food Hydrocolloids 33, 289–296.
Guerrero, P., Garrido, T., Leceta, I., Caba, K., 2013. Films based on proteins and polysaccharides: preparation and physical-chemical characterization. Eur. Polym. J. 49, 3713–3721.
Guilbert, S., Gontard, N., 2005. Agro-polymers for edible and biodegradable films. Review of agricultural polymeric materials, physical and mechanical characteristics. In: Innovating Food Packaging, pp. 263–276.

Guo, J., Ge, L., Li, X., Mu, C., Li, D., 2014. Periodate oxidation of xanthan gum and its crosslinking effects on gelatin-based edible films. Food Hydrocolloids 39, 243–250.

Han, J.H., 2005. New technologies in food packaging: overview. In: Han, J.H. (Ed.), Innovations in Food Packaging. Elsevier Ltd, London, pp. 3–11.

Haq, M.A., Hasnain, A., Jafri, F.A., Akbar, M.F., Khan, A., 2016. Characterization of edible gum cordia film: effects of beeswax. LWT—Food Sci. Technol. 68, 674–680.

Hernandez-Izquierdo, V.M., Krochta, J.M., 2008. Thermoplastic processing of proteins for film formation—a review. J. Food Sci. 73, 30–39.

Hosseinnejad, M., 2014. Active packaging for food applications—a review. Int. J. Adv. Biol. Biomed. Res. 2, 1174–1180.

Janjarasskul, T., Tananuwong, K., Kongpensook, V., Tantratian, S., Kokpol, S., 2016. Shelf life extension of sponge cake by active packaging as an alternative to direct addition of chemical preservatives. LWT—Food Sci. Technol. 72, 166–174.

Jarvis, K.L., Evans, P.J., Cooling, N.A., Vaughan, B., Habsuda, J., Belcher, W.J., et al., 2017. Comparing three techniques to determine the water vapour transmission rates of polymers and barrier films. Surf. Interfaces 9, 182–188.

Jiang, X., Li, H., Luo, Y., Zhao, Y., Hou, L., 2016. Studies of the plasticizing effect of different hydrophilic inorganic salts on starch/poly (vinyl alcohol) films. Int. J. Biol. Macromol. 82, 223–230.

Jouki, M., Yazdi, F.T., Mortazavi, S.A., Koocheki, A., 2014. Quince seed mucilage films incorporated with oregano essential oil: physical, thermal, barrier, antioxidant and antibacterial properties. Food Hydrocolloids 36, 9–19.

Kadam, D.M., Thunga, M., Wang, S., Kessler, M.R., Grewell, D., Lamsal, B., et al., 2013. Preparation and characterization of whey protein isolate films reinforced with porous silica coated titania nanoparticles. J. Food Eng. 117, 133–140.

Kaewprachu, P., Osako, K., Tongdeesoontorn, W., Rawdkuen, S., 2017. The effects of microbial transglutaminase on the properties of fish myofibrillar protein film. Food Packag. Shelf Life 12, 91–99.

Kavas, N., Kavas, G., 2016. Physical-chemical and antimicrobial properties of Egg White Protein Powder films incorporated with orange essential oil on Kashar Cheese. Food Sci. Technol. 36, 672–678.

Kavoosi, G., Rahmatollahi, A., Mohammad Mahdi Dadfar, S., Mohammadi Purfard, A., 2014. Effects of essential oil on the water binding capacity, physico-mechanical properties, antioxidant and antibacterial activity of gelatin films. LWT—Food Sci. Technol. 57, 556–561.

Kerry, J.P., O'Grady, M.N., Hogan, S.A., 2006. Past, current and potential utilisation of active and intelligent packaging systems for meat and muscle-based products: a review. Meat Sci. 74, 113–130.

Kim, J.E., Lee, D.U., Min, S.C., 2014. Microbial decontamination of red pepper powder by cold plasma. Food Microbiol. 38, 128–136.

Kirtil, E., Kilercioglu, M., Oztop, M.H., 2016. Modified atmosphere packaging of foods. Ref. Module Food Sci. 1–6.

Kokoszka, S., Debeaufort, F., Lenart, A., Voilley, A., 2010. Water vapour permeability, thermal and wetting properties of whey protein isolate based edible films. Int. Dairy J. 20, 53–60.

Krepker, M., Shemesh, R., Danin, Y., Kashi, Y., Vaxman, A., Segal, E., 2017. Active food packaging films with synergistic antimicrobial activity. Food Control 76, 117–126.

Kuorwel, K.K., Cran, M.J., Orbell, J.D., Buddhadasa, S., Bigger, S.W., 2015. Review of mechanical properties, migration, and potential applications in active food packaging systems containing nanoclays and nanosilver. Compr. Rev. Food Sci. Food Saf. 14, 411–430.

Kurt, A., Toker, O.S., Tornuk, F., 2017. Effect of xanthan and locust bean gum synergistic interaction on characteristics of biodegradable edible film. Int. J. Biol. Macromol. 102, 1035–1044.

Kwon, S., Chang, Y., Han, J., 2017. Oregano essential oil-based natural antimicrobial packaging film to inactivate *Salmonella enterica* and yeasts/molds in the atmosphere surrounding cherry tomatoes. Food Microbiol. 65, 114–121.

Lai, K.L., Fang, Y., Han, H., Li, Q., Zhang, S., Li, H.Y., et al., 2018. Orally-dissolving film for sublingual and buccal delivery of ropinirole. Colloids Surf. B Biointerfaces 163, 9–18.

Lee, D.S., 2016. Carbon dioxide absorbers for food packaging applications. Trends Food Sci. Technol. 57, 146–155.

Lee, K.Y., Lee, J.H., Yang, H.J., Song, K.B., 2016a. Production and characterisation of skate skin gelatin films incorporated with thyme essential oil and their application in chicken tenderloin packaging. Int. J. Food Sci. Technol. 51, 1465–1472.

Lee, J., Yang, H., Lee, K., Song, K.B., 2016b. Physical properties and application of a red pepper seed meal protein composite film containing oregano oil. Food Hydrocolloids 55, 136–143.

Liu, B., Xu, H., Zhao, H., Liu, W., Zhao, L., Li, Y., 2017. Preparation and characterization of intelligent starch/PVA films for simultaneous colorimetric indication and antimicrobial activity for food packaging applications. Carbohydr. Polym. 157, 842–849.

López-De-Dicastillo, C., Gómez-Estaca, J., Catalá, R., Gavara, R., Hernández-Muñoz, P., 2012. Active antioxidant packaging films: development and effect on lipid stability of brined sardines. Food Chem. 131, 1376–1384.

López-De-Dicastillo, C., Rodríguez, F., Guarda, A., Galotto, M.J., 2016. Antioxidant films based on cross-linked methyl cellulose and native Chilean berry for food packaging applications. Carbohydr. Polym. 136, 1052–1060.

Mahmoud, R., Savello, P.A., 1992. Mechanical properties of and water vapor transferability through whey protein films. J. Dairy Sci. 75, 942–946.

Majeed, T., Wani, I.A., Hussain, P.R., 2017. Effect of dual modification of sonication and γ-irradiation on physicochemical and functional properties of lentil (*Lens culinaris L.*) starch. Int. J. Biol. Macromol. 101, 358–365.

Majid, I., Nayik, G.A., Dar, S.M., Nanda, V., 2016. Novel food packaging technologies: innovations and future prospective. J. Saudi Soc. Agric. Sci. 17 (4), 454–462.

Meenatchisundaram, S., Chandrasekar, C.M., Udayasoorian, L.P., Kavindapadi Rajasekaran, R., Kesavan, R.K., Srinivasan, B., et al., 2016. Effect of spice-incorporated starch edible film wrapping on shelf life of white shrimps stored at different temperatures. J. Sci. Food Agric. 96, 4268–4275.

Meira, S.M.M., Zehetmeyer, G., Werner, J.O., Brandelli, A., 2016. A novel active packaging material based on starch-halloysite nanocomposites incorporating antimicrobial peptides. Food Hydrocolloids 63, 561–570.

Moreno, O., Pastor, C., Muller, J., Atarés, L., González, C., Chiralt, A., 2014. Physical and bioactive properties of corn starch—buttermilk edible films. J. Food Eng. 141, 27–36.

Morent, R., De Geyter, N., Desmet, T., Dubruel, P., Leys, C., 2011. Plasma surface modification of biodegradable polymers: a review. Plasma Process. Polym. 8, 171–190.

Mu, C., Guo, J., Li, X., Lin, W., Li, D., 2012. Preparation and properties of dialdehyde carboxymethyl cellulose crosslinked gelatin edible films. Food Hydrocolloids 27, 22–29.

Mujtaba, M., Salaberria, A.M., Andres, M.A., Kaya, M., Gunyakti, A., Labidi, J., 2017. Utilization of flax (*Linum usitatissimum*) cellulose nanocrystals as reinforcing material for chitosan films. Int. J. Biol. Macromol. 104, 944–952.

Muller, J., Quesada, A.C., González-Martínez, C., Chiralt, A., 2017. Antimicrobial properties and release of cinnamaldehyde in bilayer films based on polylactic acid (PLA) and starch. Eur. Polym. J. 96, 316–325.

Muriel-Galet, V., Cran, M.J., Bigger, S.W., Hernández-Muñoz, P., Gavara, R., 2015. Antioxidant and antimicrobial properties of ethylene vinyl alcohol copolymer films based on the release of oregano essential oil and green tea extract components. J. Food Eng. 149, 9–16.

Nisa, I., Ashwar, B.A., Shah, A., Gani, A., Gani, A., Masoodi, F.A., 2015. Development of potato starch based active packaging films loaded with antioxidants and its effect on shelf life of beef. J. Food Sci. Technol. 52, 7245–7253.

Nugraha, B., Bintoro, N., Murayama, H., 2015. Influence of CO_2 and C_2H_4 adsorbents to the symptoms of internal browning on the packaged "Silver Bell" pear (*Pyrus communis* L.). Agric. Agric. Sci. Procedia 3, 127–131.

Oh, Y.A., Roh, S.H., Min, S.C., 2016. Cold plasma treatments for improvement of the applicability of defatted soybean meal-based edible film in food packaging. Food Hydrocolloids 58, 150–159.

Orsuwan, A., Shankar, S., Wang, L.F., Sothornvit, R., Rhim, J.W., 2016. Preparation of antimicrobial agar/banana powder blend films reinforced with silver nanoparticles. Food Hydrocolloids 60, 476–485.

Ortega-Toro, R., Jiménez, A., Talens, P., Chiralt, A., 2014. Effect of the incorporation of surfactants on the physical properties of corn starch films. Food Hydrocolloids 38, 66–75.

Ortiz, C.M., Mauri, A.N., Vicente, A.R., 2013. Use of soy protein based 1-methylcyclopropene-releasing pads to extend the shelf life of tomato (*Solanum lycopersicum* L.) fruit. Innov. Food Sci. Emerg. Technol. 20, 281–287.

Otoni, C.G., Espitia, P.J.P., Avena-Bustillos, R.J., McHugh, T.H., 2016. Trends in antimicrobial food packaging systems: emitting sachets and absorbent pads. Food Res. Int. 83, 60–73.

Pagno, C.H., Costa, T.M.H., De Menezes, E.W., Benvenutti, E.V., Hertz, P.F., Matte, C.R., et al., 2015. Development of active biofilms of quinoa (*Chenopodium quinoa* W.) starch containing gold nanoparticles and evaluation of antimicrobial activity. Food Chem. 173, 755–762.

Peng, Y., Li, Y., 2014. Combined effects of two kinds of essential oils on physical, mechanical and structural properties of chitosan films. Food Hydrocolloids 36, 287–293.

Piñeros-Hernandez, D., Medina-Jaramillo, C., Alex, L.C., Goyanes, S., 2017. Edible cassava starch films carrying rosemary antioxidant extracts for potential use as active food packaging. Food Hydrocolloids 63, 488–495.

Piyada, K., Waranyou, S., Thawien, W., 2013. Mechanical, thermal and structural properties of rice starch films reinforced with rice starch nanocrystals. Int. Food Res. J. 20, 439–449.

Poças, M.F.F., Delgado, T.F., Oliveira, F.A.R., 2008. Smart packaging technologies for fruits and vegetables. In: Kerry, J., Butler, P. (Eds.), Smart Packaging Technologies for Fast Moving Consumer Goods. John Wiley & Sons Ltd, New Jersey, pp. 151–166.

Prakash, B., Kedia, A., Mishra, P.K., Dubey, N.K., 2015. Plant essential oils as food preservatives to control moulds, mycotoxin contamination and oxidative deterioration of agri-food commodities—potentials and challenges. Food Control 47, 381–391.

Qin, Y.Y., Yang, J.Y., Lu, H.B., Wang, S.S., Yang, J., Yang, X.C., et al., 2013. Effect of chitosan film incorporated with tea polyphenol on quality and shelf life of pork meat patties. Int. J. Biol. Macromol. 61, 312–316.

Realini, C.E., Marcos, B., 2014. Active and intelligent packaging systems for a modern society. Meat Sci. 98, 404–419.

Reis, L.C.B., De Souza, C.O., Da Silva, J.B.A., Martins, A.C., Nunes, I.L., Druzian, J.I., 2015. Active biocomposites of cassava starch: the effect of yerba mate extract and mango pulp as antioxidant additives on the properties and the stability of a packaged product. Food Bioprod. Process. 94, 382–391.

Ren, L., Yan, X., Zhou, J., Tong, J., Su, X., 2017. Influence of chitosan concentration on mechanical and barrier properties of corn starch/chitosan films. Int. J. Biol. Macromol. 105, 1636–1643.

Rhim, J.-W., Park, H.-M., Ha, C.-S., 2013. Bio-nanocomposites for food packaging applications. Prog. Polym. Sci. 38, 1629–1652.

Ribeiro-Santos, R., Andrade, M., Melo, N.R., de, Sanches-Silva, A., 2017. Use of essential oils in active food packaging: recent advances and future trends. Trends Food Sci. Technol. 61, 132–140.

Robertson, G.L., 2013. Food Packaging: Principles and Practice. CRC Press, Boca Raton, New York.

Rodrigues, D.C., Caceres, C.A., Ribeiro, H.L., de Abreu, R.F.A., Cunha, A.P., Azeredo, H. M.C., 2014. Influence of cassava starch and carnauba wax on physical properties of cashew tree gum-based films. Food Hydrocolloids 38, 147–151.

Romani, V.P., Prentice-Hernández, C., Martins, V.G., 2017. Active and sustainable materials from rice starch, fish protein and oregano essential oil for food packaging. Ind. Crops Prod. 97, 268–274.

Romani, V.P., Machado, A.V., Olsen, B.D., Martins, V.G., 2018a. Effects of pH modification in proteins from fish (*Whitemouth croaker*) and their application in food packaging films. Food Hydrocolloids 74, 307–314.

Romani, V.P., Hernández, C.P., Martins, V.G., 2018b. Pink pepper phenolic compounds incorporation in starch/protein blends and its potential to inhibit apple browning. Food Packag. Shelf Life 15, 151–158.

Rubilar, J.F., Cruz, R.M.S., Silva, H.D., Vicente, A.A., Khmelinskii, I., Vieira, M.C., 2013. Physico-mechanical properties of chitosan films with carvacrol and grape seed extract. J. Food Eng. 115, 466–474.

Saberi, B., Chockchaisawasdee, S., Golding, J.B., Scarlett, C.J., Stathopoulos, C.E., 2017. Characterization of pea starch-guar gum biocomposite edible films enriched by natural antimicrobial agents for active food packaging. Food Bioprod. Process. 105, 51–63.

Sai-Ut, S., Benjakul, S., Rawdkuen, S., 2015. Retardation of lipid oxidation using gelatin film incorporated with longan seed extract compared with BHT. J. Food Sci. Technol. 52, 5842–5849.

Salarbashi, D., Tajik, S., Ghasemlou, M., Shojaee-Aliabadi, S., Noghabi, M.S., Khaksar, R., 2013. Characterization of soluble soybean polysaccharide film incorporated essential oil intended for food packaging. Carbohydr. Polym. 98, 1127–1136.

Salgado, P.R., López-Caballero, M.E., Gómez-Guillén, M.C., Mauri, A.N., Montero, M.P., 2013. Sunflower protein films incorporated with clove essential oil have potential application for the preservation of fish patties. Food Hydrocolloids 33, 74–84.

Salgado, P.R., Ortiz, C.M., Musso, Y.S., Giorgio, L.D., Mauri, A.N., 2015. Edible films and coatings containing bioactives. Curr. Opin. Food Sci. 5, 86–92.

Schmid, M., Merzbacher, S., Müller, K., 2018. Time-dependent crosslinking of whey protein based films during storage. Mater. Lett. 215, 8–10.

Seligra, P.G., Medina Jaramillo, C., Famá, L., Goyanes, S., 2016. Biodegradable and nonretrogradable eco-films based on starch-glycerol with citric acid as crosslinking agent. Carbohydr. Polym. 138, 66–74.

Shah, U., Naqash, F., Gani, A., Masoodi, F.A., 2016. Art and science behind modified starch edible films and coatings: a review. Compr. Rev. Food Sci. Food Saf. 15, 568–580.

Shahabi-Ghahfarrokhi, I., Khodaiyan, F., Mousavi, M., Yousefi, H., 2015. Effect of γ-irradiation on the physical and mechanical properties of kefiran biopolymer film. Int. J. Biol. Macromol. 74, 343–350.

Shakila, R.J., Jeevithan, E., Varatharajakumar, A., Jeyasekaran, G., Sukumar, D., 2012. Comparison of the properties of multi-composite fish gelatin films with that of mammalian gelatin films. Food Chem. 135, 2260–2267.

Shojaee-aliabadi, S., Hosseini, H., Amin, M., 2013. Characterization of antioxidant-antimicrobial κ-carrageenan films containing *Satureja hortensis* essential oil. Int. J. Biol. Macromol. 52, 116–124.

Singh, R.K., Singh, N., 2005. Quality of packaged foods. In: Han, J.H. (Ed.), Innovations in Food Packaging. Academic Press, London, pp. 24–44.

Siripatrawan, U., Noipha, S., 2012. Active film from chitosan incorporating green tea extract for shelf life extension of pork sausages. Food Hydrocolloids 27, 102–108.

Siripatrawan, U., Vitchayakitti, W., 2016. Improving functional properties of chitosan films as active food packaging by incorporating with propolis. Food Hydrocolloids 61, 695–702.

Song, Y., Zhang, K.F., Wang, Z.R., Diao, F.X., Yan, Y.N., Zhang, R.J., 2000. Coupled thermo-mechanical analysis for plastic thermoforming. Polym. Eng. Sci. 40, 1736–1746.

Song, N.B., Song, H.Y., Jo, W.S., Song, K.B., 2013. Physical properties of a composite film containing sunflower seed meal protein and its application in packaging smoked duck meat. J. Food Eng. 116, 789–795.

Soni, A., Kandeepan, G., Mendiratta, S.K., Shukla, V., Kumar, A., 2016. Development and characterization of essential oils incorporated carrageenan based edible film for packaging of chicken patties. Nutr. Food Sci. 46, 82–95.

Sothornvit, R., Pitak, N., 2007. Oxygen permeability and mechanical properties of banana films. Food Res. Int. 40, 365–370.

Souza, A.C., Goto, G.E.O., Mainardi, J.A., Coelho, A.C.V., Tadini, C.C., 2013. Cassava starch composite films incorporated with cinnamon essential oil: antimicrobial activity, microstructure, mechanical and barrier properties. LWT—Food Sci. Technol. 54, 346–352.

Stuchell, Y.M., Krochta, J.M., 1994. Enzymatic treatments and thermal effects on edible soy protein films. J. Food Sci. 59, 1332–1337.

Sun, Q., Xiong, C.S.L., 2014. Functional and pasting properties of pea starch and peanut protein isolate blends. Carbohydr. Polym. 101, 1134–1139.

Teixeira, B., Marques, A., Pires, C., Ramos, C., Batista, I., Saraiva, J.A., et al., 2014. Characterization of fish protein films incorporated with essential oils of clove, garlic and origanum: physical, antioxidant and antibacterial properties. LWT—Food Sci. Technol. 59, 533–539.

Tongnuanchan, P., Benjakul, S., Prodpran, T., 2013. Physico-chemical properties, morphology and antioxidant activity of film from fish skin gelatin incorporated with root essential oils. J. Food Eng. 117, 350–360.

Ünalan, İ.U., Arcan, I., Korel, F., Yemenicio, A., 2013. Application of active zein-based fi lms with controlled release properties to control *Listeria monocytogenes* growth and lipid oxidation in fresh Kashar cheese. Innov. Food Sci. Emerg. Technol. 20, 208–214.

Valencia-Sullca, C., Vargas, M., Atarés, L., Chiralt, A., 2018. Thermoplastic cassava starch-chitosan bilayer films containing essential oils. Food Hydrocolloids 107–115.

Vanderroost, M., Ragaert, P., Devlieghere, F., De Meulenaer, B., 2014. Intelligent food packaging: the next generation. Trends Food Sci. Technol. 39, 47–62.

Wang, H.J., An, D.S., Rhim, J.-W., Lee, D.S., 2015. A multi-functional biofilm used as an active insert in modified atmosphere packaging for fresh produce. Packag. Technol. Sci. 23, 253–266.

Wang, K., Wang, W., Ye, R., Liu, A., Xiao, J., Liu, Y., et al., 2017a. Mechanical properties and solubility in water of corn starch-collagen composite films: effect of starch type and concentrations. Food Chem. 216, 209–216.

Wang, K., Wu, K., Xiao, M., Kuang, Y., Corke, H., Ni, X., et al., 2017b. Structural characterization and properties of konjac glucomannan and zein blend films. Int. J. Biol. Macromol. 105, 1096–1104.

Wihodo, M., Moraru, C.I., 2013. Physical and chemical methods used to enhance the structure and mechanical properties of protein films: a review. J. Food Eng. 114, 292–302.

Wittaya, T., 2012. Protein-based edible films: characteristics and improvement of properties. Struct. Funct. Food Eng. 43–70.

Woggum, T., Sirivongpaisal, P., Wittaya, T., 2015. Characteristics and properties of hydroxypropylated rice starch based biodegradable films. Food Hydrocolloids 50, 54–64.

Yu, S.H., Tsai, M.L., Lin, B.X., Lin, C.W., Mi, F.L., 2014. Tea catechins-cross-linked methylcellulose active films for inhibition of light irradiation and lipid peroxidation induced β-carotene degradation. Food Hydrocolloids 44, 491–505.

Zavareze, E.D.R., Pinto, V.Z., Klein, B., Halal, S.L.M., Elias, M.C., Prentice-Hernández, C., et al., 2012. Development of oxidised and heat-moisture treated potato starch film. Food Chem. 132, 344–350.

Zhang, Y., Ma, Q., Critzer, F., Davidson, P.M., Zhong, Q., 2015. Physical and antibacterial properties of alginate films containing cinnamon bark oil and soybean oil. LWT—Food Sci. Technol. 64, 423–430.

Zhao, Y., Xu, H., Mu, B., Xu, L., Yang, Y., 2016. Biodegradable soy protein films with controllable water solubility and enhanced mechanical properties via graft polymerization. Polym. Degrad. Stab. 133, 75–84.

Zinoviadou, K.G., Koutsoumanis, K.P., Biliaderis, C.G., 2009. Physico-chemical properties of whey protein isolate films containing oregano oil and their antimicrobial action against spoilage flora of fresh beef. Meat Sci. 82, 338–345.

Further reading

Paula, A., Resem, D., Prentice, C., 2015. Development of an intelligent enzyme indicator for dynamic monitoring of the shelf-life of food products. Innov. Food Sci. Emerg. Technol. 30, 208–217.

Vermeiren, L., Devlieghere, F., Van Beest, M., De Kruijf, N., Debevere, J., 1999. Developments in the active packaging of foods. Trends Food Sci. Technol. 10, 77–86.

Optimized cold chain to save food

7

Samuel Mercier[1,2], *Martin Mondor*[3], *Ultan McCarthy*[4], *Sebastien Villeneuve*[3], *Graciela Alvarez*[5] and *Ismail Uysal*[1]
[1]Department of Electrical Engineering, University of South Florida, Tampa, FL, United States, [2]Department of Chemical and Biotechnological Engineering, Université de Sherbrooke, Sherbrooke, QC, Canada, [3]Saint-Hyacinthe Research and Development Centre, Agriculture and Agri-Food Canada, Saint-Hyacinthe, QC, Canada, [4]School of Science & Computing, Department of Science, Waterford Institute of Technology, Waterford, Ireland, [5]Refrigeration Process Engineering Research Unit, IRSTEA, Antony, France

Chapter Outline

7.1 Introduction

Perishable foods, such as fruits and vegetables, dairy, fish, and meat products, have a limited shelf life after harvest or production. The delay before they become unmarketable or inedible depends on the food product itself and a number of environmental factors. These environmental factors include the storage temperature, pressure and relative humidity, and composition and velocity of surrounding gas.

The temperature generally has the greatest impact on the shelf life of perishable food (Hertog et al., 2014; Nunes et al., 2014). A temperature too high increases the rate of respiration and the growth of microorganisms, which can spoil some food

Saving Food. DOI: https://doi.org/10.1016/B978-0-12-815357-4.00007-9

products in a few hours or days (Giannakourou and Taoukis, 2003; Hertog et al., 2014; Gwanpua et al., 2015). A temperature too low can create cold injuries and render the food product unmarketable (Heap, 2006; Aghdam and Bodbodak, 2014). In contrast, perishable food products kept at the proper temperature can generally remain of high quality for multiple days or weeks, or even multiple months or years in the case of frozen food (Mercier et al., 2017).

Therefore, refrigeration plays a critical role in food loss. This is especially true in this current state of globalization, with fresh produce continuously traveling long distances between countries and continents to meet consumers' expectation of a having a wide range of fresh produces available year-round. As an example, the supply of Mexico-grown blackberries to the United States, which includes stages of precooling, transportation to the distribution center (DC), storage at the DC, transportation to retail, and storage at retail, can typically take from 5 to 15 days (Nunes et al., 2014). Yet, the shelf life of berries can be below 1 day when field heat is not removed and below 1 week when kept at 10°C (Hertog et al., 1999). As such, it is critical that the perishable food remains at the proper storage temperature during all stages of the supply chain, to prevent decay at a rate that would make the product unmarketable before retail and create food loss.

A range of refrigeration technologies, of varying efficiency, cost, and environmental impact, is available to preserve the temperature of the food in the desired range during each stage of the supply chain. When refrigeration is applied along the supply chain to improve food preservation, the supply chain is called a cold chain. This chapter first provides a high-level overview of the different stages found along a typical cold chain. In the following sections, the different refrigeration technologies available at each stage of the cold chain are discussed, along with their impact on food loss and waste. A section will then be dedicated to the optimization of the cold chain, that is, how we can identify the proper refrigeration technology and operating conditions to reduce food loss and waste, while also limiting cost and environmental impact. Finally, notable differences between cold chains in different regions of the world are discussed, notably regarding the refrigeration capacities and reported food loss, waste, safety, and security.

7.2 Overview of the cold chain

The stages found along a typical cold chain of a chilled product are presented in Fig. 7.1. When the product is harvested (for fresh fruits and vegetables, FFVs) or processed (for processed fruit, vegetable, meat, and dairy products), its temperature is generally above the optimal range for the preservation of its quality and safety. As such, the food is first cooled to the desired range in a precooling facility, or in some cases, directly within a refrigerated container or vessel. The food product is then transported by land to one or a series of DCs, where pallets are sorted and sent to the proper client based on product demand and a predetermined management system. The food product can also transit through a number of cross-docking sites,

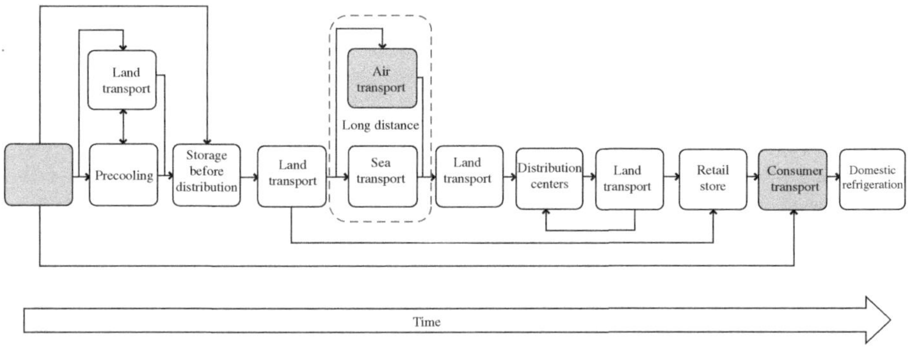

Figure 7.1 Overview of the main stages along a typical cold chain. Shaded area denote stages where no refrigeration is generally applied.

to combine shipments of small suppliers and reduce transportation cost (Mack et al., 2014). Depending on the distance-to-market, transportation can be by land or a combination of land and sea or air transportation. After transiting through the last DC, the food product is transported to a retailer, where it is stored in a backroom or in a refrigerated display cabinet until it is sold to consumers. For small growers in developing countries, the food product can often bypass the distribution stages of the cold chain and be sent directly from harvest to retail, generally nearby farmers' markets (Sibomana et al., 2016). Once the food product has been bought by consumers, it is transported home and stored in a domestic refrigerator until consumption.

7.2.1 Precooling

Precooling is commonly defined as the critical postharvest procedure that immediately follows harvest to quickly lower the temperature of FFVs to minimize spoilage and maximize shelf life and product quality. Poor precooling after harvest creates food waste and loss, endangers food safety, and represents a major economic problem to the produce industry. Delaying precooling by 6 hours at ambient temperatures can increase decay incidence by more than 25%, while a 50% increase in water loss upon arrival at the DC can be induced by a simple 4-hour delay between harvest and precooling (Nunes et al., 2005; Pelletier et al., 2011). The consequences of poor precooling are usually not apparent until later during distribution, and only a fraction of the decay or water loss demonstrated above will result in rejected loads, lost sales, and consumer dissatisfaction, resulting in wasted food (Mercier et al., 2017).

Mainly there are five different ways for precooling fresh produce—room cooling, forced-air cooling, water or hydrocooling, ice cooling, and vacuum cooling—specifically recommended based on the characteristics of the product. These characteristics include the sensitivity to chilling injuries, scale of the harvest, the minimum/maximum time required for precooling as well as product packaging. Each method has its own advantages and disadvantages with a detailed comparative analysis provided by Kader and Rolle in their book titled *The Role of Post-Harvest Management in Assuring the Quality and Safety of Horticultural Produce* (Kader and Rolle, 2004).

Room cooling is a relatively slow procedure suitable for products that generally do not decay rapidly and may be more sensitive to moisture (Thompson, 2004). As the name implies, the products are placed in a refrigerated room (which can either be a storage facility or a temperature-controlled transportation container) where there is constant circulation of cold air with the help of cooling fans. It is important to note that the air circulation in room cooling is mild and occurs around the produce instead of going through them. Room cooling is generally recommended for produce such as zucchini and squash, as well as potatoes, tomatoes, and cabbage. Among the five different precooling techniques, room cooling is both the most energy efficient and the slowest one and also requires careful placement and stacking of produce to maximize airflow between pallets.

Forced-air cooling is similar to room cooling except the cold air is "forced" to circulate through the produce instead of the ambience as in the former (Ferrua and Singh, 2009). It is by far the most effective and widely used precooling technique as it is used to cool palletized produce at industrial scales. The airflow can be horizontal or vertical depending on the placement of pallets and cooling equipment. Rows of pallets of fresh produce are covered with a plastic tarp to ensure controlled airflow from one side to the other. A cooling tunnel consists of two or more rows of produce pallets with an air intake fan in the middle that draws colder ambient air outside the tunnel through the warm produce to cool them down quickly. This air is recirculated back to the room after it goes through thermal equipment to bring it back down to desired product temperatures. Challenges remain, however, especially in regard to the uniformity of product temperature distribution as well as over cooling (freeze damage) and under cooling (loss of shelf life) of produce. Forced-air cooling is generally recommended for produce such as beans, berries (including strawberries, blueberries, and raspberries), carrots. and leafy greens.

In water or hydrocooling, cold water is applied to produce in a variety of ways such as submersion in a tank to sprinkling/spraying or passing the palletized products through constant streaming water. Additionally, the water can be treated with chemicals to control bacteria, which can cause spoilage (Reina et al., 1995). Hydrocooling is desired for situations where faster cooling is necessary to preserve product quality as water possesses a higher cooling capacity than air. However, it is also important to note that not all products can be cooled with water, which causes more spoilage in sensitive produce such as berries. Hydrocooling is typically recommended for products such as oranges, peaches, sweet corn, and cucumbers.

Ice cooling is similar to hydrocooling in that instead of cooled water, crushed, slurry ice is used to rapidly cool down the produce (Siegel et al., 2012). Ice is injected into the produce package to provide not only faster precooling times but also to extend the amount of time the product stays cooled down. It is important that ice is neither contaminated with bacteria, nor includes any chemical that might be harmful to the produce or human consumption. Ice cooling is typically recommended for products such as broccoli, onions, and parsnips.

Finally, in vacuum precooling the produced is placed in a chamber where the air is pumped out to create an environment where the atmospheric pressure is reduced immensely compared with normal atmospheric pressure (Dongquan et al., 2002). Hence, the boiling point of water is lowered, which results in significantly faster evaporation and ultimately cooling down of the produce. While, vacuum cooling is one of the fastest ways to lower produce temperature, it has its own challenges mainly in controlling the increased water loss and energy costs in running the vacuum chambers. Vacuum cooling is mostly recommended for leafy vegetables like lettuce and creates the most uniform temperature distribution among all precooling techniques.

7.2.2 Commercial transportation

7.2.2.1 Sea transportation

Refrigerated sea transportation is an important link in the food supply chain and is an important mode of transportation for dairy products, fish products, fruits, meats, vegetables, and other food products. The two most commonly used sea transportation methods for refrigerated shipments are transport in specialized refrigerated containers (also known as reefers) incorporating a mechanical vapor compression refrigeration unit into an insulated container, or in the bulk holds of dedicated refrigerated cargo ships. Wild (2009) reported that in 2000 each transportation method was accounting for approximately 50% of the world's sea refrigerated cargo transportation. Refrigerated ships were mainly dedicated to the transportation of frozen meat and bananas. Today, these ships have been partially replaced by refrigerated containers that only require an external source of electricity to operate the refrigeration system and remove heat from the container's internal environment (Fitzgerald et al., 2011). The shift from bulk to refrigerated containers is in large part due to the longer shelf life that is made possible by faster delivery for refrigerated containers, which results in greater cost-efficiency, especially for small shipments (Jedermann et al., 2014; Arduino et al., 2015). For both specialized refrigerated ships and refrigerated containers, the delivery of cold air through floor gratings is the most typical airflow pattern. The air then passes vertically through the produce and returns to the cooling unit along the ceiling (Smale, 2004). In addition to the airflow pattern, factors that will significantly influence the temperature distribution inside ships and refrigerated containers are the operation and design of the container or the ships, packaging, stacking mode, and the properties of the food product (Tanner and Smale, 2005). During the shipment of kiwis from New Zealand to Belgium, Tanner and Amos (2003) monitored the temperature inside a specialized ship and a refrigerated container. They observed significant temperature variability both spatially across the width of the container and temporally making the temperature control system inefficient since it was operated based on the basis of single-position temperature measurement in the container. A decrease of the delivery air temperature down to as low as −5°C for short periods and −2.5°C for extended periods was observed in response to a temperature measured by the sensor

0.5°C above the set point, increasing the likelihood of fruit freezing injuries. Although the number of kiwis outside the recommended temperature range was significant, they reported a more uniform temperature for the transport inside the cargo of the ship than for refrigerated containers. Amador et al. (2009) performed the temperature mapping of a load of crownless pineapples inside a refrigerated container shipped from Costa Rica to the United States (Florida). Their results suggest that the pineapples near the bottom of the pallets were more at risk of chilling injuries since the temperature was lower by as much as 3°C–4°C than near the top of the pallets. This difference was explained by the vertical airflow pattern inside the refrigerated containers. Defraeye et al. (2016) simulated an ambient loading protocol for sea transportation of oranges in refrigerated containers. A cooling period of 21 days was selected to mimic a period of forced-air precooling of 3 days followed by an 18-day transportation period. Performance of the standard ambient loading method was compared with that of the channeling configuration, which reduced airflow bypass between pallets, and the horizontal configuration, which forced the air horizontally across the pallets. Results indicated that the standard ambient loading protocol and the channeling configuration exhibited similar cooling behavior and were able to cool the produce within about 3 days to the seven-eighths cooling time. However, the oranges were of better quality and lasted longer in shelf life conditions for the channeling configuration due to less moisture loss. The horizontal configuration performed worse on all aspects. Mai et al. (2012) performed the temperature mapping for three air and three sea shipments of fresh fish transported from Iceland to the United Kingdom or France. They observed that the temperature was less stable during air transport than during sea transport, as transportation by sea using refrigerated containers reduced the number of handling operations during which the pallets could be exposed to high ambient temperatures. However, they also observed that the predicted remaining shelf life is shorter for sea transport than for an air shipment with precooled product due to the long transportation time required for the sea shipment. This indicates that several factors are to be considered when selecting the transportation mode including quality and safety of the produce, time to reach destination, as well as transportation cost. In their work, Kan et al. (2017) have employed computational fluid dynamics (CFD) technique to model and simulate the influence of cargo stacking on temperature distribution of a 20-ft. THERO-KING standard reefer container commonly used for sea transportation. They studied the impact of the height and length of the cargo stack, as well as of the space between cargo stack and the sidewall surface. They found that with the increase of the stack's height, the return air channel gradually narrows down resulting in an increase of the uneven temperature inside the containers. They also observed that an increase in the stack's length results in the apparition of enlarging high-temperature zone near the door, which also results in an increase of the uneven temperature inside the containers. Concerning the impact of the space between cargo stack and the sidewall surface, they observed that the heat transfer between the air and the cargo in the container improves with an increase of this space reducing the presence of uneven temperature profile inside the containers. They conclude that CFD technique can provide useful information in making a decision on the dimensions and stacking of the cargo to optimize the temperature distribution in the sea refrigerated containers.

7.2.2.2 Rail transportation

Transportation of containers by rail is a natural extension of sea container transport and a link of intermodal transport, it may also be an independent transportation mode. While rail transport carries huge amounts of corn syrup, French fries, canned goods, FFVs, corn and soybean oil, frozen chickens, sugar, and pasta (Foodlogistics.com), it is far from being the preferred mode of transport for the food industry because it takes a lot of time compared with all-road transport for the transportation of perishable goods (Sommar and Woxenius, 2007). As such, documentation on the temperature conditions observed during rail transport is scarce. However, there is significant growth potential owing to some current developments including high-speed trains reserved for medium-sized loads transportation, with maximum reduction of intermediate times (Fronda, 2013), and the promotion and development of fast intermodal transport solutions (Inbound Logistics, 2010; Sandberg Hanssen and Mathisen, 2011). Rail transportation also has low external costs (cost of accidents, congestion, air pollution, greenhouse gases, and noise), when compared with those of a general freight truck (Forkenbrock, 2001).

7.2.2.3 Intermodal transportation

Intermodal transportation is defined as the movement of goods in a single loading unit by a sequence of at least two transportation modes (road, sea, rail or air), the transfer from one mode to the next being performed at an intermodal terminal. Many varieties of intermodal containers are in use along the food supply chain but the standard dry-freight containers (20′, 40′, 45′ length; 8′6″ height) are the most common, while refrigerated containers (reefers) used for transportation of perishable goods (5%) represent a growing segment (Rodrigue, 2013). The main advantage of intermodal transportation is its low transportation costs compared with transportation by freight truck since usually the most suitable transport mode is used along the food supply chain (Sandberg Hanssen and Mathisen, 2011). On the other hand, intermodal transportation requires movements that may damage fragile goods, and it relies on time-consuming transshipment compared with all-road transportation, which may result in a decrease in the remaining shelf life of perishable food products and thus a decrease of its value. For example, Lervåg et al. (2001) reported that a delay of 48 hours in the transportation of fresh fish results in a price reduction of between 20% and 25%. In addition, even if the refrigerated containers slow down the decay of perishable food products, they do not completely eliminate it. As it is the case for other loading units, they can also be subject to potential temperature abuses that impact the remaining shelf life of the products. For bananas, Haass et al. (2015) reported that an increase in the reefer temperature from 15°C to 20°C may increase the daily ripening rate of bananas by as much as 75%.

In their work, Dulebenets et al. (2016) proposed a novel optimization model to minimize the total cost associated with the transportation and decay of perishable food products during intermodal transportation. The intermodal freight network for the import of seafood perishable products to the United States has been used to conduct comprehensive numerical experiments to identify important managerial insights. The numerical cases show that product decay cost significantly affects the transportation modes and the associated total transportation time and distribution costs. For example, an increase in the product decay cost was found to reduce the total miles traveled by rail while increasing the use of road transportation, which provides faster delivery and decreases the total product decay. However, the use of road transportation is more costly in terms of distribution costs. They also showed that the total miles traveled by sea was reduced with increasing product decay cost. This type of model can help in making a decision on the transportation modes to be used to minimize the total cost associated with intermodal transportation.

7.2.3 Storage at the distribution center

In the late 19th and early 20th centuries the introduction of cold chain principles across food systems was slowly being adopted. At the time this was not a fully welcomed development and it received much negative attention. This, in part, was due to a lack of trust on the part of the consumer given the fact that they were now being offered food stuffs that traditionally were only available at selected times of the year (due to local harvest seasons). This, new, year-round offering challenged social thinking at the time both locally and nationally and was a topic of many negative conversations (Freidberg, 2015). It also presented a number of new unforeseen challenges to existing legislation and know-how governing logistics at the time. The adoption of cold chain at that time in history would undoubtedly have required significant fundamental social and operational restructuring as well as all the associated financial investment.

Nowadays cold chain is widely adopted and it is difficult to present a counter argument against it. Social attitudes have evolved and people do not have the same negative views or distrust for refrigerated food supply chains once held. In fact one of the key issues causing distrust of the traditional cold chain (the year round product offering) is now a basic requirement of the modern day consumer. Cold chain, as a direct result of this year round offering, has also helped provide a more sustainable business case for produce as these business cases are generally harder to build (Liu et al., 2018). Cold chain adoption has also had direct positive implications on food security and waste and loss (McCarthy et al., 2018). Cold chain adoption has undoubtedly transformed the "food lives" of the modern day consumer, given the fact that it, in partnership with other advances, has transformed food supply chains from adopting a localized "narrow" trading span to now offering product and services across geographies, time zones, language zones, and cultures. By virtue of its name the cold chain is regulated and audited through its ability to maintain a specific temperature at which food stuff must be transported and stored to avoid premature spoilage.

Modern day DCs are central to cold chain applications and play a crucial role in the safe, efficient, and responsible carriage and storage of food stuffs and it is for this reason alone their importance cannot be overlooked. Food DCs serve as a point in time where the food can be visually inspected, assessed, and accounted for through human contact/visualization. Modern day organizations use food DCs as a hub for central/regional storage and holding sites that are strategically located at equally distance between food stores to facilitate timely and cost effective store replenishment. In theory this means that carefully placed DCs allow for all types of product (produce, dried, canned, apparel, etc.) to be within a set distance of each store nationally, and even globally, thereby reducing shelf replenishment time and out of stock occurrences. Such careful placement of DCs also helps reduce the occurrence of product loss and waste during transport. With respect to product loss and waste there is an important distinction to be made between loss and waste. Food loss is the loss of food before it reaches retail, that is, food that gets spilled or spoilt before it reaches its final product or retail stage. Food waste is considered to be food that is fit for human consumption but is not consumed and left to spoil, that is, food that is left to spoil or discarded by retailers or consumers (FAO, 2011). Collectively food waste and food loss have been reported to amount to about 1.3 billion tons per year or one-third of food produced for human consumption. While DCs alone cannot be held accountable for this 1.3 billion tones, the correct and efficient operation of DCs will certainly help reduce this figure.

To further represent the challenges faced by DC managers it is a well-established fact that not all foods have the same optimum storage temperature. As a consequence, modern day cold chain trading between DCs has been carefully tailored to the product as opposed to the process, that is, each product is and should be handled in accordance with its best practice (optimum temperature). It is no longer acceptable to simply establish trading links, and distribute food at any given temperature without considering the type and nature of the product being shipped. Improving global food supply chain and food handling processes can help address current levels of food waste, which have far-reaching impacts on global food security, resource efficiency, conservation, and climate change. The increasingly detailed regulations established in developed countries over the last few years reflect these product-specific handling requirements (Fig. 7.2). The product-specific optimal handling and storage conditions undoubtedly add complexity when it comes to food distribution and DC management and have resulted in modern day DCs being designed to accommodate a large variety of food types, shapes, sizes and weights with each requiring different storage requirements. Therein lies the cold chain challenge, that is, the development of an efficient supply network that will base decisions on product-specific information and distribute this information across the complete producer/consumer supply axis in real time (McMurray et al., 2013).

There are currently a large variety of technologies available to help improve the cold chain management each with the primary aim of reducing product losses and increasing profitability (Badia-Melis et al., 2018). Such systems, to function correctly, must consider both the incoming and outgoing flow of goods. Irrespective of

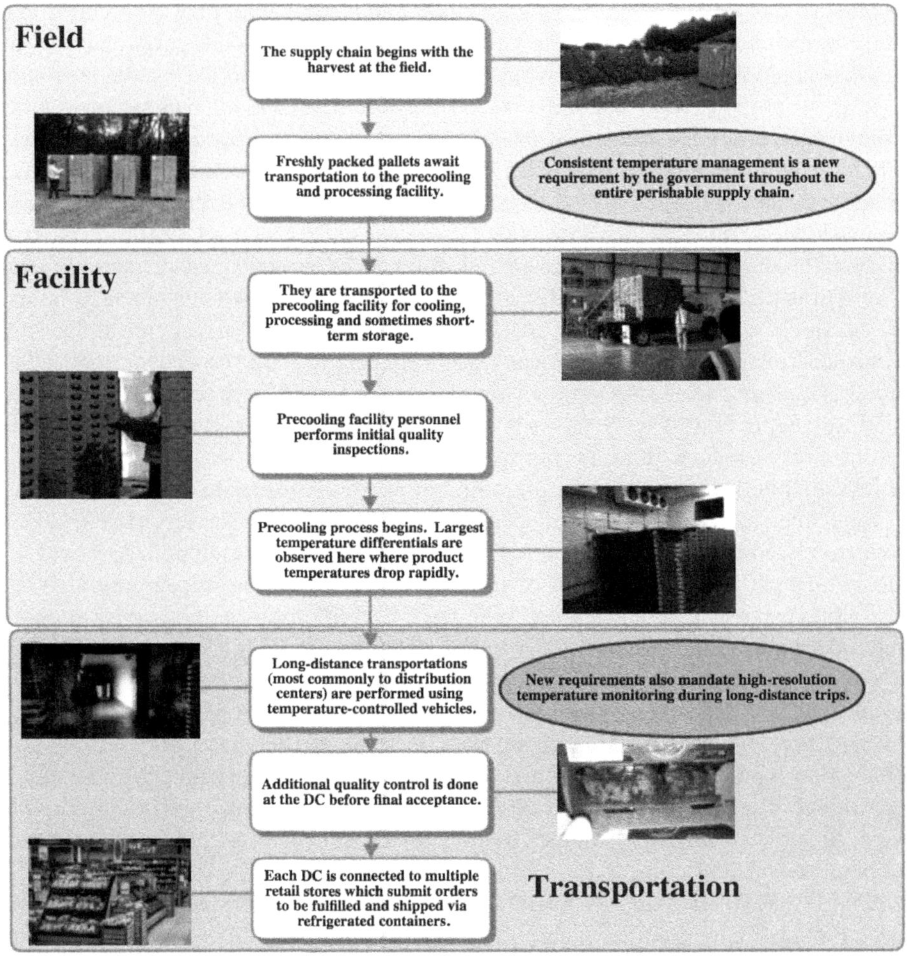

Figure 7.2 Stages along the strawberry cold chain where good practices need to be improved to comply with recent United States Food and Drug Administration (FDA) regulations.

function, form factor, functionality, cost, and level of integration these systems have the primary aim of increasing transparency of the food product during transport and storage. These systems essentially have the ability to act as custodians of the product. To be in a position to do this the systems must be able to facilitate and monitor the transit of physical product, with cyber (business/product) information (Smith, 2008). This complexity of carriage combined with where the product has been sourced in addition to where it is going relative to its current state of freshness is not a trivial task and not one that can be made in autonomy as, every decision, no matter how local, can have significant effect on the product, retailer, and the profitability of the company. Therefore there is a strong requirement that these systems function in totality across supply chains, DCs, products, time zones, and geographies.

As previously mentioned varying food types possess different ideal storage temperatures, which, as a general rule, may be grouped into three common commercial groupings, that is, frozen, chilled, and/or ambient, with each group having internal variation (i.e., ambient storage temperature may vary for different product types). In an ideal theoretical supply chain all foods (irrespective of grouping) being transported from DC-A to DC-B are loaded onto a single transport unit (truck, plane, ship, etc.) and delivered from DC-A to destination DC-B. The reality is however that many transporting units do not have an ability to mix groupings (i.e., frozen and ambient); requiring the use of a second transport (truck, plane, ship, etc.) unit traveling in unison with the first, thus doubling transport costs between DC-A and DC-B.

The science of cold chain management requires an understanding of the chemical process of food spoilage, its environmental triggers and inhibitors, its technological guardianship, and monitoring driven by business process compatibility. As a result of this decision making requires a multidisciplinary approach in a dynamic environment. DC and food logistics managers must consider a number of critical influencers prior to execution that individually and collectively are designed to reduce food waste and/or loss during transit and/or immediately on arrival at destination. Some of these decision influencers include:

1. the deterioration rate of the product at optimum conditions;
2. the state in which the food has arrived to the DC (has any negative occurrence taken place that may reduce the expected shelf life);
3. the conditions under which it must be kept at your DC;
4. product demand local and nationally;
5. the remaining shelf life of the product;
6. travel time to destination; and
7. the distance it has to travel to retail relative to remaining shelf life (Hertog et al., 2014).

Also, many food DC managers are faced with the challenge of making these decisions in the context of business process efficiency. This is due to the fact that the unfortunate reality is that, in either scenario, the negative effects of product spoilage are significant given the fact that product that spoils during transit is money wasted on transit costs and secondly product that spoils on DC arrival is, again, money wasted on transport costs, and also storage and disposal costs.

The role and importance of food DCs cannot and must not be overlooked. As food companies modernize and tend towards vertical ownership of their value chains, combined with advances in communication technologies, there are obvious efficiencies being achieved at DCs year on year. Within each company correct DC placement and management can have a direct positive impact on the social, financial, and security elements of food supply thus reducing food loss and waste across our supply networks. It is also important to note that efficient management of food DCs cannot be done in autonomy and decision making must account for the full supply network as opposed to being made at the local level.

7.2.4 Display at retail

After transiting through the last DC, the food products are transported to a retailer, where they are stored in a backroom or in a refrigerated display cabinet until they are sold to consumers. Different types of refrigerated display cabinet are used worldwide, including vertical open cabinets, vertical reach-in cabinets, tub-type cabinets, and service deli cabinets. The most common type is the vertical open-refrigerated display cabinet. This type of cabinet is equipped with a recirculating air curtain, which provides a barrier between the interior conditioned area of the cabinet and the ambient air in the retail space, with concomitant easy access to the food products by the consumers (Jouhara et al., 2017). However, they often fail to provide the temperature necessary for proper storage of the food products and they are not energy efficient. Numerous research works have been carried out to identify and optimize the variables involved in the effective design of air curtains (Bhattacharjee and Loth, 2004; Navaz et al., 2005; Yu et al., 2009).

In their work, Willcox et al. (1994) reported temperature performance for a vertical three-deck cabinet and for an horizontal one-deck cabinet used to display minimally processed, modified atmosphere packed vegetables. Both cabinets were cooled by fan-assisted cold air. For the vertical cabinet, the mean temperature measured during 1 week varied between 6.9°C and 12.3°C for the deck 1 and 3 located at the bottom and at the top of the display cabinet, respectively. Only the temperature performance of deck 1 complied with the Belgian legal requirement. For the horizontal display cabinet, the temperature distribution was more uniform and in average at 7.0°C or less, which complied with the requirement. Temperature performance of both display cabinets was influenced by the ambient air temperature as well as by the day/night regime. Nunes et al. (2009) have monitored the temperature inside 27 refrigerated and nonrefrigerated retail display cabinets. Refrigerated display cabinets were open stand-up or low cases, and were not movable, while the nonrefrigerated cabinets were also open but in some stores they were movable allowing for their positioning with the department. For the refrigerated cabinets, temperature setting was between 2°C and 4°C. The temperatures varied from −1.2°C to 19.2°C inside refrigerated retail displays, for middle shelves and "before" lower shelves, respectively, while it varied between 7.6°C and 27.7°C inside nonrefrigerated retail displays for bottom shelves and upper shelves, respectively. Maximum temperature performance reported for the refrigerated retail display cabinets was too high for produce requiring low temperature, and the minimum temperature performance too low for chilling-sensitive produce. For most of the produce stored in nonrefrigerated displays, the temperatures were too high. This was particularly problematic at the middle and top of the display. Morelli et al. (2012) studied the performance of open horizontal refrigerated display cabinets used for bakeries, pork butchers/delicatessens, and cheese/dairy products. They monitored the temperature of both food products and air in the refrigerated retail display cabinets and reported that 70% of time/temperature food product profiles were above 7°C and were thus unsatisfactory. These unsatisfactory results were attributable in equal part to poor refrigerated retail display design and poor professional practices. Zeng et al. (2014) monitored the temperature profiles of bagged salads in display cases at nine supermarkets for 2 months during summer (July to September) and for 2 months during the subsequent winter (January to March). The bagged salads were typically held for 1–3 days and were displayed for a maximum of 3 days. A total of 3799 temperature profiles were obtained. Temperature profiles indicated minimum and maximum temperature peaks ranging from 1.0°C to 14.1°C and a distribution of mean measured temperatures ranging between −1.1°C and 9.7°C. This demonstrates that temperature abuse is likely to occur during retail display. In the same work, Zeng et al. (2014) demonstrated that such abuse increased the growth probabilities for both *Escherichia coli* O157:H7 and *Listeria monocytogenes* used to inoculate bagged salads. Kou et al. (2015) monitored the spatial and temporal temperature variations within two commercial open-refrigerated display cases, consisting in three 4-ft. sections and five modular shelves that are with flexible placement, filled with a total of 72 spinach packages. Temperature was found to be the highest (average 6.5°C) in the front rows of the display cabinet while temperature in the back was the lowest (average −0.8°C), and was below freezing, which resulted in spinach damage. To solve this problem, insulating foam boards were installed. Temperature variation between the front rows and the ones in the back was successfully reduced by 3.5°C and enabled spinach packages in the front rows to remain at temperature below 5°C as recommended by the FDA. Brown et al. (2016) have monitored the temperature in nine 8-ft. display cases containing fresh-cut leafy greens. Monitors were placed in the lower bin and at the front and back of shelves. High-temperature abuse was recorded in all display cases while low-temperature abuse was recorded in five of the nine display cases. For at least 5% of the time, 40.2% of the sensors recorded temperature above 7.22°C while 17.2% of the sensors recorded temperature below −0.17°C. Temperatures were often too high at the top and too low at the bottom of the display cases. As reported in the aforementioned works, a wide range of temperature during display at retail has been observed. It can be due to several factors but the heterogeneity of temperature according to the position inside the display cabinet is certainly one of the main factors.

7.2.5 Transportation and storage by consumers

The last stages of the cold chain are under the control of consumers. Once the food product has been bought, it is generally transported home without any refrigeration, and thus at a temperature too high for proper preservation. Derens et al. (2006) measured the temperature of yogurt and meat products during transportation by consumers and reported that 85% of the products reached a temperature above 6.0°C. Morelli and Derens-Bertheau (2009) and Gogou et al. (2015) reported that the average temperature of smoked salmon and meat products during transportation by consumers was of 13.0°C and 9.8°C, respectively. The aforementioned studies reported an average duration of transportation by consumers between 40 and 75 minutes. Although it is hard to accurately quantify the impact of transportation by consumers on the amount of food waste, 1 hour at a temperature above 10°C can be sufficient to impact food quality and safety. As an example, for strawberries, 1 hour at 12°C is equivalent to nearly 4 hours in shelf life loss (Hertog et al., 1999). Similarly, James and Evans (1992) estimated that *Pseudomonas* can grow by up to two generations on a variety of perishable food products during domestic transportation in summer conditions. However, the growth can be maintained below 0.5 generation using an insulated box. As such, protecting perishable food products using insulated bags or containers is a sound practice to adopt when the ambient temperature is warm or the delay between retail and storage in the domestic refrigerator is significant.

Before the perishable food is consumed, it is generally stored in a domestic refrigerator. Three main types of domestic refrigerators are used: icebox refrigerators containing a box-plate evaporator, larder refrigerators with a backplate evaporator, and fridge-freezers with a backplate evaporator (James et al., 2017). Many studies have investigated the average temperature and the temperature distribution inside domestic refrigerators (Mercier et al., 2017). Some studies suggest that icebox refrigerators (Evans et al., 1991; Janjić et al., 2016) and refrigerators of smaller size (Laguerre et al., 2002) have a better performance and a more uniform temperature, while other studies have not found a significant difference in performance between refrigerator types (James et al., 2017). Most studies have reported the presence of significant temperature heterogeneity inside domestic refrigerators, regardless of the type. James and Evans (1992) and Bakalis et al. (2003) reported that the temperature difference between the warmest and coldest locations was above 5°C in most refrigerators. The warmest region is generally located inside the door: as an example, Janjić et al. (2016) reported that the temperature inside refrigerators in Serbia was on average 2.0°C higher when it was measured inside the door in comparison to the bottom shelf. However, conflicting results have been obtained regarding the temperature distribution along the height of refrigerators. Some studies have reported that the temperature is generally higher at the top shelf (Evans et al., 1991; New Zealand Foodsafe Partnership, 2004), while other studies have found that the temperature is generally higher at the bottom shelf (Godwin et al., 2007). Laguerre and Flick (2004) have also reported that the shelf with the highest temperature can change over time.

Nearly every study investigating domestic refrigerators has reported that the temperature inside the majority of refrigerators is too high. The recommended temperature inside refrigerators varies between countries, but is generally below 5°C (James et al., 2017). Yet, the majority of studies suggest that the average temperature inside domestic refrigerators is between 6°C and 7°C (Mercier et al., 2017). Brown et al. (2014) estimated the impact of the high temperature in domestic refrigerators on the amount of food waste in the United Kingdom. The authors assumed that the average temperature inside domestic refrigerators is currently of 7°C and that the amount of food saved would be proportional (the proportionality constant depending on the food product) to the increase in shelf life achieved from operating the refrigerators at a lower temperature. Based on these hypotheses, the authors estimated that maintaining refrigerators at a temperature of 4°C would save in the United Kingdom about 71,000 tons, or £162.9 million, of food annually. The authors also confirmed that the amount of energy and CO_2 saved from the consumption of these 71,000 tons of food would significantly exceed the amount of energy and CO_2 required to operate refrigerators 3°C below their current average temperature.

7.3 The cold chain around the world

7.3.1 Refrigeration capacities

The availability of proper infrastructure to preserve perishable food within the desired temperature range along the cold chain varies significantly across regions of the world. Table 7.1 presents the refrigerated warehouse capacity for the top 12 countries in 2014. While developing countries such as India, China, Brazil, and Indonesia have some of the largest refrigerated warehouse capacities in the world, their capacities are much smaller than most developed countries when considering the size of their population. On a per capita basis, the refrigerated warehouse capacity in India, China, Brazil, and Indonesia is below one-third of the capacity in most developed countries. In accordance with the difference in refrigerated warehouse capacity between developing and developed countries on a per capita basis, Bresolin et al. (2018) estimated that, for instance, current refrigerated warehouse capacities in Brazil only meet 29% of the country's cold chain needs.

While breaks along cold chains in developed countries are generally only observed during waiting times and transfer between stages of the cold chain, entire stages can be performed without refrigeration in developing countries, with great consequences on food loss and safety. In Africa and China, many small or medium size dairy or fruit and vegetable farms do not have access to precooling facilities because of their high capital and operating costs (Torres-Toledo et al., 2018;

Table 7.1 Refrigerated warehouse capacity (total and per capita) for the top 15 countries in 2014 (GCCA, 2014)

Country	Total refrigerated warehouse capacity (Mm^3)	Refrigerated warehouse capacity per capita (m^3 per habitant)
India	130.7	0.1
United States	114.8	0.4
China	76.1	0.1
Japan	32.7	0.3
Great Britain	25.0	0.4
Germany	24.0	0.3
Brazil	16.1	0.1
Netherlands	16.0	0.9
France	15.5	0.2
Iran	14.0	0.2
Indonesia	12.3	0.1
Canada	8.9	0.3

Zhao et al., 2018). The absence of precooling accelerates food decay, can trigger pathogens growth, and can result in the rejection of the food at the collection center, leading to substantial food and income losses (Torres-Toledo et al., 2018). Perishable food products are also frequently transported without refrigeration because of a deficit in refrigerated truck capacities. In India, it is estimated that more than five times the current number of refrigerated trucks would be required to meet cold chain needs (NCDD, 2016). In China, it is estimated that 85% of perishable food is transported in regular trucks, without refrigeration (USDA, 2008; Zhao et al., 2018). In contrast, about 90% of perishable food is transported in refrigerated trucks in developed countries (USDA, 2008). Significant differences can also be observed between the locations were fresh produce is bought by consumers in developed and developing countries. In developing countries, a large fraction of food products is sold in farmers' markets: as an example, more than 70% of agricultural products are sold in farmers' markets in China (Zhao et al., 2018). Farmer's markets have significant advantages, notably by removing intermediaries between the producers and the buyers and improving the accessibility to locally grown and low-cost fresh produce. However, the absence of proper refrigeration equipment, poor sanitation practices, and limited food quality inspection can create a significant amount of food waste at the retail level in developing countries (Torres-Toledo et al., 2018; Zhao et al., 2018).

Within developing countries, substantial differences can also be observed between small-scale and large-scale growers. Refrigeration can be completely absent along supply chains from small-scale growers to retail (Torres-Toledo et al., 2018). The absence of refrigeration creates food safety risks, and increases product rejection rate and food loss. The shorter shelf life in the absence of refrigeration also limits selling opportunities to markets located close to the production site, which can substantially reduce profit. In contrast, more modern infrastructure, closer to the infrastructure found along cold chains in developed countries, is often observed for large-scale growers selling their products to supermarkets (Torres-Toledo et al., 2018). The improved postharvest management observed for large-scale growers is stimulated by the stricter food quality requirements of supermarkets, and the postharvest management practices required by international standards when food products are exported (Torres-Toledo et al., 2018).

Another significant problem observed in some regions is the inability of consumers to preserve perishable food in the desired temperature range at home. In developed countries, nearly every household has a refrigerator. In North America, about 25% of households even have two or more (EIA, 2009; Statistics Canada, 2009). However, the household penetration rate of refrigerators is much smaller in many developing countries. In China, 91.7% of households have a refrigerator in urban areas, but the figure drops to 77.6% in rural areas (Zhao et al., 2018). In South Africa, 68.4% of households have a refrigerator (Lesame, 2014). In Indonesia, the percentage of households with a refrigerator is of 55.5% in urban areas, and only of 24.7% in rural areas (Statistics Indonesia, 2012). The major reasons for the absence of a refrigerator are the high capital cost and the absence of a reliable electricity source (Aste et al., 2017).

It is, however, important to note that progress is seen in many developing countries. The total refrigerated warehouse capacity in the world increased by 8.6% from 2014 to 2016 (GCCA, 2016). The most significant increases are observed in developing countries: for instance, the refrigerated warehouse capacity in Brazil, China, Mexico, and India increased by more than 25% annually from 2008 to 2014 (ITA, 2016). New technological developments also represent a key driver for the improvement of the cold chain in developing countries. Examples of promising technological developments include low-cost and sustainable refrigeration technologies powered by renewable energy sources, such as on-farm precooling and refrigeration systems powered by biogas or solar energy (Islam and Morimoto, 2014; Aste et al., 2017; Torres-Toledo et al., 2018). The development of accessible refrigeration technologies based on renewable energy sources is critical to the improvement of the cold chain, especially given that more than a quarter of the population living in rural regions still does not have access to electricity (The World Bank, 2014).

7.3.2 Food loss and waste in different countries

Given the larger refrigeration capacities found in developed countries, one could expect the proportion of food lost and wasted to be smaller than in developing countries. However, the opposite is actually observed: according to the 2011 report from the Food and Agriculture Organization of the United Nations (FAO, 2011), the annual mass of food loss and waste is approximately 280–300 kg per habitant in North America and Europe, in comparison to 120–170 kg per habitant in Sub-Saharan Africa and South/Southeast Asia. Differences in consumer behavior is a major factor explaining the higher amount of food loss and waste in North America and Europe. In North America and Europe, consumers waste annually, on average, 95–115 kg of food, about 10 times the amount of food wasted by consumers in Sub-Saharan Africa and South/Southeast Asia. Another factor is the higher rejection rate of food in developed countries resulting from stricter quality standards. For instance, approximately 20% of FFVs are lost at the agricultural level in North America and Europe. A large portion of these fruits and vegetables are discarded right after harvest, often only because of small defects in appearance or shape (FAO, 2011).

Nevertheless, the impact of inadequate refrigeration infrastructure in developing countries on food loss and waste is significant. More than 25% of fruits and vegetables are lost at the postharvest and processing stages in Sub-Saharan Africa and South/Southeast Asia, in large part because of insufficient refrigeration capacities and technical knowledge (FAO, 2011). In contrast, the amount of fruits and vegetables lost at these stages is less than 5% in North America and Europe. For dairy products, the amount of food lost during the postharvest and processing stages is approximately 10% in Sub-Saharan Africa and South/Southeast Asia, about twice the amount observed in North America and Europe (FAO, 2011). The International Institute of Refrigeration estimated in 2009 that, if the level of refrigeration used in developed countries was applied in developing countries, more than 200 million tons of perishable food would be saved annually, corresponding to approximately 14% of their annual consumption (IIR, 2009).

As such, the amount of food loss and waste in developed and developing countries represent two problems with distinct primary causes and solutions. In developed countries, major reduction of food loss and waste will require an increased awareness of the value of food and of the substantial consequences that poor food management at the consumer scale has at the global scale. In developing countries, supporting the cold chain with proper refrigeration infrastructure and improving technical knowledge are critical to the reduction of food loss and waste.

7.4 The cold chain in northern communities

While the improvement of refrigeration technologies and practices to decrease the temperature of food below the environmental temperature has been an issue actively researched and discussed, the problem of temperature management of food along the cold chain in cold climates has not received the same level of attention. Communities located in cold climates generally share a number features affecting the accessibility to fresh and high-quality food: many of these communities are located far from major cities, ports and DCs, have a low density and a high poverty level, face extreme environmental conditions, and have limited possibilities for local food production (Mercier et al., 2018). As an example, Emond et al. (2003) monitored the temperature and the quality of a mixed shipment of perishable food products from Montreal to Nain, a community located in the northern part of Labrador. The shipment was first transported for approximately 60 hours in a refrigerated truck from Montreal to a first community in Labrador. The shipment was then unloaded at the airport and stored overnight. The following day, the shipment was loaded inside a small airplane, and placed on a 2-hour flight to Nain. The shipment was finally unloaded and delivered to retailers using sleighs driven by snowmobiles. The delivery took from 10 to 40 minutes, during which no temperature control was applied and the food was exposed to environmental conditions such as wind, snow, and rain. The total duration of the cold chain from Montreal to Nain was between 3 and 4 days. Inspection of the shipment at arrival revealed that multiple products had become unmarketable because of chilling injuries, water loss, bruises, and browning. The low quality of the shipment was attributed to poor temperature management during storage and transportation, mechanical damages due to vibrations, and the exposure of the food to harsh weather conditions.

The cold chain delivering food to northern communities thus faces several challenges, notably:

1. the preservation for multiple days of mixed loads of fresh produce with different optimal temperatures;
2. limiting cold injuries during land transportation using vehicles without temperature control or protection against environmental conditions;
3. supplying communities at a sufficient rate to promote year-round accessibility to a range of fresh and healthy produce; and

4. reducing the cost of the cold chain, to provide affordable food reflecting the income level of the communities. The establishment of cold chains delivering high-quality food to northern communities with limited food loss requires a concerted effort to improve refrigeration of food after harvest and processing, develop low-cost and product-specific packaging protecting food against a range of environmental conditions, and establish management strategies reflecting each communities' specific needs and characteristics (Mercier et al., 2018).

Temperature management of food along cold chains in northern communities is a topic that should not be overlooked, given the major food security issues observed in many of these communities (Rosol et al., 2011; Skinner, 2013; Council of Canadian Academies, 2014).

7.5 Conclusion

Refrigeration plays a key role in the preservation of perishable food. The literature review revealed the consistent amount of food wasted or lost across regions of the world, although for different reasons: waste and loss in developed countries can be traced back in majority to poor handling of the food, while waste and loss in developing countries can be attributed to lack of refrigeration capacities, absence of electricity, and poor food handling. The review also highlighted the concerning stability of food waste and loss figures over the last two decades, despite the significant improvements in refrigeration technologies.

It is acknowledged that significant reduction of food loss and waste is required given the consistent increase of the world population and the saturation of land resources. Some of the main paths towards that objective are:

1. building a precise knowledge of food loss and waste for all categories of food products and regions of the world, to enable a proper comparison of the improvements against the reduction targets;
2. influencing a change of behavior at the microscale (households, restaurants, retailers), given that poor food management at these stages sums up to massive amount of food loss and waste, especially in developed countries; and
3. increasing refrigeration capacities in developing countries and remove/northern communities, to enable the year-round accessibility to healthy and fresh food in a sustainable manner.

References

Aghdam, M.S., Bodbodak, S., 2014. Postharvest heat treatment for mitigation of chilling injury in fruits and vegetables. Food Bioprocess 7, 37–53.

Amador, C., Emond, J.-P., do Nascimento Nunes, M.C., 2009. Application of RFID technologies in the temperature mapping of the pineapple supply chain. Sens. & Instrumen. Food Qual. 3, 26–33.

Arduino, G., Carrillo Murillo, D., Parola, F., 2015. Refrigerated container versus bulk: evidence from the banana cold chain. Marit. Policy Manag. 42, 228–245.

Aste, N., Del Pero, C., Leonforte, F., 2017. Active refrigeration technologies for food preservation in humanitarian context – a review. Sustain. Energy Technol. Assess. 22, 150–160.

Badia-Melis, R., McCarthy, U., Ruiz-Garcia, L., Garcia-Hierro, J., Robla-Villalba, J.I., 2018. New trends in cold chain monitoring applications – a review. Food Control 86, 170–182.

Bakalis, S., Giannakourou, M.C., Taoukis, P., 2003. Effect of domestic storage and cooking conditions on the risk distribution in ready-to-cook meat products. In: Proceedings of the 9th International Congress on Engineering and Food (ICEF9), Montpellier, March 7–11.

Bhattacharjee, P., Loth, E., 2004. Entrainment by a refrigerated air curtain down a wall. Trans. ASME J. Fluids Eng. 126 (5), 871–879.

Bresolin, C.S., Rego, R., Bandarra Filho, E.P., Smith Schneider, P., 2018. Brazilian protein cold chain panorama. Int. J. Refrig. Under review.

Brown, T., Hipps, N.A., Easteal, S., Parry, A., Evans, J.A., 2014. Reducing domestic food waste by lowering home refrigerator temperatures. Int. J. Refrig. 40, 246–253.

Brown, W., Ryser, E., Gorman, L., Steinmaus, S., Vors, K., 2016. Temperatures experienced by fresh-cut leafy greens during retail storage and display. In: ISHS Acta Horticulturae 1141: III International Conference on Fresh-Cut Produce: Maintaining Quality and Safety Proceedings, pp. 103–107.

Council of Canadian Academies, 2014. Aboriginal food security in Northern Canada: an assessment of the state of knowledge. Available from <http://www.scienceadvice.ca/uploads/eng/assessments%20and%20publications%20and%20news%20releases/food%20security/foodsecurity_fullreporten.pdf> (accessed 31.01.18.).

Defraeye, T., Nicolai, B., Kirkman, W., Moore, S., van Niekerk, S., Verboven, P., et al., 2016. Integral performance evaluation of the fresh-produce cold chain: a case study for ambient loading of citrus in refrigerated containers. Postharv. Biol. Technol. 112, 1–13.

Derens, E., Palagos, B., Guilpart, J., 2006. The cold chain of chilled products under supervision in France. In: Proceedings of the IUFOST, 13th World Congress of Food Science & Technology: Food is life, Nantes, France, pp. 51–64.

Dongquan, D., Heng, S., Youming, X., 2002. Current situation and development on the vacuum pre-cooling technology [J]. Sci. Technol. Food Indus. 7, 036.

Dulebenets, M.A., Ozguven, E.E., Moses, R., Ulak, M.B., 2016. Intermodal freight network design for transport of perishable products. Open J. Optim. 5 (4), 120–139.

EIA, 2009. 2009 RECS Survey Data. Available from: <https://www.eia.gov/consumption/residential/data/2009/> (accessed 31.01.18.).

Emond, J.-P., Mercier, F., Laurin, É., Nunes, M.C.N., 2003. Cold chain management of perishable distribution in northern communities of Canada. In: International Congress of Refrigeration, Washington, DC, ICR0334, 7 pages.

Evans, J.A., Stanton, J.I., Russell, S.L., James, S.J., 1991. Consumer Handling of Chilled Foods: A Survey of Time and Temperature Conditions. MAFF Publications, London.

FAO, 2011. Global food losses and food waste. Available from: <http://www.fao.org/doc-rep/014/mb060e/mb060e00.pdf> (accessed 31.01.18.).

Ferrua, M.J., Singh, R.P., 2009. Modeling the forced-air cooling process of fresh strawberry packages, Part I: Numerical model. Int. J. Refrig. 32 (2), 335–348.

Fitzgerald, W.B., Howitt, O.J.A., Smith, I.J., Hume, A., 2011. Energy use of integral refrigerated containers in maritime transportation. Energy Policy 39, 1885–1896.

New Zealand Foodsafe Partnership, 2004. One in three fridges could make you sick. 23 November 2004. New Zealand food safe partnership media release. Available from: <http://www.foodsafety.govt.nz/elibrary/industry/Three_Fridges-Around_Third.htm> (accessed 31.01.18.).

Forkenbrock, D.J., 2001. Comparison of external costs of rail and truck freight transportation. Transport. Res. Part A Policy Practice 35 (4), 321–337.

Fronda A. 2013. High-speed rail: the future of freight. Available from: <http://www.worldfinance.com/infrastructure-investment/high-speed-railthe-future-of-freight> (accessed 08.03.18.).

Freidberg, S., 2015. Moral economies and the cold chain. Histor. Res. 88, 125–137.

GCCA, 2014. Capacity and growth of refrigerated warehousing by country. Available from: <https://www.gcca.org/wp-content/uploads/2014/12/SelectCharts_Media_2014IARWCapacityReport.pdf> (accessed 31.01.18.).

GCCA, 2016. Global cold storage capacity report. Available from: <http://www.gcca.org/resources/publications/white-papers-reports/global-cold-storage-capacity/> (accessed 31.01.18.).

Giannakourou, M.C., Taoukis, P.S., 2003. Application of a TTI-based distribution management system for quality optimization of frozen vegetables at the consumer end. J. Food Sci. 68, 201–209.

Godwin, S.L., Chen, F.C., Chambers IV, E., Coppings, R., Chambers, D., 2007. A comprehensive evaluation of temperatures within home refrigerators. Food Prot. Trends 27, 16–21.

Gogou, E., Katsaros, G., Derens, E., Alvarez, G., Taoukis, P.S., 2015. Cold chain database development and application as a tool for the cold chain management and food quality evaluation. Int. J. Refrig. 52, 109–121.

Gwanpua, S.G., Verboven, P., Leducq, D., Brown, T., Verlinden, B.E., Bekele, E., et al., 2015. The FRISBEE tool, a software for optimising the trade-off between food quality, energy use, and global warming impact of cold chains. J. Food Eng. 148, 2–12.

Haass, R., Dittmer, P., Veigt, M., Lütjen, M., 2015. Reducing food losses and carbon emission by using autonomous control – a simulation study of the intelligent container. Int. J. Prod. Econ. 164, 400–408.

Heap, R.D., 2006. Cold chain performance issues now and in the future. In: The Proceedings of the *Innovative Equipment and Systems for Comfort and Food Preservation* Conference, Auckland, New-Zealand, pp. 75–87.

Hertog, M.L.A.T.M., Boerrigter, H.A.M., Van Den Boogaard, G.J.P.M., Tijskens, L.M.M., Van Schaik, A.C.R., 1999. Predicting keeping quality of strawberries (cv. 'Elsanta') packed under modified atmospheres: an integrated model approach. Postharv. Biol. Technol. 15, 1–12.

Hertog, M.L.A.T.M., Uysal, I., McCarthy, U., Verlinden, B.M., Nicolaï, B.M., 2014. Shelf life modelling for first-expired-first-out warehouse management. Philos. T. Roy. Soc. A 372, art. no20130306.

IIR, 2009. The role of refrigeration in worldwide nutrition. Available from: <http://www.iifiir.org/userfiles/file/publications/notes/NoteFood_05_EN.pdf> (accessed 31.01.18.).

Inbound Logistics, 2010. How to leverage rail/intermodal for refrigerated freight. Available from: <http://www.inboundlogistics.com/cms/article/how-to-leverage-rail-intermodal-for-refrigerated-freight/> (accessed 08.03.18.).

Islam, M.P., Morimoto, T., 2014. A new zero energy cool chamber with a solar-driven adsorption refrigerator. Renew. Energy 72, 367–376.

ITA, 2016. 2016 Top markets report cold chain. Available from: <https://www.trade.gov/topmarkets/pdf/Cold_Chain_Executive_Summary.pdf> (accessed 31.01.18.).

James, S.J., Evans, J., 1992. Consumer handling of chilled foods: temperature performance. Int. J. Refrig. 15, 299–306.

James, C., Onarinde, B.A., James, S.J., 2017. The use and performance of household refrigerators: a review. Comprehens. Rev. Food Sci. Food Safety 16, 160–179.

Janjić, J., Katić, V., Ivanović, J., Bošković, M., Starčević, M., Glamočlija, N., et al., 2016. Temperatures, cleanliness and food storage practices in domestic refrigerators in Serbia, Belgrade. Int. J. Consumer Stud. 40, 276–282.

Jedermann, R., Praeger, U., Geyer, M., Lang, W., 2014. Remote quality monitoring in the banana chain. Philos. Trans. R. Soc. A Math. Phys. Eng. Sci. 372, 20130303.

Jouhara, H., Nannou, T., Ghazal, H., Kayyali, R., Tassou, S.A., Lester, S., 2017. Temperature and energy performance of open refrigerated display cabinets using heat pipes shelves. Energy Proc. 123, 273–280.

Kader, A.A., Rolle, R.S., 2004. The Role of Post-Harvest Management in Assuring the Quality and Safety of Horticultural Produce, vol. 152. Food & Agriculture Org.

Kan, A., Hu, J., Guo, Z., Meng, C., Chao, C., 2017. Impact of cargo stacking modes on temperature distribution inside marine reefer containers. Int. J. Air-Condit. Refrig. 25 (3), 750020.

Kou, L., Luo, Y., Ingram, D.T., Yan, S., Jurick II, W.M., 2015. Open-refrigerated retail display case temperature profile and its impact on product quality and microbiota of stored baby spinach. Food Control 47, 686–692.

Laguerre, O., Derens, E., Palagos, B., 2002. Study of domestic refrigerator temperature and analysis of factors affecting temperature: a French survey. Int. J. Refrig. 25, 653–659.

Laguerre, O., Flick, D., 2004. Heat transfer by natural convection in domestic refrigerators. J. Food Eng. 62, 79–88.

Lervåg, L.-E., Meland, S., Wahl, R., 2001. Utvikling av NEMO/REGO - Parameterverdier - The Development of NEMO/REGO - Parameter Values, SINTEF Bygg og miljø, Trondheim, Norway [in Norwegian].

Lesame, Z., 2014. The South African digital access index. Mediter. J. Soc. Sci. 5, 331–341.

Liu, H., Pretorius, L., Jiang, D., 2018. Optimization of cold chain logistics distribution network terminal. EURASIP J. Wireless Commun. Netw. 2018, 158.

Mack, M., Dittmer, P., Veigt, M., Kus, M., Nehmiz, U., Kreyenschmidt, J., 2014. Quality tracing in meat supply chains. Philos. Trans. R. Soc. A: Math. Phys. Eng. Sci. 372, 20130308.

Mai, N.T.T., Margeirsson, B., Margeirsson, S., Bogason, S.G., Sigurgisladottir, S., Arason, S., 2012. Temperature mapping of fresh fish supply chains – air and sea transport. J. Food Proc. Eng. 35, 622–656.

McCarthy, U., Uysal, I., Badia-melis, R., Mercier, S., O'donnell, C.P., Ktenioudaki, A., 2018. Global Food Security – Issues, Challenges and Technological Solutions. Trends in Food Science & Technology.

McMurray, G., Arruda, C., Britton, D., Eidenberger, T., Evans, S., Gibney, M., et al., 2013. Food Security: A Systems Approach. EU Science: Global Challenges and Global Collaboration. European Parliament, Brussels, 4th to 8 th March 2013. https://globalsciencecollaboration2013.sched.com/event/Wxl6Vl/food-security-and-safety-a-systems-approach.

Mercier, S., Villeneuve, S., Mondor, M., Uysal, I., 2017. Time–temperature management along the food cold chain: a review of recent developments. Comprehens. Rev. Food Sci. Food Safety 16, 647–667.

Mercier, S., Mondor, M., Villeneuve, S., Marcos, B., 2018. The Canadian food cold chain: a legislative, scientific, and prospective overview. Int. J. Refrig. 88, 637–645.

Morelli, E., Derens, E., 2009. Evolution des temperatures du saumon′ fume au cours des circuits logistiques. Revue Générale du Froid 1090, 51–56.

Morelli, E., Noel, V., Rosset, P., Poumeyrol, G., 2012. Performance and conditions of use of refrigerated display cabinets among producer/vendors of foodstuffs. Food Control 26, 363–368.

Navaz, H.K., Amin, M., Dabiri, D., Faramarzi, R., 2005. Past, present, and future research toward air curtain performance optimization. ASHRAE Trans. 111 (1), 1083–1088.

NCDD, 2016. Report on cold-chain (rationalising concept & requirements). Available from: <http://www.nccd.gov.in/PDF/ReportCold-chain2016.pdf> (accessed 31.01.18.).

Nunes, M.C.N., Morais, A.M.M.B., Brecht, J.K., Sargent, S.A., Bartz, J.A., 2005. Prompt cooling reduces incidence and severity of decay caused by *Botrytis cinerea* and *Rhizopus stolonifer* in strawberry. HortTechnology 15 (1), 153–156.

Nunes, M.C.N., Émond, J.P., Rauth, M., Dea, S., Chau, K.V., 2009. Environmental conditions encountered during typical consumer retail display affect fruit and vegetable quality and waste. Postharv. Biol. Technol. 51, 232–241.

Nunes, M.C., Nicometo, M., Emond, J.P., Melis, R.B., Uysal, I., 2014. Improvement in fresh fruit and vegetable logistics quality: Berry logistics field studies. Philos. Trans. R. Soc. A 372, art no. 20130307.

Pelletier, W., Brecht, J.K., do Nascimento Nunes, M.C., Émond, J.P., 2011. Quality of strawberries shipped by truck from California to Florida as influenced by postharvest temperature management practices. HortTechnology 21 (4), 482–493.

Reina, L.D., Fleming, H.P., Humphries, E.G., 1995. Microbiological control of cucumber hydrocooling water with chlorine dioxide. J. Food Protect. 58 (5), 541–546.

Rodrigue J.P., 2013. World container production, 2007. Available from: <https://web.archive.org/web/20130704071409/http:/people.hofstra.edu/geotrans/eng/ch3en/conc3en/containerproduction.html> (accessed 08.03.18.).

Rosol, R., Huet, C., Wood, M., Lennie, C., Osborne, G., Egeland, G.M., 2011. Prevalence of affirmative responses to questions of food insecurity: International Polar Year Inuit Health Survey, 2007-2008. Int. J. Circumpol. Heal. 70, 488–497.

Sandberg Hanssen, T.E., Mathisen, T.A., 2011. Factors facilitating intermodal transport of perishable goods – transport purchasers viewpoint. Eur. Transp./Trasp. Eur. 49, 75–89.

Sibomana, M.S., Workneh, T.S., Audain, K., 2016. A review of postharvest handling and losses in the fresh tomato supply chain: a focus on Sub-Saharan Africa. Food Security 8, 389–404.

Siegel, R., Maté, J., Watson, G., Nosaka, K., Laursen, P.B., 2012. Pre-cooling with ice slurry ingestion leads to similar run times to exhaustion in the heat as cold water immersion. J. Sports Sci. 30 (2), 155–165.

Smale, N.J., 2004. Mathematical Modelling of Airflow in Shipping Systems: Model Development and Testing (Doctoral dissertation). Massey University, Palmerston North, New Zealand. Available from: <http://encore.massey.ac.nz/iii/encore/record/C__Rb1816458?lang = eng> (accessed 01.03.18.).

Smith, B.G., 2008. Developing sustainable food supply chains. Philos. Trans.: Biol. Sci. 363, 849–861.

Sommar, R., Woxenius, J., 2007. Time perspectives on intermodal transport of consolidated cargo. Eur. J. Transp. Infrastruct. Res. 7, 163–182.

Skinner, K., 2013. Prevalence and Perceptions of Food Insecurity and Coping Strategies in Fort Albany First Nation, Ontario (Ph.D. thesis). University of Waterloo.

Statistics Canada, 2009. Selected dwelling characteristics and household equipment (Household appliances and telephones). Available from: <http://www.statcan.gc.ca/tables-tableaux/sum-som/l01/cst01/famil09b-eng.htm> (accessed 31.01.18.).

Statistics Indonesia, 2012. Indonesia demographic and health survey 2012. Available from: <https://dhsprogram.com/pubs/pdf/fr275/fr275.pdf> (accessed 31.01.18.).

Tanner, D.J., Amos, N.D., 2003. Temperature variability during shipment of fresh produce. Acta Hortic. 599, 193–203.

Tanner, D.J., Smale, N., 2005. Sea transportation of fruits and vegetables: an update. Stewart Postharv. Rev. 1, 1–9.

The World Bank, 2014. Access to electricity, rural (% of rural population). Available from: <https://data.worldbank.org/indicator/EG.ELC.ACCS.RU.ZS> (accessed 31.01.18.).

Thompson, J.F., 2004. Pre-cooling and storage facilities. In: The Commercial Storage of Fruits, Vegetables, and Florist and Nursery Stocks, Agriculture Handbook Number 66. USDA, ARS, Beltsville. p. 11.

Torres-Toledo, V., Hack, A., Mrabet, F., Salvatierra-Rojas, A., Müller, J., 2018. On-farm milk cooling solution based on insulated cans with integrated ice compartment. Int. J. Refrig. 90, 22–31.

USDA. 2008. China's cold chain industry. Available from: <apps.fas.usda.gov/gainfiles/200903/146347650.doc> (accessed 31.01.18.).

Wild, Y., 2009. Refrigerated Containers and CA Technology. Container Handbook. The German Insurance Association, Berlin.

Willocx, F., Hendrickx, M., Tobback, P., 1994. A preliminary survey into the temperature conditions and residence time distribution of minimally processed MAP vegetables in Belgian retail display cabinets. Int. J. Refrig. 17 (7), 436–444.

Yu, K., Ding, G., Chen, T., 2009. A correlation model of thermal entrainment factor for air curtain in a vertical open display cabinet. Appl. Therm. Eng. 29 (14–15), 2904–2913.

Zeng, W., Vorst, K., Brown, W., Marks, B.P., Jeong, S., Pérez-Rodriguez, F., et al., 2014. Growth of *Escherichia coli* O157:H7 and *Listeria monocytogenes* in packaged fresh-cut Romaine mix at fluctuating temperatures during commercial transport, retail storage, and display. J. Food Protect. 77 (2), 197–206.

Zhao, H., Liu, S., Tian, C., Yan, G., Wang, D., 2018. An overview of current status of cold chain in China. Int. J. Refrig. 88, 483–495.

Further reading

Derens-Bertheau, E., Osswald, V., Laguerre, O., Alvarez, G., 2015. Cold chain of chilled food in France. Int. J. Refrig. 52, 161–167.

Optimized food supply chains to reduce food losses

8

Despoudi Stella
Aston Business School, Aston University, Birmingham, United Kingdom

Chapter Outline

8.1 Introduction

An important way to increase food supply and decrease the environmental consequences of current food production is to reduce food losses (Godfray et al., 2010). Reducing food losses can increase food availability without requiring additional production resources (Hodges et al., 2010). Foresight (2011) stated that food and drink loss is a significant issue for economic, environmental, and food security reasons. Although food loss arises at every stage of the food supply chain, the causes of food loss vary greatly depending on the stage of the supply chain. Almost the 50% of food produced is wasted along the supply chain and does not reach consumers. Food waste is waste of resources used in production (e.g., land, water, energy, crops). The production of food that is not being consumed not only pollutes the environment, but also it is a loss of economic value (FAO, 2011).

Food is lost or wasted throughout the supply chain, from the initial agricultural production down to final household consumption (Gustavsson et al., 2011). The authors suggest that food losses and waste in developing low-income countries are related to the upstream supply chain (producer to processor), whereas the losses in the affluent world are related to the downstream supply chain (retailer to final

Saving Food. DOI: https://doi.org/10.1016/B978-0-12-815357-4.00008-0

consumer). Food losses are not only a waste of food, but they also represent a similar waste of human effort, farm inputs, livelihoods, investments, and natural resources such as water. Food losses have an impact on food security for poor people, on food quality and safety, on economic development, and on the environment. The need for food losses reduction is not a recent issue. According to Foresight (2011) in 1974 at the World's Food Conference it was decided to reduce food losses to 50% by 1985 and a special action program for the prevention of food losses was established by the FAO with a technological focus in storage and farm on reduction in losses of durable grain. After that there is no recorded progress on food loss reduction until 2008 when Lundqvist et al. (2008) called for action to reduce food losses from producers to consumers by 50% to be achieved by 2025.

Reducing food loss is one of the prominent goals in the current research, which has also been set by the United Nations to achieve a more sustainable world by 2030. Given that previous studies mainly examined causes for food waste generation related to consumers, this chapter aims to provide an overview on losses in the food industry. In addition, ways to reduce food losses by optimizing supply chains are discussed in this chapter.

8.2 Definition of food loss

There are different definitions about food loss in terms of where in the food supply chain it is happening. Postharvest food loss (PHFL) and food waste are commonly used as synonyms to food loss in the literature (Kader, 2005; WRAP, 2009; Hodges et al., 2010; Atanda et al., 2011). The World Economic Forum (2011) defines PHFL as upstream loss in agriculture and transport prior to processing, and food waste as food fit for human consumption that is wasted in all further downstream parts of the supply chain. In some cases, food waste is termed as food loss occurring at the end of the food supply chain (FAO, 2012). Food loss refers to the decrease of edible food mass throughout the supply chain from farm to fork or from production to consumption (Sharma and Singh, 2011).

Food loss in this chapter is defined as the decrease of edible food mass that occurs from producers until reaching consumers and includes all the edible food that was lost either intentionally or unintentionally (FAO, 2011). Food waste is a type of food loss that is related to intentional spillage of edible food mass and could happen from the producers and after harvesting until postconsumption stages (Parfitt et al., 2010). Food waste is generated due to a conscious decision to discharge food. The highest rates of food waste are at the retailer and consumer stages of the supply chain as they intentionally throw food away. Whereas, in other stages of the supply chain (e.g., production, processing) food is usually unavoidably lost.

According to FAO (2010) food loss falls into three categories: (1) physical losses resulting from spoilage where the product is diminished by weight and/or quality, (2) opportunity or monetary losses where sales might be lost or only be made in a lower value market, and (3) external losses that fall on both the value chain participants and the rest of the society (e.g., where the chemical pesticides used to protect grain impact on the environment or human health).

8.3 Overview of food losses in the food industry

In the food industry, food loss occurs across the supply chain, that is, from production to consumption. The exact causes of food losses vary throughout the world and are very much dependent on the specific conditions and local situation in a given country (Lupien, 2008). In broad terms, food losses may be influenced by crop production choices and patterns, internal infrastructure and capacity, distribution channels, and consumer purchasing, and food use practices (Hodges et al., 2010). Food losses in the food industry can be categorized in two elements: upstream supply chain (producer to processor), and downstream supply chain (retailer to final consumer). Parfitt et al. (2010) indicated that food losses and waste in developing low-income countries are related to the upstream supply chain, whereas the losses in the affluent world are related to the downstream supply chain. Table 8.1 shows the estimates of food losses in EU for both upstream and downstream supply chain (EU FUSIONS, 2016).

8.3.1 Food losses in the upstream supply chain

Losses in the upstream supply chain arise from the challenges experienced in harvesting techniques, storage and cooling facilities in difficult climatic conditions, logistics, warehousing infrastructure, packaging, and marketing systems (Lupien, 2008). Developing countries were found to have the highest percentage of upstream food losses (Parfitt et al., 2010), although variations in wastage rates exist for different types of food and it is difficult to estimate the actual loss (Premanandh, 2011).

Table 8.1 Estimates of food waste in EU-28 in 2012 from this quantification study; includes food and inedible parts associated with food

Sector	Food waste (million tonnes) with 95% CI[a]	Food waste (kg per person) with 95% CI[a]
Primary production	9.1 ± 1.5	18 ± 3
Processing	16.9 ± 12.7	33 ± 25
Wholesale and retail	4.6 ± 1.2	9 ± 2
Food service	10.5 ± 1.5	21 ± 3
Households	46.5 ± 4.4	92 ± 9
Total food waste	87.6 ± 13.7	173 ± 27

[a]Confidence interval. https://www.eu-fusions.org/phocadownload/Publications/Estimates%20of%20European%20food%20waste%20levels.pdf.

8.3.1.1 Food losses in primary production

Food losses during production are usually due to spillage during growing of the produce, harvest operation (e.g., threshing, crop picking), or mechanical damage (treatment of the produce). Uncontrollable factors such as temperature and weather variations are the main causes of food losses at the production stage (Lupien, 2008). Insect infestation and improper handling could also cause food losses at this stage. The heterogeneity of the primary production sector in terms of the different products that are produced makes it difficult to quantity the food losses.

Further assessing the crop quality based on the climate behavior and the weather events, Ahmed and Stepp (2016) indicated that the quality of the agricultural produce is affected by the atmospheric changes in precipitation, the carbon dioxide levels that are on the rise because of the emission of greenhouse gases, and temperature changes through the ongoing climate change. The climatic variables affected both the quantity and the quality of crops that are produced based on the antioxidant activities that result from the changes in the climate. In addition, the changes in the quality and quantity of crops that are produced can be noted in the variations of grades of the crops. This alteration of the crops has in turn affected the agricultural strategies that are used to produce the crops.

With the climatic changes affecting on agriculture, it can be deduced that there are implications on the operations with regards to the farming processes. Gornall et al. (2010) indicated that the changes in the temperatures caused losses of up to $5 billion on maize, wheat, and barley in the United States. Lobell et al. (2014) stated that changes in the weather patterns occurred in a way that has balanced the losses in some countries due to other countries increasing their production. The researchers cite the case of soybean and rice where the countries with gains balanced those with losses. Coumou and Rahmstorf (2012) further indicated that in 2010, the Pakistani flooding and the Moscow heat wave not only led to the loss of lives, but the losses of up to 30% of the grain harvest, which led the government to ban exports.

With reference to the change in climate and the extreme weather events, it can be noted that there is a notable difference in the crop quality. According to Lobell et al. (2011), climatic change affected the food availability due to the changes in the weather in the farming land. The growing seasons in most countries are affected by the changes in temperature that impact the quality of the crops produced. Further, Schlenker and Roberts (2008) state that the emission of greenhouse gases has affected the production of crops, especially in the United States. According to the research, the change in the climatic conditions is noted to have implications on the quality of crops being produced, which is reflected on the types of crops produced. Hence, deterioration of crop quality leads to further food losses. Mirza (2003) stated that the vulnerability of the agricultural land also plays a role in the losses incurred with the extremities in climatic and weather events.

8.3.1.2 Food losses in postharvest handling, storage, processing, and distribution

Food losses in handling and storage occur due to spillage and degradation during handling, lack of storage facilities, and transportation between farm and distribution (Akkerman and Van Donk, 2008). In storage, considerable quantitative losses can be attributed to pests and microorganisms. While, losses in processing of food products include losses due to spillage and degradation during industrial or domestic processing, for example, juice production, canning, and bread baking. Food losses may occur when crops are sorted out if not suitable to process or during washing, peeling, slicing, and boiling or during process interruptions and accidental spillage. Food losses in distribution occur due to lack of appropriate transportation methods, time constraints, and power relationships.

The food losses incurred in the logistics can be seen in terms of the changes in quantity and quality of exports (Coumou and Rahmstorf, 2012). Rosenzweig et al. (2001) pointed out that the deterioration in food production in the United States led to taxpayers giving approximately $3 billion in relief to corn farmers. The changes in the climate affect the production as there are constraints in the production of food.

8.3.2 Food losses in the downstream supply chain

Food losses in the downstream supply chain refer to food losses at the retailer and consumer stages. In the developed countries the majority of the food losses occur due to intentional spillage of the food, which is called food waste. Fig. 8.1 shows the top US food groups in terms of annual food loss at both retailer and consumers level measured by amount value and calories (USDA, 2010). In terms of the food

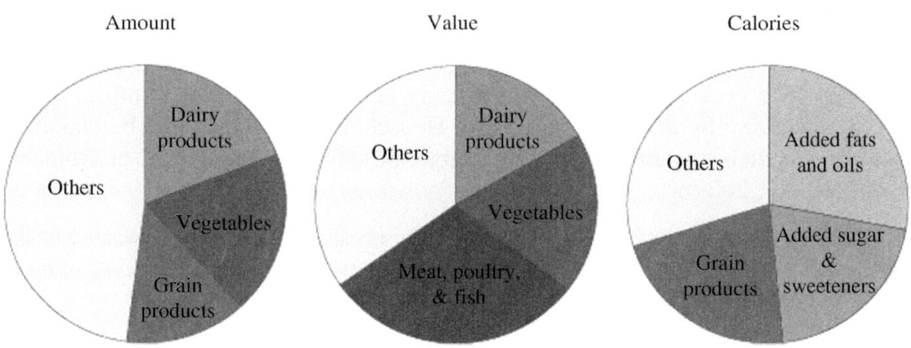

Figure 8.1 The top three US food groups in terms of annual food loss at the retail and consumer levels vary depending on if measured by amount, value, or calories.
Source: USDA, 2010. Economic research service loss-adjusted food availability data. <https://www.ers.usda.gov/amber-waves/2014/june/food-loss-questions-about-the-amount-and-causes-still-remain/>.

amount wasted all three categories, that is, dairy, vegetables, and grain were found to have similar amount of losses. The "other" category represented almost half of the wasted food products. The value "other" and the meat, poultry, fish category were found to have the higher losses in terms of value. The caloric losses were the same for all categories. Thus, some food product categories seem to have a higher amount of losses than others.

8.3.2.1 Food losses of retailers

From the retailers' perspective food loss is created mainly due to poor demand forecasting and inventory management, which results in overproduction. Some other reasons for food loss are temperature sensitivities, weather conditions during transportation, disposal of unsold food, and inappropriate packaging (Defra, 2009). Retailers also contribute to waste as a result of their contractual arrangements with suppliers. Failure to supply agreed quantities renders producers or processors liable to have their contracts canceled. Therefore, producers need to plan to produce more than actually required to meet the contract requirements.

Retailers throw away significant quantities of food that have reached best before, sell-by, or use-by dates. Even at the consumer level the food industry plays a crucial role in influencing consumers' behavior towards food losses (Defra, 2006). Consumers throw away food that was not expired due to misinterpreting the product labeling. Retailers have been also accused of creating food waste through their in-store promotions. This is because consumers perceive it as a bargain and they buy more food products that they need.

8.3.2.2 Consumer and postconsumer food losses

Food losses at the consumer level arise due to many reasons such as the individual shopper's psychology, lack of awareness regarding the negative implications of food losses, lack of knowledge regarding efficient food use, cultural perceptions regarding food consumption, lack of shopping planning, packaging confusion (best before versus use-by dates; Defra, 2009). Household food losses can be classified according to their avoidability into avoidable, possibly avoidable, and unavoidable waste (WRAP, 2009):

- Avoidable waste is food and drink thrown away because it is no longer wanted or has been allowed to go past its best. The vast majority of avoidable food is composed of material that was, at some point prior to disposal, edible, even though a proportion is not edible at the time of disposal due to deterioration.
- Possibly avoidable is food and drink that some people eat, and others do not (e.g., bread crusts), or that can be eaten when prepared in one way but not in another (e.g., potato skins). As with "avoidable" waste, "possibly avoidable" waste is composed of material that was, at some point prior to disposal, edible.
- Unavoidable waste is the waste from food preparation that was not edible under any circumstance.

8.4 Ways to reduce food losses

Effective food loss management will benefit all the food supply chain entities. By reducing raw material usage and increasing recycling and reusage activities, cost reduction, increased performance, and increased sustainability could be achieved. The US EPA (2011) proposed the food waste recovery hierarchy model (Fig. 8.2). Instead of three R's (Reduce, Reuse, and Recover), the EPA suggested that reducing the amount of food waste being generated is the most important aspect in food loss reduction. After that feeding people, feeding animals, industrial use of food waste, and composting follow.

All the food supply chain entities should consider the EPA's model in their food loss reduction efforts to achieve zero waste. First the more proactive approaches to food loss reduction should be actioned by the food supply chain entities and then the food loss treatment methods should be considered. In the sections that follow different approaches to food loss reduction are suggested, which are categorized as follows: primary production solutions, solutions at handling, storage, processing and distribution stage, solutions at retailer stage, and supply chain solutions.

8.4.1 Primary production solutions

Primary production involves the producers and their respective relationships with business partners to whom they sell their produce. The role of producers in reducing food losses is highlighted by Food Agricultural Organization (FAO, 2012). The majority of the food losses are happening at the producers' stage (FAO, 2011). Different ways are suggested below to reduce food losses at this stage of the food supply chain.

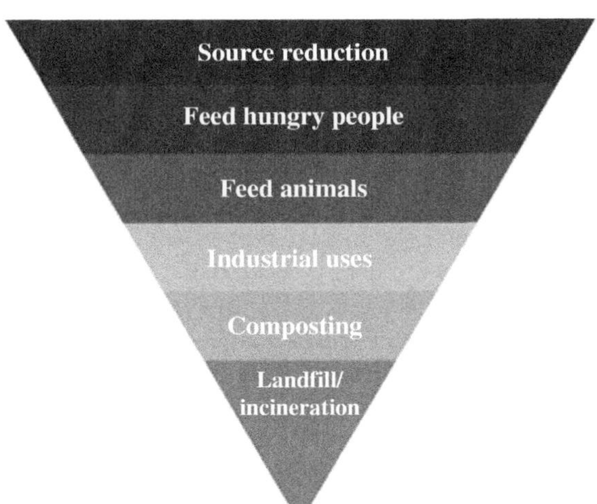

Figure 8.2 Food waste recovery hierarchy (EPA, 2011).

8.4.1.1 Focus on collaboration and collective action at producers' stage

Recent research found that higher levels of collaboration between producers and agricultural cooperatives could reduce food losses (Despoudi, 2016; Despoudi et al., 2018). Forms of collective action at the producers' stage can significantly support producers in sharing best practices, resources, and information to enable food loss reduction. Agricultural cooperatives in the form of producer organizations (POs) can be further developed as tools for producers to improve their competitiveness and strengthen their bargaining position towards retailers (Kaditi and Nitsi, 2010). Following the guidelines of the reformed CAP will enable producers to build strong POs that could compete internationally (Paisiadis, 2013). Through the POs improvements in the quality of the fresh produce is expected to be seen by adopting international quality certifications, improved packaging and labeling, adoption of new management techniques, and by highlighting the quality and the recognition of the food products. Market institution development and collective marketing generally improve the marketing system (Lupien, 2008; Kader, 2010). Formation of collective marketing groups to process unsold food is proposed as a way to reduce food losses.

The structure of the supply chain also influences the price of the product, as the more intermediaries are involved, the more the payments and the greater the spoilage of the product (Kamenidis, 2004). Elimination of the intermediaries involved is essential for small-scale producers to not only provide better quality of products, but also to get better prices and increase bargaining power. Through the development of cooperatives and the POs intermediaries could be eliminated and better prices with the retailers could be achieved. Collaboration between producers and wholesalers or retailers could also be considerably improved by having specific contractual agreements.

8.4.1.2 Training of the producers

The changes in the food industry environment require partners to develop and acquire new skills and capabilities. Food producers need to adopt new farming methods through seminars and by sharing best practices among them. Producers not only need to improve their technical skills, but they also need to be better organized, act collectively, an acquire stronger group business and marketing skills. The more recent emphasis on market-oriented approaches and on "linking producers to markets" has been fundamental for understanding the constraints and lack of incentives for food loss improvements (The World Bank, 2011). Producers need to be educated and informed about new production methods, the different food regulations, and the changing consumers' needs and wants. In this way producers will become more resilient and they will be able to respond appropriately to the changing market needs with new high-value products that consumers require.

8.4.1.3 Focus on value-added and high-quality products

Consumers look for branded and high-quality food products. Producers should focus on exceptional quality and high value-added products (e.g., Protected Destination of

Origin, Geographical Indication, organic farming). In this way the confidence of the consumers will be increased and the demand for the products and the income of the producers will increase too. There is a need for an efficient marketing strategy in the EU fresh produce sector (Fotopoulos and Krystallis, 2003). The EU producers need to focus on product differentiation and take advantage of the land and climate characteristics that enable this differentiation. Producers need to put more effort in the development of branded and certified products.

Also, producers need to develop their knowledge in terms of using unmarketable crops. Value-added products such as puree, juice, and marmalades could be created by using any crops that are not in perfect shape or are damaged and cannot be sold in the market (FAO, 2011). Creation of value-added activities means waste elimination either by preventing waste or by converting waste into another product. These value-added products could increase producers' income and at the same time reduce food losses.

8.4.1.4 Agroecology

Agroecology is an approach that takes into account natural ecosystems and uses local knowledge to plant a diversity of crops that boost the sustainability of the farming system as a whole (Moore, 2016). It helps to deliver contextualized solutions to global issues. There is a need for contextual solutions to address food losses. Instead of trying to provide generic solutions to unsustainable agricultural systems, the system's adaptive capacity and autonomy needs to be enhanced. According to FAO (2018) there are 10 guiding principles of agroecology: diversity, synergies, efficiency, resilience, recycling, cocreation and sharing of knowledge, human and social values, culture and food traditions, responsible governance, circular and solidarity economy. The focus is on social and economic aspects of the food systems related to local producers, youth, and women. An example is the improvement of soil and plant quality through available biomass and biodiversity instead of using chemical inputs.

8.4.2 Solutions at handling, storage, processing, and distribution stage

8.4.2.1 Postharvest storage and handling solutions

Appropriate storage technologies should be implemented such as evaporative coolers, and storage bags. Investment in cold chain facilities is essential due to the perishability of the products. Governmental support or close collaboration with the supplier is recommended for supply chain entities who the lack cold chain facilities due to financial issues.

8.4.2.2 Postharvest processing and transport solutions

In the processing stage, food losses could be reduced through improved packaging solutions that increase the shelf-life of the product and optimize the portion size.

Packaging protects food from damaging and preserves its freshness. However, sometimes inappropriate packaging can lead to considerable food waste (Williams and Wikström, 2011). Close collaboration with the suppliers is fundamental to understand the product and customer requirements and identify the right packaging solutions. In an effort to reduce food waste Tesco and M&S are using new packaging to extend fruits' and vegetables' shelf-life (Smithers, 2010). Intelligent container technologies could be used to reduce food losses. Some examples of intelligent container technologies are (GoSupplyChain, 2018):

- The time temperature indicator (TTI): This is useful for determining if foods have been temperature abused. An irreversible change (such as color change) will occur if the TTI experiences abusive conditions.
- Gas indicators: Food can respire and may therefore change its own atmosphere when inside a package. Gas indicators monitor the composition of gases inside a package and typically signal presence or absence of oxygen or carbon dioxide.
- Biosensors: Foodborne pathogens are of great concern to the food industry. A biosensor can detect a substance (a pathogen in this case) and then transmit the information in a quantifiable manner.

In terms of the transportation methods and lack of proper road infrastructure especially in developed countries alternative routes to markets should be identified in collaboration with the supply chain partners. Appropriate use of the different distribution hubs can reduce food losses. A First Expiry First Out strategy could be employed to first transport products with short expiry date. Also, supply chain demand decisions need to be considered in relation to seasonality, and changes in weather patterns. In addition, any food product losses due to spillage or degradation at this stage should be recycled or redeveloped to a byproduct.

8.4.3 Solutions at retailers stage

At this point of the supply chain the food loss is rather called food waste as it is food that is wasted intentionally. The perishability of the food products, the short expiration dates, the not perfectly looking food products, the unsold products, and the overordering of food products are some of the causes of food loss at this stage of the food supply chain. To reduce food losses at the retailers' stage different solutions are suggested that include modification of the product labeling, change of consumers' food waste behavior, donations and recycling, and technological investments.

8.4.3.1 Modification of product labeling

Product labeling contributes significantly to food losses. The "best by" dates are usually misinterpreted by consumers resulting in unsold food. Modifications in the product labeling to enable consumers to understand when the products should be discharged could reduce food losses. For example, a "sell by" label and a "best by if used" label could be used for the retailers and consumers understanding of the expiration dates respectively (Kor et al., 2017). Providing more information about the ingredients of the products and its freshness could help consumers make more informed decisions about when the products are not safe for consumption.

8.4.3.2 Change of consumers' food waste behavior

Retailers' food product promotions have been blamed for food losses at the consumer's point of the supply chain. This is because consumers overbuy products and then they discharge them. Educating consumers about the issue of food loss and its economic, social, and environmental implications could reduce food losses. Many supermarkets started to have food waste reduction campaigns to educate consumers and to change their mindset. Retailers could also promote food loss reduction through food waste reduction tips and recipes for utilizing leftover food. This could be done using online platforms and forums for sharing waste reduction ideas to engage the younger generation too. Organizing special campaigns about food loss reduction through the engagement of consumers could also be used as a way to make consumers aware of the food loss issue. Creation of online shopping lists and matching them to recipes could be another way to help consumers towards buying only what they need.

The high cosmetic standards in the food sector are one of the major causes of food losses at the retailers' stage. Consumers need to be educated regarding the appearance of the food products. The perception of having perfectly looking fruit and vegetables should be changed. Some retailers already promote this product category of products as "wonky" vegetables and they explain to consumers that they are perfect for consumption. This is an effort to educate consumers that fruit and vegetables cannot all have the same appearance.

8.4.3.3 Donations, recycling, and compost

Any surplus or unsold food with short expiration date should be donated to food banks or other similar charitable organizations. Some retailers cooperate with charitable organizations or food banks to distribute food and advise consumers how to use food that will be wasted otherwise (Kaye, 2011). While others are prevented from doing this in case of any food contamination issues. Special agreements with charities should be put in place to have a systematic process for distribution the food products. This would not only contribute to food loss reduction, but it could also increase the retailers' corporate social responsibility efforts. Store-to-distribution strategies could be used to move unsafe for consumption food products for recycling or recovery. Any leftovers of food could be used for composting. Small-scale anaerobic digestion could be used at retail stores to recover materials for energy production.

8.4.3.4 Technological investments

Investing in new technologies to increase transparency in the supply chain and improve the time to market of the food products is essential to food loss reduction (Kor et al., 2017). This requires also better collaboration with suppliers to eliminate any intermediaries. The benefits of this technology would be increased visibility in the supply chain and thus better management of the inventory of the food products.

Also, special smart sensors to check the quality and freshness of food products could also be placed on the retailers' trucks or on the packaging of the products. Any change in the food product quality during its transport to the retailers' store or within the store could be noticed before the product's quality deteriorates. Besides, consumers will be able check the product freshness in real time. According to Capgemini (2017) retailers can invest in applications that can match the supermarket supply with the demands of food banks. Using this app the food banks can adjust their food needs in real time and donors can check the availability of the required products.

8.4.4 Supply chain solutions

Although solutions for food loss reduction at the different stages of the supply chain are essential, there are supply chain wide solutions that should be implemented too. In this section different supply chain solutions for food loss reduction and thus supply chain optimization are suggested. These are namely awareness of the changing food standards and regulations, collaboration, across the supply chain, formation of communities of practice, technological and infrastructural solutions, lean and total quality management (TQM), sustainability across the supply chain, and developing a resilient supply chain.

8.4.4.1 Awareness of the changing food standards and regulations

The surrounding policy and regulatory framework might affect the ability of the supply chain entities to reduce food loss levels (HLPE, 2014). When food safety rules are well designed, they will enable food loss reduction (HLPE, 2014). The main regulations that food supply chain members need to comply and adopt are food safety regulations, food quality regulations, food labeling and packaging regulations, food traceability regulations, food transport and handling regulations, and organic food regulations. According to Waarts et al. (2011), in Europe private food safety regulations are the main reason of food loss occurrence. This is because food products are getting rejected due to noncompliance to the private food safety standards that are required from buyers in other EU countries. According to Despoudi et al. (2015) producers perceived that there are no specific guidelines on what food regulations they need to adopt and comply with.

Adoption and compliance with food safety and quality standards can help to reduce food losses (Lupien, 2008). For example, for a producer who wants to export his products in another country and his products do not comply with the food safety standards in this country (e.g., banned pesticides), the products will be rejected, and all the crops will get wasted. Upstream chain members, that is, producers, processors, and retailers need to be aware of the different international food regulations to prevent any noncompliance. Establishing a common communication channel among supply chain members to communicate order requirements in terms of food regulations could significantly reduce food losses.

8.4.4.2 Collaboration across the supply chain

Chapman (2010) referred to food loss as a shrinkage problem and characterized it as a "complex" problem that needs to be addressed in a collaborative way involving a wide range of stakeholders to get different perspectives and deliver holistic solutions. Recent research showed that better producer–buyer and supplier–retailer relations and collaborative action could possibly reduce food losses (WRAP, 2011; Despoudi, 2016; Despoudi et al., 2018). Other research suggested that better and closer collaboration between suppliers and retailers can be the starting point to deal with the majority of root causes of food losses (Mena et al., 2011). There are many benefits for supply chain partners achieving collaboration, some of which are the following: information exchange, improved planning and support, joint problem solving, sharing resources gain of competitive advantage, reduced costs, and reduction of negative bullwhip effect (Daugherty, 2011).

Supply chain collaboration (SCC) can be achieved in different forms such as vertical and/or horizontal and external and/or internal collaboration (Barratt, 2004). Vertical collaboration involves internal and external collaboration with customers and suppliers, respectively. Horizontal collaboration involves internal collaboration, but also external collaboration with competitors and other organizations. Internal collaboration refers to an organization's collaborative culture (e.g., existence of elements of trust and commitment). A common case with internal collaboration is the dilemma arising between decisions to be made for the interest of all chain partners and/or the individual firm (Simatupang and Sridharan, 2002). External downstream collaboration involves customer relationship management, while external upstream collaboration involves supplier management. Each entity in the supply chain might collaborate in different levels; not all partner relationships need to be involved in high levels of SCC (Holweg et al., 2005). There are different types/levels of SCC such as transaction collaboration, cooperative collaboration, and cognitive collaboration (Whipple and Russell, 2007). Transaction collaboration involves simple communication and partners exchanging data, while cooperative collaboration involves partners sharing data, processes, and setting common supply chain objectives. Cognitive collaboration requires higher levels of involvement as partners work together in joint planning and decision making.

Collaboration requires resources and effort from all partners (Whipple and Russell, 2007). Food supply chain entities do not need to collaborate closely with everyone in their supply chain; they rather focus on a small number of strategic partners (De Leeuw and Fransoo, 2009). However, there is a dilemma with whom and in what level to collaborate with partners; collaborating internally, with customers, with suppliers, with competitors, with governments and/or other institutions. To determine what level of SCC is needed for a specific chain or a specific problem first the current levels of SCC need to be assessed and after that seek for ways to improve collaborative efforts/practices (Simatupang and Sridharan, 2002). However, achieving collaboration with partners does not always have the expected benefits (Kampstra et al., 2006). There are a number of challenges mentioned in the literature as impediments in achieving collaboration. The main barriers associated with SCC are the following: difficulties in implementation, overreliance on technological solutions of collaboration, failure to differentiate with whom to collaborate with, and lack of trust between trading partners (Barratt, 2004).

8.4.4.3 Formation of communities of practice and learning alliances

Creation of learning alliances has been proposed as a way to reduce food losses (World Bank, 2006). Learning alliances are about identifying, sharing, and adapting good practices in research and development in specific contexts between research organizations, development agencies, policymakers, and private business. FAO's (2010) workshop on reducing food losses in Africa proposed a strategy for developing communities of practice about food losses to facilitate information exchange and share knowledge about new technologies and strategies to manage crops. It is essential to develop strategies that promote coordination, collaboration and information flow among all actors in the chain.

8.4.4.4 Technological and infrastructural solutions

Investments in technology and technology transfer are considered to be essential for better processing of food and better management of processed food and avoid food losses (Hodges et al., 2010). For example, collaborative planning forecasting could be implemented across the supply chain to better forecast the product demand. Technological advancements in the processing and transportation of the products could diminish food losses (Caixeta-Filho, 1999). This could involve new packaging solutions and/or innovations in cold chain logistics. Development of better infrastructure is a crucial step for reducing food losses including creation of better warehouses and logistics development such as cold chain facilities and handling equipment (Kader, 2010).

The nature of the agricultural products requires them to be distributed on time and to be stored under the right conditions (Zanoni and Zavanella, 2012). The lack of cold chain facilities or any delay in cooling of the products can result in quality deterioration or quality losses (Nunes et al., 2009). Temperature control during processing of the crops is a challenging task and fluctuating temperatures have an effect on product's quality (Brecht et al., 2003). Inadequate and improper management of cold chains leads to food loss (Atanda et al., 2011). Perishability, shelf-life, and quality variations are significantly influencing food loss levels (Mena et al., 2011). Both technological and infrastructural improvements are needed to enable food loss reduction and their absence seems to be a major obstacle to achieve it.

8.4.4.5 Total quality management and lean

TQM is a management philosophy and a set of accompanying quality continuous improvement techniques (Slack et al., 2013). By applying TQM philosophy and techniques, businesses undertake continuous improvement across all operations by seeking to discover the reasons for poor quality performance and customer service and implementing methods to reduce and/or eliminate the causes of poor quality. TQM is an effective system for integrating the quality development, quality maintenance, and quality improvement efforts of the various groups in an organization to enable production and service at the most economical levels that allow for full customer satisfaction. Food loss reduction could be achieved by implementing TQM across the supply chain.

There are different aspects in TQM implementation. The first aspect of TQM is about meeting the needs and expectations of customers. Customer centricity is essential to meet the customer's perception of quality and the changing needs and wants of customers (Slack et al., 2013). The customer's voice should be translated into quality objectives to increase customer satisfaction. TQM is about covering all parts of an organization and involving everyone, that is, each department, each activity, each person, and each level need to work together. Everyone is a customer and supplier within an organization. Service level agreements or contracts are usually used within an organization, for example, for response times and range of services. Another aspect of TQM is that all the costs of quality need to be considered. The different costs of quality are prevention costs, appraisal costs, internal failure costs, and external failure costs. Systems and procedures to support quality improvement should be implemented. For example, ISO 9001 could be implemented to provide guidance and tools for organizations who want to ensure that their products and services consistently meet customer requirements, and that quality is consistently improved.

Developing a continuous process of improvement is a core principle of TQM. Lean philosophy can be used to determine value-adding and nonvalue-adding activities at every stage of the food supply chain. Lean refers to approaches that focus on the elimination of waste in all forms, and smooth and efficient flow of materials and information throughout the supply chain to obtain faster customer response, higher quality, and lower costs. The different types of waste need to be identified first and then ways of eliminating them should be implemented. Food loss and its different waste implications need to be identified first. Then, different lean tools such as value stream mapping and 5S could be implemented in the food supply chain to eliminate food losses across the supply chain.

8.4.4.6 Implement sustainability across the supply chain

Climatic changes are impacting yields, altering weather patterns, increasing the uncertainty and likelihood of disruption (Bereuter et al., 2014). Food supply chains are not always able to respond appropriately and on time resulting in lost production, resources, and sales. SustainAbility (2011) defined a sustainable food supply chain as a reliable, resilient, and transparent, one that produces food within ecological limits, empowers food producers, and ensures accessible and nutritious food for all. A sustainable food supply chain:

1. produces safe and healthy products,
2. supports the viability and diversity of the communities,
3. enables sustainable livelihoods by respecting human rights and fair trade terms, and
4. sustains the available resources and minimizes inputs (Defra, 2006).

A sustainable food supply chain must meet the world's need for food and avoid adverse environmental impacts (Defra, 2006). Hence, by increasing the sustainability efforts across the supply chain food losses could be reduced. However, to achieve supply chain sustainability efforts from all supply chain entities are needed.

In the supply chain from the product's perspective, sustainability can be illustrated through the product stewardship concept. Product stewardship can be defined as the shared responsibilities that all the participants in a product's lifecycle have for minimizing its environmental and health impacts (Product Stewardship Institute, 2011). The producers' responsibilities lie from the downstream (customer end of the supply chain) in the supply chain to the upstream (supplier end of the supply chain supply chain). The responsibilities in a supply chain do not end when the product is delivered to its customers. Companies that produce goods are responsible for the whole lifecycle of their products, from raw material extraction to use and disposal. Thus, there is a need for increasing the awareness regarding the shared responsibility of all the food supply chain entities to increase sustainability.

8.4.4.7 Developing a resilience supply chain

Lal et al. (2014) referred to resilience as the state of having the natural ecosystem to withstand environmental changes based on the ability to have resistance to certain forms of disturbances. Shenggen et al. (2014) defined resilience in the food and agricultural sector as the ability to prevent crises and disasters by being able to anticipate, engross, and accommodate the effects of the disaster and creating an efficient and timely solution that will manage any potential issues in a timely manner. The authors further indicate that the protection, restoration, and improving on the systems of agriculture will assist in improving food security.

Almas and Campbell (2012) indicated that the effectiveness of the agricultural strategies in the management of sustainability will be able to determine the effectiveness of the resilience strategy that has been set. In addition, the authors indicate that the policies that are set aside by the governing body will assist in the management of resilience in the organization. The measurement of resilience and its supply chain processes can be assessed through the assessment of the differences in operation during and after the crisis (Barthel and Isendahl, 2012). Having no decrease in operations in food production, resilience can be noted as being effective in the management of the food supply chain.

According to Christopher and Peck (2004) resilience in the supply chain process is a method that creates effective operations in terms of the distribution of the food products. A resilient supply chain is able to manage the consumer needs without risking the delivery of the products to the end use. The resilience of the supply chain is based on the development of a process that is able to withstand the volatility and turbulence in the operating environment. In addition, the supply chain processes in place should be able to ensure that any changes in the operating environment factors of the organization such as politics, the economy will have limited impact on the supply chain. A resilient supply chain may be also managed through redundancy management and having flexibility in the supply chain process (Sheffi and Rice, 2005).

The selection of the different strategies in the management of the food chain supply chain is effective in controlling the losses that may be incurred in the supply process (Ceryno et al., 2013). Whipple and Russell (2007) suggested that the use of the collaborative approaches such as having a collaborative management of transactions, and collaborating management of events and processes assists in the effective creation of a resilient supply chain process. Hudson (2009) stated that the management of the environmental footprint will assist in ensuring that the supply chain process is effectively managed. Sonnino and Marsden (2006) stated that the supply chain process should be built derived from the implications that the actions will have on the rural and agricultural development.

In the assessment of the resilient food supply chain, Sonnino and Marsden (2006) indicated that the relationships in the operating environment should be managed effectively. Having sustainable relationships between the key players in the supply chain will assist in building a strategy that will have minimum negative implications. The food production systems should be mirrored on the management of the processes, which should be driven by the goals of the retailer, those of the consumer, and the producers (LeBlanc et al., 2014). Having a resilient food supply chain further incorporates the participation of different actors in the agricultural industry, who will assist in the attainment of the set goals in the supply chain. The current food supply chain processes have been affected by the complexity of the operations in the market. According to the findings, the social and environmental implications on the processes and logistics have affected the resilience of the operations in the market. The systems created in the different strategies of maintaining resilience have affected the operations in the market and in turn food supply.

The management of a resilient food chain is dependent on the availability of resources in the market. Manning and Soon (2016) stated that the availability of resources will assist in ensuring that the supply chain in food production is conducted effectively. The use of the resources available will assist in the assessment of the different tools that will be used to measure the best and most effective strategy to be applied. According to Maslaaric et al. (2013), the management of the resources that are invested in the supply chain processes will be effective in ensuring that there is availability of food in the market. The management of the costs of operations will assist in ensuring that there is no compromise in the quality of the food being produced, which may have adverse effects if costs are decreased.

Min et al. (2005) asserted that collaboration efforts in the supply chain processes are one of the most effective means in the management of resources. Having collaborative efforts in the food industry is important in ensuring that the people have access to the adequate food supply. Brabeck-Letmathe (2016) stated that the efficiency of resources can be managed through the mitigation of the climate changes by the different food production organizations. In the assessment, the maintenance of stability is a critical aspect that will have an implication on the operations in the food market. Foley et al. (2011) mentioned that the environmental changes affected the yields of the crops, hence the need to manage the current resources to ensure that there is sufficient food supply in the global growing population.

8.5 Conclusion

This chapter provided an overview of the food losses in the food supply chain. Also, solutions to reduce food losses by optimizing supply chains are discussed. The occurrence of food losses in the supply chain is described in terms of the upstream and downstream supply chain. Food losses occur across the supply chain and there are different causes at each stage. The different ways to reduce food losses are discussed based on the different supply chain entities and on the whole supply chain. Different solutions for each supply chain entity need to be considered. However, to enable food loss reduction both actor specific and supply chain solutions are needed. Overall supply chain optimization suggestions can provide holistic solutions to the food loss issue.

References

Ahmed, S., Stepp, J.R., 2016. Beyond yields: climate change effects on specialty crop quality and agroecological management climate effects on specialty crop quality. Elementa: Sci. Anthropocene 4, https://www.elementascience.org/articles/10.12952/journal.elementa.000092/.

Akkerman, R., Van Donk, D.P., 2008. Development and application of a decision support tool for reduction of product losses in the food-processing industry. J. Cleaner Product. 16 (3), 335–342.

Almas, R., Campbell, H., 2012. Rethinking Agricultural Policy Regimes: Food Security, Climate Change and the Future Resilience of Global Agriculture. Emerald Group Publishing.

Atanda, S.A., Pessu, P.O., Agoda, S., Isong, I.U., Ikotun, I.I., 2011. The concepts and problems of post-harvest losses in perishable crops. Afr. J. Food Sci. 5 (11), 603–613.

Barratt, M., 2004. Understanding the meaning of collaboration in the supply chain. Supply Chain Manage. Int. J. 9 (1), 30–42.

Barthel, S., Isendahl, C., 2012. Urban gardens, agriculture, and water management: sources of resilience for long-term food security in cities. Ecol. Econ. 224–234.

Brabeck-Letmathe, P., 2016. Climate Change, Resource Efficiency and Sustainability. Springer Science + Business Media, New York.

Brecht, J., Chau, K., Fonseca, S., Oliviera, F., Silva, F., Nunes, M., et al., 2003. Maintaining optimal atmosphere conditions for fruits and vegetables throughout the postharvest handling chain. Postharv. Biol. Technol. 27 (1), 87–101.

Bereuter, D., Glickman, D., Nelson, G.C., 2014. Advancing Global Food Security in the Face of a Changing Climate. Chicago Council on Global Affairs.

Caixeta-Filho, J.V., 1999. Losses in the transportation of fruits and vegetables: a Brazilian case study. Int. J. Logis. Res. Appl. 2 (3), 325–341.

Capgemini (2017): https://www.capgemini.com/wp-content/uploads/2017/10/food-waste-management_web.pdf

Chapman, P., 2010. Reducing product losses across the food supply chain. In: Mena, C., Graham, S. (Eds.), Delivering Performance in Food Supply Chains. Woodhead Publishing.

Christopher, M., Peck, H., 2004. Building the resilient supply chain. Int. J. Logis. Manage. 15 (2), 1–13.

Ceryno, P.S., Scavarda, L.P., Klingebiel, K., Yuzgulec, G., 2013. Supply chain risk management: a content analysis approach. Int. J. Industr. Eng. Manage. (IJIEM) 4 (3), 141–150.

Coumou, D., Rahmstorf, S., 2012. A decade of weather extremes. Nat. Ology 1–6.

Daugherty, P.J., 2011. Review of logistics and supply chain relationship literature and suggested research agenda. Int. J. Phys. Distrib. Logis. Manage. 41 (1), 16–31.

Defra, 2006. Food industry sustainability strategy. <http://www.defra.gov.uk/publications/files/pb11649-fiss2006-060411.pdf> (accessed 10.09.14.).

Defra, 2009. UK Food SecurityAssessment: our approach. <http://archive.defra.gov.uk/foodfarm/food/pdf/food-assess-approach-0908.pdf> (accessed 25.01.12.).

De Leeuw, S., Fransoo, J., 2009. Drivers of close supply chain collaboration: one size fits all? Int. J. Operat. Prod. Manage. 29 (7), 720–739.

Despoudi, S., Papaioannou, G., Saridakis, G., Dani, S., 2018. Does collaboration pay in agricultural supply chain? An empirical approach. Int. J. Prod. Res. Available from: https://doi.org/10.1080/00207543.2018.1440654.

Despoudi, S., 2016. An investigation of the collaboration – postharvest food loss relationship and the effect of the environmental turbulence factors, Loughborough University. <https://dspace.lboro.ac.uk/dspace-jspui/bitstream/2134/21785/1/Thesis-2016-Despoudi.pdf> (accessed 18.02.18.).

Despoudi, S., Papaioannou, G., Dani, S., 2015. An investigation of the environmental turbulence factors and their sources in the collaboration-post-harvest food loss relationship. Farm Fork: Int. J. Innovat. Res. Practice 2 (1), 1–6.

EPA, 2011. Generators of food waste. <http://www.epa.gov/osw/conserve/materials/organics/food/fd-gener.htm#food-hier> (accessed 18.02.12.).

EU FUSIONS, 2016. Reducing waste through social innovation. <https://www.eu-fusions.org/phocadownload/Publications/Estimates%20of%20European%20food%20waste%20levels.pdf> (accessed 10.06.18.).

FAO, 2018. The ten elements of agroecology. <http://www.fao.org/3/I9037EN/i9037en.pdf> (accessed 10.09.18.).

FAO, 2010. FAO – World Bank workshop on reducing post-harvest losses in grain supply chain in Africa. <http://www.fao.org/fileadmin/user_upload/ags/publications/FAO_WB_ph_web.pdf> (accessed 19.03.12.).

FAO, 2011. Global food losses and waste. <http://www.fao.org/fileadmin/user_upload/ags/publications/GFL_web.pdf> (accessed 10.01.12.).

FAO, 2012. Top production – peaches and nectarines – 2012. <http://faostat.fao.org/DesktopDefault.aspx?PageID = 339&lang = en&country = 84> (accessed 04.06.14.).

Foley, J., et al., 2011. Solutions for a cultivated planet. Nature 478, 1–6.

Foresight, 2011. Foresight Project on Global Food and Farming Futures. Synthesis Report C7: Reducing Waste. The Government Office for Science, London.

Fotopoulos, C., Krystallis, A., 2003. Quality labels as a marketing advantage: the case of the 'PDO Zagora' apples in the Greek market. Eur. J. Market. 37 (10), 1350–1374.

Gornall, J., Betts, R., Burke, E., Clark, R., et al., 2010. Implications of climate change for agricultural productivity in the early twenty-first century. Philos. Trans. R. Soc. B 365, 2976–2992.

Gustavsson, J., Cederberg, C., Sonesson, U., Otterdijk, R., Meyberg, A., 2011. Global food losses and waste. <http://www.fao.org/fileadmin/user_upload/ags/publications/GFL_web.pdf> (accessed 10.01.12.).

Godfray, H.C., Crute, I.R., Haddad, L., Lawrence, D., Muir, J.F., Nisbett, N., et al., 2010. The future of Global Food System. Philos. Trans. R. Soc. B 365, 2769–2777.

Go Supply Chain, 2018. Nine ways to reduce food waste in the supply chain. <https://www.gosupplychain.com/blog/9-ways-to-reduce-food-waste-in-the-supply-chain.html> (accessed 05.09.18.).

HLPE, 2014. Food losses and waste in the context of sustainable food systems. A report by the High Level Panel of Experts on Food Security and Nutrition of the Committee on World Food Security, Rome 2014.

Hodges, R., Buzby, J.C., Benett, B., 2010. Postharvest losses and waste in developed and less developed countries: opportunities to improve resource use. J. Agric. Sci. Cambridge Univ. 37–45.

Holweg, M., Disney, S., Holmström, J., Småros, J., 2005. Supply chain collaboration: making sense of the strategy continuum. Europ. Manage. J. 23 (2), 170–181.

Hudson, R., 2009. Resilient regions in an uncertain world: wishful thinking or a practical reality? Cambridge J. Econ. Soc.

Kader, A.A., 2005. Increasing food availability by reducing postharvest losses of fresh produce. In: Proceedings 5th International Postharvest Symposium.

Kader, A.A., 2010, Handling of horticultural perishables in developing vs. developed countries. In: Proceedings 6th International Postharvest Symposium. <http://ucce.ucdavis.edu/files/datastore/234-1875.pdf> (accessed 10.01.12.).

Kaditi, E.A., Nitsi, E.I., 2010. The Agricultural Sector in Greece. Centre of Planning and Economic Research (KEPE), Athens (in Greek).

Kamenidis, H., 2004. Agricultural Marketing. Charis EPE, Athens (in Greek).

Kampstra, R.P., Ashayeri, J., Gattorna, J.L., 2006. Realities of supply chain collaboration. Int. J. Logis. Manage. 17 (3), 312–330.

Kaye, L., 2011. Food retailers must do more to reduce food waste. <http://www.guardian.co.uk/sustainable-business/food-retailers-must-reduce-waste?INTCMP = ILCNETTXT3487> (accessed 03.02.12.).

Kor, Y., Prabhu, J., Esposito, M., 2017. How large food retailers help solve the food waste crisis. <https://hbr.org/2017/12/how-large-food-retailers-can-help-solve-the-food-waste-crisis> (accessed 10.06.18.).

Lal, L., Singh, B., Mwaseba, D.L., Kraybill, K., Hansen, D.O., Eik, L.O., 2014. Sustainable Intensification to Advance Food Security and Enhance Climate Resilience in Africa. Springer, New York.

LeBlanc, J.R., Conner, D., McRae, G., Darby, H., 2014. Building resilience in nonprofit food hubs. J. Agric. Food Syst. Commun. Develop. 121–135. Advance online publication.

Lobell, D.B., Schlenker, W., Costa-Roberts, J., 2011. Climate trends and global crop production since 1980. Science 333 (2011), 616–620.

Lobell et al (2014): https://www.theguardian.com/vital-signs/2014/oct/29/diet-climate-maple-syrup-coffee-global-warming.

Lupien, J.R., 2008. Small and medium-size food producers and processors: potential in national development, international trade, and role in solving nutrition problems. <http://www.iufost.org/publications/books/documents/Lupien.pdf> (accessed 10.01.12.).

Lundqvist, J., De Fraiture, C., Molden, D., Berndes, G., Berntell, A., Falkenmark, M., 2008. Saving Water: From Field to Fork – Curbing Losses and Wastage in the Food Chain, SIWI Policy Brief. SIWI.

Manning, L., Soon, J.M., 2016. Building strategic resilience in the food supply chain. Br. Food J. 118 (6), 1477–1493.

Maslaaric, M., et al., 2013. Assessing the trade-off between lean and resilience through supply chain risk management. Int. J. Industr. Eng. Manage. (IJIEM) 4 (4), 229–236.

Mena, C., Adenso-Diaz, B., Yurt, O., 2011. The causes of food waste in the supplier-retailer interface: evidences from UK and Spain. Resour. Conserv. Recycl. 55, 648–658.

Min, S., Roath, A.S., Daughertyu, P.J., Genchey, S.E., Chen, H., Arndt, A.D., 2005. Supply chain collaboration: what's happening? Int. J. Logis. Manage. 16 (2), 237–256.

Mirza, M.Q., 2003. Climate change and extreme weather events: can developing countries adapt? Clim. Policy 3 (2003), 233–248.

Nunes, M., Emond, J., Rauth, M., Dea, S., Chau, K., 2009. Environmental conditions encountered during typical consumers retailer distribution affect fruit and vegetable quality and waste. Postharv. Biol. Technol. 51 (2), 232–241.

Moore, H., 2016. Can agroecology feed the world and save the planet? <https://www.theguardian.com/global-development-professionals-network/2016/oct/09/agroecological-farming-feed-world-africa> (accessed 10.09.18.).

Paisiadis, S., 2013. In progress is the reform of the fruit and vegetables in the EU small interest from Greece. Agropypos magazine. <http://www.agrotypos.gr/index.asp?mod-= articles&id = 77494> (accessed 10.03.13.) (in Greek).

Parfitt, J., Bartherl, M., Macnaughton, S., 2010. Food waste within food supply chains: quantification and potential for change to 2050. Philos. Trans. R. Soc. B 365, 3065–3081.

Premanandh, J., 2011. Factors affecting food security and contribution of modern technologies in food sustainability. J. Sci. Food Agric. 91 (15), 2707–2714.

Product Stewardship Institute, 2011. <http://productstewardship.us/displaycommon.cfm?an = 1&subarticlenbr = 55> (accessed 17.04.11).

Rosenzweig, C., Iglesius, A., Yang, X.B., Epstein, P.R., Chivian, E., 2001. Climate Change and Extreme Weather Events – Implications for Food Production, Plant Diseases, and Pests. NASA Publications, Paper 24.

Schlenker, W., Roberts, M.J., 2008. Nonlinear temperature effects indicate severe damages to U.S. crop yields under climate change. Proc. Natl. Acad. Sci. 106 (37), 15594–15598.

Sharma, G., Singh, S.P., 2011. Economic analysis of post-harvest losses in marketing of vegetables in Uttarakhand. Agric. Econ. Res. Rev. 24, 309–315.

Sheffi, Y., Rice, J.B., 2005. A supply chain view of resilient enterprise. Manage. Sloan Rev. 47, 41–51.

Shenggen, F., Pandya-Lorch, R., Yosef, S., 2014. Resilience for Food and Nutrition Security. International Food Policy Resistance Institution.

Sonnino, R., Marsden, T., 2006. Beyond the divide: rethinking relationships between alternative and conventional food networks in Europe. Journal of Econ Geography 6 (2), 181–199.

Smithers, R., 2010. GM and farming technology 'key to fighting climate change. <http://www.guardian.co.uk/environment/2010/feb/24/gm-precision-farming-environment-agency?INTCMP = ILCNETTXT3487> (accessed 15.02.12).

Simatupang, T.M., Sridharan, R., 2002. The collaborative supply chain. Int. J. Logis. Manage. 13 (1), 13–50.

Slack, N., Chambers, S., Johnston, R., 2013. Operations Management, seventh ed. Prentice Hall, Harlow.

Sustainability, 2011. Appetite for change. <http://www.sustainability.com/library/appetite-for-change#.Tzzj5Vy15WU> (accessed 03.02.12).

The World Bank, 2011. Missing food: the case of postharvest grain losses in Sub-Saharan Africa. <http://siteresources.worldbank.org/INTARD/Resources/MissingFoods10_web.pdf> (accessed 26.12.12).

USDA, 2010. Food loss—questions about the amount and cause still remain. <https://www.ers.usda.gov/amber-waves/2014/june/food-loss-questions-about-the-amount-and-causes-still-remain/> (accessed 10.06.18).

Waarts, Y., Eppink, M.M., Oosterkamp, E.B., Hiller, S., Van Der Sluis, A.A., Timmermans, A.J.M., 2011. Reducing Food Waste: Obstacles and Experiences in Legislation and Regulations. Rapport LEI.

Whipple, J.M., Russell, D., 2007. Building supply chain collaboration: a typology of collaborative approaches. Int. J. Logis. Manage. 18 (2), 174–196.

Williams, H., Wikström, F., 2011. Environmental impact of packaging and food losses in a life cycle perspective: a comparative analysis of five food items. J. Cleaner Prod. 19 (1), 43–48.

World Bank, 2006. Enhancing agricultural innovation: how to go beyond the strengthening of research systems. <http://siteresources.worldbank.org/INTARD/Resources/Enhancing_Ag_Innovation.pdf> (accessed 10.01.12).

World Economic Forum, 2011. Driving sustainable consumption: value chain waste. <https://members.weforum.org/pdf/sustainableconsumption/DSC%20Overview%20Briefing%20-%20Value%20Chain%20Waste.pdf> (accessed 10.01.12).

WRAP, 2009. Household food and drink waste in the UK. <http://www.wrap.org.uk/sites/files/wrap/Household_food_and_drink_waste_in_the_UK_-_report.pdf> (accessed 10.01.12).

WRAP, 2011. Reducing food waste through retail supply chain collaboration. <http://www.wrap.org.uk/sites/files/wrap/WRAP_IGD_supply_chain_report.pdf> (accessed 26.02.12).

Zanoni, S., Zavanella, L., 2012. Chilled or frozen? Decision strategies for sustainable food supply chains. Int. J. Prod. Econ. 140 (2), 731–736.

Measuring food losses in the supply chain through value stream mapping: a case study in the dairy sector

9

Joshua Wesana[1,2], Xavier Gellynck[1], Manoj K. Dora[3], Darian Pearce[1] and Hans De Steur[1]
[1]Department of Agricultural Economics, Faculty of Biosciences Engineering, Ghent University, Ghent, Belgium, [2]School of Agricultural and Environmental Sciences, Mountains of the Moon University, Fort Portal, Uganda, [3]College of Business, Arts & Social Sciences, Brunel Business School, Brunel University, London, United Kingdom

Chapter Outline

9.1 Introduction

9.1.1 Background

The terms food loss and food waste are often utilized synonymously, but they do in fact differ based on where they occur along the supply chain. Food losses take place during production, harvest, processing, and distribution; unlike food waste, which occurs at the retail and consumer levels of the chain (Parfitt et al., 2010; Richter and Bokelmann, 2016; Willersinn et al., 2015). Nonetheless, both elements point to

Saving Food. DOI: https://doi.org/10.1016/B978-0-12-815357-4.00009-2

a certain quantity of food, calories, or nutrients that are intentionally or unintentionally disappearing from the food supply chain before consumption. Food loss and waste (FLW) is an endemic and growing global problem, estimated at over 30% of produced food that is not consumed (Gustavsson et al., 2011). Within vulnerable regions, FLW contributes to the dire state of food insecurity at a time when increased food production, as a solution, is costly and exploits scarce productive agricultural land and water (Godfray et al., 2010; Phalan et al., 2011).

There is an increasing interest in promoting efforts leveraging FLW reduction as a means of assuring adequate and equitable food availability, if surplus food could be redistributed appropriately to the hungry (Garrone et al., 2014). Tackling FLW in both developed and developing countries is associated with positive outcomes especially on food prices, thus increasing economic access to food among people likely to experience hunger (Buzby and Hyman, 2012; Rutten, 2013). Thereby, actions that minimize FLW in food systems directly support their sustainability, contributing to food security to offset pressure on increased food production (Munesue et al., 2015; Smith, 2013; West et al., 2014). The fight against FLW is reinforced by Sustainable Development Goal (SDG) target 12.3, which aims at halving food waste at retail and consumer levels, whilst simultaneously reducing food losses along production and supply chains (Hanson, 2017). SDG 12.3 primarily targets quantifiable losses or wastes, equivalent to a quarter of available calories that are missed and never consumed (Pangaribowo et al., 2013). Such a loss would ideally feed close to 10% of the current 821 million undernourished people in developing countries (FAO et al., 2018; Munesue et al., 2015). However, strategies to reduce FLW in developing countries are hindered by an absence of reliable data on FLW that occurs within different food value chains (Affognon et al., 2015). The few studies that do link FLW with macro- or micronutrients lost from the supply chain are also limited to developed countries (Cooper et al., 2018; Love et al., 2015; Spiker et al., 2017). This absence could hinder evidence-based follow-up of SDG 12.3 indicators especially in countries experiencing food and nutrition insecurity (Barrett et al., 2010; Francis et al., 2012; Gil et al., 2006).

There exist additional obstructive factors to FLW data acquisition. FLW definitions and measurements methods are inconsistently used, exacerbating identification and quantification problems that ultimately affect mitigation efforts (Chaboud and Daviron, 2017; Redlingshöfer et al., 2017). The lack of harmonized or integrated FLW assessment is a historical problem limiting acquisition of reliable and comparable FLW data. This is partly the reason for inconsistencies in the approximation of the magnitude of FLW around the world (Xue et al., 2017). To solve this problem, the FLW protocol was developed as a standard for accurate accounting and reporting of FLW (Hanson, 2016). It facilitates comparison across regions, countries, and between other smaller entities like companies and organizations. It also covers the entire food chain, distinguishes food loss from food waste, considers (in-)edible food parts, as well as possible destinations of FLW (Hanson, 2017). The protocol is based on the idea that what gets measured can also be managed and hence crucial to the design and development of appropriate FLW mitigation strategies. Although the FLW protocol proposes 10 FLW quantification methods, it does not address the need for complementary approaches for identification of FLW hotspots. Such identification approaches could form the basis to successfully apply standards, like the FLW protocol, hence strengthening FLW measurements and improving subsequent mitigating efforts along the supply chain, while considering a life cycle perspective of FLW (Corrado et al., 2017). Because the supply chain constitutes various hotspots where FLW occurs, a life cycle assessment (LCA) further lays the foundation and facilitates holistic analysis of products, processes, or activities (Roy et al., 2009). As such, approaches that transverse the entire supply chain should consider stakeholder awareness creation, interest development, and establishment of strategic actor partnerships so as to increase success (Aschemann-Witzel et al., 2017; Muriana, 2017; Parmar et al., 2017; Richter and Bokelmann, 2016).

9.1.2 Stakeholder adoption of lean manufacturing practices for food loss and waste assessment and mitigation

Although comprehensive assessments of FLW are still a challenge in many contexts, efforts to minimize FLW contribute to the realization of nutrition sensitive agriculture, prioritized to sustainably address global hunger by 2030 (Keding et al., 2013). As such, a recent study by Wesana et al. (2018) provides evidence that value chain actors support initiatives to reduce FLW and subsequently promote nutrition sensitive food systems. Findings in this study are based on the theory of organizational readiness to change, modeled to evaluate actor willingness to adopt lean manufacturing practices along the dairy supply chain to reduce FLW. The theory specifically associates change valence to change commitment on one hand and implementation capability to change efficacy on the other hand. Further, the concept of the multiactor approach was linked to both change commitment and efficacy.

Interviews conducted among farmers, processors, retailers, and distributors of milk products provide evidence that unmarketable products are normally discarded and so constitute FLW. The study affirms that FLW occurs at multiple stages of the supply chain and further justifies the need for multiactor collaboration to tackle this problem (De Steur et al., 2016; Göbel et al., 2015). It is thus indicated that actors who value the adoption of lean manufacturing and are also positive about a multiactor approach exhibit an increased commitment for implementing lean practices to reduce FLW. In addition, a positive perception of the resources required to reduce FLW using lean practices is more likely to increase an actor belief of having the ability to successfully implement advocated initiatives. This study contributes to this limited body of evidence related to perceptions of value chain actors towards nutrition sensitive agriculture (Allen and de Brauw, 2017; Jaenicke and Virchow, 2013). Current policy dialogues are also in favor of this approach so as to improve expected impacts of agriculture on nutrition outcomes (Hodge et al., 2015; Van Den Bold et al., 2015). This could potentially increase the success of strategies that are nutrition sensitive if an enabling policy environment is established following recommended frameworks (Gillespie et al., 2013; Pingali and Sunder, 2017).

9.1.3 Value stream mapping as a hot spot identification approach for food loss and waste assessments

Value stream mapping (VSM) has been proposed as a method that can be used to identify and map hotspots of FLW along the agrifood value chains (De Steur et al., 2016). It is part of lean manufacturing, a management philosophy that was developed to eliminate wastes in production systems (Womack et al., 1990). Since its inception in the automobile sector, it has been utilized in other sectors including the agrifood industry (Dora et al., 2014; Zokaei and Simons, 2006). As a lean tool, VSM, a method that facilitates systematic documentation of flow of materials and information in producing a product, is becoming particularly popular in the agrifood sector (Panwar et al., 2015). This approach involves mapping the production configuration to identify lean wastes categorized as defects, overproduction, inappropriate processing, unnecessary inventory, unnecessary motion, transport and waiting (Dal Forno et al., 2014; Womack, 2006).

A study by De Steur et al. (2016) was the first to systematically aggregate evidence on the potential of using VSM to identify FLW along food supply chains. This study compiled evidence from the available literature that applied VSM in an agrifood context to confirm VSM's adaptability to efforts targeting reduction of FLW. Thereby, 24 studies dealing with lean manufacturing aspects or concepts (i.e., VSM, lean management, lean philosophy, lean thinking, lean principles, lean practices, and lean tools) as well as food related aspects (i.e., food, food supply chain, agrifood chain, food industry, food sector, and agriculture) were selected as they tackled loss and waste identification and or minimization. Information that was considered relevant to FLW was obtained including supply chain actors, types of food products, VSM related aspects (state maps, other lean tools, and lean metrics), and types and reasons for waste.

In this study, it was observed that VSM was either used in a single- or multiactor setting in various agrifood contexts including factories, warehouses, hospital kitchens, primary producers, and distributors (i.e., wholesaler and retailers). Different food products were studied including bread, ready to eat foods, peaches, wine, mango juice, yogurt, ketchup, biscuits, snacks, coffee, tea, nougat, soups, vegetables, beef, lamb, pork, and edible oil while some studies targeted restaurants and warehouses with no specific type of food mentioned. This study further illustrated that VSM is used in various ways. Thereby, a graphical mapping technique was commonly applied including both the current and future state maps while the description of these states was less used. From case studies that used VSM, lead time and the number of operators were the commonest indicators used to determine performance improvement. Subsequently, future state maps were characterized by other lean improvement tools including Kaizen, Just-In-Time, Kanban, and Cellular Manufacturing.

Table 9.1 provides an overview of lean related waste occurrences at different stages of the supply chain that were linked to FLW in an agrifood setting. Findings showed that there were two categories of FLW including discarded food and nutrient losses. Discarded food was mainly associated with defects in products,

Table 9.1 Hotspots and wastes and their causes derived from agrifood studies applying value stream mapping, split up according to stage

Hot spot	Form of loss/waste	Lean waste	Cause of waste	References
Primary production	Discard	Unnecessary inventory	Uncertainty in supply of raw material	Seth et al. (2008)
			Use of push production system	Taylor (2005, 2006)
		Defect in product	Nonconformance to specifications[a]	Taylor (2005)
Processing	Discard	Defect in product	Nonconformance to specifications[a]	Folinas et al. (2015), Goriwondo et al. (2011), Jiménez et al. (2012), Noorwali (2013), Sathiyabama and Dasan (2013), Seth et al. (2008), Shobha and Subramanya (2012), Taylor (2005, 2006), Vlachos (2015)
			Short shelf life due to microbial spoilage	Darlington and Rahimifard (2006), Francis et al. (2008), Melvin and Baglee (2008)
		Inappropriate processing	Poor and overtopping, overbaking, variation in size/shape	Sathiyabama and Dasan (2013)
			Poor timing of slicing operation	Goriwondo et al. (2011)
			Food loss due to forming and loss of processing materials	Kennedy et al. (2013)
		Overproduction	Poor demand forecast	Darlington and Rahimifard (2006), Noorwali (2013)
		Unnecessary inventory	Excess stock of either raw materials or finished products	Jiménez et al. (2012), Lehtinen and Torkko (2005), Noorwali (2013), Shobha and Subramanya (2012), Tanco et al. (2013), Taylor (2005)

(*Continued*)

Table 9.1 (Continued)

Hot spot	Form of loss/waste	Lean waste	Cause of waste	References
	Nutrient loss	Defect in product	Nonconformance to specifications[a]	Sathiyabama and Dasan (2013)
		Inappropriate processing	Overbaking	Sathiyabama and Dasan (2013)
			Inappropriate peeling, washing, and pasteurization	Folinas et al. (2015)
Storage	Discard	Defect in product	Short shelf life due to microbial spoilage	Glover et al. (2014)
Foodservice/ consumption	Discard	Defect in product	Wrong meal service	Ahmed et al. (2015)
			Mismatch with customized needs of consumers	Rahimnia et al. (2009)
		Overproduction	Poor demand forecast	Engelund et al. (2009)

[a]Including incorrect weight and fat levels, poor/overtopped products, variation in size/shape, breakages, scrap, and/or poor quality.

Source: From De Steur, H., Wesana, J., Dora, M.K., Pearce, D., Gellynck, X. (2016). Applying value stream mapping to reduce food losses and wastes in supply chains: a systematic review. Waste Manage. 58, 359–368.

unnecessary inventory, inappropriate processing, and overproduction wastes in lean manufacturing. Main attributing factors were nonconformance with specifications, short shelf life, rejected meals, uncertain supply of raw materials, push production, poor processing outcomes, and poor demand forecast. Similarly, nutrient loss was associated with defects in products as well as inappropriate processing, where nonconformance to specifications, overbaking, inappropriate peeling, washing, and pasteurization were identified as key causes for this type of loss. The processing stage of production was by far considered the main hot spot for the occurrences of lean related FLW along the food supply chain. This gives an indication of the potential of VSM to systematically identify FLW together with hotspots where they occur, which creates an avenue for quantification of losses and wastes in a holistic manner as well as facilitates information sharing among stakeholders. This opens up opportunities to use other lean manufacturing tools to minimize the occurrence of FLW. As such, this review pointed to the need of a practical application of VSM specifically targeting losses and wastes that occur in a specific food supply chain.

Therefore, the aim of this chapter is to apply VSM analysis at chain level, while integrating the FLW standard. This is expected to lead to a reliable and systematic mapping of hotspots to facilitate FLW measurement and reporting. As a consequence, mitigating approaches could be initiated along food supply chains. Furthermore, these methods are applied within the confines of the product life cycle approach (Corrado et al., 2017). There are few studies conducted following the FLW standard (Tostivint et al., 2017), and none have used a systematic mapping approach in an agrifood chain of a nutrient-rich food product. This study used the dairy value chain in Uganda as a case.

9.2 Methodology

Data were collected in August 2017, using a case study approach at a dairy company (not named because of confidentiality), located in the western region of Uganda. The company operates a dairy farm, a processing plant, and various distribution channels. This set-up formed a value chain that was suitable for the application of the VSM methodology to conduct a holistic assessment of FLW whilst adhering to the FLW protocol (Hanson, 2016; Womack, 2006). With reference to the guidelines of the protocol and having established the purpose of this case study, the scope of this study included the period of data collection, specified target type of material [i.e., only edible (milk) products] as well as setting boundaries for data collection [i.e., three stages of the supply chain, one dairy company, (milk) products]. Destinations of lost or wasted products were observed during data collection and were reported as findings. Interviews were conducted with different personnel that worked at the three supply chain levels of the dairy company. In addition, observations of processes were made so as to confirm key informants' responses. In case of inconsistencies in responses, the observed situation took precedence.

A semistructured questionnaire was used to guide the data collection. Its development was based on the principles of lean manufacturing and comprised three sections (Hines et al., 2004; Womack, 2006). The first included general information about the stage of the supply chain, the process name, and the constituent step. The second sought information on the cycle time (i.e., time a process takes from start to finish), waiting, or nonvalue adding time and the number of operators managing a process. The third section was used to detail losses and wastes observed along the different stages of the supply chain and included types of loss/waste based on the seven lean wastes (i.e., defects, waiting, transport, overprocessing, motion, overproduction, and unnecessary inventory). This information facilitated the creation of a "current state map" depicting the present situation along the dairy supply chain with an emphasis on steps, processes, and occurrence of FLW. Microsoft Visio 2016 was used to design the current state map. Lastly, lead time was also calculated using the cycle time and the waiting/nonvalue adding time observed by following operations along the supply chain. The quantity of FLW was calculated following the load tracking method developed by the Food and Agriculture Organisation (FAO, 2016). Thereby, the quantity of milk or its products was recorded before and after an activity, from which the difference constituted the quantity of FLW.

9.3 Results

9.3.1 *Characteristics of the dairy supply chain examined in the case study*

Table 9.2 shows findings from observations made along the value chain and the interviews conducted with key personnel operating within the company. Observations made at three levels of the value chain (farm, processor, and distribution) indicate that production follows specified steps, each made up of at least two operations. During data collection, the focal farm had 51 lactating cows that were milked twice a day. This was also reported as the average number in the previous 6 months. The farm is run by a farm manager and an accountant who are employed on a long-term basis, in addition to over 15 personnel on short term employment basis (mostly milk men and other casual laborers). The farm on average produces 200 L of milk a day but also acts as a collection center for farmers in the neighborhood. Therefore, farm records indicated that approximately 1400 L of milk were normally collected every 3 days for delivery to the processing plant during dry seasons. However, the quantity of milk collected was reported to be higher in the wet seasons of the year. The processor mainly operates on orders made from customers (i.e., wholesalers and retailers) and so generally uses a pull system to produce milk products.

During fieldwork for this case study, it was observed that the processor was supplied with 20,000 L of raw milk, based on a past order from farms in the region. Although the processing plant was directly linked to the focal farm and its partner

Table 9.2 Characteristics of the dairy supply chain

Supply chain stage	Steps	Operation	Capacity/units handled/processed	Operators
Farmer	Milking	• Preparation • Hand milking • Measurement • Pouring milk into cans	51 cows	5
	Collection and storage	• Transfer cans to cooling center • Delivery of milk from other farms	2000 L	2
	Distribution	• Milk quality testing • Transfer milk from cooling tanks to trucks • Delivery to the processing plant	1400 L/3 days	4
Processor	Milk reception	• Milk quality testing • CIP of inlet, pasteurizer, and tanks • Milk inlet • Pasteurization • Cooling • CIP of inlet and pasteurizer	50,000 L	2
	Yogurt			
	Mixing	• CIP of mixture • Milk transfer to mixture • Add milk powder and sugar	3000 L	3
	Pasteurization + homogenization	• CIP of tube pasteurizer and homogenizer • Pasteurization and homogenization	3000 L	3
	Fermentation	• CIP of fermentation tank • Milk inlet • Add culture, flavor, and color • Start fermentation • Test pH	3000 L	3

(*Continued*)

Table 9.2 (Continued)

Supply chain stage	Steps	Operation	Capacity/units handled/processed	Operators
	Cooling	• CIP of cooling tank • Milk inlet • Start cooling	3000 L	3
	Packaging	• Prepare packaging machine • Prepare packaging material (cups + seals) • Calibrate machine with real product • Channel yogurt to machine • Pack and seal • Print manufacture and expiry dates • Arrange sealed cups in boxes	72 cups/min	15
	Storage	• Place boxes on pellets • Transfer pellets to store	25 boxes/pellet >200 m^2	3
	UHT			
	Homogenization + sterilization	• CIP of sterilizer • Pasteurization • Transfer to deaerator • Homogenization • Milk inlet into sterilizer • Sterilization • CIP of sterilizer	6000 /h	2
	Aseptic tank holding	• CIP of aseptic tank • Milk inlet from sterilizer • CIP of aseptic tank	6000 L	2

	Packaging	• Prepare tetra packaging machine + CIP • Prepare packaging material (tetra pack + caps) • Calibrate machine • Channel milk from aseptic tank to tetra packing • Print manufacture and expiry dates • Apply top caps • Arrange sealed tetra packs in boxes • CIP of tetra packing machine	6000 L/h	15
	Storage	• Place boxes on pellets • Transfer pellets to store	15 boxes/pellet >200 m^2	3
Distributor	Loading and transportation	• Transfer stock from storage to truck	Depends on order	4
		• Truck journey to Kampala		2
	Unloading & storage	• Transfer stock from truck to store		4
		• Distribution to customers		4

farmers, it was also supplied by other farmers in the region to reach its storage capacity of 50,000 L of milk. This made it possible to receive 20,000 L of milk or more whenever there was a need. The processor currently makes yogurt and ultra-high temperature (UHT) milk, and mainly distributes these products for sale to wholesale and retail outlets in Kampala and neighboring towns. The plant has a capacity to process 3000 L of pasteurized milk into yogurt while 6000 L of pasteurized milk can be processed into UHT milk at a time. Line production is used and there are two separate lines for yogurt and UHT processing. A batch system is utilized and all pasteurized milk contained in storage tanks is normally processed so that the next delivery of milk is not mixed up with old stock of milk that would still be kept in the tanks. There is a milk laboratory, stationed between the two production lines, where all quality tests are carried out to ensure that recommended standards are met. In addition, the plant is equipped with two separate types of packaging machinery for yogurt and UHT. Packing material is supplied from Nairobi, Kenya on a monthly basis. The plant has two storage facilities located adjacent to the packaging areas of both lines, with each connected to a loading area. There are close to 30 personnel working at the processing plant including the Chief Executive Officer, process manager, marketing manager, technicians, laboratory analysts, food technologist, and other staff responsible for packing and storage of finished products. Yogurt and UHT milk are periodically transported to an additional and separate storage facility located elsewhere in Kampala to replenish old stock before final distribution to wholesalers and retailers, or for sale to end consumers.

9.3.2 Current state map for production of yogurt and ultra-high temperature milk

Fig. 9.1 outlines the dairy company's production processes for yogurt and UHT milk. Below, the findings are described for each stage of the supply chain.

9.3.2.1 Farmer level

Focal farm and partner farms: The process of production starts at this level with milking of cows. At the focal farm, this takes place in a milking parlor, which accommodates around 10 cows at a time, while the rest are held in a nearby paddock awaiting their turn. Each cow is restrained before being hand milked with buckets by one of four men, each milking one cow at a time. Once the cow's udders are emptied, the milk is measured and then poured into a 50-L milk can. It was noted that this process of milking each of the 51 cows took approximately 3 hours to be completed.

Collection center: Following milking, the cans are transferred to the collection center for cold storage. Other farmers also deliver their milk to this center. At the storage center, there are two employees that receive milk from farmers and store it in a 2000-L tank; this process takes on average 2 hours. The process of transferring the milk is manual and so delivery to the center is done either with the assistance of

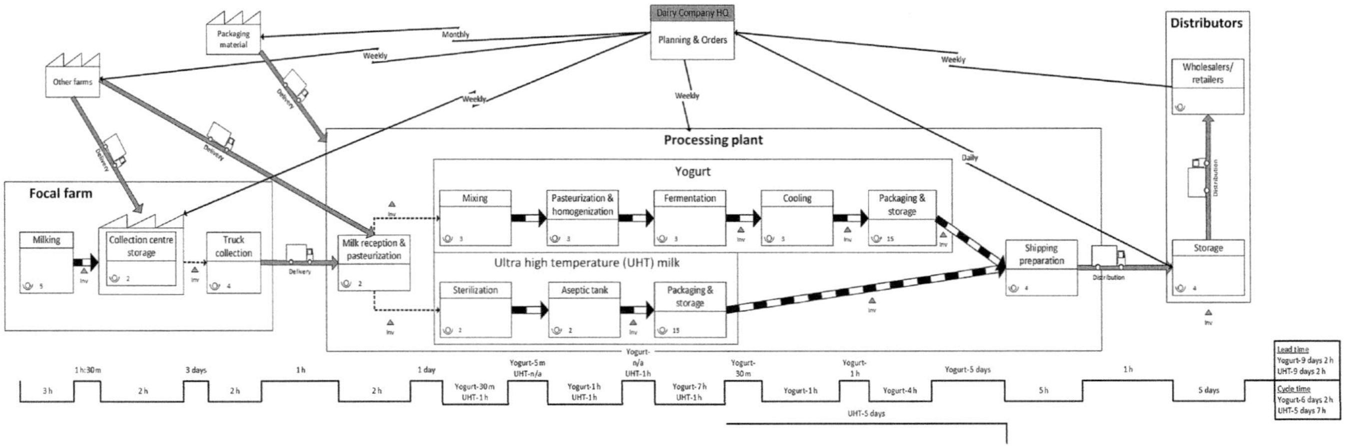

Figure 9.1 Current state map of the dairy supply chain of yogurt and UHT milk.

a wheelbarrow or with a bicycle or motorcycle. At the center, the milk is decanted into small 10 L buckets, allowing the 50-L can to be easily lifted and emptied into the storage tank. The cooling center also uses a generator as a source of power for cooling and this is run for 30 minutes in every 24 hours.

Distribution: Every third day, a truck collects milk from the center for delivery to the processing plant. Collectors first have to test the quality of the milk before it can be loaded into the truck. The process of transferring milk into the truck is manually done by four persons and normally takes 2 hours to complete. A pipe is connected from the cooling tank to the truck. Milk is first measured using a 50-L can, and so 50 L of milk are poured into the truck following each measurement. This process continues until the truck is full or milk in the cooling tank is finished, the former situation occurring more often.

9.3.2.2 Processor level

Orders for milk delivery are placed weekly with these orders initiating milk processing activities at the plant.

Milk reception: This is the first activity conducted at the plant. On the day that milk was delivered, it was observed that a sample of milk was first tested to determine its quality and assess if it would be suitable for processing as yogurt or UHT milk. Thereafter, CIP of the inlet system was conducted, followed by the actual input of milk into the plant. As milk is pumped into the system, pasteurization immediately starts before milk is channeled to the cooling storage tanks. At the start of the milk inlet, there is a milk-milk push through the system but at the end, water is used to push pasteurized milk into storage tanks. When all the milk is received and stored, CIP of the inlet system and the pasteurization tubes is conducted. The whole process of milk reception was done by two personnel and took 2 hours to receive 20,000 L of milk that were delivered by two trucks.

9.3.2.2.1 Yogurt

Mixing: The actual processing of pasteurized milk into yogurt starts when a mixture is made with sugar and milk powder. On the day this process was observed, two batches of yogurt were produced (i.e., plain and mango flavored yogurt). Plain yogurt was produced first, with 2800 L of pasteurized milk being channeled into the mixer from one of the storage tanks. Then 160 kg of skimmed milk powder and 128 kg of sugar were poured into the mixer. This was performed by three workers and the mixer ran for exactly 30 minutes, before the product was channeled to the pasteurizer and homogenizer. The same process was followed for the next batch of mango flavored yogurt, which only started when the first batch was already at the next step of processing.

Pasteurization + homogenization: The product from the mixer is pasteurized again before it is channeled to the next step. The pasteurizer also acts as a temporary storage element and this is facilitated by its structure (i.e., a series of holding tubes). Pasteurization takes place first and homogenization immediately commences but some milk remains in the tubes. Milk sent to the homogenizer pushes out water that would have remained during CIP conducted earlier, into the drainage system. Since the process here is continuous, drainage of water is closely observed, with the outlet valve being manually closed after the output is presumed to be milk and not water. In this case, milk is used to push out water but the opposite occurs at the end when water is instead used to push out the remaining milk that could not be sent to the next step, to the drainage system. Both processes take around 1 hour and are managed by three persons. This process was also repeated for the second batch of mango flavored yogurt.

Fermentation: Once pasteurization and homogenization are complete, milk is then sent to the fermentation tank. There are two tanks, each with a capacity of 3000 L, which makes it possible to handle two batches almost concurrently. At this step, it was however difficult to determine how much of the product was sent to the tank. This could be determined later during packaging. It was also at this step that only culture (thermophilic bacteria) was added in case of plain yogurt. For mango yogurt, flavor and color were also added. There is also a heat treatment that is applied that facilitates the fermentation process. Fermentation took 7 hours to complete and it was monitored to maintain the pH at 4.2–4.5, a lower pH being detrimental to the expected quality of the product. However, it was reported that the duration may be longer than 7 hours if the desired acidity is not yet reached. Two personnel were responsible for this process.

Cooling: At the end of the fermentation process, yogurt is sent to one of two cooling tanks. A valve is opened, and yogurt instantaneously moves to the cooling tank. The main purpose is to inactivate thermophilic bacteria so that fermentation stops. As earlier noted, the start of this process was delayed by 30 minutes for both batches, yet the preceding process had completed. Cooling takes around 1 hour and is managed by two workers. They also had to observe a yellowish-orange change in color of yogurt in the pipes because the mango flavored batch was later channeled to the other cooling tank.

Packaging: Before this commences, the packaging machine has to be prepared with all of the necessary packing material (i.e., cups and seals) and a date printer. Additionally, at least 15 people have to be positioned along the packing conveyer belt to arrange finished products in boxes, ready for storage. Therefore, cooled plain yogurt was channeled directly to the packaging machine and it was packed in 450 g cups, which was later, followed by mango flavored yogurt. In the end, there were 5659 cups with plain yogurt and 6055 cups with mango flavored yogurt that were appropriately packaged, with the whole process lasting 4 hours.

Storage: This is done concurrently with packaging. Boxes each with 12 cups are arranged on a pallet and then a plastic wrap is applied around it. Each pallet could accommodate 24 boxes, which were subsequently transferred to the storage area using a hand pallet jack. Products were arranged according to the date of production; hence new stock was not mixed with old stock.

9.3.2.2.2 Ultra-high temperature milk

Sterilization + homogenization: Before this process, 9900 L of milk in storage tanks were first repasteurized using the pasteurizer of the yogurt line. The double pasteurized milk was then directly sent to the UHT production line, pushing out water to the drainage area in the process. The temperature of milk was then raised and maintained between 70°C and 75°C and then it was channeled to the deaerator, also kept at the same temperature. Before sterilization at 132°C–140°C, milk was first homogenized so that fats could be broken down. Milk was held at sterilization temperature for 3–5 seconds and it took around 90 minutes to process the 9900 L of milk.

Aseptic tank holding: Prior to sterilization, the aseptic tank was first prepared to receive milk in a condition that significantly reduces the risk of microbial growth. This was done using steam at a temperature of 147°C and cooled down using sterile air. Milk that is sterilized was then sent to the aseptic tank for temporary storage before it was packed. This process lasted for 1 hour.

Packaging: Preparation of the tetra brik aseptic packing machine was done at the same time sterilization was initiated. This involved CIP and placing the packing materials into the machine. As already noted, milk in the aseptic tank was not immediately packed. This was because the packing machine was still being prepped even though this had been started earlier. Once ready, milk was then sent from the aseptic tank for packing. The first tetra packs were used to calibrate the machine so as to reduce errors on packages. Good packs were labeled with dates and as they moved on the conveyer belt, top covers were applied using a precise cap applicator. UHT milk was then packed in boxes, each containing 10 one-liter tetra packs. There were 15 personnel who were engaged in the whole process of packaging, and this lasted for 1 hour.

Storage: UHT milk in 1-L packs were arranged in a box with a capacity of 12 packs. Sealed boxes were then placed on pallets, wrapped with a plastic, and transferred to storage with a hand pallet jack. Care was taken that newly produced UHT milk was not mixed with old stock, hence pallets were arranged according to the date of production.

9.3.2.3 Distribution

Finished products (yogurt and UHT milk) are periodically transported to the storage facility in Kampala. Products on pallets were loaded into a truck using a hand pallet jack. The truck travels a distance of 300 km to deliver products to the storage facility in Kampala. Once there, products are offloaded and stored according to the dates of arrival. Thereafter, the same process of loading, transportation, and off-loading is followed when products are distributed to customers.

9.3.3 Identification of food loss and waste and their destinations along the dairy value chain, with a link to lean manufacturing

Findings in Table 9.3 illustrate losses and wastes identified and are also explained later according to the stage of the supply chain.

Table 9.3 Food losses, lean waste linkage, and their destinations along the dairy value chain

Supply chain stage	Steps	Food losses	Lean waste linkage	Destination
Farmer	Milking	Spillage of milk	Defect	Discard
		Milk kept in open cans for long periods	Inventory Defect	Discard
	Collection and storage	Spillage of milk	Defect	Discard
		Milk in cooling tank before distribution	Inventory	
	Distribution	Spillage of milk	Defect	Discard, given to employees, ghee and milk powder production
		Poor quality milk rejected		
		Uncollected milk in tank	Overproduction Inventory	
Processor	Milk reception	Spillages of milk	Defect	Discard, ghee and milk powder production
		Unpasteurized milk sent to the drain		
		Milk in trucks not pumped		
		Poor quality milk rejected		
	Yogurt			
	Mixing (2 × 2800 L)	Spillage of milk powder and sugar	Defect	Discard
	Pasteurization + homogenization	Milk mixed with water	Defect	Discard
		Heat labile micronutrient degradation	Overprocessing	Packaged
	Fermentation(Plain and mango)	Yogurt with very low pH (sour taste) rejected	Defect Overprocessing	Discard
		Yogurt unnecessarily kept longer in tank	Inventory	
	Cooling(Plain and mango)	Yogurt drained out during batch change-over	Defect	Discard
		Yogurt kept unnecessarily longer in tank	Inventory	Packaged

(*Continued*)

Table 9.3 (Continued)

Supply chain stage	Steps	Food losses	Lean waste linkage	Destination
	PackagingPlain: 5659 cupsMango: 6055 cups	Yogurt with incorrect weight rejected	Defect	Discard, given to employees
		Yogurt in damaged cups rejected		
		Yogurt with seal leakage rejected		
		Yogurt with error/unclear dates rejected		
		Yogurt with mixed flavor rejected		
	Storage	Old stock used as buffer	Inventory Overproduction	Distributed, discard
	UHT			
	Sterilization(with pasteurization + homogenization)9900 L	Milk drained out during removal of water	Defect	Discard
		Heat labile micronutrient degradation	Overprocessing	Packaged
	Aseptic tank holding	Sterilized milk awaiting packaging	Inventory	Packaged
	Packaging8532 tetra packs	Tetra pack with weak seal rejected	Defect	
		Tetra pack with design error rejected		
		Tetra pack with pin hole rejected		Discard
		Tetra pack with no applied cap rejected		
		Tetra pack with wrong/unclear dates rejected		
	Storage	Old stock used as buffer	Inventory Overproduction	Distributed, discard
Distributor	Loading, transportation, unloading, and storage	Damage on packaging	Defect	Discard
		Delivered products not distributed immediately	Inventory	Distributed

9.3.3.1 Farmer level

Focal farm and partner farms: During milking, it was observed that a portion of the milk was normally spilled on the floor. This takes place during hand milking and when milk is poured from a bucket into cans. The main causes of spillages that were identified are inattentiveness of milk men when performing a task and also the restlessness of the cows. For the latter, there was also an increased risk that the cow would kick the bucket, causing a bigger loss of milk. Spilled milk becomes a product defect that cannot be recovered and hence can be categorized as discarded milk. There is also a practice of keeping milk in open cans located in the milking parlor for prolonged periods of time. Flies were observed hovering over the cans and coming into contact with the milk. This exposed the milk to microbial contamination and hence increased the likelihood of deterioration. The loss attributed to this can occur in subsequent stages of processing when milk goes bad due to poor handling practices at a preceding task, hence being rejected and or discarded. As far as lean manufacturing is concerned, this practice constitutes a defect that additionally results in an accumulation of inventory, limiting the start of the next step in the production system.

Collection center: The system of transportation used exposes milk to spillages if cans are not properly covered and its occurrence is exacerbated by bumpy roads en route to the center. As milk was poured into the cooling tank, it was also easily spilled on the floor and on top of the tank. Spilled milk is considered a defect since it cannot be used anymore. During storage, it is also presumed that the tank is capable of maintaining cold temperatures, built up during the first 30 minutes of cooling. This is problematic since there was no control that was observed on the tank to monitor changes in temperature. Hence, there is also a high chance of milk deterioration due to microbial growth especially if there was some form of microbial contamination at an earlier stage. Moreover, milk that is stored in the cooling tank for days without being distributed results in an accumulation of inventory.

Distribution: If milk in the cooling tank is of low quality, it may be rejected by distribution trucks. Sometimes the collection center receives a lower price for low quality milk. Alternatively, such milk ends up with processors who produce dairy products that do not depend a lot on the quality of the milk. An example that was reported was the production of ghee and milk powder from such milk, where it could be used as a raw material. Another possible destination reported was that rejected milk is sometimes given to farm employees or thrown away since it cannot further be used for any purpose. While milk is loaded into the truck using 50-L cans, a lot of spillages normally occur. This results in a considerable loss of product, for example, on the day 1400 L of milk were collected, it was observed that 28 cans were loaded into the truck. Each can spilled around 100 mL of milk, which was approximately 3 L of milk lost at that stage. Further, the truck was unable to load all of the milk contained in the cooling tank. This was the case because it had already made rounds from other centers arriving at the focal collection center last. This was reported to be the usual routine followed by the truck. Therefore, almost 500 L of milk remained in the tank, and this balance can be considered an overproduction waste. This also leads to a situation where the remaining milk is easily mixed with fresh milk that is received from farmers on subsequent days. Hence, increasing the chances of cross-contamination and possible rejection of milk during the next truck pick-up.

9.3.3.2 *Processor level*

Milk reception: During milk inlet, spillages were observed around the truck. However, the connection to the plant inlet valve was tight enough and no spillages were observed. It was reported that milk of poor quality was always rejected and was not used at the plant. Such milk was distributed to other processors for the production of ghee and milk powder. An observation was made that a proportion of unpasteurized milk remained in the whole system after pasteurization and storage. This milk was pushed out of the reception unit into the drainage system using a force provided by water that is automatically pumped into the system once pasteurized milk is stored. There was also milk that remaining in the trucks that could not be pumped into the processing plant. This milk is disposed of while trucks are cleaned for the next delivery.

9.3.3.2.1 Yogurt

Mixing: Because the whole mixture is sent to the next step, losses at mixing were minimal. It was only the ingredients added (i.e., skimmed powder milk and sugar) that were spilled on the working surfaces and floor. Spilled ingredients could not be reused and were discarded into drainage as scrap.

Pasteurization + homogenization: During pasteurization and homogenization, milk is lost twice into the drainage. First is when incoming milk is used to push out water from the system. The outlet valve is only closed once the personnel think that it is only milk in the system. This is done manually; it is very hard to tell and a subjective decision is always made. Therefore, a certain quantity of milk is allowed to drain out together with water. The second time is when a new batch has to be processed and the system has to be cleared of any milk. All the remaining milk is pushed out by water into drainage. It was estimated that between 20 and 50 L of milk are lost at this step. It should be noted that losses at this level continue to occur even with the next batch because the same principle applies. Milk is also pasteurized the second time since it was delivered to the plant. This increases the likelihood that thermal labile micronutrients are affected in terms of quality.

Fermentation: During the fermentation phase, the main threat as far as losses are concerned is increased acidity of yogurt (i.e., pH below 4.2). Once this occurs, yogurt develops a sour taste that is irreversible, and the product is discarded; hence the whole batch is lost. Another possible loss was with the ingredients added that were seen spilled on top of the tanks. When fermentation was complete, there was an observed 30-minute lag before yogurt was channeled to the next process, hence product held at this step became inventory. This also provided an opportunity for thermophilic bacteria to continue the breakdown of yogurt, which could further lower the pH.

Cooling: By the time the first batch was sent through pipes to one of the cooling tanks, another batch was almost completing its fermentation. At this point, no CIP was performed, the cooling tank was sealed, and it was noted that plain yogurt that remained in the pipes was pushed out into drainage by incoming mango flavored yogurt. The operator reported that approximately 12 L of yogurt are drained out of the system per minute. As such, it took around 5 minutes for almost 60 L of plain yogurt to drain out of the system. As earlier mentioned, a yellowish-orange color change in the pipes was the signal operators used to initiate closure of the drainage valve. It may also be suggested that inventory accumulates at this stage, especially when the next process was not prepared in time.

Packaging: During packaging, there were 68 plain, 17 mango, and 132 mixed flavored yogurt cups that had defects. These defects included incorrect weight, damaged cups, seal leakages, errors in dates printed, and unclear dates. Many of the defects occurred at the very start of packing when the machine was being calibrated. Such products were separated and not packed for distribution to customers. Additionally, before mango flavored yogurt was packed, the product that first came out of the system was not purely mango (i.e., mixed flavor). It was clearly observed that the first product had a very light yellowish color, which indicated a mix with plain yogurt. The operator in charge also highlighted that it is even worse if another flavor such as pineapple is also produced. Therefore, cups with this mix were also separated from those with a consistent yellowish-orange color typical of mango flavored yogurt. It was also observed that the surfaces of working tables were slippery and without end-stops, in that three sealed cups were knocked over by workers who were arranging them in boxes. Most products damaged during packing were thrown away and a few were given to employees. When the number of cups packed are converted into liters, 2472.4 and 2645.4 L of plain and mango flavored yogurt, respectively, were eventually packed and suitable for distribution to customers. Compared with 2800 L of pasteurized milk that were used as raw material for each batch, overall, there was a 327.8 L (11.7%) and 154.6 L (5.5%) loss of marketable milk product from mixing to packaging stage for plain and mango flavored yogurt, respectively.

Storage: There was no sign of packed yogurt loss observed during storage. However, old stock was observed in storage that was reported to act as buffer whenever an urgent order was made at times when production was not planned. This constitutes both accumulating inventory and overproduction and issues may arise if the old stock is not distributed in good time before specified expiry dates.

9.3.3.2.2 Ultra-high temperature

Sterilization + homogenization: At this step, the process of pushing out water from the system using incoming milk was the source of loss. The operator had to wait and ensure that all water had been drained. This required that some amount of milk be concurrently disposed of in the process. It was observed that almost 400 L of milk were lost to drainage at this point. In addition, exposing milk to a second pasteurization process increases the likelihood that heat labile micronutrients are compromised, hence affecting the nutritional value of the final product.

Aseptic tank holding: There was no sign of physical loss observed at this step. However, since the next process did not start promptly, there was an accumulation of inventory.

Packaging: Losses of milk were immediately observed when the tetra brik machine was being calibrated. The first packs that came out had a lot of errors and it took many attempts to come up with an acceptably packaged product. Observed errors included weak package seals, design errors, pin holes, wrong application of the cap, and wrong/unclear dates. It was both reported and observed that such milk would not be reused and all of it was discarded. Midway through packing, the same errors happened and still more milk was discarded. There were 8532 tetra packs that were appropriately packed. This was equivalent to 8532 L of milk since each pack contained 1 L. When compared with 9900 L channeled from the storage tanks, a loss of 1368 L of milk or 13.8% was identified.

Storage: No loss was observed during storage. But the delay to distribute finished products was associated with an accumulation of inventory. It was also highlighted that the old stock present was used as buffer in case an urgent order is made at times of no production. This gave an indication that although the processing plant mainly operates on orders, it also produces more products than ordered. This in a way may be rational, but the plant also runs a risk of loss if such a buffer is not distributed on time before its expiry date.

9.3.3.3 Distribution

No losses were observed at the time of data collection. However, workers reported having experienced losses during loading, off-loading, and transportation. This gave an indication of additional hotspots where losses, in terms of physical damage to packages, could occur if enough care is not taken. There is also accumulation of inventory at the second storage facility since distribution to customers is normally not done immediately.

9.4 Discussion

This case study applied VSM, following guidelines of the FLW protocol to map hotspots for food related losses and wastes along three stages of a dairy value chain in Uganda, thereby confirming that VSM is applicable in an agrifood context to make a comprehensive overview of the nature of losses and wastes that occur along the supply chain (De Steur et al., 2016), while following specific lean manufacturing practices adapted to the dairy sector (Malmbrandt and Åhlström, 2013). As a foundation for value chain analysis, the current state map of the dairy value chain indicates that the production of milk products constitutes a series of dependent steps and operations that are potential hotspots for losses and wastes. Although the majority of losses and wastes were noted to occur at the processing stage, unsatisfactory handling practices at the farmer level increased the likelihood of milk rejection and subsequent disposal upstream. The issue of losses and wastes instigated at earlier stages of the chain has also been reported in a study about food losses and reduction strategies in Switzerland (Beretta et al., 2013). Unfortunately, awareness among actors of what happens down or upstream is limited and is hardly observed because of barriers that hinder integration along the supply chain (Taylor and Fearne, 2009), thereby reinforcing the need for targeted awareness creation to promote implementation of collective strategies at all chain levels with input from various actors (Göbel et al., 2015; Halloran et al., 2014).

With respect to loss or waste types, results indicated that milk products were oftentimes discarded, while some supply chain operations were linked to nutrient losses. Product defects were by far the main reason for discarding milk products, a finding that supports previous literature (Halloran et al., 2014; Muriana, 2017). Selectively discarding products that fail to match quality standards expected by consumers is common practice among producers as a way of increasing or sustaining market share of their products (Willersinn et al., 2015). There were also instances of accumulated inventory along the chain, and subsequent poor handling could in a way render milk unacceptable for further processing, hence being discarded. The same was true for overproduction of milk products that were not transferred upstream at the same rate as they were produced. Although production of food is increasingly affected by uncertain demand forecasts, producers continue to use push strategies that result in either accumulation of inventory or stock (Buzby and Hyman, 2012; Silvennoinen et al., 2015). Perishability of dairy products such as UHT milk and yogurt hence underlines the need to adopt lean production based pull strategies such as just-in-time production (Lyonnet and Toscano, 2014; Mackelprang and Nair, 2010). This has the potential to reduce losses due to unnecessary inventory and overproduction. Overprocessing was also identified as a factor affecting the integrity of milk products as far as nutrient quality of final products is concerned. Although the practice of double pasteurization at high temperatures has merits linked to the microbial safety assurance of food products, it potentially results in nutritional losses (Qi et al., 2015; Shewfelt, 2017). This also applies to other nonheat operations such as washing and other physical treatments applied to food, with vitamins being most susceptible (Atungulu and Pan, 2014; Francis et al., 2012), thereby confirming that the processing stage is an important hot spot for nutrient losses.

Limited standardization of operations especially at the farmer level could have been the underlying casual factor for the losses that were observed at this stage (Papargyropoulou et al., 2014; Parfitt et al., 2010). At the processing plant, there were some established process controls, but these were insufficient to prevent almost 14% or 5%–12% of observed losses along the UHT and yogurt production lines, respectively. Nonetheless, it remains important to establish and continuously improve controls, traversing the entire value chain, as a way of promoting collaborative efforts against losses and wastes (Mena et al., 2014). This practice could facilitate continuous improvement, a principle in lean manufacturing that promotes efficiency and lowers production costs (Rivera and Chen, 2007).

Lean metrics such as lead time play a key role to justify processes that need improvement. Our results indicate that the production process of a given batch of milk products take approximately 10 days from the farmer level to the point of distribution. Given the perishability of most milk products (Kaipia et al., 2013), improvements in production efficiency are needed so that consumption is not limited by the shortened shelf life of edible products (De Treville et al., 2004). Future research should therefore consider investigating and confirming the causal association between process standardization and control with the occurrence of FLW at different stages of the food chain. In addition, food producers should strive to improve production efficiency to lower the time it takes to have a finished product ready for consumer use.

Although discarding of unmarketable milk products was a popular destination, there are times when such products are given to employees. This supports findings from a study on the adoption of lean manufacturing to reduce FLW among dairy chain actors (Wesana et al., 2018). While this is a good practice, it can only be implemented to a limited extent because not all rejected products can be absorbed by available employees. In developing country contexts like Uganda, with a considerable part of the population facing hunger especially due to compromised economic access to milk or other nutrient-rich food products, there is a need to develop effective mechanisms by which unmarketable but edible products can be effectively redistributed beyond employees to the needy. This can be in the form of organized charity distributions, like those that have been implemented in other countries (Richter and Bokelmann, 2016; Schneider, 2013). Governments can take initial steps to foster an enabling environment for actors in the food industry and charity organizations to interact and promote effective collaboration as far as FLW is concerned (Garrone et al., 2016). This also could provide a suitable platform to create critical awareness and promote collective problem diagnosis to design alternative uses and destinations of products that would otherwise be discarded from the supply chain.

Even though identification of FLW hot spots along the three supply chain levels was possible while following the principles of VSM and the FLW protocol, quantifying the magnitude was not straightforward, as also reported in previous studies (Affognon et al., 2015; Chaboud and Daviron, 2017; Elimelech et al., 2018). There were observable efficiency differences in operations and equipment used at different stages of the supply chain, a limitation also identified by Corrado et al. (2017). Findings from the case study point to the absence of automation at the farmer and distribution/storage levels relative to the processor level. The organization of operations during processing of yogurt and UHT milk to a given extent facilitated FLW quantification. By comparing the amount of raw material used at the start of processing with the final product at the end, the magnitude of loss during the processing level was determined. However, there are some process components (i.e., drainage outlets) that ideally would require future investment in innovative technologies with quantification capabilities. This would be complementary to advocated improvement of production efficiency as a way of mitigating FLW (Parfitt et al., 2010; Shafiee-Jood and Cai, 2016).

9.5 Conclusion

This study has implications for the agrifood industry with regard to the systematic identification of hotspots along the value chain where FLW occurs. Applying VSM could help value chain actors to holistically establish the magnitude of FLW, by comparing the amount of material used at the start of a production process to the final quantity of product emerging at the end of the processes. Wherever possible, this should be done for every activity along the value chain. Efforts to minimize FLW should emphasize adoption of this practice more at the processor level of the chain, as this has shown to be a key stage where most losses occur, while also promoting actor integration and collaboration along the supply chain. Given the complexity of food production systems, establishing suitable controls to monitor FLW may be hindered by associated costs if new equipment needs to be installed, especially in resource constrained country settings. However, recent evidence shows that actors in the dairy value chain are more likely to adopt lean manufacturing strategies to reduce food losses if they are aware of associated benefits and are able to collaborate with other actors for a common purpose. Therefore, food producers should continuously be engaged and informed about the potential of lowering production costs following the adoption of lean waste reduction strategies along supply chains. As a consequence, the availability of nutrient-rich foods like dairy products is enhanced in a sustainable way without necessarily investing more in increased food production that has proven to be a costly venture. Future studies should extend this work and apply VSM to other agrifood value chains and further justify the potential of lean manufacturing strategies, integrated with other established accounting and reporting guidelines or approaches for FLW assessments and subsequent minimization.

References

Affognon, H., Mutungi, C., Sanginga, P., Borgemeister, C., 2015. Unpacking postharvest losses in sub-Saharan Africa: a meta-analysis. World Develop. 66, 49–68.

Ahmed, M., Jones, E., Redmond, E., Hewedi, M., Wingert, A., Gad El Rab, M., 2015. Food production and service in UK hospitals. Int. J. Health Care Qual. Assur. 28, 40–54.

Allen, S., de Brauw, A., 2017. Nutrition sensitive value chains: theory, progress, and open questions. Glob. Food Secur. 16, 22–28.

Aschemann-Witzel, J., De Hooge, I.E., Rohm, H., Normann, A., Bossle, M.B., Grønhøj, A., et al., 2017. Key characteristics and success factors of supply chain initiatives tackling consumer-related food waste—a multiple case study. J. Clean. Prod. 155, 33–45.

Atungulu, G.G., Pan, Z., 2014. Rice industrial processing worldwide and impact on macro- and micronutrient content, stability, and retention. Ann. N.Y. Acad. Sci. 1324, 15–28.

Barrett, D.M., Beaulieu, J.C., Shewfelt, R., 2010. Color, flavor, texture, and nutritional quality of fresh-cut fruits and vegetables: desirable levels, instrumental and sensory measurement, and the effects of processing. Crit. Rev. Food Sci. Nutr. 50, 369–389.

Beretta, C., Stoessel, F., Baier, U., Hellweg, S., 2013. Quantifying food losses and the potential for reduction in Switzerland. Waste Manage. 33, 764–773.

Buzby, J.C., Hyman, J., 2012. Total and per capita value of food loss in the United States. Food Policy 37, 561–570.

Chaboud, G., Daviron, B., 2017. Food losses and waste: navigating the inconsistencies. Glob. Food Secur. 12, 1–7.

Cooper, K.A., Quested, T.E., Lanctuit, H., Zimmermann, D., Espinoza-Orias, N., Roulin, A., 2018. Nutrition in the bin: a nutritional and environmental assessment of food wasted in the UK. Front. Nutr. 5, 19.

Corrado, S., Ardente, F., Sala, S., Saouter, E., 2017. Modelling of food loss within life cycle assessment: from current practice towards a systematisation. J. Clean. Prod. 140, 847–859.

Dal Forno, A.J., Pereira, F.A., Forcellini, F.A., Kipper, L.M., 2014. Value Stream Mapping: a study about the problems and challenges found in the literature from the past 15 years about application of Lean tools. Int. J. Adv. Manuf. Technol. 72, 779–790.

Darlington, R., Rahimifard, S., 2006. A responsive demand management framework for the minimization of waste in convenience food manufacture. Int. J. Comput. Integr. Manufact. 19, 751–761.

De Steur, H., Wesana, J., Dora, M.K., Pearce, D., Gellynck, X., 2016. Applying value stream mapping to reduce food losses and wastes in supply chains: a systematic review. Waste Manage. 58, 359–368.

De Treville, S., Shapiro, R.D., Hameri, A.-P., 2004. From supply chain to demand chain: the role of lead time reduction in improving demand chain performance. J. Operat. Manage. 21, 613–627.

Dora, M., Van Goubergen, D., Kumar, M., Molnar, A., Gellynck, X., 2014. Application of lean practices in small and medium-sized food enterprises. Br. Food J. 116, 125–141.

Elimelech, E., Ayalon, O., Ert, E., 2018. What gets measured gets managed: a new method of measuring household food waste. Waste Manage. 76, 68–81.

Engelund, E.H., Breum, G., Friis, A., 2009. Optimisation of large-scale food production using Lean Manufacturing principles. J. Foodserv. 20, 4–14.

FAO, 2016. Food Loss Analysis: Causes and Solutions, Case Studies in the Small-Scale Agriculture and Fisheries Subsectors. FAO, Rome.

FAO, IFAD, UNICEF, WFP, WHO, 2018. The State of Food Security and Nutrition in the World 2018: Building Climate Resilience for Food Security and Nutrition. FAO, Rome.

Folinas, D.K., Aidonis, D., Karayannakidis, P., 2015. Greening the canned peach production. J. Agric. Inform. 6, 24–39.

Francis, G., Gallone, A., Nychas, G., Sofos, J., Colelli, G., Amodio, M., et al., 2012. Factors affecting quality and safety of fresh-cut produce. Crit. Rev. Food Sci. Nutr. 52, 595–610.

Francis, M., Simons, D., Bourlakis, M., 2008. Value chain analysis in the UK beef foodservice sector. Supply Chain Manage. Int. J. 13, 83–91.

Garrone, P., Melacini, M., Perego, A., 2014. Opening the black box of food waste reduction. Food Policy 46, 129–139.

Garrone, P., Melacini, M., Perego, A., Sert, S., 2016. Reducing food waste in food manufacturing companies. J. Cleaner Prod 137, 1076–1085.

Gil, M.I., Aguayo, E., Kader, A.A., 2006. Quality changes and nutrient retention in fresh-cut versus whole fruits during storage. J. Agric. Food Chem. 54, 4284–4296.

Gillespie, S., Haddad, L., Mannar, V., Menon, P., Nisbett, N., Maternal and Child Nutrition Group, 2013. The politics of reducing malnutrition: building commitment and accelerating progress. Lancet 382, 552–569.

Glover, W.J., Poopunsri, T., Hurley, R., 2014. Applying Lean to non-profit organizations: a food bank case study. In: IIE Annual Conference. Proceedings. Institute of Industrial Engineers-Publisher, p. 274.

Göbel, C., Langen, N., Blumenthal, A., Teitscheid, P., Ritter, G., 2015. Cutting food waste through cooperation along the food supply chain. Sustainability 7, 1429–1445.

Godfray, H.C.J., Beddington, J.R., Crute, I.R., Haddad, L., Lawrence, D., Muir, J.F., et al., 2010. Food security: the challenge of feeding 9 billion people. Science 327, 812–818.

Goriwondo, W.M., Marecha, A., Mhlanga, S., 2011. Use of the value stream mapping tool for waste reduction in manufacturing. Case Study Bread Manufact. Zimbabwe. Proceedings of (IEOM2011). 236–241.

Gustavsson, J., Cederberg, C., Sonesson, U., van Otterdijk, R., Meybeck, A., 2011. Global food losses and food waste: extent, causes and prevention. Report by the Swedish Institute for Food and Biotechnology for the Food and Agriculture Organization of the United Nations, Rome.

Halloran, A., Clement, J., Kornum, N., Bucatariu, C., Magid, J., 2014. Addressing food waste reduction in Denmark. Food Policy 49, 294–301.

Hanson, C., 2016. Food Loss and Waste Accounting and Reporting Standard. Food Loss and Waste Protocol. WRI, Washington, DC.

Hanson, C., 2017. Guidance on Interpreting Sustainable Development Goal Target 12.3. WRI, Washington, DC.

Hines, P., Holweg, M., Rich, N., 2004. Learning to evolve: a review of contemporary lean thinking. Int. J. Operat. Prod. Manage. 24, 994–1011.

Hodge, J., Herforth, A., Gillespie, S., Beyero, M., Wagah, M., Semakula, R., 2015. Is there an enabling environment for nutrition-sensitive agriculture in East Africa? Stakeholder perspectives from Ethiopia, Kenya, and Uganda. Food Nutr. Bullet. 36, 503–519.

Jaenicke, H., Virchow, D., 2013. Entry points into a nutrition-sensitive agriculture. Food Secur. 5, 679–692.

Jiménez, E., Tejeda, A., Pérez, M., Blanco, J., Martínez, E., 2012. Applicability of lean production with VSM to the Rioja wine sector. Int. J. Prod. Res. 50, 1890–1904.

Kaipia, R., Dukovska-Popovska, I., Loikkanen, L., 2013. Creating sustainable fresh food supply chains through waste reduction. Int. J. Phys. Distrib. Logist. Manage. 43, 262–276.

Keding, G.B., Schneider, K., Jordan, I., 2013. Production and processing of foods as core aspects of nutrition-sensitive agriculture and sustainable diets. Food Secur. 5, 825–846.

Kennedy, I., Plunkett, A., Haider, J., 2013. Implementation of lean principles in a food manufacturing company. Advances in Sustainable and Competitive Manufacturing Systems. Springer, pp. 1579–1590.

Lehtinen, U., Torkko, M., 2005. The Lean concept in the food industry: a case study of a contract manufacturer. J. Food Distrib. Res. 36, 57.

Love, D.C., Fry, J.P., Milli, M.C., Neff, R.A., 2015. Wasted seafood in the United States: quantifying loss from production to consumption and moving toward solutions. Global Environ. Change 35, 116–124.

Lyonnet, B., Toscano, R., 2014. Towards an adapted lean system—a push-pull manufacturing strategy. Prod. Plan. Control 25, 346–354.

Mackelprang, A.W., Nair, A., 2010. Relationship between just-in-time manufacturing practices and performance: a meta-analytic investigation. J. Operat. Manage. 28, 283–302.

Malmbrandt, M., Åhlström, P., 2013. An instrument for assessing lean service adoption. Int. J. Operat. Prod. Manage. 33, 1131–1165.

Melvin, A., Baglee, D., 2008. Value stream mapping: a dairy industry prospective. In: Engineering Management Conference, IEMC Europe, pp. 1–5.

Mena, C., Terry, L.A., Williams, A., Ellram, L., 2014. Causes of waste across multi-tier supply networks: cases in the UK food sector. Int. J. Prod. Econ. 152, 144–158.
Munesue, Y., Masui, T., Fushima, T., 2015. The effects of reducing food losses and food waste on global food insecurity, natural resources, and greenhouse gas emissions. Environ. Econ. Policy Stud. 17, 43–77.
Muriana, C., 2017. A focus on the state of the art of food waste/losses issue and suggestions for future researches. Waste Manage. 68, 557–570.
Noorwali, A., 2013. Apply lean and taguchi in different level of variability of food flow processing system. Proc. Eng. 63, 728–734.
Pangaribowo, E.H., Gerber, N., Torero, M., 2013. Food and Nutrition Security Indicators: A Review. Center for Development Research (ZEF), University of Bonn, Bonn.
Panwar, A., Nepal, B.P., Jain, R., Rathore, A.P.S., 2015. On the adoption of lean manufacturing principles in process industries. Prod. Plan. Control 26, 564–587.
Papargyropoulou, E., Lozano, R., Steinberger, J.K., Wright, N., bin Ujang, Z., 2014. The food waste hierarchy as a framework for the management of food surplus and food waste. J. Clean. Prod. 76, 106–115.
Parfitt, J., Barthel, M., Macnaughton, S., 2010. Food waste within food supply chains: quantification and potential for change to 2050. Philos. Trans. R. Soc. B Biol. Sci. 365, 3065–3081.
Parmar, A., Hensel, O., Sturm, B., 2017. Post-harvest handling practices and associated food losses and limitations in the sweet potato value chain of southern Ethiopia. NJAS-Wageningen J. Life Sci. 80, 65–74.
Phalan, B., Onial, M., Balmford, A., Green, R.E., 2011. Reconciling food production and biodiversity conservation: land sharing and land sparing compared. Science 333, 1289–1291.
Pingali, P., Sunder, N., 2017. Transitioning toward nutrition-sensitive food systems in developing countries. Annu. Rev. Resour. Econ. 9, 439–459.
Qi, P.X., Ren, D., Xiao, Y., Tomasula, P.M., 2015. Effect of homogenization and pasteurization on the structure and stability of whey protein in milk1. J. Dairy Sci. 98, 2884–2897.
Rahimnia, F., Moghadasian, M., Castka, P., 2009. Benchmarking legality in mass services: the case of a fast food restaurant chains in Iran. Benchmark. Int. J. 16, 799–816.
Redlingshöfer, B., Coudurier, B., Georget, M., 2017. Quantifying food loss during primary production and processing in France. J. Clean. Prod. 164, 703–714.
Richter, B., Bokelmann, W., 2016. Approaches of the German food industry for addressing the issue of food losses. Waste Manage. 48, 423–429.
Rivera, L., Chen, F.F., 2007. Measuring the impact of Lean tools on the cost–time investment of a product using cost–time profiles. Robot. Comput.-Integr. Manufact. 23, 684–689.
Roy, P., Nei, D., Orikasa, T., Xu, Q., Okadome, H., Nakamura, N., et al., 2009. A review of life cycle assessment (LCA) on some food products. J. Food Eng. 90, 1–10.
Rutten, M.M., 2013. What economic theory tells us about the impacts of reducing food losses and/or waste: implications for research, policy and practice. Agric. Food Secur. 2, 13.
Sathiyabama, C.N., Dasan, P., 2013. Optimum resource allocation and eliminating waste inside food industry. Centr. Eur. Bus. Rev. 2, 31.
Schneider, F., 2013. The evolution of food donation with respect to waste prevention. Waste Manage. 33, 755–763.
Seth, D., Seth, N., Goel, D., 2008. Application of value stream mapping (VSM) for minimization of wastes in the processing side of supply chain of cottonseed oil industry in Indian context. J. Manufact. Technol. Manage. 19, 529–550.

Shafiee-Jood, M., Cai, X., 2016. Reducing food loss and waste to enhance food security and environmental sustainability. Environ. Sci. Technol. 50, 8432–8443.
Shewfelt, R.L., 2017. How does food processing change the nutritional value of foods? Defense of Processed Food. Springer, pp. 107–123.
Shobha, N., Subramanya, K., 2012. Application of value stream mapping for process improvement in a food manufacturing industry—a case study. In: International Conference on Challenges and Opportunities in Mechanical Engineering, Industrial Engineering and Management Studies.
Silvennoinen, K., Heikkilä, L., Katajajuuri, J.-M., Reinikainen, A., 2015. Food waste volume and origin: case studies in the Finnish food service sector. Waste Manage. 46, 140–145.
Smith, P., 2013. Delivering food security without increasing pressure on land. Global Food Secur. 2, 18–23.
Spiker, M.L., Hiza, H.A., Siddiqi, S.M., Neff, R.A., 2017. Wasted food, wasted nutrients: nutrient loss from wasted food in the United States and comparison to gaps in dietary intake. J. Acad. Nutr. Diet. 117, 1031–1040. e1022.
Tanco, M., Santos, J., Rodriguez, J.L., Reich, J., 2013. Applying lean techniques to nougat fabrication: a seasonal case study. Int. J. Adv. Manufact. Technol. 68, 1639–1654.
Taylor, D.H., 2005. Value chain analysis: an approach to supply chain improvement in agri-food chains. Int. J. Phys. Distrib. Logis. Manage. 35, 744–761.
Taylor, D.H., 2006. Strategic considerations in the development of lean agri-food supply chains: a case study of the UK pork sector. Supply Chain Manage. Int. J. 11, 271–280.
Taylor, D.H., Fearne, A., 2009. Demand management in fresh food value chains: a framework for analysis and improvement. Supply Chain Manage. Int. J. 14, 379–392.
Tostivint, C., de Veron, S., Jan, O., Lanctuit, H., Hutton, Z.V., Loubière, M., 2017. Measuring food waste in a dairy supply chain in Pakistan. J. Clean. Prod. 145, 221–231.
Van Den Bold, M., Kohli, N., Gillespie, S., Zuberi, S., Rajeesh, S., Chakraborty, B., 2015. Is there an enabling environment for nutrition-sensitive agriculture in South Asia? Stakeholder perspectives from India, Bangladesh, and Pakistan. Food Nutr. Bullet. 36, 231–247.
Vlachos, I., 2015. Applying lean thinking in the food supply chains: a case study. Prod. Plan. Control 26, 1351–1367.
Wesana, J., De Steur, H., Dora, M.K., Mutenyo, E., Muyama, L., Gellynck, X., 2018. Towards nutrition sensitive agriculture. Actor readiness to reduce food and nutrient losses or wastes along the dairy value chain in Uganda. J. Clean. Prod. 182, 46–56.
West, P.C., Gerber, J.S., Engstrom, P.M., Mueller, N.D., Brauman, K.A., Carlson, K.M., et al., 2014. Leverage points for improving global food security and the environment. Science 345, 325–328.
Willersinn, C., Mack, G., Mouron, P., Keiser, A., Siegrist, M., 2015. Quantity and quality of food losses along the Swiss potato supply chain: stepwise investigation and the influence of quality standards on losses. Waste Manage. 46, 120–132.
Womack, J.P., 2006. Value stream mapping. Manuf. Eng. 136, 145–156.
Womack, J.P., Jones, D.T., Roos, D., 1990. Machine That Changed the World. Simon and Schuster.
Xue, L., Liu, G., Parfitt, J., Liu, X., Van Herpen, E., Stenmarck, Å., et al., 2017. Missing food, missing data? A critical review of global food losses and food waste data. Environ. Sci. Technol. 51, 6618–6633.
Zokaei, K., Simons, D., 2006. Performance improvements through implementation of lean practices: a study of the UK red meat industry. Int. Food Agribus. Manage. Rev. 9, 30–53.

Food waste valorization

10

Stella Plazzotta and Lara Manzocco
Department of Agricultural, Food, Environmental and Animal Sciences, University of Udine, Udine, Italy

Chapter Outline

10.1 Introduction and main definitions

In the last years, food waste has been increasingly considered not a critical discard, but rather a valuable biomass that can be successfully converted into profitable products. In fact, food waste presents interesting characteristics: it is a source of different compounds including carbohydrates, proteins, lipids, and bioactive molecules (Ravindran and Jaiswal, 2016), it is renewable (i.e., available on a continuous basis), and cheap (Elmekawy et al., 2013). Based on these characteristics, food waste has been exploited to obtain not only energy and biofuels, but also enzymes, antioxidant extracts, novel biodegradable materials, and many other derivatives with a commercial value. In other words, food waste presents a great potential for valorization in a holistic biorefinery view. The term *biorefinery* refers to the production of bioenergy/biofuels and value-added derivatives from renewable biomass sources, with the final aim of substituting fossil fuels and reducing the

Saving Food. DOI: https://doi.org/10.1016/B978-0-12-815357-4.00010-9

depletion of other natural sources (Cherubini and Ulgiati, 2010). To this aim, different valorization strategies can be applied to food waste (Galanakis, 2012). A general definition of *valorization* can be based on the definition of *recovery* given by the Waste Framework Directive (2008/98/EC, 2008). Recovery refers to any operation whose result is waste serving a useful purpose by replacing other materials, which would otherwise have been used to fulfill a specific function in the plant or in the wider economy. Recovery can be based both on the direct exploitation of waste or after a modification of its characteristics. However, it should be added that valorization aims not only at recovering a waste material, but also at maximally exploiting its features, reaching the highest possible added value. In this regard, the term *food by-product* is increasingly used among researchers in the sector of food waste valorization instead of *food waste*. This term posits waste as raw material for the recovery of food and nonfood compounds intended for the development of novel value-added products (Laufenberg et al., 2003). However, it must be underlined that the legal definition of *by-product*, given by the European Commission, requires meeting conditions other than the possibility to develop a commercial value from a waste substance. In fact, according to Directive 2008/98/EC, a substance or object resulting from a production process, the primary aim of which is not the production of that item, may be regarded as by-product and not as waste only if:

1. the further use of that substance is certain;
2. the substance can be directly used, without any further processing other than normal industrial practice;
3. the substance is produced as integral part of the production process; and
4. further substance use fulfills legal requirements in terms of environment, safety, and quality.

The same Directive also defines the "end-of-waste" status: in this case, waste shall cease to be a waste if it has been somehow recovered and is in accordance with the following conditions:

1. the substance is commonly used for specific purposes;
2. a market or demand exists for such a substance;
3. the substance meets technical requirements, existing legislation and standards; and
4. the recovery will not lead to adverse environmental or human health impacts.

Food waste valorization strategies are often addressed to substances that do not meet conditions set by this Directive. In fact, many of them require a deep modification of waste before using it as a new marketable product. In addition, they are often based on the pioneering exploitation of food waste to produce outputs whose added value, legal classification, and impact on the food supply chain sustainability is unknown. In this work, the term *food waste* will be used to generally indicate substances discarded from a food process, which have a potential for valorization.

Based on these considerations, the present chapter discusses traditional and innovative valorization strategies of waste deriving from different food processing. In addition, a guideline for the development and implementation of new valorization strategies is proposed.

10.2 Sources and targets of food waste valorization

Food waste can be classified based on different criteria. The first one relies on the point in the food supply chain in which it is generated (harvesting, postharvest, industrial process, distribution, retail, consumption). For implementing a valorization strategy, the availability of substrates presenting a homogeneous composition and a segregated localization is crucial. In fact, the absence of these features hinders the implementation of valorization strategies, since additional collection and separation processes would be required. This would pose significant cost issues and, besides, would also increase the time needed between food waste collection and processing intended for its valorization. This aspect is of critical importance, since food waste is extremely perishable, presenting a high concentration of water and nutrients that make it a perfect substrate for the growth of pathogens (Galanakis, 2012). Currently, the collection of homogeneous wastes, concentrated in rather few locations, is possible only in the initial steps of the food supply chain (harvesting, postharvest, and industrial processing) while, in most cases, food waste generated during distribution, retailing, and consumption is a mixture of heterogeneous and not-segregated materials (Ravindran and Jaiswal, 2016). For these reasons, food waste intended for valorization is commonly collected in the first steps of the food supply chain and, among them, in the postharvest and industrial processing steps. In industrialized countries, in fact, food waste is mainly generated during these steps of the food supply chain, due to different causes, including postharvest evaluation of raw materials based on quality standards requested by retailers and programmed overproduction (FAO, 2011; Segrè and Falasconi, 2011).

A second possible classification criterion for food waste sources relies on the food sector origin. Based on this criterion, food wastes can be classified in two main groups represented by animal- and plant-origin wastes, which can be further declined into subcategories. Animal-origin waste includes materials discarded from meat, fish and seafood, and dairy industries; plant-origin waste originates from cereal, root and tuber, oil crop and pulse, and fruit and vegetable industries.

10.3 Defining priorities in food waste valorization

A possible criterion for prioritizing food waste valorization can be represented by the amount of waste generated along the food supply chain. In this regard, most studies have highlighted that perishable food items account for the highest proportion of food waste. Based on these considerations, valorization strategies should focus their attention on such waste, mainly deriving from fruit and vegetables, dairy products, meat, and fish (Pekcan et al., 2006; Morgan, 2009; WRAP, 2010; Thönissen, 2009).

Another criterion based on relative waste amount compares animal- and plant-derived waste. The latter has been reported to represent a higher proportion (about 63%, on wet basis) of food waste produced in industrialized countries in comparison to the former (Pfaltzgraff et al., 2013).

However, it must be underlined that different food products have different environmental impact, as clearly stated by the Climate Smart Agriculture Sourcebook (FAO, 2013). Based on this consideration, the Barilla Centre for Food and Nutrition (BCFN) elaborated the so defined *Double Pyramid* model, which classifies foods based not only on their nutritional supply but also on their environmental impact. The latter was evaluated based on the life cycle assessment (LCA) methodology, which considers the impact on environment of all the phases of the production process, including waste generation and management. This pyramid could thus be used to define priorities in food waste valorization. In this regard, the Double Pyramid model neatly highlights that foods for which the highest consumption is recommended (fruit and vegetables) also present the lowest environmental impact. Similarly, animal products, for which a lower weekly consumption is recommended, are characterized by the highest environmental impact (BCFN, 2010).

Alternatively, valorization strategies can be prioritized based on the so-called Waste Hierarchy, which is independent of the waste characteristics and origin. According to this classification, stated by the directive 2008/98/EC on waste, waste management strategies should be prioritized in the following order: prevention and minimization, reuse, recycle, energy recovery, and disposal. If the scope is obtaining valuable products from waste, applicable management strategies are those based on reuse, recycle, and energy recovery. In fact, prevention would allow preventing waste generation, eliminating the need for its valorization, while disposal is not recommended, since it simply consists in the landfilling of waste, with significant transport and disposal costs as well as high environmental impact. In particular, reuse indicates valorization strategies based on the use of waste materials after no or minor modification of waste characteristics (e.g., checking, cleaning, repairing); recycle refers to valorization strategies based on a deep modification of waste materials, through which they are reprocessed into products, materials, and substances whether for the original or other purposes; energy recovery indicates those interventions performed to recover energy contained in the waste material (Kothari et al., 2010).

Finally, valorization strategies can be compared based on the produced added value. In this regard, when food waste is used as a feedstock to produce energy (e.g., through incineration and anaerobic digestion), fuels (e.g., conversion into bioethanol, biohydrogen, biomethanol), animal feeding and fertilizers (e.g., composting), its interesting functional molecules are lost or, at best, underutilized (Pfaltzgraff et al., 2013). The latter are instead maximally exploited when food waste serves as a source of bioactive compounds, functional ingredients, and biocompatible materials to be exploited for human consumption. Thereby, it has been recently proven that the conversion of food waste to bulk chemicals is about 3.5 and 7.5 times more profitable than its conversion to animal feed or transportation fuel respectively, confirming the marginal economic value of these first-generation valorization strategies (Papargyropoulou et al., 2014). Thus herein the attention is focused on valorization strategies aiming at the maximal exploitation of food waste potentialities.

10.4 Valorization and sustainability

To understand the role of food waste valorization on food supply chain sustainability, the first step is surely the definition of the sustainability concept. According to the Brundtland Commission, "sustainability is the development that meets the needs of the present without compromising the ability of future generations to meet their own needs" (Brundtland, 1987). Even if this definition puts the accent on the fact that a sustainable development should not compromise resources for future generations, it gives no indications to companies about what should be done and what should be avoided for achieving a sustainable development. Such indications can be found in the "corporate social responsibility" definition, according to which, organizational sustainability must include not only economic performances (profit) but also environmental and social impact: only by balancing these three components, long-term sustainability can be achieved (Elkington, 1994). If specifically referring to the food supply chain, sustainability can be achieved by a proper "product stewardship", that is, "minimizing the health, safety, environmental, and social impacts of a product and its packaging throughout all lifecycle stages, while also maximizing economic benefits". The manufacturer, or producer, of the product has the greatest ability to minimize adverse impacts, but other stakeholders, such as suppliers, retailers, and consumers, also play a role. Consequently, food supply chain sustainability depends on all the stakeholders and affects economic, social, and environmental frameworks (Product Stewardship Institute, 2011).

Food waste valorization could improve all the three aspects of sustainability. The economic sustainability would be increased both directly, by reducing waste management costs and creating new value-added products from a cheap and always available source and indirectly, by giving companies the possibility to build an eco-friendly image.

Social sustainability may also be improved by food waste valorization, since it would increase food production and optimize the resource used as food to feed an ever-increasing world population.

Environmental sustainability would be directly improved by food waste valorization, due to a reduction in waste to be disposed of and to the conversion of waste materials into new renewable resources, to be used in production processes without depleting other primary resources.

Nevertheless, data available on the real impact of food waste valorization on the sustainability of food supply chain are still limited and this topic is worthy of further investigation. For example, despite that the extraction of bioactive compounds from food waste using classical solvent procedures is simple and cheap, it has been demonstrated to have a negative impact on the environment, due to huge solvent amounts to be disposed of. Similarly, the application of novel extraction technologies can result in high investment costs, and is not sustainable from an economic point of view (Chemat et al., 2012).

10.5 Valorization of animal waste

10.5.1 Definition and quantification

10.5.1.1 Meat and poultry

Waste produced by meat industries consists of the portions of slaughtered animals that cannot be processed and sold as meat or meat products and are not intended for direct human consumption (Jayathilakan et al., 2012). This definition thus excludes all those edible products such as fat, lard, or internal organs (kidney, liver, tripe, brains, spleen) that fall under the definition of by-products given by the Directive 2008/98/EC. Rather, meat-derived waste includes heterogeneous materials such as bones, tendons, skin, portions of the gastrointestinal tract and other internal organs, and blood. Waste generated by slaughtering of cattle, pigs, and lambs represents more than 50% of the live animal weight (Russ and Meyer-Pittroff, 2004). Animal blood is the first product obtained after slaughter and constitutes 3%–5% of the live weight of animals. According to the different animal species, hides/skins and bones could represent respectively 4%–11% and 15%–22% of the live animal weight (Halliday, 1973; Wanasundara et al., 2003).

10.5.1.2 Seafood

Waste generated by seafood processing typically ranges from 20% to 60% of the initial raw material, with an average of 20 million tons globally (Suresh and Prabhu, 2013; Pangestuti and Kim, 2014). Fish and seafood waste mainly consists (27% of the fish) of offal, head, and tails collected by the eviscerating, cutting, and filleting processes. A second major residue is also represented by skins, fishbones, and blood (25% of the fish) (AWARENET, 2004).

10.5.1.3 Dairy

Dairy and cheese processing plants generate large volumes of liquid waste, mainly represented by cheese whey. The latter is the waste liquid portion produced during cheese-making or during coagulation of the milk casein process. In particular, to make 1 kg of cheese approximately 9 kg of whey is generated (Audic et al., 2003), leading to a current total worldwide production of whey of about 190 million tons/year (Yadav et al., 2015).

10.5.2 Valorization strategies

Animal waste derivatives have been recovered for centuries to serve a wide range of applications: animal blood, skin, and bones have been used for the production of foods such as sausages, snacks, soups; hides and feathers have been used since prehistoric times for shelters, clothing, containers, filler, adornment, and forage (Jayathilakan et al., 2012); cheese whey has been used for animal feeding (Audic et al., 2003; Schingoethe, 1976). Moreover, in the past, animal waste (especially animal blood and cheese whey) was disposed of in the environment by spraying on agricultural fields or composting. However, these strategies present several issues. For example, strict legal requirements limit the use of meat waste in food and feed, due to possible health and hygienic issues (Ofori and Hsieh, 2014). Similarly, land spreading of animal waste is not allowed or strongly hindered, due to the environmental problems caused by its physicochemical characteristics such as high solid content, elevated biochemical and chemical oxygen demand, and production of landfill leachate and polluting gases (Nowak and von Mueffling, 2006; Del Hoyo et al., 2007; Prazeres et al., 2012). Finally, composting can absorb only a reduced amount of animal waste, as it is a slow process, requiring space and regular maintenance to avoid environmental pollution and off-odor generation (Mittal, 2006).

Since animal waste represents a cheap source of high-quality proteins and nutrients, alternative strategies can be applied to maximally valorize this waste (Table 10.1).

10.5.2.1 Meat and poultry

Blood, hides/skins, and bone meals are prepared from meat waste by the rendering process. The latter consists in cooking, defatting, and grinding of meat waste to obtain two separate fractions of fat and flour, generally referred to as meat and bone meals. These substrates have been commonly used in animal feed and as ingredients for pet food, due to their high concentration of available essential amino acids, minerals, and vitamin B12. However, as a result of the bovine spongiform encephalopathy (BSE) crisis in the European beef industry, the use of meat and bone meal for cattle feed has not been allowed since 2000. Therefore, particularly interesting is the valorization strategy based on the incineration of meat and bone meals, leading to ashes rich in phosphate and cadmium. The latter have been used as source for phosphoric acid production, phosphate source for industry, agricultural soil enrichment, heavy metals immobilization in soil or water, and for the development of phosphate rich materials and for biosorbents to be used in wastewater treatment (Hodson et al., 2000; Deydier et al., 2005).

Moreover, both hides and bones contain large quantities of collagen, which is the most abundant animal structural protein, representing about 30% of the total animal protein content (Pal et al., 2015). The controlled hydrolysis of collagen produces gelatin. Both collagen and gelatin are widely used in various industrial sectors: collagen can be used as emulsifier and filler in meat products, while gelatin is added to a wide range of foods, forming a major ingredient in jellied desserts, aspics, and as a stabilizer for ice cream and other frozen desserts. Gelatin is also widely used in the pharmaceutical industry, to produce capsules, medicated tablets, and pastilles and can also be turned into sterile sponges used in surgery. In addition, based on the excellent emulsifier and stabilizing properties of gelatin, it is used in cosmetic emulsions and foams (Mokrejs et al., 2009). Even if currently the primary sources of commercial collagen are bovine and porcine skin, bones, and hides, in recent decades, the production of land-based animal's collagen has decreased, due to the concerns about outbreaks of BSE, foot mouth disease, and other prion

Table 10.1 Targets of valorization strategies applied to animal waste and their possible use

Waste origin	Waste material	Target	Use/possible use	References
Meat and poultry	Skins, hides, bones	Ashes	Source of phosphoric acid and phosphate; components of bioadsorbent materials	Deydier et al. (2005), Hodson et al. (2000)
		Collagen	Emulsifier, foaming, filler, gelling agents for food, pharmaceutical, medical and engineering applications	Pal et al. (2015)
		Hydroxyapatite	Biomaterial for craniofacial, oral, maxillofacial and orthopedic applications	Kowalski et al. (2008)
	Blood	Proteins	Emulsifying agents, protein supplements, fat replacers; growth factors of microbial culture media, reactants of biological assays	Viana et al. (2005), Kurbanoglu and Kurbanoglu (2004)
		Plasma	Ingredients for food and pet food with high-quality protein content and palatability	Coffey and Cromwell (2001), Polo et al. (2005), Fernández-Michel et al. (2006), Yousif et al. (2003)
		Hemoglobin	Ingredients for food iron fortification	Walter et al. (1993)
		Fibrinogen, fibrinolysin, serotonin, immunoglobulins, and plasminogen	Chemical and medical uses	Young and Lawrie (2007)
		Plasma flour	Ingredients for protein fortification of bakery products	
	Internal glands and organs	Cholesterol, melatonin, bile, vitamin B12, progesterone, estrogen, heparin, insulin	Substrate for vitamin D3 synthesis; active substances in drugs for treatment of several diseases	Jayathilakan et al. (2012)
	Feathers	Keratin	Biomaterials for medical and bioengineering applications	Ji et al. (2014)

Seafood	Mixed seafood waste	Fish meal	Pet food, animal feed and fertilizer	Ferraro et al. (2010)
		Fish oil	Ingredient in shortenings and margarines; component in soaps and creams	Ferraro et al. (2010)
	Skins, scales, bones, fins	Collagen	Emulsifier, foaming, filler, gelling agent for food, pharmaceutical, medical and engineering applications	Pal and Suresh (2016)
		Hydroxyapatite	Biomaterial for craniofacial, oral, maxillofacial and orthopedic applications	Huang et al. (2011), Muhammad et al. (2016)
	Liver and flesh	PUFA	Ingredients in the formulation of nutritional supplements	Sahena et al. (2009)
		Free amino acids	Ingredients in the formulation of food supplements, energy drinks, infant formulas, drugs	Kang et al. (2009)
Seafood	Shells	Chitin and chitosan	Components in drugs (delivery systems, hypocholesterolemic agents); food adjuvants (beverage clarification), antimicrobial and emulsifying agents, fat mimetics, components of edible films	Cahú et al. (2012)
		Carotenoids	Antioxidants and immune system enhancers; natural food colorants	Sánchez-Machado et al. (2006)
	Blood	Antifreeze proteins	Adjuvants in cryopreservation and cryosurgery; ingredients in frozen and low-fat foods	Feeney and Yeh (1998)
	Internal organs	Enzymes (proteases, lipases, chitinolytic)	Additives in polyester production and wastewater treatment; adjuvants in meat tenderization, protein hydrolysate production, enzyme-assisted extraction	Gildberg (2004)

(*Continued*)

Table 10.1 (Continued)

Waste origin	Waste material	Target	Use/possible use	References
Dairy	Whey	Whey powder, whey protein concentrate, whey protein isolate	Food emulsifying, foaming, water binding, and gelling agents; components of edible and active films and coatings	Yadav et al. (2015), Ramos et al. (2012)
		α-Lactalbumin	Regulator of serotonin, lipid oxidation and mineral absorption; antimicrobial activity and immunomodulatory agents	De Wit (1990)
		β-Lactoglobulin	Emulsifying and gelling agent; carrier for fat-soluble vitamins and lipids	De Wit (1990)
		Bovine serum albumin	Carrier for fatty acids	Korhonen (2009)
		Immunoglobulins	Antimicrobial, antitoxins and antiviruses agents; milk replacers in infant formula	Mohanty et al. (2016)
		Lactoferrin and lactoperoxidase	Antimicrobial and antifungal agents	Hernández-Ledesma et al. (2011)
		Bioactive proteins	Ingredients in feed and pet food formulation; antimicrobial agents	Mohanty et al. (2016)
		Biopolymers (exopolysaccharides, xanthan gum)	Food viscosants and gelling agents	Prazeres et al. (2012)
		Enzymes (e.g., lipase, α-amylase)	Adjuvants in biotechnological applications	Yadav et al. (2015)
	Whey permeate	Lactose	Source of lactulose and lactitol for foods and pharmaceutical formulations	Audic et al. (2003)

diseases. Besides, use of mammalian collagen is a hurdle in the development of kosher and halal products, due to religious factors (Wang et al., 2014).

Animal bones have been recently exploited as a source of natural hydroxyapatite. This mineral is the primary constituent of bones, calcified cartilage tissues and teeth and finds large use in craniofacial, oral, maxillofacial, and orthopedic applications. Since hydroxyapatite is currently mainly derived from chemical synthesis, its extraction from animal waste would offer the possibility of obtaining a biomaterial with interesting stability, biocompatibility, and inertness features (Kowalski et al., 2008).

Plasma is the blood liquid fraction, thus excluding the cellular fraction (red and white blood cells and platelets) and is rich in proteins, which are used as natural binders in whole meat processing, emulsifiers in meat systems, protein supplements, fat replacers in meat products and enzymatic inhibitors in some fish derived products presenting high endogenous protease activity (Viana et al., 2005). Moreover, blood protein derivatives can be used as growth factors for culture media and as reactants of different biological assays (Kurbanoglu and Kurbanoglu, 2004). Plasma has also been used in the formulation of food for both pets and farm animals, to increase its nutritional quality, texture, and palatability (Coffey and Cromwell, 2001; Polo et al., 2005). Similarly, flour derived from plasma has been exploited in the formulation of protein-rich bakery products (Fernández-Michel et al., 2006; Yousif et al., 2003).

By contrast, the use of cellular fraction of blood is limited due to the undesirable sensory properties imparted to the final product in terms of color, odor, and taste (Duarte et al., 1999). Nevertheless, this fraction is exploited for the extraction of hemoglobin, to be used for food iron fortification (Walter et al., 1993).

Finally, many blood components such as fibrinogen, fibrinolysin, serotonin, kalikreninsa, immunoglobulins, and plasminogen are isolated for chemical or medical uses (Young and Lawrie, 2007).

Internal organs and glands are a source of different compounds such as cholesterol and hormones. Cholesterol can be extracted from brains, nervous systems, and spinal cords and is used as raw material for the synthesis of vitamin D3 and as emulsifier in cosmetics. The pineal gland is a source of the hormone melatonin that is used for the treatment of different diseases such as schizophrenia and insomnia. Similarly, the gall bladder is used for the extraction of bile, used for the treatment of indigestion, constipation, and gastrointestinal tract disorders. Moreover, liver extract is used as a source of vitamin B12 and heparin, used to treat anemia and coagulation disorders. Progesterone and estrogen, hormones used to treat reproductive problems in women, can be extracted from pig ovaries, while the pancreas provides insulin and glucagon, which are used in the treatment of diabetes and low sugar diseases, respectively, as well as chymotrypsin and trypsin, used to improve healing after surgery or injury (Jayathilakan et al., 2012).

Feathers from the poultry industry have been recently used as a source of keratin. It is known, in fact, that feathers contain usually more than 70% (w/w) of this structural protein that has important applications in different sectors, including the production of biomaterials, flocculants, and adhesives (Ji et al., 2014).

10.5.2.2 Seafood

Most seafood waste is currently processed in fish meal plants, where fish meal and fish oil are produced. Fish meal production consists in mincing, cooking, and pressing fish waste to separate the solid cake (the fish meal) from a liquid phase, which is centrifuged to obtain the fish oil. While fish meal is used as animal feed, pet food, or plant fertilizer due to its rich composition in protein and minerals, fish oil can be exploited for both food and nonfood uses according to its composition. Among food uses, the production of margarine and shortenings is the most common use of fish oil; nonfood applications include production of soap, glycerol, fertilizers, and substrates for fermentations (Ferraro et al., 2010).

Skin, scales, fins, and bones deriving from seafood waste can represent a valuable alternative to meat waste to produce collagen and gelatin, not presenting issues related to prion diseases and religious factors (kosher and halal products). Moreover, with respect to animal gelatin, seafood gelatin presents analogous functional properties, associated to an enhanced digestibility (Woodard et al., 2007).

Similarly, seafood bone can represent an alternative to animal waste to produce biocompatible materials such as hydroxyapatite (Yamamura et al., 2018).

The liver and residual flesh of some fishes (cod fish, mackerel) are important natural sources of polyunsaturated fatty acids (PUFA), which are used in the production of PUFA concentrates and nutritional supplements. PUFA, in fact, are well known to be associated to numerous biological and physiological functions in the human body (Zuta et al., 2003).

Fish waste can also be considered a source of free amino acids, such as taurine and creatine, which are largely used for producing sport drinks, food supplements, infant formulae, and drugs. Free amino acids, in fact, present different biological activities: for example, taurine is involved in renal functionality and antiinflammatory activity, while creatine is responsible for skeletal and muscle regeneration and contraction. Nowadays these free amino acids are mainly produced by chemical synthesis. However, the final products contain process contaminants and by-products that can have negative health effects. For these reason, a lot of research is being dedicated to the possibility of extracting these amino acids form fish flesh. In particular, raw mussels, fresh clams, and raw fish flesh are particularly rich in taurine, while herring, salmon, and cod are valuable sources of creatine (Kang et al., 2009; Ferraro et al., 2010).

Shell waste from crabs, shrimps, and krill are the main sources for the extraction of chitin and chitosan. These ubiquitous marine polysaccharides are used in food, pharmaceutical, and health industries. In particular, chitin, chitosan and their oligomers are used in nutraceutical formulation for their role as dietary fibers, in lipid absorption reduction and hypocholesterolemia effect. In the food industry they are used as additives for beverage clarification, as texturing and emulsifying agents, fat mimetics, and to produce edible films. Chitinous materials are also used as antimicrobial agents and in drugs as delivery systems and helpers in wound healing. Other applications include water purification from dyes, pesticides, and phenols and laboratory application as chromatographic separation agents and enzyme immobilizers (Kim et al., 2008; Kumar, 2000).

The production of chitinous materials from seafood shell waste is a well-established and profitable process, especially if it includes also the recovery of pigments such as carotenoids (e.g., astaxanthin). In this regard, crustacean cells represent a major source for the recovery of these compounds, which present interesting characteristics for use in food and medical applications. In particular, they are used as natural colorants and powerful antioxidants. Medical uses exploit their protection activity against chemically induced cancers and age-related macular degeneration, as well as their enhancement of the immune system (Sánchez-Machado et al., 2006).

Seafood waste is also a very promising source of antifreeze proteins, which prevent fish from freezing and are thus present in fish adapted to very cold sea waters, such as cod. Antifreeze proteins find large application in frozen foods, low-fat products, cryopreservation, cryosurgery, and aquaculture. In frozen foods they allow lowering the freezing point, thus reducing cellular damage and maintaining texture. Their structuring ability has also been exploited in low-fat ice cream production (Feeney and Yeh, 1998; Ferraro et al., 2010).

Aquatic invertebrates as well as the internal organs of fish and the shells of crustaceans constitute natural sources of enzymes. At present, proteases constitute the dominant group of marine enzymes with a commercial value. They mainly include gastric, intestinal, and hepatopancreas proteinases but also nonproteolytic enzymes, such as transglutaminase, lipases, and chitinolytic enzymes. These enzymes present a wide range of well-established or possible applications in food (e.g., extraction of pigments, production of protein hydrosilates, meat tenderization) and nonfood sectors (wastewater treatment, polyester production) (Kim and Dewapriya, 2014).

10.5.2.3 Dairy

According to Galanakis (2012), whey derivatives represent the majority of the labeled products deriving from food waste. Whey proteins and their derivatives, in fact, have found a wide range of applications, due to their excellent functional and nutritional properties. To transform whey into value-added products, two main procedures are usually carried out. The first one is processing of whey to obtain whey powder, whey protein concentrate, whey protein isolate, reduced lactose whey, lactose, and other protein fractions (Yadav et al., 2015). Whey powder, whey protein concentrate, and isolate can be used for food and feed formulation, due to their high nutritional value. Moreover, they are largely used as emulsifying, foaming, water binding, gelling, and texturizing agents (Morr and Ha, 1993).

Viable materials have been successfully produced from whey proteins. Based on their wide-range mechanical properties, such materials can be used either as primary coatings or as packaging films; in addition, their ability to serve other functions, such as carriers of antimicrobials, antioxidants, or other nutraceuticals, will add value for eventual further commercial applications (Ramos et al., 2012).

In addition, the isolation of specific proteins from whey produces highly functional compounds. β-Lactoglobulin is an excellent emulsifying and gelling agent and can also be used as carrier for fat-soluble vitamins and lipids. α-Lactalbumin helps serotonin regulation, and thus leads to positive effects on cognitive performance, mood, and sleep; in addition, it helps the regulation of lipid oxidation and mineral absorption and possesses antimicrobial activity and immunomodulatory effects. Bovine serum albumin is often used as carrier for fatty acids, while immunoglobulins present antimicrobial, antitoxin, and antiviral activity and are used in infant formula as milk replacers (Korhonen, 2009; Mohanty et al., 2016; De Wit, 1990). Lactoferrin and lactoperoxidase are mainly used for medical applications, due to their antimicrobial and antifungal activity (Hernández-Ledesma et al., 2011).

A second approach for whey valorization involves biotechnological processing, where whey is used as substrate for various microbial/enzymatic processes to obtain valuable products. The latter include bioactive proteins such as nisin (bacteriocin) and peptides with a desired aminoacidic profile for feed and pet food formulation, biopolymers such as exopolysaccharides and xanthan gum to be used as food viscosants/gelling agents, and enzymes (e.g., lipase, α-amylase) for biotechnological applications (Siso, 1996; Prazeres et al., 2012).

The residual fraction from the production of whey protein derivatives is called whey permeate and is used for lactose recovery. In this regard, Peters (2005) assessed that the transformation of whey into whey protein concentrate, isolate, and/or single protein fractions generates a large stream of whey permeate, which needs further processing to make the valorization process profitable. Lactose, in fact, represents the raw material to produce functional derivatives such as lactulose, widely used in pharmaceutical and feed applications, and lactitol, used in ester emulsifier production (Audic et al., 2003).

10.6 Valorization of plant-origin waste

10.6.1 Definition and quantification

10.6.1.1 Cereals

Cereals are a primary human food source, being the staple food for a large sector of the world population. Cereal waste is produced during the milling process, in which bran and germ are eliminated. The latter are rich sources of dietary fibers, phenolics, vitamins, and minerals (Elmekawy et al., 2013). According to the cereal, the amount of generated waste can be different. For example, it is estimated that during the production of white wheat flour, 150 million tons of wheat bran are produced per year worldwide (Prückler et al., 2014). Similarly, rice production generates about 29.3 million tons of rice bran annually (Sharif et al., 2014).

10.6.1.2 Roots and tubers

Among the several roots and tubers, potato is the largest crop worldwide, while cassava is very popular in South Asia and America (FAO, 2009). Processing of potatoes is conducted mainly for the production of chips or French fries and corresponding solid wastes consist of peels or cull potatoes (Schieber et al., 2001). The amount of the waste depends on the used peeling process, reaching on average 27% of the initial weight, accounting for more than 100 million tons in 2013 (Guechi and Hamdaoui, 2016).

Similarly, cassava peels constitute 10% wet weight of the roots and represent the second waste generated during processing of cassava, following the so defined cassava pulp or pomace, representing 15%–20% of initial cassava weight (Jamal et al., 2011).

10.6.1.3 Oil crops and pulses

Olive and olive oil production is mostly located in Mediterranean countries, where more than 98% of the world's olive oil is produced (an estimated 2.5 million tons/year). Upon oil extraction, a solid fraction called olive pomace (olive cake) containing olive pulp, skin, and stones, and a liquid one (olive mill wastewater) are produced. According to the production process, 1 ton of olives generates about 200 kg of olive oil, 400–600 kg of solid waste, and 600–1200 kg of wastewater (Azbar et al., 2004).

In the case of oil crops and pulses, sunflower and soybean are the dominating crops in Europe and North America, respectively (Galanakis, 2012). Sunflower is cultivated for the high oil content, which represents up to 80% of its economic value. Sunflower oil extraction typically has a yield of 85%, while the remaining 15% is represented by residual press cake (Evon et al., 2009).

In the case of pulses, soybeans have been cultivated in Asian civilizations for thousands of years and are one of the most important food crops globally today. Okara is the Japanese term referring to the soy pulp, that is, the ground soybean insoluble residue remaining after filtering the water-soluble fraction during soymilk and soybean curd production. Since about 1.1–1.2 kg of okara is produced from every kilogram of soybeans, huge quantities of okara are produced annually, especially in Asian countries with high soybean consumption. The amount of okara generated from the soybean manufacturing sector is about 800,000 tons in Japan, 310,000 tons in Korea, and 2,800,000 tons in China. The amount of okara produced annually in Singapore alone is at least 10,000 tons, comparable to that produced in Canada (Vong and Liu, 2016).

10.6.1.4 Fruit and vegetables

Fruit and vegetable processing can generate different waste amount, due to the removal of peels, stones, and other inedible parts (Ajila et al., 2012). For example, fresh-cut processing generates at least 25%–30% of waste, up to, in some cases, 50% (Sagar et al., 2018; Plazzotta et al., 2017), while during juice production, generated waste can amount up to 10%–20% (Argun and Dao, 2017). Overall, a total of 42 and 70 million tons of annually produced waste has been estimated for fruits and vegetables, respectively (Sagar et al., 2018).

10.6.2 Valorization strategies

Traditional uses of plant waste include soil amendment, composting, bioenergy recovery, and animal feed. However, soil amendment practice, which is based on the ability of organic waste to immobilize trace metals and metalloids, is often difficult to put into practice, due to the high biological instability of plant waste, responsible for pathogen growth risk and off-odors generation (Clemente et al., 2015). Moreover, in some cases, such as olive mill wastewater, pretreatments are required to remove specific phytotoxic components (Dermeche et al., 2013). Similarly, despite that composting is an ancient eco-friendly method to convert organic waste into organic fertilizer, it is well established that anaerobic digestion is a more attractive strategy to produce fertilizers from plant waste, due to the energy recovery as biogas. In this regard, it must be noted that a significant portion of some crops, such as soy and cereals, is not intended for food use, but for producing biofuels, as a biobased alternative to fossil sources (Milazzo et al., 2013). However, codigestion of plant waste with other organic waste with a richer composition (e.g., animal manure) is usually recommended to increase process yield (Sharma et al., 2000). Plant-based waste can also be exploited to formulate animal feeds with increased nutritional value. However, these waste materials are not always suitable for animal feed, due to the high water content, the low protein concentration, and the presence of indigestible compounds (especially insoluble fibers) or antinutritional factors (such as trypsin inhibitors present in soy-derived waste) (Stanojevic et al., 2013). Moreover, composition of vegetable products varies according to season, forcing manufacturers to often change feed formulations (San Martin et al., 2016).

Beside these issues, traditional recovery strategies are not able to maximally valorize functional compounds of plant food waste, leading to the need for alternative valorization strategies (Table 10.2).

10.6.2.1 Cereals

Cereal milling process produces bran, which is a by-product so as defined by European legislation, that is, presenting a common use in food sector. Bran includes the outer shell of the seed, which is particularly rich in fiber and protein. The most common types are corn, rice, and wheat bran, which are commonly used as ingredients to increase nutritional value of different foods, including bakery products and breakfast cereals (Elmekawy et al., 2013).

Biovalorization of cereal waste based on microbial fermentation has been reviewed by Elmekawy et al. (2013). According to the cereal type, used microorganisms, and fermentation mechanisms, different outputs can be obtained. They include organic acids, such as lactic, citric and succinic acid, which are widely used in food, pharmaceutical, leather, and textile industries. Also, enzymes can be obtained by cereal waste fermentation. They include α-amylase, β-glucosidase, cellulase, glucoamylase, and proteases, which find wide uses in food and nonfood sectors, such as pharmaceutical, paper, textile, detergent, and tanning industries.

Table 10.2 Targets of valorization strategies applied to plant waste and their possible use

Waste origin	Waste material	Target	Use/possible use	Reference
Cereals	Corn, rice, and wheat bran	Dietary fiber	Ingredients for bakery products and breakfast cereals with enhanced fiber content	Elmekawy et al. (2013)
		Antioxidant compounds	Natural antioxidants for food and pharmaceutical applications	Yen and Razak (2014)
	Corn mill waste	Organic acids (lactic, citric, succinic)	Additives and adjuvants in food, pharmaceutical, leather, and textile industries	Li et al. (2006)
		Pullulan	Component of biodegradable packaging films, adhesives, molded articles, coatings, fibers	Sharma et al. (2013)
	Wheat mill waste, rice bran	Enzymes α-amylase, β-glucosidase, cellulase, glucoamylase and proteases	Additives and adjuvants in food, pharmaceutical, paper, textile, detergent, and tanning industries	Kammoun et al. (2008), Ng et al. (2010)
	Corn husk	Antibiotics and biosurfactants	Components in drugs, ingredients in food emulsified systems	Mahalaxmi et al. (2010)
		Arabinoxylans, xylitol,	Low calorific value sweetener	Zhang et al. (2014)
		Furfural	Component in the synthesis of different solvents, lubricants, medicines and adhesives	Sánchez-Bastardo et al. (2017)
	Corn, rice mill waste, and wheat bran	Vanillin	Flavoring and antimicrobial agent in food industry; intermediate in herbicides, drugs and antifoaming agents; component of household products (polishes and air fresheners)	Di Gioia et al. (2007)
	Brewer's spent grain	Xylitol	Sweetener for food and pharmaceutical formulations	Mussatto and Roberto (2005)

(*Continued*)

Table 10.2 (Continued)

Waste origin	Waste material	Target	Use/possible use	Reference
	Rice bran	Polyhydroxyalkanoate	Component of biodegradable packaging materials	Saranya Devi et al. (2012)
	Oat mill waste	β-Glucan	Gelling agent	Patsioura et al. (2011)
Roots and tubers	Potato peel	Reducing sugars with antioxidant properties	Food preservatives and cosmetic sector	Pathak et al. (2018)
		Lactic acid	Component in polylactic acid production	Liang et al. (2014)
		Tissue structure	Biosorbents for the treatment of effluents or other contaminated sources containing dyes, pigments, and metals	Tiwari et al. (2015)
	Cassava	Enzymes (cellulolytic and amylolytic)	Additives in food, pharmaceutical, detergent, and tanning industries	Silva et al. (2009), Ofuya and Nwajiuba (1990)
		Organic acids	Preservatives, acidulants, flavoring agents	John et al. (2006)
		Xanthan	Viscosant in jams, puddings, sauces, and canned and frozen food and drinks; stabilizer in emulsions, suspensions, and foam products; additive in textile industries, paint and automotive oils, ceramic coatings	Padmaja and Jyothi (2016)
Roots and tubers	Cassava	Pullulan	Component of biodegradable packaging films, adhesives, molded articles, coatings, fibers	Padmaja and Jyothi (2016)
Oil crops and pulses	Olive pomace	Fat-balanced flour	Highly nutritional feed	Servili et al. (2015)
		Cell wall polysaccharides	Microcrystalline or powdered cellulose, gelling agents and fat replacers	Galanakis et al. (2010)

	Olive stone	Porous structure	Biosorbents for water purification and other decontamination processes	Spahis et al. (2008)
	Olive wastewater	Polyphenols	Antioxidant, antiinflammatory and antimicrobial agents for food and pharmaceutical applications	Servili et al. (2015)
	Sunflower oil waste	Proteins	Emulsifying, foaming, and whipping agents	Pickardt et al. (2015), Rodrigues et al. (2012)
	Okara from soy processing	Dietary fiber, proteins	Ingredient for functional food production (snacks, bakery and meat products)	Park et al. (2015), Olga and Etelka (2013), Grizotto et al. (2012)
			Source of nanofibers	Fung et al. (2010)
		Bioactive peptides and free amino acids	Ingredients of food supplements, energy drinks, infant formulas, drugs	Vong and Liu (2016)
		Isoflavones	Ingredients of functional food formulation	Preece et al. (2017)
		Oil	Nutritional ingredient for food and pharmaceutical formulations	Quintana et al. (2018)
Fruit and vegetables	Fruit and vegetable waste	Antioxidant phenolic compounds	Natural antioxidants for food and pharmaceutical applications	Laufenberg et al. (2003)
		Carotenoids, anthocyanins, betanin and chlorophylls	Natural colorants	Bridle and Timberlake (1997)
		Dietary fiber	Flour intended for bakery product fortification	Plazzotta et al. (2018c), Ferreira et al. (2015)
		Tissue structure	Biosorbents for wastewater treatment; templates for solvent imbibition	Plazzotta et al. (2018b), Azouaou et al. (2008)
		Water	Water for use in industries facilities	Anon (2017)

(Continued)

Table 10.2 (Continued)

Waste origin	Waste material	Target	Use/possible use	Reference
	Apple, orange, carrot, waste	Pectin	Gelling agent for fruit derivatives and bakery fillings	Perussello et al. (2017), Balu et al. (2012)
	Citrus peels	Essential oils	Flavoring agents in foods; anti-inflammatory and antibacterial in drugs; component of toilet soaps, perfumes, cosmetics	Boukroufa et al. (2015)
	Grape waste	Seed oil	Food oil rich in linoleic acid and polyphenols	Gowe Chala (2015)

Vanillin can also be obtained by wheat, corn, and rice waste. In fact, ferulic acid, which is the main vanillin precursor, can be released from these substrates by proper physicochemical and enzymatic treatments and bioconverted into vanillin. The latter is used as flavoring and antimicrobial agent in the food industry, as intermediate in herbicides, drugs, and antifoaming agents, as a component of household products (polishes and air fresheners) (Di Gioia et al., 2007).

Biopolymeric materials have been also produced from cereal waste. Brewer's spent grain has been used as a medium for producing xylitol, corn waste as nutrient for pullulan production, and hydrolyzed rice bran as substrate for polyhydroxyalkanoate production. Xylitol is used in the food and pharmaceutical industries as a low-calorific-value sweetener (Mussatto and Roberto, 2005). Due to its unique structure, biodegradable nature, and characteristic physical properties, pullulan has a wide range of industrial applications such as packaging films, adhesives, molded articles, coatings, and fibers. Similarly, polyhydroxyalkanoates bear similar physicochemical properties to conventional polymers such as polyethylene (PE/LDPE) and polypropylene (PP). These waste-derived materials present applications in pharmaceutical, food, and cosmetic industries and represent attractive alternative to petrochemically derived ones, since their use can minimize the detrimental impact of persistent plastics on the environment (Saranya Devi et al., 2012).

Finally, specific microorganisms can ferment cereal waste and produce antibiotics (e.g., rifamycin) and biosurfactants, leading to value-added compounds of natural origin, able to substitute chemically synthesized ones and increasingly appreciated by consumers (Elmekawy et al., 2013).

Beside bioconversion, cereal waste can also be exploited for selective extraction of specific compounds. In this regard, oat mill waste has been suggested for the extraction of β-glucan with advanced gelling properties (Patsioura et al., 2011).

Similarly, rice and wheat bran have been used for the extraction of antioxidant compounds (Yen and Razak, 2014).

Moreover, arabinoxylans can be extracted by hydrothermal methods, chemical treatments, enzymatic extractions, and mechanical processes, hydrolyzed into xylose and arabinose and then further hydrogenated to obtain xylitol. Another product to be produced from xylose is furfural, which is employed in the synthesis of different solvents, lubricants, medicines, and adhesives (Sánchez-Bastardo et al., 2017).

10.6.2.2 Roots and tubers

A widely studied valorization strategy for potato peels is the extraction of reducing sugars intended for antioxidant extract preparation, possibly applicable in food and cosmetic sectors (Al-Weshahy and Rao, 2012; Pathak et al., 2018). In this regard, the efficiency as a food preservative of an ethanolic extract from potato peels was tested on fish fillets and soybean oil (Amado et al., 2014).

Carbohydrates from potato peels were also exploited for microbial bioconversion into useful organic compounds such as acetic and lactic acid, long chain fatty acid, alcohols, or hydrocarbons. In this regard, scale-up studies were conducted on the bioconversion of potato peel waste in lactic acid intended for production of polylactic acid, an added-value biodegradable plastic (Liang et al., 2014).

Over the past few years, one of the major applications of potato peels has been the development of bioadsorbents for the treatment of effluents or other contaminated sources containing dyes, pigments, and metals. For this application, potato waste has been submitted to different treatments such as pyrolysis, acid treatment, and hydrothermal (Tiwari et al., 2015).

Around 50%–60% of the unextracted starch from cassava tubers goes along with the residue, and hence this biowaste is a cellulose-starch product, that can be thus exploited for efficient bioconversion into commercially important compounds such as cellulolytic and amylolytic enzymes, organic acids (lactic, citric) and polysaccharides such as xanthan and pullulan (Ofuya and Nwajiuba, 1990; Silva et al., 2009). These compounds find a wide range of applications in food and nonfood sectors. Organic acids are widely used as preservatives, acidulants, and flavoring agents (John et al., 2006). Xanthan is employed as a stabilizer in emulsions, suspensions, and foam products and is also used as an additive in textile industries, paint and automotive oils, and ceramic coatings. It is used in various food products such as jams, puddings, sauces, and canned and frozen food and drinks. Pullulan, as anticipated, represents a biodegradable alternative to traditional plastic materials (Padmaja and Jyothi, 2016).

By contrast, cassava peel valorization is limited due to its high content of toxic cyanogenic glucosides, which make it not suitable for direct soil fertilization. However, if properly fermented, cassava peel can serve as a potential resource for animal feeds contributing up to 10% or more to feed composition (Jamal et al., 2011).

10.6.2.3 Oil crops and pulses

A proper valorization of olive oil pomace and vegetation wastewater can be obtained by different strategies. Olive pomace can be dried and stoned to obtain highly nutritive feed, presenting interesting characteristics in terms of fat content and balanced composition of PUFA. These features have been shown to exert positive effects on milk and meat quality in terms of reduction of saturated fatty acids, increase in monounsaturated fatty acids and vitamin E and improvement of oxidative stability (Servili et al., 2015).

Cell wall polysaccharides can also be recovered from olive pomace. These compounds include cellulosic, hemicellulosic, and pectic carbohydrates and have been proposed as a source of microcrystalline or powdered cellulose, gelling agents, and fat replacers (Galanakis et al., 2010).

Moreover, the peculiar porous structure of olive stone has been exploited for its absorption capacity of different contaminants (e.g., dyes, metals), finding thus applications in the field of water purification and other decontamination processes (Spahis et al., 2008).

The extraction of phenolic bioactive compounds is a possible strategy for valorizing oil vegetative wastewaters. In this regard, during extra virgin olive oil extraction, only 2%–3% of bioactive phenolic compounds is transferred into the oil. The remaining portion is retained in waste portions (Servili et al., 2015). Different technologies have been proposed to this aim, including enzymatic preparation, solvent extraction, supercritical fluid extraction using carbon dioxide, and high-energy ultrasounds. The obtained phenolic extracts hold promising potential as antioxidant, antiinflammatory, and antimicrobial agents, exploitable as antioxidant agents in food and pharmaceutical applications.

Press cakes generated from sunflower oil extraction are promising sources of food proteins, due to their widespread availability, the high protein content, and the low amounts of antinutritive compounds. In particular, sunflower proteins have been proposed as alternative to soy-derived ones, which present allergy issues. Overall, literature data on the functional properties of sunflower proteins are contradictory due to the diversity of employed methods and pretreatments, as well as the large variety of investigated products. However, sunflower proteins have shown emulsifying properties comparable to those of soy proteins and formed stable foams, suggesting them as emulsifying and whipping agents (Pickardt et al., 2015).

Being rich in dietary fiber and protein, okara (waste from soybean processing) can be directly used in food formulations to enhance their nutritional profile. In this regard, Préstamo et al. (2007), reported okara as an effective weight-loss supplement with a potential prebiotic effect. For these reasons, wet okara has been incorporated in different snacks and bakery products (Park et al., 2015). Unfortunately, the use of fresh okara in the food industry is limited because of its high water content (70%–80%), which makes it prone to quick spoilage, even under refrigerated conditions (Radočaj and Dimić, 2013). To avoid these limitations, drying has been exploited to obtain okara flour, which has been proposed for the production of functional foods (Li et al., 2013). Recent studies have focused on applying okara flour in the production of bakery and meat products presenting enhanced fiber, protein, and bioactive content and a concomitant lower fat amount (Grizotto et al., 2012). However, drying process of okara is costly and energy intensive, due to the high amount of water to be removed.

An alternative valorization strategy is based on okara fermentation, with the purpose of producing extractable bioactive compounds or using the fermented foodstuff as functional ingredient. Microbial biotransformation of okara may offer some important advantages. Firstly, fermented okara has improved digestibility; moreover, the bioconversion of high molecular weight okara proteins to smaller ones may increase the solubility of protein isolates, and generate bioactive peptides and amino acids. Trypsin inhibitors and fatty acids may also be degraded by microorganisms, improving the nutritional value of okara and producing more desirable aroma compounds, respectively (Vong and Liu, 2016).

Extraction of specific compounds could also represent a possible strategy for okara valorization. This strategy is based on the extraction of target compounds such as fibers, carbohydrates, proteins, and bioactive compounds such as isoflavones. The latter belong to a group of polyphenols believed to be partially responsible for the health benefits of soy (Preece et al., 2017). In this regard, fibers for nanofiber production, antioxidant carbohydrates, β-glucosidase for functional food preparation, and oil fraction for cosmetic and pharmaceutical usage have been extracted from okara (Fung et al., 2010; Mateos-Aparicio et al., 2010; Li et al., 2013; Quintana et al., 2018).

10.6.2.4 Fruit and vegetables

Fruit and vegetable processing wastes are the most widely investigated substrates for the extraction of several types of antioxidants and dietary fibers. This is due to the fact that soft tissues of fruit and vegetable waste are rich in both ingredients, which allow their simultaneous extraction in two separate streams (Laufenberg et al., 2003).

Antioxidant phenols intended for pharmaceutical and food applications are a common target in fruit and vegetable waste valorization strategies since they have been associated with a reduction in the incidence of diseases such as cancer, heart disease, hepatic injury, and neurodegenerative disorders. Their antioxidant activities make them suitable for food applications to prevent rancidity and oxidation instead of chemical antioxidants with documented toxicity, such as BHA and BHT (Peschel et al., 2006).

Similarly, carotenoids, anthocyanins, betanin, and chlorophylls, which are present in large quantities in fruit and vegetable waste, can also be exploited as natural colorants (Bridle and Timberlake, 1997; Rodriguez-Amaya, 2016).

Fruit and vegetable waste also presents high dietary fiber content. In this regard, fruit and vegetable waste can be turned into flour by means of drying and grinding. Fruit and vegetable waste flour has been used as ingredient for the formulation of fiber-enriched foods (Ferreira et al., 2015; Plazzotta et al., 2018c). Drying of fruit and vegetable waste has also been exploited to produce fibrous materials with high contact surface intended as bioadsorbents for the removal of pollutants such as dyes and heavy metals from water and ground and as templates for solvent absorption (Azouaou et al., 2008; Plazzotta et al., 2018b).

Among the different compounds, pectin represents an interesting polysaccharide, which can be extracted from different fruit and vegetable waste, including apple pomace, orange peel, carrot steam peels, green beans cutting waste, leek cutting waste, and celeriac stem peels. In this regard, pectin is primarily known as a gelling agent and is extensively applied in the production of jams and jellies, fruit juice, confectionery products, and bakery fillings (Christiaens et al., 2015; Perussello et al., 2017; Balu et al., 2012).

Citrus waste represents a source of essential oils that can be used in food as flavoring ingredients in drinks, ice creams, and other food products as well as in pharmaceutical industries for its antiinflammatory and antibacterial effect. In addition, substantial quantity of this oil is also used in the preparation of toilet soaps, perfumes, cosmetics, and other home care products or as green solvent (Boukroufa et al., 2015).

Seeds of grapes intended for wine production constitute a valuable source of seed oil, which is produced all over Europe and is highly appreciated due to its rich content of unsaturated fatty acids (particularly linoleic acid) and phenolic compounds (Gowe Chala, 2015).

Water can also be considered a valuable output of a recovery strategy. In this regard, patented or patent-pending systems able to convert organic material into water are already applied in companies, supermarkets, and restaurants. They are based on the hyperacceleration of aerobic decomposition through the activity of naturally occurring microorganisms with enhanced degradation capabilities under tightly controlled environmental conditions (Anon, 2017).

10.7 Development and implementation of food waste valorization strategies

Food waste can be considered a cheap source of energy, water, and valuable ingredients/products. To maximally exploit these potentialities, an integrated approach to waste management should be developed by selecting, and eventually combining, the most efficacious recovery strategies. Such approach results from the application of a rational four-step procedure, including waste characterization, output definition, process design, and feasibility study.

10.7.1 Waste characterization

The first step for developing a rational valorization strategy involves an accurate characterization of the food waste material, in terms of legal classification, amount, and composition. In particular, the reclassification of a processing discard as a by-product or the recognition of its end-of-waste status are possible only if meeting specific requirements of safety, quality, marketability, and sustainability (Dir. 2008/98/CE).

Amount data are often already available to companies, based on the knowledge of company sources flow, transport, and disposal costs.

Secondarily, the identification of process steps mainly involved in waste generation should be carried out. It is likely that in these steps, in fact, waste materials presenting a more homogeneous composition could be available. This would favor and simplify the implementation of a valorization strategy, avoiding or shortening collection and separation processes.

Wastes should be then accurately characterized not only in terms of actual homogeneity and composition but also in terms of long-term variability and perishability. This would allow identifying waste features possibly exploitable in a valorization strategy (e.g., high fiber or protein content). This step would also allow identifying critical features: for example, a high compositional variability would hinder the possibility of standardizing the valorization strategy and a high perishability would pose the need for a quick transformation of the waste.

10.7.2 Output definition

Based on the key properties identified in step "Waste characterization", possible final products of food waste valorization can be hypothesized. Beside valorization strategies based on the extraction of bioactive compounds, in this step, research and development expertise should be exploited to hypothesize innovative solutions to valorize the waste material. In this regard, it should be underlined that the same waste material could offer a wide range of valorization possibilities. For example, lettuce waste, despite its poor composition, has been shown to be a possible source of phenols but can also be turned into functional flour and innovative porous materials exploitable for solvent loading (Llorach et al., 2004; Plazzotta et al., 2018a,b,c). In this step, a holistic biorefinery approach should be adopted, identifying all the possible outputs, with the final aim of reducing the waste to zero. This can be attained by applying a multiple-step approach. For example, after extracting bioactive molecules from a vegetable matrix, the residual waste could be further exploited to produce flour or serve as water source for company facilities. This step should result in a clear definition of the possible amounts and compositional features of the waste derivatives, as well as in their classification in terms of possible use (ingredients, products, adjuvants, additives, materials) purpose (increase food functionality, material biodegradability, or biocompatibility), and sector (e.g., food, engineering, biomedical, packaging).

10.7.3 Process design

Production processes required to obtain the outputs defined in step "Output definition" are designed. Despite the extremely wide variability in characteristics according to different food waste kinds, Galanakis (2012) developed the 5-Stages Universal Recovery Process, potentially applicable to every kind of food waste. According to this universal approach, food waste recovery could be accomplished in five distinct stages: macroscopic pretreatment (adjustment of the water, solids and lipid content, activation or deactivation of enzymes, reduction of the microbial load, increase in the permeability of the matrix); macro- and micromolecules separation (separation of antioxidants, acids, or ions from biopolymers); extraction (solubilization of free molecules and dissociation of bound ones); purification (clarification of the target compounds from coextracted impurities); and product formation (encapsulation or drying to obtain a stable product).

This recovery strategy has the advantage of being potentially applicable to all kinds of food waste materials, since it can be tailored by omitting some stages. In the simplest case, reuse is possible and thus only pretreatment operations such as cleaning, washing, mincing, and partial dehydration are needed. However, this only applies to traditional recovery strategies such as animal feeding, soil amendment, and composting, which, as already underlined, do not attain a proper valorization of functional compounds present in food waste. In most cases, all steps are required and can be accomplished with different conventional or emerging technologies, both of which present specific advantages and pitfalls that should be accurately considered. Equipment and know-how for the application of traditional technologies are already available and thus they are easy to use and characterized by low or null investment costs. Such technologies include, for example, air-drying, classical solvent-assisted extraction, and alcohol precipitation. However, they are often energy intensive and can damage the treated matrix producing overheating, structural, and functionality modifications. In addition, they usually require huge amounts of solvents, which represent an economic and environmental burden for companies. Novel technologies, such as microwave, ultrasounds, pulsed electric fields, high pressure homogenization, and supercritical drying represent a suitable alternative, since generally reducing thermal effect and matrix damaging, while maintaining or increasing yield efficacy. Such technologies are commonly referred to as "green technologies" because they reduce solvent and energy consumption, due to higher extraction efficacy. By contrast, since their industrial application is still limited, high investment costs for equipment and dedicated expertise are required for process scaling-up from laboratory to industrial plants.

10.7.4 Feasibility study

The final step for the development and implementation of a food waste valorization strategy is the determination of its feasibility from an economic, environmental and social point of view. In other words, the sustainability of the proposed strategy must be assessed. In this regard, different aspects should be taken into consideration, including capital and operating costs, consumer response and acceptance, environmental impact, and adherence to increasingly stringent legal requirements on food and material safety.

The most important issues, possibly hindering the implementation of a food waste valorization strategy, include:

1. Food waste characteristics: the high perishability and inhomogeneity of food waste materials can make their valorization economically unsustainable.
2. Valorization know-how: the lack of information about functional and bioactive compounds present in food waste materials would make necessary a massive investment not only in equipment but also in research and development expertise. In this regard, food science and technology as well as medical expertise would be of crucial importance in defining new possibilities.
3. Consumer response: commercial and marketing knowledge regarding consumer response to food waste derivatives is nowadays still limited. Different attitudes can affect consumer acceptance of such innovative products. Innovations in the food industry suffer a high market failure rate, partly due to a phenomenon known as "neophobia," which is the rejection that some people express towards new or unfamiliar foods (Barrena and Sánchez, 2013). By contrast, consumers are increasingly concerned about food supply chain sustainability and recent surveys have demonstrated a positive reaction to product labels reporting sustainability claims on them, even using specific terms related to food waste. Moreover, issues related to specific food waste should be considered. For instance, consumer fear of contracting BSE, potential allergic reactions to blood proteins, and the belief that products obtained from animals contains harmful microorganisms, toxins, and metabolites, militates against efforts to fully utilize blood proteins as a food and feed source (Ofori and Hsieh, 2014).
4. Possibility to standardize and scale up the process: much of food waste applications are not effective and are only described in the scientific literature. This is due to the massive investment cost often required for scaling-up and validating valorization processes on industrial scale.

5. Fulfillment of regulatory framework: in the process of valorization and of new products development, there is a need for companies to comply with the stipulated governmental environmental regulatory guidelines and legal concerns of the public so that, eventually, all food processing industries contribute directly to sustainable utilization of natural resources and consequent sustainable development of society.

Only by an accurate consideration of all these aspects, the impact of a food waste valorization strategy on economic, environmental and social sustainability can be assessed. This requires the use of a holistic approach, integrating a wide set of expertise in economy and marketing, in social sciences and in Life Cycle Assessment (LCA), which evaluates the environmental impact of a product in all its life stages (Kim and Kim, 2010).

10.8 Conclusions

Valorization of food waste provides a sustainable solution for solving the existing waste disposal problem. In fact, it offers a boon to food processing industries for augmentation of resources available and, with them, for product diversification and innovation to meet increasing consumer demand for novelty. However, most valorization strategies are nowadays only described in the scientific literature and food waste materials are thus mainly unexploited or used as low-value sources of energy, animal feed, and fertilizers. Valorization strategies can be considered the eco-design of an integrated production scheme, involving expertise in food science and technology, economy, marketing, engineering, environmental, and social sciences. Only legislation and integrated studies following well-funded research and development programs will eventually attain the goal of optimized food waste processing, to successfully create a global food waste biorefinery, with the final aim of realizing a zero-waste food supply chain.

References

2008/98/EC, Directive 2008/98/EC of the European Parliament and of the Council of 19 November 2008 on waste and repealing certain directives. 2008. Offic. J. Eur. Union. 312, 3–30.

Ajila, C.M., et al., 2012. Sustainable solutions for agro processing waste management: an overview. In: Malik, A., Grohmann, E. (Eds.), Environmental Protection Strategies for Sustainable Development. Springer, Netherlands, pp. 65–109.

Al-Weshahy, A., Rao, V., 2012. Potato peel as a source of important phytochemical antioxidant nutraceuticals and their role in human health – a review. In: Rao, V. (Ed.), Phytochemicals as Nutraceuticals. InTech, Rijeka, pp. 207–225.

Amado, I.R., et al., 2014. Optimization of antioxidants extraction from potato peel wastes of *Solanum tuberosum* by surface response methodology. Food Chem. 165 (290–299), 290–299.

Anon, 2017. Waste to water. <http://www.wastetowater.com.au/> (accessed 26.05.18.).

Argun, H., Dao, S., 2017. Biohydrogen production from waste peach pulp by dark fermentation: effect of inoculum addition. Int. J. Hydrogen Energy 42 (4), 2569–2574.

Audic, J., Chaufer, B., Daufin, G., 2003. Non-food applications of milk components and dairy co-products: a review. Lait 83, 417–438.

AWARENET, 2004. Handbook for the prevention and minimization of waste and valorisation of by-products in European agro-food industries, agro-food waste minimization and reduction network – Grow Programme, European Commision.

Azbar, N., et al., 2004. A review of waste management options in olive oil production. Crit. Rev. Environ. Sci. Technol. 34 (3), 209–247.

Azouaou, N., Sadaoui, Z., Mokaddem, H., 2008. Removal of cadmium from aqueous solution by adsorption on vegetable wastes. J. Appl. Sci. 8 (24), 4638–4643.

Balu, A.M., et al., 2012. Valorisation of orange peel residues: waste to biochemicals and nanoporous materials. Chem. Sustain. Energy Mater. 5 (9), 1694–1697.

Barrena, R., Sánchez, M., 2013. Neophobia, personal consumer values and novel food acceptance. Food Qual. Pref. 27 (1), 72–84.

BCFN, 2010. Double Pyramid: healthy food for people, sustainable food for the planet, pp. 1–75. <http://scholar.google.com/scholar?hl = en&btnG = Search&q = intitle: Double + Pyramid + : + healthy + food + for + people + , + sustainable + food + for + the + planet#0> (accessed 26.05.18.).

Boukroufa, M., et al., 2015. Biorefinery of orange peels waste: a new concept based on integrated green and solvent free extraction processes using ultrasound and microwave techniques to obtain essential oil, polyphenols and pectin. Ultrason. Sonochem. 24, 72–79.

Bridle, P., Timberlake, C.F., 1997. Anthocyanins as natural food colours—selected aspects. Food Chem. 58 (1–2), 103–109.

Brundtland, G.H., 1987. Our Common Future. Report of the World Commission on Environment and Development. Oxford University Press, Oxford.

Cahú, T.B., et al., 2012. Recovery of protein, chitin, carotenoids and glycosaminoglycans from Pacific white shrimp (*Litopenaeus vannamei*) processing waste. Process Biochem. 47 (4), 570–577.

Chemat, F., Vian, M.A., Cravotto, G., 2012. Green extraction of natural products: concept and principles. Int. J. Molec. Sci. 13 (7), 8615–8627.

Cherubini, F., Ulgiati, S., 2010. Crop residues as raw materials for biorefinery systems – a LCA case study. Appl. Energy 87 (1), 47–57.

Christiaens, S., et al., 2015. Pectin characterisation in vegetable waste streams: a starting point for waste valorisation in the food industry. LWT – Food Sci. Technol. 61 (2), 275–282.

Clemente, R., et al., 2015. Food byproducts as amendments in trace elements contaminated soils. Food Res. Int. 73, 176–189.

Coffey, R.D., Cromwell, G.L., 2001. Use of spray-dried plasma in diets for weaning pigs. Pig News Info 47N, 22–39.

De Wit, J.N., 1990. Thermal stability and functionality of whey proteins. J. Dairy Sci. 73 (12), 3602–3612.

Del Hoyo, P., et al., 2007. Demineralization of animal blood plasma by ion exchange and ultrafiltration. Meat Sci. 76 (3), 402–410.

Dermeche, S., et al., 2013. Olive mill wastes: biochemical characterizations and valorization strategies. Process Biochem. 48 (10), 1532–1552.

Deydier, E., et al., 2005. Physical and chemical characterisation of crude meat and bone meal combustion residue: "waste or raw material?". J. Hazard. Mater. 121 (1–3), 141–148.

Di Gioia, D., et al., 2007. Production of biovanillin from wheat bran. Enzyme Microb. Technol. 41 (4), 498–505.

Duarte, R.T., Carvalho Simões, M.C., Sgarbieri, V.C., 1999. Bovine blood components: fractionation, composition, and nutritive value. J. Agric. Food Chem. 47 (1), 231–236.

Elkington, J., 1994. Towards the sustainable corporation. Calif. Manage. Rev. Winter 90–100.

Elmekawy, A., et al., 2013. Valorization of cereal based biorefinery byproducts: reality and expectations. Environ. Sci. Technol. 47 (16), 9014–9027.

Evon, P., et al., 2009. Aqueous extraction of residual oil from sunflower press cake using a twin-screw extruder: feasibility study. Industr. Crops Prod. 29 (2–3), 455–465.

FAO, 2009. FAOSTAT. Harvested sheets. FAOSTAT. <http://faostat.fao.org/site/444/DesktopDefault.aspx?PageID¼444#ancor> (accessed 26.0518.).

FAO, 2011. Global food losses and food waste - extent, causes and prevention. <www.fao.org/docrep/014/mb060e/mb060e00.pdf> (accessed 26.05.18.).

FAO, 2013. Climate-smart agriculture sourcebook. Sourcebook on Climate-Smart Agriculture, Forestry and Fisheries. <http://www.fao.org/docrep/018/i3325e/i3325e00.htm> (accessed 26.05.18.).

Feeney, R.E., Yeh, Y., 1998. Antifreeze proteins: current status and possible food uses. Trends Food Sci. Technol. 9 (3), 102–106.

Fernández-Michel, S., Montfort, G.R., Vazquez-Moreno, L., 2006. Physicochemical and microbiological characteristics and sensory analysis of chocolate cakes added with porcine serum proteins. Revista Científica de Veterinaria 16 (4), 420–427.

Ferraro, V., et al., 2010. Valorisation of natural extracts from marine source focused on marine by-products: a review. Food Res. Int. 43 (9), 2221–2233.

Ferreira, M.S., et al., 2015. Formulation and characterization of functional foods based on fruit and vegetable residue flour. J. Food Sci. Technol. 52 (2), 822–830.

Fung, W.Y., Yuen, K.H., Liong, M.T., 2010. Characterization of fibrous residues from agrowastes and the production of nanofibers. J. Agric. Food Chem. 58 (13), 8077–8084.

Galanakis, C.M., 2012. Recovery of high added-value components from food wastes: conventional, emerging technologies and commercialized applications. Trends Food Sci. Technol. 26 (2), 68–87.

Galanakis, C.M., Tornberg, E., Gekas, V., 2010. A study of the recovery of the dietary fibers from olive mill wastewater and the gelling ability of the soluble fiber fraction. LWT Food Sci. Technol. 43 (7), 1009–1017.

Gildberg, A., 2004. Enzymes and bioactive peptides from fish waste related to fish silage, fish feed and fish sauce production. J. Aquat. Food Prod. Technol. 13 (2), 3–11.

Gowe, C., 2015. Review on potential use of fruit and vegetables by-products as a valuable source of natural food additives. Food Sci. Qual. Manage. 45 (November), 47–61.

Grizotto, R.K., et al., 2012. Physical, chemical, technological and sensory characteristics of Frankfurter type sausage containing okara flour. Food Sci. Technol. 32, 538–546.

Guechi, E., Hamdaoui, O., 2016. Biosorption of methylene blue from aqueous solution by potato (*Solanum tuberosum*) peel: equilibrium modelling, kinetic, and thermodynamic studies. Desalin. Water Treatm. 57 (22), 10270–10285.

Halliday, D.A., 1973. Blood – a source of proteins. Process Biochem. 8, 15–17.

Hernández-Ledesma, B., del Mar Contreras, M., Recio, I., 2011. Antihypertensive peptides: production, bioavailability and incorporation into foods. Adv. Colloid Interf. Sci. 165 (1), 23–35.

Hodson, M.E., Valsami-Jones, É., Cotter-Howells, J.D., 2000. Bonemeal additions as a remediation treatment for metal contaminated soil. Environ. Sci. Technol. 34 (16), 3501–3507.

Huang, Y.C., Hsiao, P.C., Chai, H.J., 2011. Hydroxyapatite extracted from fish scale: effects on MG63 osteoblast-like cells. Ceram. Int. 37 (6), 1825–1831.

Jamal, P., et al., 2011. Valorization of cassava peels by the white rot fungus *Panus tigrinus* M609RQY. Austr. J. Basic Appl. Sci. 5 (7), 808–816.

Jayathilakan, K., et al., 2012. Utilization of byproducts and waste materials from meat, poultry and fish processing industries: a review. J. Food Sci. Technol. 49 (3), 278–293.

Ji, Y., et al., 2014. Extraction of keratin with ionic liquids from poultry feather. Separat. Purif. Technol. 132, 577–583.

John, R.P., Madhavan Nampoothiri, K., Pandey, A., 2006. Simultaneous saccharification and fermentation of cassava bagasse for l-(+)-lactic acid production using Lactobacilli. Appl. Biochem. Biotechnol. 134 (3), 263–272.

Kammoun, R., Naili, B., Bejar, S., 2008. Application of a statistical design to the optimization of parameters and culture medium for α-amylase production by *Aspergillus oryzae* CBS 819.72 grown on gruel (wheat grinding by-product). Bioresour. Technol. 99 (13), 5602–5609.

Kang, K., et al., 2009. Optimization of amino acids production from waste fish entrails by hydrolysis in sub and supercritical water. Can. J. Chem. Eng. 79 (1), 65–70.

Kim, I., et al., 2008. Chitosan and its derivatives for tissue engineering applications. Biotechnol. Adv. 26 (1), 1–21.

Kim, M.H., Kim, J.W., 2010. Comparison through a LCA evaluation analysis of food waste disposal options from the perspective of global warming and resource recovery. Sci. Total Environ. 408 (19), 3998–4006.

Kim, S.K., Dewapriya, P., 2014. Enzymes from fish processing waste materials and their commercial applications. In: Kim, S.K. (Ed.), Seafood Processing By-Products. Springer, New York, NY, pp. 183–196.

Korhonen, H., 2009. Milk-derived bioactive peptides: from science to applications. J. Funct. Foods 1 (2), 177–187.

Kothari, R., Tyagi, V.V., Pathak, A., 2010. Waste-to-energy: a way from renewable energy sources to sustainable development. Renew. Sustain. Energy Rev. 14 (9), 3164–3170.

Kowalski, Z., et al., 2008. The possibilities of obtaining hydroxyapatite from meat industry. Molec. Cryst. Liquid Cryst. 486 (1), 282–290.

Kumar, R.M.N.V., 2000. A review of chitin and chitosan applications. React. Funct. Polym. 46 (1), 1–27.

Kurbanoglu, E.B., Kurbanoglu, N.I., 2004. Utilization as peptone for glycerol production of ram horn waste with a new process. Energy Convers. Manage. 45 (2), 225–234.

Laufenberg, G., Kunz, B., Nystroem, M., 2003. Transformation of vegetable waste into value added products: (A) the upgrading concept; (B) practical implementations. Bioresour. Technol. 87 (2), 167–198.

Li, Z., et al., 2006. L-Lactic acid production by *Lactobacillus casei* fermentation with corn steep liquor-supplemented acid-hydrolysate of soybean meal. Biotechnol. J. 1 (12), 1453–1458.

Li, S., et al., 2013. Soybean curd residue: composition, utilization, and related limiting factors. ISRN Industr. Eng. 2013, 1–8.

Liang, S., McDonald, A.G., Coats, E.R., 2014. Lactic acid production from potato peel waste by anaerobic sequencing batch fermentation using undefined mixed culture. Waste Manage. 45, 2022–2027.

Llorach, R., Tomás-Barberán, F.A., Ferreres, F., 2004. Lettuce and chicory byproducts as a source of antioxidant phenolic extracts. J. Agric. Food Chem. 52 (16), 5109–5116.

Mahalaxmi, Y., et al., 2010. Corn husk as a novel substrate for the production of rifamycin B by isolated *Amycolatopsis* sp. RSP 3 under SSF. Process Biochem. 45 (1), 47–53.

Mateos-Aparicio, I., et al., 2010. Multifunctional antioxidant activity of polysaccharide fractions from the soybean byproduct okara. Carbohydr. Polym. 82 (2), 245–250.

Milazzo, M.F., et al., 2013. Sustainable soy biodiesel. Renew. Sustain. Energy Rev. 27, 806–852. Available from: http://www.sciencedirect.com/science/article/pii/S1364032113004760.

Mittal, G.S., 2006. Treatment of wastewater from abattoirs before land application—a review. Bioresour. Technol. 97 (9), 1119–1135.

Mohanty, D.P., et al., 2016. Milk derived bioactive peptides and their impact on human health – a review. Saudi J. Biol. Sci. 23 (5), 577–583.

Mokrejs, P., et al., 2009. Extraction of collagen and gelatine from meat industry by-products for food and non food uses. Waste Manage. Res. 27 (1), 31–37.

Morgan, E., 2009. Fruit and vegetable consumption and waste in Australia, Victoria, Australia. <https://www.vichealth.vic.gov.au/media-and-resources/publications/fruit-and-vegetable-consumption-and-waste-in-australia> (accessed 26.05.18.).

Morr, C.V., Ha, E.Y.W., 1993. Whey protein concentrates and isolates: processing and functional properties. Crit. Rev. Food Sci. Nutr. 33 (6), 431–476.

Muhammad, N., et al., 2016. Extraction of biocompatible hydroxyapatite from fish scales using novel approach of ionic liquid pretreatment. Separat. Purif. Technol. 161, 129–135.

Mussatto, S.I., Roberto, I.C., 2005. Acid hydrolysis and fermentation of brewer's spent grain to produce xylitol. J. Sci. Food Agric. 85 (14), 2453–2460.

Ng, I.S., et al., 2010. High-level production of a thermoacidophilic β-glucosidase from *Penicillium citrinum* YS40-5 by solid-state fermentation with rice bran. Bioresour. Technol. 101 (4), 1310–1317.

Nowak, B., von Mueffling, T., 2006. Porcine blood cell concentrates for food products: hygiene, composition, and preservation. J. Food Protect. 69 (9), 2183–2192.

Ofori, J.A., Hsieh, Y.P., 2014. Issues related to the use of blood in food and animal feed. Crit. Rev. Food Sci. Nutr. 54 (5), 687–697.

Ofuya, C.O., Nwajiuba, C.J., 1990. Fermentation of cassava peels for the production of cellulolytic enzymes. J. Appl. Bacteriol. 68 (2), 171–177.

Olga, R., Etelka, D., 2013. Valorization of wet okara, a value-added functional ingredient, in a coconut-based baked snack. Cereal Chem. 90 (3), 256–262.

Padmaja, G., Jyothi, A.N., 2016. Root and tuber. In: Chandrasekaran, M. (Ed.), Valorization of Food Processing By-Products. Taylor & Francis Group: CRC Press.

Pal, G.K., Suresh, P.V., 2016. Sustainable valorisation of seafood by-products: recovery of collagen and development of collagen-based novel functional food ingredients. Innov. Food Sci. Emerg. Technol. 37 (Part B), 201–215.

Pal, G.K., Nidheesh, T., Suresh, P.V., 2015. Comparative study on characteristics and in vitro fibril formation ability of acid and pepsin soluble collagen from the skin of catla (*Catla catla*) and rohu (*Labeo rohita*). Food Res. Int. 76, 804–812.

Pangestuti, R., Kim, S.K., 2014. Bioactive materials derived from seafood and seafood processing by-products. In: Noomhorm, A., Ahmad, I., Anal, A.K. (Eds.), Functional Foods and Dietary Supplements: Processing Effects and Health Benefits. Wiley-Blackwell, USA, pp. 139–158.

Papargyropoulou, E., et al., 2014. The food waste hierarchy as a framework for the management of food surplus and food waste. J. Cleaner Product. 76 (August), 106–115.

Park, J., Choi, I., Kim, Y., 2015. Cookies formulated from fresh okara using starch, soy flour and hydroxypropyl methylcellulose have high quality and nutritional value. LWT – Food Sci. Technol. 63 (1), 660–666.

Pathak, P.D., et al., 2018. Valorization of potato peel: a biorefinery approach. Crit. Rev. Biotechnol. 38 (1), 218–230.
Patsioura, A., Galanakis, C.M., Gekas, V., 2011. Ultrafiltration optimization for the recovery of β-glucan from oat mill waste. J. Membr. Sci. 373 (1–2), 53–63.
Pekcan, G., et al., 2006. Household Food Wastage in Turkey. FAO, Rome.
Perussello, C.A., et al., 2017. Valorization of apple pomace by extraction of valuable compounds. Compreh. Rev. Food Sci. Food Safety 16 (5), 776–796.
Peschel, W., et al., 2006. An industrial approach in the search of natural antioxidants from vegetable and fruit wastes. Food Chem. 97 (1), 137–150.
Peters, R.H., 2005. Economic aspects of cheese making as influenced by whey processing options. Int. Dairy J. 15 (6), 537–545.
Pfaltzgraff, L.A., et al., 2013. Food waste biomass: a resource for high-value chemicals. Green Chem. 15 (2), 307–314.
Pickardt, C., et al., 2015. Pilot plant preparation of light-coloured protein isolates from de-oiled sunflower (*Helianthus annuus* L.) press cake by mild-acidic protein extraction and polyphenol adsorption. Food Hydrocol. 44, 208–219.
Plazzotta, S., Manzocco, L., Nicoli, M.C., 2017. Fruit and vegetable waste management and the challenge of fresh-cut salad. Trends Food Sci. Technol. 63, 51–59.
Plazzotta, S., Calligaris, S., Manzocco, L., 2018a. Application of different drying techniques to fresh-cut salad waste to obtain food ingredients rich in antioxidants and with high solvent loading capacity. LWT – Food Sci. Technol. 89 (November 2017), 276–283.
Plazzotta, S., Calligaris, S., Manzocco, L., 2018b. Innovative bioaerogel-like materials from fresh-cut salad waste via supercritical-CO2-drying. Innov. Food Sci. Emerg. Technol. 47, 485–492.
Plazzotta, S., Sillani, S., Manzocco, L., 2018c. Exploitation of lettuce waste flour to increase bread functionality: effect on physical, nutritional, sensory properties and on consumer response. Int. J. Food Sci. Technol. 53 (10), 1–8.
Polo, J., et al., 2005. Functional properties of spray-dried animal plasma in canned pet food. Anim. Feed Sci. Technol. 122 (3), 331–343.
Prazeres, A.R., Carvalho, F., Rivas, J., 2012. Cheese whey management: a review. J. Environ. Manage. 110, 48–68.
Preece, K.E., et al., 2017. Intensified soy protein extraction by ultrasound. Chem. Eng. Process. Process Intensif. 113, 94–101.
Préstamo, G., et al., 2007. The effects of okara on rat growth, cecal fermentation, and serum lipids. Eur. Food Res. Technol. 225 (5), 925–928.
Product Stewardship Institute, 2011. Product Stewardship. <http://productstewardship.us/displaycommon.cfm?an = 1&subarticlenbr = 55> (accessed 30.05.158.).
Prückler, M., et al., 2014. Wheat bran-based biorefinery 1: Composition of wheat bran and strategies of functionalization. LWT – Food Sci. Technol. 56 (2), 211–221.
Quintana, G., Gerbino, E., Gómez-Zavaglia, A., 2018. Valorization of okara oil for the encapsulation of *Lactobacillus plantarum*. Food Res. Int. 106, 81–89.
Radočaj, O., Dimić, E., 2013. Valorization of wet okara, a value-added functional ingredient, in a coconut-based baked snack. Cereal Chem. 90 (3), 256–262.
Ramos, Ó.L., et al., 2012. Edible films and coatings from whey proteins: a review on formulation, and on mechanical and bioactive properties. Crit. Rev. Food Sci. Nutr. 52 (6), 533–552.
Ravindran, R., Jaiswal, A.K., 2016. Exploitation of food industry waste for high-value products. Trends Biotechnol. 34 (1), 58–69.

Rodrigues, I.M., Coelho, J.F.J., Carvalho, M.G.V.S., 2012. Isolation and valorisation of vegetable proteins from oilseed plants: methods, limitations and potential. J. Food Eng. 109 (3), 337–346.
Rodriguez-Amaya, D.B., 2016. Natural food pigments and colorants. Curr. Opin. Food Sci. 7, 20–26.
Russ, W., Meyer-Pittroff, R., 2004. Utilizing waste products from the food production and processing industries. Crit. Rev. Food Sci. Nutr. 44 (1), 57–62.
Sagar, N.A., et al., 2018. Fruit and vegetable waste: bioactive compounds, their extraction, and possible utilization. Compreh. Rev. Food Sci. Food Safety 17, 512–531.
Sahena, F., et al., 2009. PUFAs in fish: extraction, fractionation, importance in health. Compreh. Rev. Food Sci. Food Safety 8 (2), 59–74.
San Martin, D., Ramos, S., Zufía, J., 2016. Valorisation of food waste to produce new raw materials for animal feed. Food Chem. 198, 68–74.
Sánchez-Bastardo, N., Romero, A., Alonso, E., 2017. Extraction of arabinoxylans from wheat bran using hydrothermal processes assisted by heterogeneous catalysts. Carbohydr. Polym. 160, 143–152.
Sánchez-Machado, D.I., Gutiérrez-Coronado, M.A., Ríos-Vázquez, N.J., 2006. Quantification of astaxanthin in shrimp waste hydrolysate by HPLC. Biomed. Chromatogr. 20 (10), 981–984.
Saranya Devi, E., Vijayendra, S.V.N., Shamala, T.R., 2012. Exploration of rice bran, an agro-industry residue, for the production of intra- and extra-cellular polymers by *Sinorhizobium meliloti* MTCC 100. Biocatal. Agric. Biotechnol. 1 (1), 80–84.
Schieber, A., Stintzing, F.C., Carle, R., 2001. By-products of plant food processing as a source of functional compounds—recent developments. Trends Food Sci. Technol. 12 (11), 401–413.
Schingoethe, D.J., 1976. Whey utilization in animal feeding: a summary and evaluation. J. Dairy Sci. 59 (3), 556–570.
Segrè, E.A., Falasconi, L., 2011. Il libro nero dello spreco in Italia: Il cibo. Ambiente Edizioni, Milan.
Servili, M., et al., 2015. New approaches to virgin olive oil quality, technology, and by-products valorization. Eur. J. Lipid Sci. Technol. 117 (11), 1882–1892.
Sharif, M.K., et al., 2014. Rice bran: a novel functional ingredient. Crit. Rev. Food Sci. Nutr. 54 (6), 807–816.
Sharma, V.K., et al., 2000. Inclined-plug-flow type reactor for anaerobic digestion of semi-solid waste. Appl. Energy 65 (1), 173–185.
Sharma, N., Prasad, G.S., Choudhury, A.R., 2013. Utilization of corn steep liquor for biosynthesis of pullulan, an important exopolysaccharide. Carbohydr. Polym. 93 (1), 95–101.
Silva, T.M., et al., 2009. Use of cassava peel as carbon source for production of amylolytic enzymes by *Aspergillus niveus*. Int. J. Food Eng. 5 (5), 1–11.
Siso, M.I.G., 1996. The biotechnological utilization of cheese whey: a review. Bioresour. Technol. 57 (1), 1–11.
Spahis, N., et al., 2008. Purification of water by activated carbon prepared from olive stones. Desalination 222 (1), 519–527.
Stanojevic, S.P., et al., 2013. Bioactive proteins and energy value of okara as a byproduct in hydrothermal processing of soy milk. J. Agric. Food Chem. 61 (38), 9210–9219.
Suresh, P.V., Prabhu, G.N., 2013. Seafood. In: Chandrasekaran, M. (Ed.), Valorization of Food Processing By-Products. CRC Press, New York, pp. 685–736.
Thönissen, R., 2009. Food waste: The Netherlands. FAO, Lund.

Tiwari, D.P., Singh, S.K., Sharma, N., 2015. Sorption of methylene blue on treated agricultural adsorbents: equilibrium and kinetic studies. Appl. Water Sci. 5 (1), 81–88.

Viana, F.R., et al., 2005. Quality of ham pâté containing bovine globin and plasma as fat replacers. Meat Sci. 70 (1), 153–160.

Vong, W.C., Liu, S.Q., 2016. Biovalorisation of okara (soybean residue) for food and nutrition. Trends Food Sci. Technol. 52, 139–147.

Walter, T., et al., 1993. Effect of bovine-hemoglobin-fortified cookies on iron status of schoolchildren: a nationwide program in Chile. Am. J. Clin. Nutr. 57 (2), 190–194.

Wanasundara, J.P.D., Pegg, R.B., Shand, P.J., 2003. Value added applications for plasma proteins from the beef processing industry. Can. Meat Sci. Assoc. 10–15.

Wang, H., et al., 2014. Physical-chemical properties of collagens from skin, scale, and bone of grass carp (*Ctenopharyngodon idellus*). J. Aquatic Food Prod. Technol. 23 (3), 264–277.

Woodard, J.R., et al., 2007. The mechanical properties and osteoconductivity of hydroxyapatite bone scaffolds with multi-scale porosity. Biomaterials 28 (1), 45–54.

WRAP, 2010. A review of waste arisings in the supply of food and drink to UK households. <http://www.wrap.org.uk/sites/files/wrap/Waste arisings in the supply of food and drink toUK households, Nov 2011.pdf> (accessed 26.05.18.).

Yadav, J.S.S., et al., 2015. Cheese whey: a potential resource to transform into bioprotein, functional/nutritional proteins and bioactive peptides. Biotechnol. Adv. 33 (6), 756–774.

Yamamura, H., et al., 2018. Physico-chemical characterization and biocompatibility of hydroxyapatite derived from fish waste. J. Mech. Behav. Biomed. Mater. 80, 137–142.

Yen, G.B., Razak, L.T., 2014. Screening of antioxidant potential from cereal wastes and fruit peels. Int. J. Eng. Res. Technol. 3 (1), 1990–1998.

Young, R.H., Lawrie, R.A., 2007. Utilization of edible protein from meat industry by-products and waste. Int. J. Food Sci. Technol. 10 (4), 453–464.

Yousif, A.M., Cranston, P., Deeth, H.C., 2003. Incorporation of bovine dry blood plasma into biscuit flour for the production of pasta. LWT – Food Sci. Technol. 36 (3), 295–302.

Zhang, Z., Smith, C., Li, W., 2014. Extraction and modification technology of arabinoxylans from cereal by-products: a critical review. Food Res. Int. 65, 423–436.

Zuta, P.C., et al., 2003. Concentrating PUFA from mackerel processing waste. J. Am. Oil Chem. Soc. 80 (9), 933–936.

Food consumption and wasted food

11

Taija Sinkko, Carla Caldeira, Sara Corrado and Serenella Sala
European Commission, Joint Research Centre (JRC), Ispra, Italy

Chapter Outline

11.1 Introduction

An increasing global population, an evolution in consumers' needs, changes in consumption models, and considerable generation of food waste pose serious challenges to the overall sustainability of food production and consumption. About one third of the food produced for human consumption is currently wasted at the global scale (FAO, 2011). This reflects the high level of inefficiency of the food supply chain, which has significant economic, social, and environmental impacts. Besides being associated to relevant economic losses, food waste exacerbates food insecurity and malnutrition, and increases pressures on climate, water, and land resources contributing to natural resources depletion and environmental pollution (Godfray et al., 2010). In this context, the United Nations (UN) has proposed 17 Sustainable Development Goals (SDGs). Within SDGs, target 12.3 focuses on food waste reduction, requiring, by 2030, to: "halve per capita global food waste at the retail and consumer levels and reduce food losses along production and supply chains, including postharvest losses" (UN, 2017). The European Commission (EC) has also committed to achieve SDGs including the 12.3 (EC, 2016). In addition, EC has identified food waste as one of the priority areas of the European Circular Economy Action Plan (EC, 2015). This plan presents a set of actions to be implemented in

Saving Food. DOI: https://doi.org/10.1016/B978-0-12-815357-4.00011-0

Europe to facilitate and promote the transition towards circular economy, "where the value of products, materials and resources is maintained in the economy for as long as possible, and the generation of waste is minimized" (EC, 2015).

In the last years, food waste quantification has aroused considerable interest, reflected by the increasing availability of data on food waste generation along the food supply chain at various geographical scales. Corrado and Sala (2018) reviewed 10 studies focused on food waste. The reviewed studies reported that food waste generation along the supply chain ranged between 194 and 389 kg per person per year at the global level and between 158 and 290 kg per person per year when referring to the European scale. The latest estimate produced by Eurostat reported 149 kg per capita food waste in EU in 2014 (EC, 2018). The highest share of food waste was in most cases produced at the consumption stage, followed by the food manufacturing stage. However, it was also observed that estimations for different stages were quite uncertain and further in-depth analysis would be advisable. Caldeira et al. (2017) observed that the share of food waste generated in each stage varies according to the food waste definition adopted and the sources of data. For example, Bräutigam et al. (2014) and FAO (2011) reported a considerable amount of food waste generation at primary production and postharvest stages (43% and 47%, respectively) which was, instead, completely excluded by Monier et al. (2010), Tisserant et al. (2017), and Alexander et al. (2017), or only partly captured in van Holsteijn et al. (2017) and FUSIONS (2016). The distribution stage was found to generate a lower amount of food waste than other stages in all the analyzed studies.

When dealing with supply chains, Life cycle assessment (LCA) is a key methodology to assess environmental impacts of products taking into account all phases during their life cycle, from raw material extraction through processing, distribution, and use until the end of life (EoL). The use of LCA to assess environmental impacts of food has been increasing over time, aiming at assessing the sustainability of the food system (Sala et al., 2017a). Most of the studies available in the literature related to the environmental impacts of food consumption have been focused on climate change, acidification, and eutrophication impacts (McClelland et al., 2018). Other environmental impacts associated to the food production increasingly studied are water consumption (e.g., Lundqvist et al., 2008), land use (e.g., Meier et al., 2014), and biodiversity loss (Wolff et al., 2017; Crenna et al., 2019).

Current food production and consumption patterns are considered unsustainable. According to FAO (2010, p.10), "Sustainable diets are those diets with low environmental impacts which contribute to food and nutrition security and to healthy life for present and future generations". Sustainable diets are protective and respectful of biodiversity and ecosystems, culturally acceptable, accessible, economically fair and affordable, nutritionally adequate, safe, and healthy, while optimizing natural and human resources. Due to current concerns, dietary guidelines set by different national bodies are now including sustainability issues in their recommendations [e.g., the Nordic Nutrition Recommendations (Brugård Konde et al., 2015)].

Different food products have different environmental impacts and food waste rates, thus foods included in the diet and food waste generation among different food types have significant impact on the environmental impacts. Generally, meat has higher impacts compared with plant-based food due to the land and other inputs needed for feed production, but also due to enteric fermentation of beef cattle (Notarnicola et al., 2017). Thus, 1 kg of animal-based food waste has often higher environmental impact compared with 1 kg of plant-based food waste. Similarly, diets containing animal-based food have a higher environmental impact compared with a vegetarian diet (van Dooren et al., 2014). However, food waste rate is typically higher in plant-based food compared with animal-based food (FAO, 2011).

Food waste can be included in LCA studies using different approaches. Corrado et al. (2017) reviewed 100 studies aiming at identifying which approaches have been used to account for food waste in LCA applied on food products. According to Corrado et al. (2017), food waste has not been defined or included systematically in LCA studies. When included, different approaches have been adopted. For example, avoidable food waste at the processing stage was explicitly reported only in few studies. Conversely, possibly avoidable and unavoidable food waste at the processing stage were reported in a higher number of studies. Depending on the process, the amount of food waste can be relevant and the modeling approach adopted can considerably influence the LCA results.

Beretta et al. (2017) quantified environmental impacts of Swiss food consumption and environmental impacts of food waste along the food supply chain. They considered 33 food categories, which represent the whole food basket in Switzerland. They also included impacts of food waste treatment. In terms of climate change, the food waste related emissions were estimated to be 25% of the total emissions of consumed food. In addition, Beretta et al. (2017) observed that food waste at the end of the food value chain (households and food services) causes almost 60% of the total climate impact of food waste, because of large food waste quantities at this stage and higher accumulated impacts per kg of product.

Eberle and Fels (2016) assessed environmental burdens of food consumption and food waste in Germany. The German food basket was differentiated between in-house consumption and out-of-home consumption. Both baskets contained the same food items but in different quantities. Food waste along the food chain had a share of 15%–21% of environmental impacts of the food basket. Eberle and Fels (2016), however, did not made any distinction between avoidable and unavoidable waste. Results also showed that animal products, such as meat and dairy, cause most of the environmental burden of food consumption and food waste, although the share of plant products was higher regarding amounts of consumption and waste.

Scherhaufer et al. (2018) estimated the potential scale of food waste related impacts based on available food waste data on European level using nine indicator products (apple, tomato, potato, bread, milk, beef, pork, chicken, and fish). Food waste was considered as the edible or inedible part of food removed from supply chain and sent to food waste treatment and disposal facilities. The share of the food waste related impacts was from 15.1% to 15.7% of the overall food consumption impacts, depending on impact category. Most of the food waste related impacts derived from the primary production.

The aim of this chapter is firstly to calculate environmental impacts of food consumption of an average EU-28 citizen in 2015, taking into account the apparent consumption of food in EU-28 countries and food waste generation along the food chain. Secondly, the aim is to assess the environmental impacts of food waste, taking into account impacts from waste treatment of wasted food and additional food production, when part of the food is wasted. Moreover, scenarios for food waste reduction and alternative diets are defined, to assess the impact of food waste reduction and diet change when compared to the overall environmental impacts of food consumption.

11.2 Materials and methods

Materials and methods used in this study are presented in this section. First, the food products included in the study and their amounts are presented, then the LCA-based method to calculate the environmental impacts of food consumption in different life cycle stages, as well as the amount of food waste used in the study. Finally, different scenarios related to food waste reduction and dietary shift are illustrated.

11.2.1 Basket of representative food products

In 2017, Notarnicola et al. (2017) published a study on the environmental impact of food consumption in Europe in 2010, focusing on a selection of food products covering 61% of food consumption. The reference flow was the amount of food consumed by an average EU-27 citizen in the reference year 2010. It consisted of a process-based life cycle inventory (LCI) model for a basket of products (BoP) that represented the most relevant food product groups, selected by importance in mass and economic value, to depict the average consumption for nutrition of EU citizens in 2010 (Notarnicola et al., 2017). The product groups in the original BoP were meat, dairy products, cereal-based products, sugar, oils, tubers, fruits, coffee, beverages, and pre-prepared meals.

The work done by Notarnicola et al. (2017) has been complemented with the new products in accordance to their relevance in terms of environmental impacts, for example, biodiversity and water use, even if their consumption amount was not so high. Also, products with high share of imports, and products representative of new trends in nutrition, for example, sources of vegetable proteins, nuts, and seeds, were added to the BoP food. In addition, the reference flow was updated to be the amount of food consumed by an average EU-28 citizen in the reference year 2015. The functional unit was defined as the average food consumption per person in the EU in terms of food categories.

The new product groups added to the basket were fish and seafood, eggs, vegetables, legumes, nuts and seeds, and confectionery products. Moreover, some products under existing products groups were added (rice, bananas, and tea). All products and their apparent consumption (=production + import−export) in 2015 are presented in Table 11.1 (Eurostat, 2018).

Table 11.1 Composition of the BoP food in terms of product groups, representative products, and related quantities referred to the reference flow new products in the basket on top of those in Notarnicola et al., 2017 are marked with*

Product group	Representative product	Per-capita consumption (kg/pers.[a]yr^{-1})	Share from the whole basket (%)
Meat	Pork meat	44.9	6.6
	Beef meat	15.2	2.2
	Poultry meat	26.3	3.9
Fish and seafood	Cod*	10.4	1.5
	Salmon*	3.5	0.5
	Shrimp*	1.5	0.2
Dairy	Milk	78.4	11.5
	Cheese	15.1	2.2
	Butter	4.4	0.6
Eggs	Eggs*	14.0	2.0
Cereal-based products	Bread	40.0	5.9
	Pasta	9.3	1.4
	Rice*	9.6	1.4
Sugar	Sugar	28.6	4.2
Oils	Sunflower oil	5.7	0.8
	Olive oil	4.7	0.7
Tubers	Potatoes[a]	68.5	10.0
Vegetables	Tomatoes*	14.5	2.1
Legumes	Beans*	2.8	0.4
	Tofu[b]*	4.3	0.6
Fruits	Apples	17.5	2.6
	Oranges	13.0	1.9
	Bananas*	11.5	1.7
Nuts and seeds	Almonds[a]*	0.6	0.1
Coffee and tea	Coffee	3.3	0.5
	Tea*	0.6	0.1
Beverages	Beer	70.0 L	10.2
	Wine	26.0 L	3.8
	Mineral water	122.3 L	17.9
Confectionery products	Biscuits*	7.1	1.0
	Chocolate*	6.0	0.9
Preprepared meals	Meat-based dishes	3.4	0.5

[a]Based on 2013 data, since 2015 data was not available.
[b]Based on EFSA (2018).

11.2.2 Calculation of environmental impact of food consumption

Environmental impacts were calculated using LCA approach and modeled with SimaPro 8.5 software. Included life cycle stages were agriculture, industrial processing, packaging, logistics, retail, use, and EoL (Table 11.2). The same stages were

Table 11.2 Summary of life cycle stages and related activities included in the BoP food

Life cycle stage	Activities included
Agriculture/breeding	Cultivation of crops Animal rearing Food waste management
Industrial processing	Processing of ingredients Slaughtering and processing Chilled or frozen storage Food waste management
Logistics	International transport of imports Transport to processing Transport to regional distribution center Transport to retailer Food waste management
Packaging	Manufacture of packaging Final disposal of packaging
Retail	Storage at retail Food waste management
Use	Transport of the products from retailer to consumer's home Refrigerated storage at home Cooking of the meal
EoL	Final disposal of food waste Wastewater treatment and auxiliary processes due to human excretion

also included in the original BoP food (Notarnicola et al., 2017). The most representative datasets for each product in the basket were identified from existing LCA literature. LCI data sources of the agriculture and production stages of the new products in BoP food are summarized in Table 11.3. Data sources for other stages can be found in the following sections. Data sources and LCI results of original products can be found in Notarnicola et al. (2017) and Castellani et al. (2017). All the agricultural datasets, taken from the literature or from databases, have been modified to adapt them to the method and assumptions reported in Notarnicola et al. (2017).

11.2.2.1 Agricultural stage

Emissions from agriculture were calculated using methodology described in Notarnicola et al. (2017), that is, N_2O emissions from managed soils, CO_2 emissions from lime and urea application, NH_3 emissions to air and nitrates leaching in the soil were estimated according to the IPCC guidelines (IPCC, 2006). It was assumed that all nitrogen that volatizes converts to ammonia, and all nitrogen that leaches is emitted as nitrate. It was also estimated that 5% of phosphorus applied through fertilizers is emitted to freshwater resources (Blonk Consultants, 2014). The emissions of pesticides during their use were also taken into account, assuming

Table 11.3 Overview of LCI datasets relative to the agriculture and processing phase of new products in the BoP food

Representative product	Activities	Geographical scope of the data source	References
Wild cod	• Fishing • Processing of cod fillets	Sweden	Svanes et al. (2011)
Farmed salmon	• Salmon aquaculture • Slaughtering • Processing of salmon fillets	Norway	Pelletier et al. (2009), Ellingsen et al. (2009)
Shrimp	• Shrimp aquaculture • Processing	China	Cao et al. (2011)
Eggs	• Laying hens	The Netherlands	Blonk Consultants (2014)
Rice	• Rice cultivation • Rice processing	Italy	Blengini and Busto (2009), Water: Chapagain and Hoekstra (2010)
Tomatoes	• Tomato cultivation	Spain	Torrellas et al. (2012)
Beans (dry)	• Bean cultivation and drying	The Netherlands	Blonk Consultants (2014)
Tofu	• Soybean cultivation	Brazil, Argentina, and The United States	Blonk Consultants (2014)
	• Production of tofu	The United States (adapted to Europe)	Mejia et al. (2017)
Bananas	• Banana cultivation • Postharvest handling	Ecuador	Iriarte et al. (2014), Water: Mekonnen and Hoekstra (2011), Dole (2011)
	• Ripening	Europe	Svanes and Aronsson (2013)
Almonds	• Almond cultivation • Almond processing	Greece	Bartzas et al. (2017), Kendall et al. (2015)
Tea	• Tea cultivation • Tea processing	Average of Kenya, India, and Indonesia, processing in the United Kingdom	Jefferies et al. (2012)

(*Continued*)

Table 11.3 (Continued)

Representative product	Activities	Geographical scope of the data source	References
Biscuits	• Wheat cultivation and flour production • Sugar beet cultivation and sugar production • Oil palm cultivation and palm oil production • Laying hens (producing eggs) • Dairy cattle breeding and milk production	Europe and Indonesia (palm oil)	Blonk Consultants (2014)
	• Baking of biscuits	Europe	Noya et al. (2018)
Chocolate	• Cocoa bean cultivation • Cocoa processing	Ghana	Ntiamoah and Afrane (2008)
	• Chocolate production	Italy	Recanati et al. (2018)

that 100% of the active pesticide ingredient is emitted to soil (de Beaufort-Langeveld et al., 2003).

LCIs of the cultivation of new plant-based products are presented in Table 11.4. LCI results of original products in BoP food can be found in Notarnicola et al. (2017) and Castellani et al. (2017). The emissions from the combustion of diesel in agricultural machinery are not reported in Table 11.4, but are considered in the inventory according to data in the agri-footprint database (Blonk Consultants, 2014).

Table 11.5 shows the LCIs of the farming phase of salmon and shrimp aquaculture, and egg production. The table reports the feed used, the water consumed, and energy inputs as well as main emissions to the air and water. Economic allocation was used to allocate burdens between eggs and meat in the farming phase. Fish processing is also producing some coproducts, for example, heads and guts, which can be used as fish feed after processing. However, in this study, no burdens were

Table 11.4 Life cycle inventories of the cultivation of plant-based products or main ingredients used in new products added to the BoP food (per cultivated ha per year), excluding inventories of products based on agri-footprint database (i.e., inventories of beans and some ingredients used in modeling new products, e.g., soybeans)

	Unit	Banana	Tomato	Rice	Cocoa beans	Almonds	Tea
Products	t	32.8	166.7	7.0	25.5	3.3	1.7
Coproducts (total weight)	t	0	0	8.4	0	0	0
Inputs							
N fertilizer	kg	295	798	130	0	180	306
P fertilizer	kg	38	506	18	818	100	728
K fertilizer	kg	131	1562	135	670	200	128
Lime	kg	22	0	0	0	0	0
Pesticides (total weight)	kg	15	46.5	16.7	210	8.2	< 0.1
Irrigation water	m^3	3083	4748	4087[a]	131	4650	136
Diesel	kg	108	0	95.5	0	539	14.3
Heat	MJ	0	0	4280	0	0	0
Electricity	kWh	10	6485	51	0	803	0
Outputs							
Emissions to air							
N_2O from fertilizers	kg	6.1	16.6	2.7	0	3.7	6.4
NH_3 from fertilizers	kg	35.9	97.0	15.8	0	21.9	37.2
CO_2 from fertilizers	kg	9.5	0	80.1	0	0	0
CH_4 from field	kg	0	0	338	0	0	0
Emissions to water							
NO_3 from fertilizers	kg	88.6	239.4	39.0	<0.1	54	91.8
P from fertilizers	kg	1.9	25.3	1.0	<0.1	5.0	36.4
Emissions to soil							
Chlorpyrifos	kg	0	3.8	0.2	9.1	0	0
Captan	kg	0	0	0	0	4.1	0
Glyphosate	kg	2.8	0	0.8	0	0	0
Mancozeb	kg	6.2	28.5	0	142	0	<0.1
Diuron	kg	0	0	0.1	0	0	0
Bispyribac-sodium	kg	0	0	0.3	0	0	0
Pretilchlor	kg	0	0	4.6	0	0	0

[a]Takes into account water uptake of the rice and available rainwater in the area. Although rice is usually cultivated in flooded field, part of this water can be captured and reused downstream (Chapagain and Hoekstra, 2010).

Table 11.5 Life cycle inventories of the farming phase of animal-derived products

	Unit	Salmon aquaculture	Shrimp aquaculture	Egg production
Products	kg	1000	1000	1000
Coproducts (total weight)	kg	–	–	67.7
Inputs				
Compound feed	kg	1103	1600	2162
Water	m^3	–	–	3.5
Heat from gas	MJ	0.1	–	–
Diesel	kg	12.8	–	–
Electricity	kWh	20.1	2550	98.0
Outputs				
CH_4, biogenic	kg	–	–	1.3
N_2O	kg	–	–	0.4
NH_3	kg	–	–	19.6
N to water	kg	41.1	66	–
P to water	kg	5.2	9	–

allocated to the coproducts of fish filleting, because according to other studies (e.g., Ellingsen et al., 2009) fish filleting coproducts have very small economic value. In addition, in the fish feed process, used as salmon feed, no burdens were allocated to raw material derived from fish coproducts.

11.2.2.2 Packaging, logistics, and retail

Table 11.6 reports the amounts of packaging inventoried for each product added to the BoP food [see inventories for original product in Notarnicola et al. (2017) and Castellani et al. (2017)]. Packaging types and amounts were mainly obtained from LCA studies presented in Table 11.3, with exception of beans and almonds, for which the information was not available. Thus, the package type and amount of beans and almonds were estimated according to other food products, assuming to be packed into plastic bags with the same weight as plastic package used in tofu packaging [according to Mejia et al. (2017)]. In addition, tomato packaging amount was based on Cellura et al. (2012) and egg packaging was based on Sonesson et al. (2008).

Logistics consists of international transportation from outside the EU, transport of raw materials to the processing site, and transport of processed goods from industry to retailing. The transport of imported products was assumed to occur from the capital of the exporting country to the city of Frankfurt, which was considered a central destination for the arrival of imports in Europe. For exporting countries directly connected to Europe by land, such as Switzerland or Belarus, only a transport by lorry was considered from the capital of the exporting country to the city of

Table 11.6 Amounts of packaging per typology, grams per 1 kg packaged product

Food product	Cardboard	Corrugated board box	Kraft paper	Cellulose fiber	Aluminum	LDPE	PS
Cod	100	–	–	–	–	–	–
Salmon	–	–	–	–	–	–	25
Shrimp	135	–	–	–	–	10.5	–
Eggs	69	–	–	–	–	–	–
Rice	50					10	
Tomatoes	–	89.2	–	–	–	–	–
Beans	–	–	–	–	–	39	–
Tofu	93	–	–	–	–	39	–
Bananas	–	112.3	–	–	–	3.3	–
Almonds	–	–	–	–	–	39	–
Tea	260	280	440	10	–	42	–
Biscuits	170	–	–	–	–	5	–
Chocolate	118	–	–	–	18	–	–

Frankfurt. For the others, the transport was considered to be composed by a transport by lorry between the capital of the exporting country and the country's main port; a transport by ship from the port of the exporting country to the main European ports of goods (Rotterdam or Marseilles), and, finally, a transport by lorry between the port of destination and the city of Frankfurt. The distances were calculated by using http://www.sea-distances.org (transport by ship) and Google maps (transport by lorry). This transport was allocated to a percentage of the final product in the LCI model, corresponding to the share of imported goods out of the total apparent consumption of that kind of product. Distances and shares of imported products of new products are reported in Table 11.7. The same information for products originally in the BoP food can be found in Castellani et al. (2017).

The use of refrigerants (both load and leakage) was included in the inventory of refrigerated/frozen transportation and storage in retail phase when applicable (fish, shrimp, eggs, bananas, tomatoes). Refrigerant R404A was considered, as it is the most commonly used refrigerant in Europe. The LCA data for the production of the refrigerants were according to Bovea et al. (2007).

11.2.2.3 Use phase and end of life

The use phase consists of transport of food to consumer home and domestic consumption. It was assumed that 30 products were bought in a single purchase, including food and nonfood products. The impact of transport was therefore allocated between the purchased products considering that each product is one of 30 items purchased (3.33% of the transport burden) (Vanderheyden and Aerts, 2014).

Refrigerated storage at home is included in the life cycle of salmon and tofu (2 days), and frozen storage for cod and shrimp (10 days). The electricity consumption of the domestic refrigerator was assumed to be 0.0023 kWh/L per day and the

Table 11.7 Summary of the share of imported food products, sea transport distances and road transport distances for products added to the BoP food

Product	Import (%)	Sea transport (tkm) per kg of product imported	Road transport (tkm) per kg of product imported
Fish (cod and salmon)	66.9	4.62	0.57
Shrimp	64.7	10.42	1.12
Eggs	0.1	2.35	1.26
Rice	28.4	9.44	1.44
Tomatoes	7.7	1.87	0.53
Beans	16.5	3.76	0.84
Soy beans for tofu	92.8	7.13	2.16
Bananas	87.4	9.37	0.79
Almonds	35.8	6.27	0.68
Tea	100	10.48	1.46
Wheat for biscuits	4.2	2.19	0.29
Palm oil for biscuits	100	12.83	1.04
Sugar for biscuits and chocolate	4.5	0.43	0.10
Cocoa beans for chocolate	100	7.26	0.99

electricity consumption of the freezer was assumed to be 0.0042 kWh/L per day (Nielsen et al., 2003).

Regarding home preparation, the following specific energy consumptions were considered (Foster et al., 2006; Jefferies et al., 2012):

- Boiling of eggs: 2 MJ of natural gas per kg eggs (50% of eggs assumed to be boiled).
- Boiling of presoaked beans: 5.5 MJ of natural gas per kg beans.
- Boiling of rice: 1.7 MJ of natural gas per kg rice.
- Frying: 7.5 MJ of natural gas per kg product [fish, tofu, eggs (50% eggs assumed to be fried)].
- Boiling of tea: 49.5 MJ of natural gas per kg tea.

In addition, water consumption in cooking was taken into account when applicable (e.g., boiling of food) and related waste water treatment (e.g., waste water from bean cooking).

The EoL stage was modeled taking into account both burdens of waste management and benefits of recycling and reuse. The end-of-life phase includes packages, treatment of food scraps and unconsumed foods, together with the human metabolism, modeled according to the method of Muñoz et al. (2007). Specifically, each food product was considered in terms of its nutritional composition (e.g., fiber/carbohydrate/protein) to account for the impacts of human excretion (Ciraolo et al., 1998).

11.2.3 Food waste quantification

In this study, food waste was considered as food that is intended for human consumption but is not consumed, that is, avoidable food waste. The estimation of food waste generated in the different production stages was mainly based on FAO (2011). However, some LCA studies, used as data sources in this study, included also food waste generation from processing (Kendall et al., 2015; Pelletier et al., 2013; Noya et al., 2018), in this case these data were used also in this study. For eggs, chocolate and biscuits, the food waste amount from the household was not available in FAO (2011), instead WRAP (2014) data was used. In case of tofu, the food waste amount was not available at any stage, thus the waste generation was assumed to be equal to that for oilseed and pulses reported by FAO (2011). In case of tea and chocolate, food waste amount in processing and retail was not available, thus FAO (2011) values of cereals (tea) and oilseed and pulses (cocoa bean processing and chocolate retail) were used.

Amounts of the avoidable food waste used in the baseline modeling are presented in Table 11.8. In most of the products, the waste amount from agricultural phase is zero, because usually agricultural losses are taken into account already in estimating the yield, that is, the inputs used in cultivation are per yield that is harvested, but the real yield could be higher if part of the yield is left to the field or lost otherwise.

Emissions from food waste treatment were calculated assuming the average waste treatment scenario for Europe based on Eurostat data (Eurostat, 2014), where 8% of industrial food waste is sent to landfill, 5% is incinerated, and 87% is sent to other recovery treatments (composting and biogas production). In case of household wastes, 59.9% is sent to landfill, 33.3% to energy recovery, and 9.8% to other recovery than energy (Eurostat, 2014). In addition to emissions and benefits from waste treatment (e.g., energy recovery from biogas production), the food waste amount was also taken into account when calculating how much food was actually produced compared with apparent consumption in EU-28. Apparent consumption was assumed to be equal to food consumed in households, and this amount was increased according to amount of wasted food, that is, in the zero waste situation all food produced would be eaten, but when food is wasted along the food chain, the wasted amount increases the food production amount in previous phases. An example of an approach can be seen in Fig. 11.1.

11.2.4 Food waste reduction and dietary shift scenarios

In addition to the average food consumption in the EU, different scenarios were developed to assess the differences in the environmental impacts when the food waste amount is reduced or the diet is changed. The United Nations Sustainable Development Goals has the target 12.3 focusing on food waste reduction. The higher reduction target is on retail and consumer level (50% reduction by 2030), but the aim is to reduce food losses also in other phases (UN, 2017). The food waste reduction scenarios assessed in this study were:

Table 11.8 Avoidable food waste used in the baseline modeling; processing phase includes also postharvest selection

	Agriculture (%)	Processing (%)	Logistics and retail (%)	Household (%)
Meat	0	5	4	11
Fish and shrimp	0	6	9	11
Milk	3.5	1.7	0.5	7
Cheese and butter	3.5	0	3.4	3.4
Eggs	0	1.1	4	6.5
Bread	0	5	2	25
Pasta	0	7	2	25
Rice	0	10	2	25
Sugar	0	0	0	17.2
Sunflower oil	10	6	1	4
Olive oil	0	0	0	13.6
Potatoes	0	0	7	17
Tomatoes	0	5	10	19
Legumes	0	5	1	4
Apples	0	15	10	19
Oranges	0	20	10	19
Bananas	0	5	10	19
Almonds	0	7.5	2	4
Coffee	0	0	0	32
Tea	0	10.5	2	33.3
Beer	0	0	0	10
Wine	0	0.3	1	10
Mineral water	–	0	0	10
Biscuits	0	2.3	2	4.3
Chocolate	0	5	1	4.4
Meat-based dishes	0	3.1	1	24

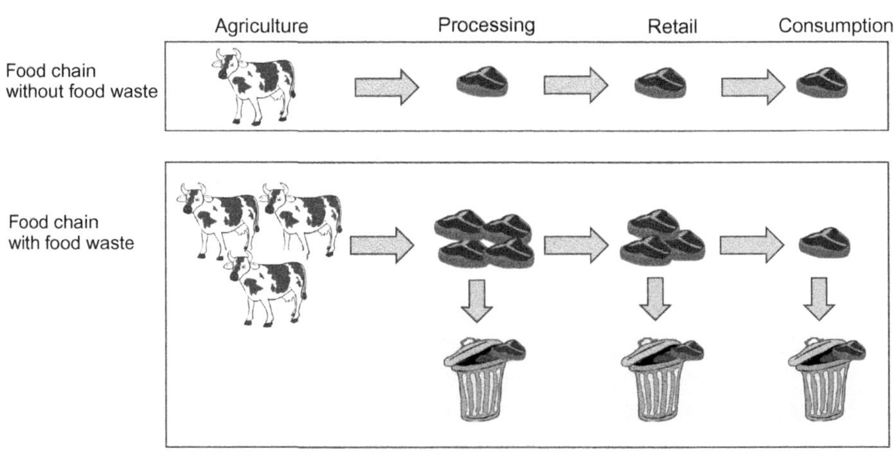

Figure 11.1 Example of the approach used in this study to calculate environmental impacts of food consumption and food waste of an average EU-28 citizen.

1. linear reduction in all phases and all product groups (food waste −50% in all products and all phases),
2. food waste reduction in retail and households (food waste −50% in retail and household phases of all products),
3. food waste reduction in animal-based products (food waste −50% in all phases of all animal-based products), and
4. food waste reduction in plant-based products (food waste −50% in all phases of all plant-based products).

Different diets were selected to see how changes in food products consumed would affect the environmental impact of diet and food waste, because different foods have different food waste intensities and different environmental impact (e.g., meat vs plant proteins). Dietary change scenarios were defined according to literature based on:

1. Swedish nutrition recommendations, which take into account both nutritional value and sustainability aspects of food (Brugård Konde et al., 2015),
2. Mediterranean diet containing less meat and more fish, fruits, and vegetables compared to other diet in Europe (van Dooren et al., 2014), and
3. Vegetarian diet without meat and fish but including eggs and dairy products (van Dooren et al., 2014).

In some cases, it was necessary to adjust the original diet proposals to keep different diets at a reasonable level in terms of energy, protein, fat, and carbohydrate intake, and because representative products included in the BoP food did not cover all recommended food. Due to nature of the different diets, the purpose was not to have exactly the same protein, fat, and carbohydrate intakes with different diets, but to keep them at a reasonable level, for example, to make sure that fat intake from food is not too high related to all food consumed, or that the amount of carbohydrates is high enough.

Swedish nutrition recommendations are based on Nordic Nutrition Recommendations, which set maximum amounts for red meat (max 500 g per week), sugar (max 10% of energy), and alcohol (max 5% of energy), and minimum amounts for fruits and vegetables without potatoes (min 500 g per day) (Brugård Konde et al., 2015). These amounts were used in this study under the assumption that equal amounts of fruits and vegetables are consumed, 250 g per day both. There are also recommended amounts for dairy products (2−5 dL per day), fish (2−3 times per week), and nuts and seeds (couple of tablespoon per day), when amount of milk and cheese was assumed to be 350 g per day (average from recommendation), fish amount 300 g per week (three times 100 g), and almonds 30 g per day. In addition, there is recommendation to use whole-grain pasta and rice, and unsweetened dairy products, which could not be included in our study, because BoP food includes only traditional pasta and rice, and all dairy products are already unsweetened. Recommendation to use healthy oils, for example, rapeseed oil, was included by removing butter from the diet. Amount of vegetable oils was not increased because energy intake from fat was already at the level it should be (35% from total energy intake). The amount of sugar was decreased by 50% compared with the baseline to achieve the target that only 10% of energy intake is from sugar. After decreasing sugar, the carbohydrate and energy intake was very low. Because of that, the amounts of bread and pasta were increased by 50%. Amount of drinks and preprepared food was kept the same as the average diet.

Mediterranean and vegetarian diets were adapted from van Dooren et al. (2014), making some small adjustments as follows. Diets reported in van Dooren et al. (2014) did not include any sugar, biscuits, chocolate, almonds, coffee, or tea, so the amount of those was mainly kept the same as in the baseline. In addition, amount of alcohol and beverages was kept constant to keep different diets more comparable. However, the energy intake was much higher in Mediterranean and vegetarian diets, so the amount of sugar, biscuits, and chocolate was halved in the end. Even after that, energy content of the Mediterranean diet was higher compared with other diets.

In the data sources of all additional diet scenarios (van Dooren et al., 2014; Brugård Konde et al., 2015), the data was not divided between different meat, fish, oils, and fruits as it was done in BoP food. The division between different meat, fish, oil, and fruits was done according to division in average EU consumption, for example, 40.8% of fruits were assumed to be apples, 27.6% oranges, and 31.6% bananas as was in the baseline diet. In addition, tomato was the only vegetable included in the BoP food, although tomatoes represent only 9.5% of vegetables eaten in the EU. The amount of tomatoes was upscaled to represent all vegetable consumption, because in diet scenarios vegetables have a significant role. Energy content of tomatoes is lower compared with many other vegetables, so also the total energy content of whole diets can be slightly lower compared with real diet with varying vegetables. Food amounts, energy, protein, fat, and carbohydrate intakes, as well as energy from sugar and alcohol related to each diet, are presented in Table 11.9.

Total mass of consumed food is very low in the baseline diet, because quite a big part of energy comes from sugar (17%). Energy intake is also lower compared with other diets, but difference from the Swedish recommended diet and vegetarian diet is not very big. Contrarily, the protein intake is highest in the baseline. In fact, protein intake is higher than recommended protein intake, 0.8 g protein per kg of weight, which means 60 g protein per day if the weight is 75 kg (Pendick, 2015), in all other diets except in vegetarian diet, where protein intake is 59.4 g per day. Fat intake is at the same level in all diets, being little higher in the Swedish recommended diet. Carbohydrate intake is much higher in the Mediterranean and the vegetarian diets compared with the other two diets.

11.3 Results

LCIs of BoP food were characterized using EF 2017 midpoint life cycle impact assessment method (Sala et al., 2017b; EC, 2017). This section represents the characterized results of the average EU-28 citizen food consumption as baseline results, and then the results of different scenarios compared with baseline results, related to either food waste reduction or dietary shift.

Table 11.9 Differences between the diet scenarios tested in this study, in terms of amounts per food product. At the end of the table the energy, protein, fat, and carbohydrate intakes as well as the percentage of energy deriving from sugar and from alcohol are reported.

Representative product	Unit	Baseline	Swedish recommendations	Mediterranean diet	Vegetarian diet
Pork meat	g/day	99.6	37.1	15.6	0
Beef meat	g/day	33.7	12.6	5.3	0
Poultry meat	g/day	58.4	21.7	9.1	0
Cod	g/day	21.1	28.9	25.0	0
Salmon	g/day	7.1	9.7	8.4	0
Shrimp	g/day	3.0	4.2	3.6	0
Milk	g/day	187.5	292.2	300	450
Cheese	g/day	37.1	57.8	15	30
Butter	g/day	10.8	0	0	0
Eggs	g/day	33.9	33.9	29	29
Bread	g/day	74.5	111.8	210	210
Pasta	g/day	16.8	25.2	50.4	30.7
Rice	g/day	16.6	16.6	49.6	30.3
Sugar	g/day	64.9	32.4	32.4	32.4
Sunflower oil	g/day	12.3	12.3	23.7	23.7
Olive oil	g/day	11.1	11.1	21.3	21.3
Potatoes	g/day	142.6	142.6	25	117
Tomatoes	g/day	276.0	250	300	200
Beans	g/day	6.9	6.9	75	11
Tofu	g/day	10.6	10.6	4	43
Apples	g/day	26.8	102	102	81.6
Oranges	g/day	18.2	69	69	55.2
Bananas	g/day	20.8	79	79	63.2
Almonds	g/day	1.4	30	1.4	1.4
Coffee	g/day	6.1	6.1	6.1	6.1
Tea	g/day	1.1	1.1	1.1	1.1
Beer	g/day	172.6	172.6	172.6	172.6
Wine	g/day	63.2	63.2	63.2	63.2
Mineral water	g/day	301.6	301.6	301.6	301.6
Biscuits	g/day	17.8	17.8	8.9	8.9
Chocolate	g/day	14.7	14.7	7.4	7.4
Meat-based dishes	g/day	6.7	6.7	0	0
Total mass	Kg/day	1.78	1.98	2.01	1.99
Total energy	Kcal/day	1948	2046	2348	2097
Proteins	g/day	84.2	80.7	77.8	59.4
Fat	g/day	76.7	79.8	73.8	75.7
Carbohydrates	g/day	204	222	312	272
Energy from sugar	%	17	10	7	8
Energy from alcohol	%	6	6	5	6

11.3.1 Baseline results

Characterized baseline results are presented in Table 11.10 divided into the emissions due to food consumption without any food waste (actual food consumption) and emissions caused by avoidable food waste. Emissions caused by food waste include both emissions and benefits from food waste treatment, and emissions due to additional production of food because part of the food is wasted. Emissions due to food waste generation are between 15% and 21% of the total emissions of food consumption of EU average citizen in the year 2015, depending on impact category. The lowest share is in the impact category human toxicity noncancer effect and the highest share is in mineral and metal resource use.

Meat consumption is the biggest contributor in almost all impact categories included in the study, except for human toxicity noncancer effect, ionizing radiation, water use, and mineral and metal resource use (Fig. 11.2). Highest contribution to the human toxicity noncancer effect is due to the consumption of dairy products, especially due to production of feeds. Drinks, including alcoholic and nonalcoholic drinks, coffee, and tea, have the highest contribution to water use, and mineral and metal resource use, because the consumed amount of drinks is high, so the amount of water in drinks and amount of materials used for packaging are also high. In addition, drinks have the highest contribution to ionizing radiation and quite high impact on fossil resource use due to high-energy consumption in the processing. In general, cereal-based products, fish, vegetables, oils, and fruits have low share of impacts in all impact categories. However, ozone depletion potential of fruits can be identified as a hotspot due to the refrigerants used in the transportation and storage of fruits. In case of fish, photochemical ozone formation and freshwater eutrophication have higher share of the impacts compared with other impact categories. High photochemical ozone formation potential is due to the high fuel consumption in cod fishing, and high eutrophication potential is due to the nutrient emissions from salmon aquaculture. Water use of cereal-based products is the second highest after drinks due to the high water consumption in rice cultivation.

In case of environmental impacts of the food waste, agriculture has the highest contribution to most of the impact categories (Fig. 11.3) because there is a need to produce more food when part of the food is wasted. In case of ozone depletion, retail has the highest contribution due to the refrigerants used in the storage of the cold or frozen products. Processing has high contribution to the ozone depletion, ionizing radiation, and resource use impacts due to the additional energy needed for the production of wasted food. In addition, packaging has quite high impact on resource use. EoL impacts, namely the treatment of the wasted food, has very low impact compared with other life cycle phases moreover, the EoL phase takes into account also benefits, for example, energy recovery from biogas process. EoL phase has the highest share in climate impact, 9% of the total climate impact of the wasted food.

11.3.2 Scenario results: food waste reduction

Results of the food waste reduction scenarios compared with the baseline scenario are presented in Fig. 11.4. Food production amount could be almost 10% lower if

Table 11.10 Characterized baseline results of the food consumption of an average EU-28 citizen in 2015 and share between impacts due to food consumption and impacts attributable to avoidable food waste

Impact category	Unit	Impacts due to actual food consumption		Impacts due to avoidable food waste		Total impacts due to food consumption including food waste	
		Value	%	Value	%	Value	%
Climate change	kg CO_2 eq.	1.90E + 3	82	4.08E + 2	18	2.31E + 3	100
Ozone depletion	kg CFC-11 eq.	2.71E−3	80	6.89E−4	20	3.40E−3	100
Human toxicity, noncancer	CTUh	1.45E−3	85	2.59E−4	15	1.71E−3	100
Human toxicity, cancer	CTUh	2.16E−5	83	4.44E−6	17	2.61E−5	100
Particulate matter	Disease incidence	1.94E−4	83	3.92E−5	17	2.33E−4	100
Ionizing radiation	kBq U^{235} eq.	4.02E + 1	80	9.86E + 0	20	5.01E + 1	100
Photochemical ozone formation	kg NMVOC eq.	3.31E + 0	81	7.83E−1	19	4.09E + 0	100
Acidification	molc H^+ eq.	2.66E + 1	83	5.39E + 0	17	3.19E + 1	100
Eutrophication, terrestrial	molc N eq.	1.12E + 2	83	2.25E + 1	17	1.35E + 2	100
Eutrophication, freshwater	kg P eq.	5.40E−1	81	1.27E−1	19	6.67E−1	100
Eutrophication, marine	kg N eq.	1.27E + 1	82	2.77E + 0	18	1.55E + 1	100
Ecotoxicity, freshwater	CTUe	5.42E + 3	82	1.20E + 3	18	6.62E + 3	100
Land use	Pt	1.75E + 5	82	3.73E + 4	18	2.12E + 5	100
Water use	m^3 water eq.	4.16E + 3	81	9.89E + 2	19	5.15E + 3	100
Resource use, fossils	MJ	1.17E + 4	81	2.77E + 3	19	1.45E + 4	100
Resource use, minerals and metals	kg Sb eq.	2.14E−3	79	5.52E−4	21	2.69E−3	100

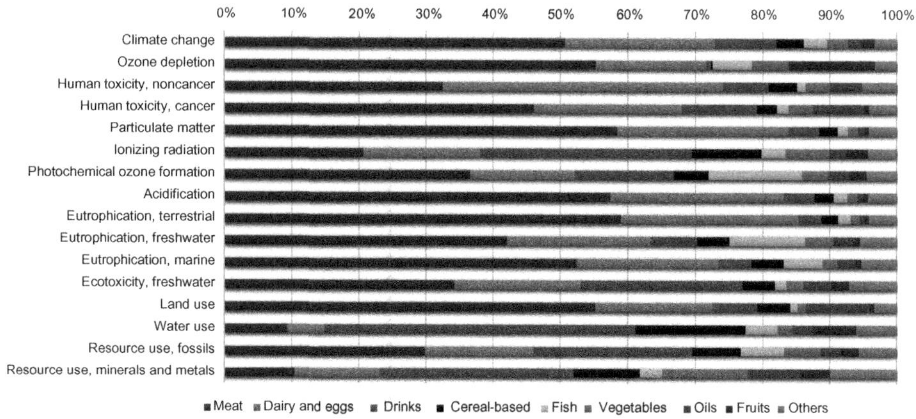

Figure 11.2 Environmental impacts of food consumption of EU-28 average citizen with the breakdown of the contribution of the different product groups.

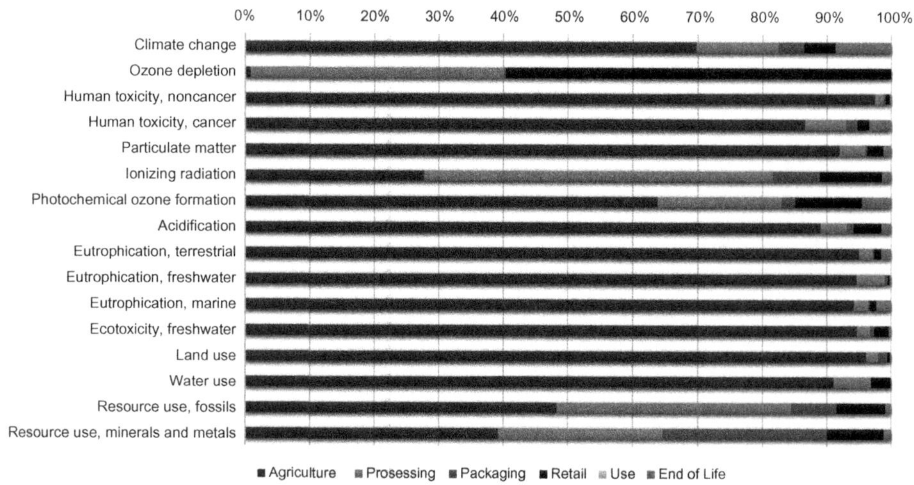

Figure 11.3 Contributions of different life cycle phases to the environmental impacts of food waste.

food waste would be reduced 50% in all product groups and all phases. If the food waste amount would be reduced 50% only in retail and consumer phases, the food production amount could be decreased 8%. Similarly, if food waste is reduced only in animal-based products, the food production amount could be only around 3% lower, whereas if the reduction of food waste is in plant-based products, the food production amount could be decreased by 7%, because plant-based products have in general higher food waste amount along the whole supply chain. Environmental impacts of food consumption by average EU-28 citizen could be decreased from 7% to almost 10%, depending on impact category, by decreasing 50% of avoidable

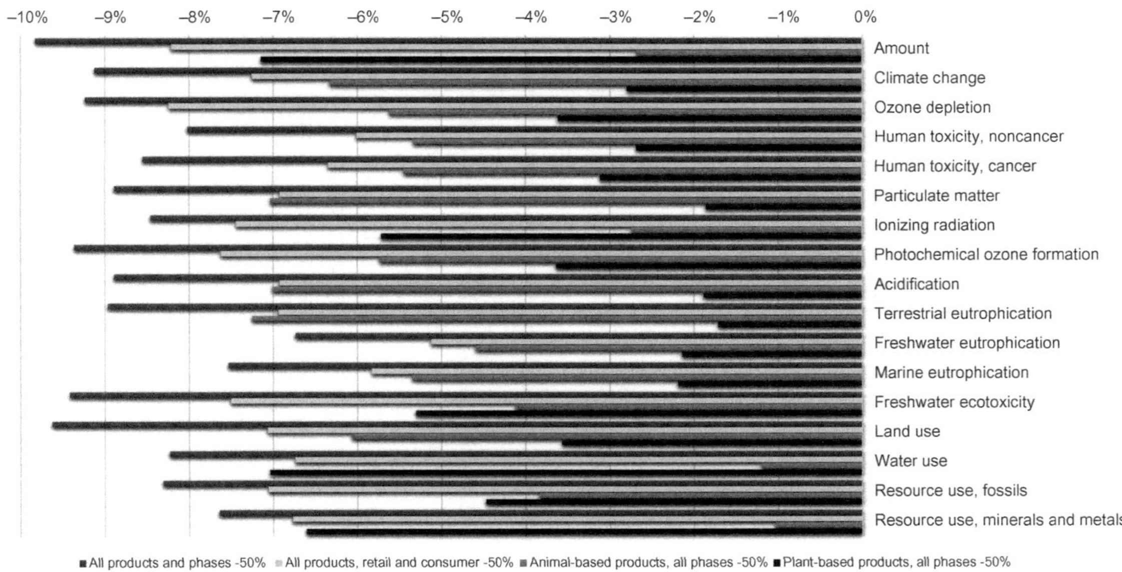

Figure 11.4 Environmental impact reduction compared with baseline when 50% of food waste is reduced under different assumptions: (i) reduced in all life cycle stages of all products; (ii) reduced in the retail and consumption phases of all products; (iii) reduce in all stages of animal-based products; (iv) reduced in all stages of plant-based products.

food waste of all products in all phases of food supply chain, and 5%–8% if the reduction is only in retail and consumer phases of all products. Highest reductions can be achieved in land use, freshwater ecotoxicity, and photochemical ozone formation.

In the majority of the impact categories, the higher impact reduction can be achieved reducing food waste in animal-based products instead of plant-based products (5%–8% and 2%–7%, respectively, depending on impact category), because in the majority of impact categories the animal-based products are the main drivers of environmental impacts. However, in case of ionizing radiation, freshwater ecotoxicity, water use, and resource use (both fossil, and minerals and metals), higher impact reduction potential is achieved when food waste is reduced in plant-based products, because in these impact categories plant-based products are the major drivers of environmental impacts. For example, in case of water use, rice and wine contribute 45% of water use impact of the average food consumption in the EU. For ionizing radiation, the highest share of impact is due to coffee consumption, in particular energy consumption in processing of coffee. In case of resource use, potatoes, wine, and beer have a high share of impacts, but also coffee and animal-based food in case of fossil resources. In case of animal-based products, the highest reduction can be achieved in ozone depletion potential due to cold or frozen storage of all animal-based products.

11.3.3 Scenario results: diet shift

Results of the different dietary scenarios compared with the baseline results are presented in Table 11.11. Environmental impacts in the majority of the impact categories are lower compared with the baseline results when alternative diets are applied. In case of all diet scenarios, the impact on ionizing radiation, water use, and mineral and metal resource use are higher compared with the baseline. In case of the recommended diet in Sweden and the Mediterranean diet, water use impact is 90% and 39% higher, respectively, compared with the average diet in EU. Almond cultivation has high water consumption, and almond consumption is much higher in the Swedish diet compared with the other diets. Rice cultivation has also quite high water use, and rice consumption is higher in the Mediterranean and the vegetarian diets compared with the average diet. Higher mineral and metal resource use in the Mediterranean and the vegetarian diets is mainly due to higher sunflower and olive oil consumption compared with the baseline, when pesticide production associated to agricultural stage of sunflower oil and olive oil packaging are the main contributing processes. Pasta and milk also have quite high resource use impacts, whose consumption amounts are also higher in all alternative diet scenarios. Ionizing radiation is higher in all diet scenarios compared with the baseline due to higher bread consumption (electricity consumption in baking of bread).

When applying the diet according to the Swedish nutrient recommendations, there is also a small increase in photochemical ozone formation, freshwater eutrophication, and fossil resource use. Increase in photochemical ozone formation is due to the higher consumption of wild cod, diesel consumption in cod fishing being

Table 11.11 Impacts due to the different diet scenarios and comparison with the baseline impacts

Impact category	Unit	Swedish dietary recommendations		Mediterranean diet		Vegetarian diet	
		Value	Diff. to baseline (%)	Value	Diff. to baseline (%)	Value	Diff. to baseline (%)
Climate change	kg CO_2 eq.	1.95E + 3	−15	1.57E + 3	−32	1.43E + 3	−38
Ozone depletion	kg CFC-11 eq.	3.35E−3	−1	2.41E−3	−29	2.24E−3	−34
Human toxicity, noncancer	CTUh	1.50E−3	−12	1.18E−3	−31	1.37E−3	−20
Human toxicity, cancer	CTUh	2.13E−5	−18	1.96E−5	−25	1.89E−5	−28
Particulate matter	Disease incidence	1.78E−4	−24	1.34E−4	−43	1.19E−4	−49
Ionizing radiation	kBq U^{235} eq.	5.52E + 1	+10	5.35E + 1	+7	5.13E + 1	+2
Photochemical ozone formation	kg NMVOC eq.	4.21E + 0	+3	3.67E + 0	−10	2.89E + 0	−29
Acidification	molc H^+ eq.	2.46E + 1	−23	1.87E + 1	−41	1.64E + 1	−49
Eutrophication, terrestrial	molc N eq.	1.01E + 2	−25	7.53E + 1	−44	6.56E + 1	−51
Eutrophication, freshwater	kg P eq.	6.95E−1	+4	6.38E−1	−4	4.79E−1	−28
Eutrophication, marine	kg N eq.	1.29E + 1	−17	1.14E + 1	−26	9.26E + 0	−40
Ecotoxicity, freshwater	CTUe	6.48E + 3	−2	5.67E + 3	−14	5.24E + 3	−21
Land use	Pt	1.72E + 5	−19	1.65E + 5	−22	1.43E + 5	−32
Water use	m^3 water eq.	9.79E + 3	+90	7.16E + 3	+39	5.95E + 3	+15
Resource use, fossils	MJ	1.50E + 4	+4	1.38E + 4	−5	1.26E + 4	−13
Resource use, minerals and metals	kg Sb eq.	3.14E−3	+17	3.13E−3	+16	2.93E−3	+9

the main contributing process. Increase in freshwater eutrophication impact is due to higher amount of salmon consumption (nutrient emissions from salmon aquaculture), and increase in fossil resource use is due to higher cheese consumption (energy consumption in cheese processing and in compound feed production).

In general, the vegetarian diet has the highest impact reduction potential compared with other diet scenarios in almost all impact categories. However, in case of human toxicity noncancer effect, the highest reduction, 31%, can be achieved with the Mediterranean diet, due to lower consumption of cheese compared with other diets. Related to the impact categories, whose impacts are higher in alternative diets than in the baseline, the increase of impacts is lower with the vegetarian diet compared with the other two diet scenarios.

When comparing environmental impacts of avoidable food waste in different diets, it can be noted that the share of food waste of total environmental impacts of the food consumption is highest in the Mediterranean diet (Fig. 11.5). This is because the food waste amount is highest for fruit (over 40% along the whole supply chain), and vegetables, bread, pasta, and rice (over 30% along the whole supply chain). Fruit, vegetables, bread, pasta, and rice have higher share in the Mediterranean diet compared with the baseline. These are also mainly higher in the Mediterranean diet compared with other diet scenarios, except fruit consumption amount is the same as the recommended diet in Sweden, and bread consumption amount is same as the vegetarian diet.

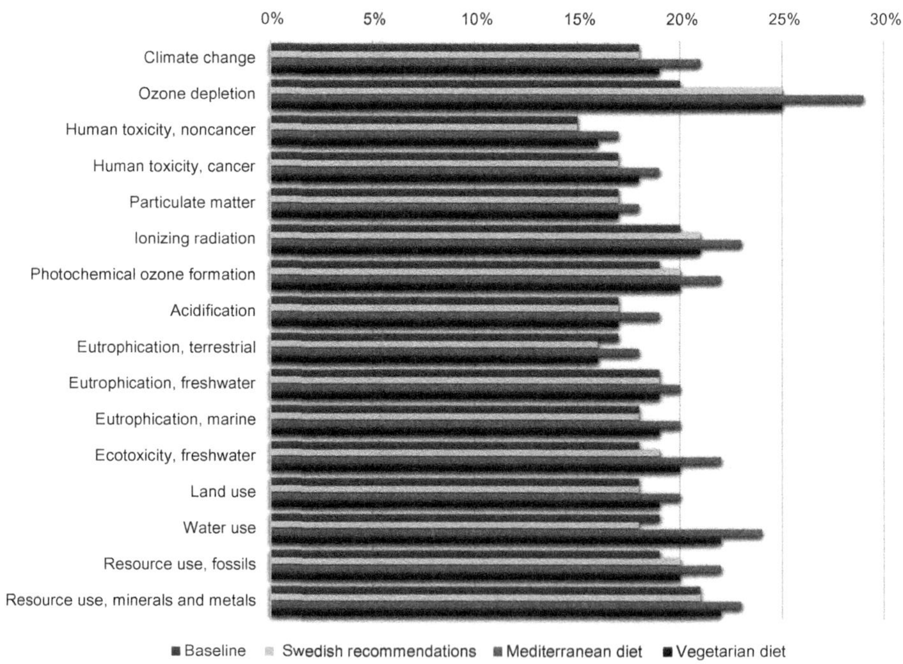

Figure 11.5 Percentage of environmental impacts due to avoidable food waste over the total environmental impacts of food consumption with different diets.

11.4 Discussion

According to this study, a considerable amount of total environmental impacts of food consumption is due to the food waste that could be avoided, ranging between 15% and 21% of the total impact of food consumption. Also Eberle and Fels (2016) and Scherhaufer et al. (2018) reported impacts in simlar ranges, 15%–21% and 15.1%–15.7% of the total food consumption, respectively. However, Eberle and Fels (2016) accounted both in-house and out-of-home food waste, whilst in this study only in-house food wast is accounted for. Environmental impacts due to only in-house food waste were between 11% and 17% in the Eberle and Fels (2016) study, being lower than impacts due to the out-of-house waste or impacts in this study. Beretta et al. (2017) reported 25% climate impact of the consumed food due to the food waste, which is significantly higher compared with this and other published studies: 18% in this study, 15.7% in Scherhaufer et al. (2018), and 15% in Eberle and Fels (2016). Comparison of shares of impacts due to food waste in the different impact categories are reported in Table 11.12. Discrepancies in the results can be due to many factors:

1. different food waste definition and thus different amounts of food waste used in the studies,
2. food products selected for running the study, which may imply different impacts, and
3. differences in the per-kg emissions of the same food products, depending on system boundaries, allocation methods, and used data sources.

Table 11.12 Share of the food waste impact from total impact of the food consumption in different studies

Impact category	This study	Eberle and Fels (2016)	Scherhaufer et al. (2018)
Climate impact (%)	18	15	15.7
Ozone depletion (%)	20	11	–
Particulate matter (%)	17	15	–
Photochemical ozone formation (%)	19	11	–
Acidification (%)	17	17[a]	15.1
Eutrophication, freshwater (%)	19	11	15.2[b]
Eutrophication, marine (%)	18	17	
Land use (%)	18	15	–
Water use (%)	19	16[c]	–
Resource use, fossils (%)	19	12[d]	–
Resource use, minerals and metals (%)	21	15[e]	–

[a]Terrestrial acidification.
[b]Eutrophication, not specified.
[c]Only agricultural water use.
[d]Fossil depletion.
[e]Metal depletion.

In case of environmental impacts of the food waste, agriculture has the highest contribution to the most of the impact categories, because there is a need to produce more food when part of the food is wasted. This is also due to the fact that agriculture is the life cycle stage with higher contributions in most of the impact categories in food LCA (Castellani et al., 2017). Also, according to EEA (2016), agricultural activities for production of food, fibers, and fuel in Europe account for 90% of ammonia emissions, 80% of methane emissions, and 50%–80% of nitrogen load in freshwater bodies. EoL impacts, that is, treatment of the wasted food, have very low contribution compared with other life cycle phases, but the EoL phase also takes into account benefits, for example, energy recovery from biogas process. Consequently the contribution could be higher without benefits.

Food production amount could be almost 10% lower, if food waste would be reduced 50% in all product groups, which is especially important when land resources are limited due to constantly growing population and pressures to use land area also for biofuel production. Similarly, environmental impacts of food consumption by the average EU-28 citizen could be decreased from 7% to almost 10% by decreasing 50% of food waste in all phases of the food supply chain. The highest reduction can be achieved in land use, photochemical ozone formation, and freshwater ecotoxicity. In the majority of the impact categories, the higher impact reduction could be achieved by reducing food waste in animal-based products instead of plant-based products, because animal-based products are the main source of environmental impacts in the majority of the impact categories, although the food waste generation rate is mainly higher with plant-based products (FAO, 2011). This can also be seen in higher reduction in food production amount if the food waste reduction is applied to plant-based products (7% reduction) instead of animal-based products (less than 3% reduction).

In the majority of impact categories, the environmental impacts are lower compared with baseline results when alternative diets are applied. However, ionizing radiation, water use, and mineral and metal resource use are lower with the average diet compared with alternative diet scenarios. In fact, the water use impact of diet according to Swedish dietary recommendations is 90% higher compared with the average diet. This is due to the fact that nutrient recommendations in Sweden contain a considerable amount of nuts and seeds. Only almonds were included in the basket, representing all nuts and seeds included in the diet. However, environmental impacts of different nuts and seeds are different. All nuts and seeds have high water consumption, as was the case with almonds, but for almonds it can be slightly higher (Barilla, 2016). This could have caused an overestimation to the water use impact of diet according to Swedish recommendations, which contains nuts and seeds 30 g per day, while average consumption in EU is only 1.4 g per day.

In general, the vegetarian diet had the highest impact reduction potential compared with other diet scenarios in almost all impact categories, except human toxicity noncancer effects, because the vegetarian diet did not include any meat-based food or fish, which are among the food products contributing most to the environmental impacts of food consumption. Contrarily, the share of avoidable food waste is not lower with the vegetarian diet compared with the baseline, because the vegetarian diet includes significant amounts of fruit and other plant-based food with high food waste amount. In fact, the average diet has the lowest or similar share of impacts with alternative diets due to the food waste in majority of the impact categories, except in water use impact.

Limitations of this study are mainly related to representative products included in the basket. In addition to almonds already discussed to represent all nuts and seeds and thus potentially affecting overestimation on water use impact of diet according to the Swedish recommendations, tomatoes were the only vegetable included in the basket. Tomatoes have the highest consumption among vegetables, but still the share of tomatoes is only 9.5% of all vegetables consumed. For example, share of carrots is 7.5% and share of onions 7.3%. However, all vegetables have in general quite low environmental impacts, but some vegetables can have specific hotspot impact areas that are not covered when using only tomatoes to represent all vegetables. Pesticide usage in the cultivation of tomatoes, in turn, can be higher compared with some other vegetables, which can cause overestimation in ecotoxicity impacts.

There are also limitations related to the accuracy of food intake and the food waste amounts. The food intake was calculated according to apparent consumption in EU-28 (production + import−export) and food waste amounts along the food supply chain. Apparent consumption was assumed to be the consumed food amount, that is, food waste amounts were added as additional food that has to be produced. However, food waste amounts were based on different studies, which includes uncertainty.

Selected data sources and assumptions made in the assessment also influences the results. Data used in the assessment were based on the literature that was assessed to be the most reliable and representative in terms of data quality and geographical scope. For example, the study by Blengini and Busto (2009) was selected as a data source on rice cultivation representing rice cultivation in Italy, which has the highest rice production in Europe. In that study, water use was 19,800 m^3/ha, based on average from different studies and reusage rate of 28%, thus rice is cultivated in flooded field and part of this water can be reused. This was much higher compared with the study by Chapagain and Hoekstra (2010), which takes into account water uptake of the rice and available rainwater in the area, that is, amount of rainwater was deducted from the irrigation water needed. Thus, the assumption of Chapagain and Hoekstra (2010) was that all "additional" water, which is not taken up by the plant, can be reused later somewhere else. According to that, water consumption was 4087 m^3/ha. The latter approach was used also in this study. This selection has clear impact on the water use results.

The main purpose of this study was to assess how much environmental impacts of food consumption will change with different dietary habits, when the amount of food waste is changing, because food waste amount of different food products is different. The potential next step would be to find the most favorable diet in terms of food waste amount, environmental impact, and nutrient intake, that is, optimization of diet related to three different factors.

11.5 Conclusions

According to this study, between 15% to 21% of total environmental impact of food consumption is due to avoidable food waste. Food production amount could be decreased by almost 10%, if food waste would be reduced 50% within all product groups and all phases of the food supply chain. Similarly, environmental impacts of food consumption by the average EU-28 citizen could be decreased from 7% to almost 10%, depending on the impact category, by decreasing 50% of the food waste in all phases. The highest reduction can be achieved in land use, photochemical ozone formation, and freshwater ecotoxicity potential. In the majority of the impact categories, the higher impact reduction could be achieved by reducing food waste in animal-based products instead of plant-based products, since in the majority of impact categories the animal-based products are the main sources of environmental impacts.

In the majority of impact categories, the environmental impacts are lower compared with the baseline results when alternative diets are applied. Environmental impacts in some impact categories (ionizing radiation, water use, and mineral and metal resource use) are higher with alternative diet scenarios. Water use impact is especially high with the diet according to the Swedish dietary recommendations, which contains a high amount of nuts and seeds which are related to an high demand of water for irrigation. In general, the vegetarian diet has the highest impact reduction potential compared with other diet scenarios in almost all impact categories, because the vegetarian diet did not include any meat-based foods, which have high environmental impacts per kg of food. Contrarily, the share of avoidable food waste is not lower with the vegetarian diet compared with the baseline, because the vegetarian diet includes significant amounts of fruit and other plant-based food with high food waste rate. In fact, the average diet had mainly lower or similar environmental impacts with alternative diets due to avoidable food waste.

References

Alexander, P., Brown, C., Arneth, A., Finnigan, J., Moran, D., Rounsevell, M.D., 2017. Losses, inefficiencies and waste in the global food system. Agric. Syst. 153, 190–200.

Barilla, 2016. BCFN database for double pyramid. Barilla Center for Food & Nutrition. <https://www.barillacfn.com/en/> (accessed 15.05.18.).

Bartzas, G., Vamvuka, D., Komnitsas, K., 2017. Comparative life cycle assessment of pistachio, almond and apple production. Inform. Process. Agric. 4, 188–198.

Beretta, C., Stucki, M., Hellweg, S., 2017. Environmental impacts and hotspots of food losses: value chain analysis of Swiss food consumption. Environ. Sci. Technol. 51 (19), 11165–11173.

Blengini, G.A., Busto, M., 2009. The life cycle of rice: LCA of alternative agri-food chain management systems in Vercelli (Italy). J. Environ. Manage. 90, 1512–1522.

Blonk Consultants, 2014. Agri-footprint 2.0. Description of data. <http://www.agri-footprint.com/wp-content/uploads/2016/08/Agri-footprint-2.0-Part-2-Description-of-data.pdf> (accessed 05.03.18.).

Bovea, M.D., Cabello, R., Querol, D., 2007. Comparative life cycle assessment of commonly used refrigerants in commercial refrigeration systems. Int. J. Life Cycle Assess. 12 (5), 299–307.

Brugård Konde, Å., Bjerselius, R., Haglund, L., Jansson, A., Pearson, M., Sanner Färnstrand, J., et al., 2015. Swedish dietary guidelines – risk and benefit management report. Livsmedelsverket. National Food Agency. Rapport 5. <https://www.livsmedelsverket.se/globalassets/rapporter/2015/rapp-hanteringsrapport-engelska-omslag–inlaga–bilagor-eng-version.pdf> (accessed 11.04.18.).

Bräutigam, K.R., Jörissen, J., Priefer, C., 2014. The extent of food waste generation across EU-27: different calculation methods and the reliability of their results. Waste Manage. Res. 32 (8), 683–694.

Caldeira, C., Corrado, S., Sala, S., 2017. Food waste accounting. Methodologies, challenges and opportunities. JRC Technical Reports. EUR 28988 EN. Publications Office of the European Union, Luxemburg.

Cao, L., Diana, J.S., Keoleian, G.A., Lai, Q., 2011. Life cycle assessment of Chinese shrimp farming systems targeted for export and domestic sales. Environ. Sci. Technol. 45, 6531–6538.

Castellani, V., Fusi, A., Sala, S., 2017. Consumer footprint. Basket of products indicator on food. JRC Technical Reports. EUR 28764 EN. Publication Office of the European Union, Luxemburg.

Cellura, M., Longo, S., Mistretta, M., 2012. Life Cycle Assessment (LCA) of protected crops: an Italian case study. J. Clean. Prod. 28, 56–62.

Chapagain, A.K., Hoekstra, A.Y., 2010. The green, blue and grey water footprint of rice from both a production and consumption perspective. Value of Water. Research Report Series No. 40. UNESCO-IHE. Institute for Water Education.

Ciraolo, L., Giaccio, M., Morgante, A., Riganti, V., 1998. Handbook of Commodity Science. Monduzzi Editore, Bologna.

Corrado, S., Ardente, F., Sala, S., Saouter, E., 2017. Modelling of food loss within life cycle assessment: from current practice towards a systematisation. J. Clean. Prod. 140, 847–859.

Corrado, S., Sala, S., 2018. Food waste accounting along global and European food supply chains: state of the art and outlook. Waste Manage. 79, 10–131.

Crenna, E., Sinkko, T., Sala, S., 2019. Biodiversity impacts due to food consumption in Europe. J. Clean. Prod. (submitted).

de Beaufort-Langeveld A.S.H., Bretz R., van Hoof G., Hischier R., Jean P., Tanner T., et al., 2003. Code of life-cycle inventory practice. SETAC. <http://www.setac.org> (accessed 01.06.16.).

Dole, 2011. Water recycling programs for banana packing. Water management. Dole corporate responsibility and sustainability. <http://dolecrs.com/sustainability/water-management/water-recycling-programs-for-banana-packing/> (accessed 11.04.18.).

Eberle, U., Fels, J., 2016. Environmental impacts of German food consumption and food losses. Int. J. Life Cycle Assess. 21 (5), 759–772.

EC, 2015. Communication from the Commission to the European Parliament, the Council, the European Economic and Social Committee and the Committee of the Regions. Closing the loop – an EU action plan for the circular economy. COM (2015) 614.

EC, 2016. Communication from the Commission to the European Parliament, the Council, the European Economic and Social Committee and the Committee on Regions. Next steps for a sustainable European future. European action for sustainability. COM (2016) 739 final.

EC, 2017. PEFCR guidance document – guidance for the development of Product Environmental Footprint Category Rules (PEFCRs), version 6.3, December 2017. <http://ec.europa.eu/environment/eussd/smgp/pdf/PEFCR_guidance_v6.3.pdf> (accessed July 2018).

EC, 2018. Communication from the Commission to the European Parliament, the Council, the European Economic and Social Committee and the Committee of the Regions on a monitoring framework for the circular economy. COM (2018) 29 final.

EEA, 2016. From production to waste: the food system. <https://www.eea.europa.eu/signals/signals-2014/articles/from-production-to-waste-food-system> (accessed 13.09.18.).

EFSA, 2018. The Comprehensive Food Consumption database. European Food Safety Authority. <http://www.efsa.europa.eu/en/food-consumption/comprehensive-database> (accessed 22.03.18).

Ellingsen, H., Olaussen, J.O., Utne, I.B., 2009. Environmental analysis of the Norwegian fishery and aquaculture industry – a preliminary study focusing on farmed salmon. Mar. Policy 33, 479–488.

Eurostat, 2018. Statistics on production of manufactured goods (Prodcom). <http://ec.europa.eu/eurostat/data/database> (accessed 08.02.18).

Eurostat, 2014. Treatment of waste. Code: env_wastrt. <http://ec.europa.eu/eurostat/web/waste/data/database> (accessed 25.11.14).

FAO, 2010. Biodiversity and Sustainable Diets United Against Hunger. International Scientific Symposium. FAO Headquarters, Rome. Available at: http://www.fao.org/ag/humannutrition/29186-021e012ff2db1b0eb6f6228e1d98c806a.pdf. (accessed 25.01.2019).

FAO, 2011. Global Food Losses and Food Waste – Extent, Causes and Prevention. FAO, Rome.

Foster, C., Green, K., Bleda, M., Dewick, P., Evans, B., Flynn, A., et al., 2006. Environmental Impacts of Food Production and Consumption: A Report to the Department for Environment, Food and Rural Affairs. Manchester Business School, Defra, London.

FUSIONS, 2016. Estimates of European food waste levels.

Godfray, H.C.J., Beddington, J.R., Crute, I.R., et al., 2010. Food security: the challenge of feeding 9 billion people. Science 327 (5967), 812–818.

IPCC, 2006. N_2O emissions from managed soils and CO_2 emissions from lime and urea application. IPCC 4, 1–54. Chapter 11.

Iriarte, A., Almeida, M.G., Villalobos, P., 2014. Carbon footprint of premium quality export bananas: case study in Ecuador, the world's largest exporter. Sci. Total Environ. 472, 1082–1088.

Jefferies, D., Muñoz, I., Hodges, J., King, V.J., Aldaya, M., Ercin, A.E., et al., 2012. Water Footprint and Life Cycle Assessment as approaches to assess potential impacts of products on water consumption. Key learning points from pilot studies on tea and margarine. J. Clean. Prod. 33, 155–166.

Kendall, A., Marvinney, E., Brodt, S., Zhu, W., 2015. Life cycle-based assessment of energy use and greenhouse gas emissions in almond production, part I. Analytical framework and baseline results. J. Industr. Ecol. 19 (6), 1008–1018.

Lundqvist, J., de Fraiture, C., Molden, D., 2008. Saving Water: From Field to Fork – Curbing Losses and Wastage in the Food Chain. SIWI Policy Brief, Stockholm.

McClelland, S.C., Arndt, C., Gordon, D.R., Thoma, G., 2018. Type and number of environmental impact categories used in livestock life cycle assessment: a systematic review. Livestock Sci. 209, 39–45.

Meier, T., Christen, O., Semler, E., Jahreis, G., Voget-Kleschin, L., Schrode, A., et al., 2014. Balancing virtual land imports by a shift in the diet. Using a land balance approach to assess the sustainability of food consumption. Germany as an example. Appetite 74, 20–34.

Mejia, A., Harwatt, H., Jaceldo-Siegl, K., Sranacharoenpong, K., Soret, S., Sabate, J., 2017. Greenhouse gas emissions generated by tofu production: a case study. J. Hunger Environ. Nutr. 1–12.

Mekonnen, M.M., Hoekstra, A.Y., 2011. The green, blue and grey water footprint of crops and derived crop products. Hydrol. Earth Syst. Sci. 15, 1577–1600.

Monier, V., Mudgal, S., Escalon, V., O'Connor, C., Gibon, T., Anderson, G., et al., 2010. Preparatory study on food waste across EU 27. Report for the European Commission [DG ENV—Directorate C].

Muñoz, I., Mila i Canals, L., Clift, R., Doka, G., 2007. A simple model to include human excretion and wastewater treatment in life cycle assessment of food products. CES Working Papers, 01/07. Centre for Environmental Strategy, University of Surrey.

Nielsen, P.H., Nielsen, A.M., Weidema, B.P., Dalgaard, R., Halberg N., 2003. LCA food database. Available at: http://www.lcafood.dk/. (accessed 16.05.18).

Notarnicola, B., Tassielli, G., Renzulli, P.A., Castellani, V., Sala, S., 2017. Environmental impacts of food consumption in Europe. J. Clean. Prod. 140, 753–765.

Noya, L.I., Vasilaki, V., Stojceska, V., Gonzalez-Garcia, S., Kleynhans, C., Tassou, S., et al., 2018. An environmental evaluation of food supply chain using life cycle assessment: a case study on gluten free biscuit products. J. Clean. Prod. 170, 451–461.

Ntiamoah, A., Afrane, G., 2008. Environmental impacts of cocoa production and processing in Ghana: life cycle assessment approach. J. Clean. Prod. 16, 1735–1740.

Pelletier, N., Tyedmers, P., Sonesson, U., Scholz, A., Ziegler, F., Flysjo, A., et al., 2009. Not all salmon are created equal: life cycle assessment (LCA) of global salmon farming system. Environ. Sci. Technol. 43, 8730–8736.

Pelletier, N., Ibarburu, M., Xin, H., 2013. A carbon footprint analysis of egg production and processing supply chains in the Midwestern United States. J. Clean. Prod. 54, 108–114.

Pendick, D., 2015. How much protein do you need every day? Harvard Health Publishing. <https://www.health.harvard.edu/blog/how-much-protein-do-you-need-every-day-201506188096> (accessed 18.04.18).

Recanati, F., Marveggio, D., Dotelli, G., 2018. From beans to bar: a life cycle assessment towards sustainable chocolate supply chain. Sci. Total Environ. 613–614, 1013–1023.

Sala, S., Anton, A., McLaren, S., Notarnicola, B., Saouter, E., Sonesson, U., 2017a. In quest of reducing the environmental impacts of food production and consumption. J. Clean. Prod. 140 (2), 387–398.

Sala, S., Benini, L., Castellani, V., Vidal Legaz, B., Pant, R., 2017b. Environmental footprint – update of life cycle impact assessment methods. Resources, water, land and particulate matter. JRC technical report.

Scherhaufer, S., Moates, G., Hartikainen, H., Waldron, K., Obersteiner, G., 2018. Environmental impacts of food waste in Europe. Waste Manage. 77, 98–113.

Sonesson, U., Cederberg, C., Flysjö, A., Carlsson, B., 2008. Livscykelanalysis (LCA) av svenska ägg (ver.2). SIK-rapport Nr 783 2008.

Svanes, E., Vold, M., Hanssen, O.J., 2011. Environmental assessment of cod (*Gadus morhua*) from autoline fisheries. Int. J. Life Cycle Assess. 16, 611–624.

Svanes, E., Aronsson, K.S., 2013. Carbon footprint of a Cavendish banana supply chain. Int. J. Life Cycle Assess. 18, 1450–1464.

Tisserant, A., Pauliuk, S., Merciai, S., Schmidt, J., Fry, J., Wood, R., et al., 2017. Solid waste and the circular economy: a global analysis of waste treatment and waste footprints. J. Industr. Ecol. 21 (3), 628–640.

Torrellas, M., Anton, A., Lopez, J.C., Baeza, E.J., Parra, J.P., Muñoz, P., et al., 2012. LCA of a tomato crop in a multi-tunnel greenhouse in Almeria. Int. J. Life Cycle Assess. 17, 863–875.

UN, 2017. Sustainable development goals. <https://sustainabledevelopment.un.org/sdg12> (accessed 09.10.17).

Vanderheyden G., Aerts J., 2014. Comparative LCA assessment of Fontinet filtered tap water vs. natural sourced water in a PET bottle. Final report 04/03/2014. Futureproofed.

van Dooren, C., Marinussen, M., Blonk, H., Aiking, H., Vellinga, P., 2014. Exploring dietary guidelines based on ecological and nutritional values: a comparison of six dietary patterns. Food Policy 44, 36–46.

van Holsteijn, F., Kemna, R., Lee, P., Sims, E., 2017. Optimal food storage conditions in refrigeration appliances. Preparatory/review study on Commission Regulation (EC) No. 643/2009 and Commission Delegated Regulation (EU) No. 1060/2010 – complementary research on Optimal food storage conditions in refrigeration appliances.

Wolff, A., Gondran, N., Brodhag, C., 2017. Detecting unsustainable pressures exerted on biodiversity by a company. Application to the food portfolio of a retailer. J. Clean. Prod. 166, 784–797.

WRAP, 2014. Household food and drink waste: a product focus. Final report. Waste & Resources Action Programme.

Suboptimal food? Food waste at the consumer–retailer interface

12

Jessica Aschemann-Witzel[1], Ilona E. de Hooge[2] and Valérie L. Almli[3]
[1]MAPP Centre, Department of Management, Aarhus School of Business and Social Sciences, Aarhus University, Aarhus, Denmark, [2]Marketing and Consumer Behaviour group, Wageningen University, Wageningen, The Netherlands, [3]Sensory and Consumer Science Department, Nofima, Ås, Norway

Chapter Outline

12.1 Why a lot of food waste is due to suboptimal food

Food waste is commonly defined as food for human consumption but due to some reason lost or wasted and potentially used for other purposes at some point in the supply chain (FAO, 2011; Fusions, 2015). When food waste occurs closer to the consumption stage, the waste more often concerns items that are ready to be eaten or used by consumers. In reports or in literature, it might quite often be mentioned that the wasted food had been "perfectly edible." However, for food supply chain actors and consumers it does not make sense to waste "perfect food." Therefore, the occurrence of food waste can be mostly attributed to a situation or food not being as optimal as desired, and/or procedures not going according to plan.

In the early stages of the supply chain and in particular in emerging or developing countries, it is primarily the procedures that are not optimal, due to shortcomings of and inefficiencies in harvesting, storage, and transportation (Parfitt et al., 2010). A share of the harvest can be lost in the field because machinery is not efficient, foods can spoil due to shortcomings in storage conditions, or can be lost during transport because the roads are in a bad state or the transport distances are too long. Examples of these are fruits that perish in the field due to being exposed to too much sunlight, or damaged during transportation in the truck (Henz, 2017).

Saving Food. DOI: https://doi.org/10.1016/B978-0-12-815357-4.00012-2

In developed countries, the technology is more optimized, but then the food itself is not optimal enough compared with the standards and requirements of the processing facilities and the esthetic expectations of retailers and customers. In addition, it can be the case that the item would be too costly to be changed into a product worth selling, given that there is sufficient alternative raw material in a better state, and given the low prices of commodities. For example, potatoes can be too small to be processed, or vegetables can have such deviating shapes that they do not meet storage and transport standards or the expectations of consumers (Stuart, 2009), such as the famous bent cucumber.

In the later stages of the supply chain, it is more likely that the procedures are not optimal, with a lack of optimal storage conditions in stores or consumer households. Also, packages get mislabeled by the processor or dented in wholesale (Raak et al., 2017), retailers order more units than required (Eriksson et al., 2017), and consumers buy items that are subsequently incorrectly treated or not used as planned. Consequently, the product that was perfectly edible at purchase has become not so perfect over time. Or, it is perfectly edible but there is an even better item: it is not as optimal in some way or other as a competing item of the same price in store, or as the item of the same type that awaits use in a household's fridge or cupboard.

12.2 Definition of suboptimal food and the consumer–retailer interface

Food being in essence perfectly edible but not as optimal as other available food or not regarded as optimal as desired by a member of the supply chain and in particular by the end-consumer is a major reason why food is wasted. Accordingly, such imperfect or *suboptimal food* can be defined as foods that "deviate from normal or optimal products" in a number of ways, without food safety or the item's intrinsic quality being affected (De Hooge et al., 2017, p. 81). This suboptimality can in particular but not exclusively be in terms of (1) appearance in for example, shape, size, or weight; in terms of (2) the time frame in which the food can still be used, determined by, for example, its state of ripeness or the current date being close to or beyond the indicated date label; or in terms of (3) the status of its packaging being, for example, mislabeled, torn, or dented.

A large share of food waste in developed countries is caused by households (EC, 2010; FAO, 2013). These countries are in a state of economic prosperity and most households show high affluence levels. There is no scarcity of food supply and there is a great diversity of foods offered. The majority of consumer households can not only secure sufficient food with their available income, but also the share of income they use for food is about a tenth of their budget only. Thus, for the occurrence of food waste it is not so crucial whether the food is actually edible or not, but rather in which status of optimality or suboptimality the food is. Given that the greater share of food waste is caused by consumer households, the decision of using or discarding a food regarded as suboptimal is to a great extent taken by consumers. However, consumer choices in terms of food waste are influenced by actions and choices of the food supply chain, and actions in the food supply chain concerning suboptimal food are influenced by consumer choices, or at least by what the supply chain anticipates the consumer would do. The place where both parts "meet" and interact is typically the retailer, that is, the supermarkets where the consumers purchase foods. This crucial interaction determining food waste up and down the supply chain is therefore the so-called *consumer–retailer interface*.

Examples of interactions between retailers and consumers are consumers selecting the fruits and vegetables with the best appearance from the piles offered in store, and retailers are consequently demanding high esthetic standards from their suppliers. The uniform and "perfect" appearance of the items offered, in turn, also shapes consumer's expectations over time, determining what consumers assume is the adequate and normal appearance of a fruit or a vegetable. Retailers offer lower prices for purchases of larger quantities and of unit sizes with a variety of pricing strategies. This leads consumers to potentially buy more than they need, resulting in food being wasted due to having passed the optimal usage stage in terms of ripeness or date. Consumers, in turn, are often price-sensitive and deal prone, and many consumers favor price-oriented shopping formats, discount-level private labels, and price promotions during shopping. These consumer behaviors motivate retailers to use such approaches. However, producers, processors, and retailers are via their decisions also setting the boundaries for what consumers can decide about in their food choices. For example, the fact that certain misshapen vegetables that the farmers grow do not reach the store in the first place, inhibits any consumer who would actually buy suboptimal foods. The decision of food processors on which date label to print on a package of, for example, pasta—6 months, or maybe 12 months?—might be decisive for the situation much later in time when a consumer checks the cupboards and throws anything past the date in the bin. Only when mislabeled packages or foods in torn or dented packages are nevertheless sold in some way or other—whether in the original store or in some alternative format—can consumers buy and use such suboptimal foods.

12.3 Types of suboptimal food and respective decisions on optimality

It follows from the definition of suboptimal food and from the occurrence of food waste as caused by consumers that there are two crucial moments where consumers make a decision on the optimality or suboptimality of food. The first is before *purchase* and in-store, when consumers make a choice on which items to purchase. The second moment is after purchase and at home, when consumers make a choice on whether and which items to use, thus on *consumption*. In the first case, any item not chosen by any consumer might end up being discarded, and the food waste is accounted for as being caused by the retailer. In the second case, any item not chosen by the consumers in the household might end up being discarded, and the burden of food waste is accounted for as consumer household food waste. The suboptimality of the food might be of any type, but the most often described and cited categories are suboptimality in terms of the *date label*, thus a textual description of the optimality, and the sensory perception of the status of the actual food in *appearance* or any other sensory perception, or in terms of the status of the *packaging* of the food (see Fig. 12.1 for an overview).

This distinction underlines that consumer–related food waste—defined as food that is wasted because of the consumer—is linked to firstly, the purchase decision, and secondly, to the consumption decision. In addition, it underlines that primarily consumer perceptions of and decisions based on date label, appearance, and packaging lead to waste of food otherwise perfectly edible.

The definition of suboptimal food assumes that food safety is not affected or that the item's intrinsic quality is not reduced to any great extent. However, consumers assess certain suboptimal foods to be unsafe even if they are not, or they somehow fear or are uncertain about the safety, and therefore rather not purchase the item or discard it. A lot of research shows that consumers have difficulties understanding and handling date labels and assessing whether foods are still edible (Van Boxstael et al., 2014). Foods that are in fact unsafe to eat are not suboptimal anymore, but have likely been at a stage previously where they were only suboptimal. An example of this is fish past the use-by date, which should not be eaten due to food safety reasons, but which had been perfectly edible before it got to that stage. A food's intrinsic quality refers to the characteristics inherent to the food, such as taste or healthiness (Grunert, 2007). Food quality is understood to be composed of a range of dimensions (Grunert, 2005; Oude Ophuis and van Trijp, 1995), and the degree of quality is assessed on a continuum.

The categories of food suboptimality and the distinction of the purchase versus the consumption situation underline an important observation: a considerable share of suboptimal food is food that has become suboptimal after it was optimal, due to choices and circumstances in the supply chain. Consumer research shows that consumers at times find themselves in the situation where they move food from a stage of suboptimality to inedibility, and then discard the food. An example of this is a banana with a spot, which might remain in the fruit bowl until it has become so intensely brown or black that it is necessary to discard it.

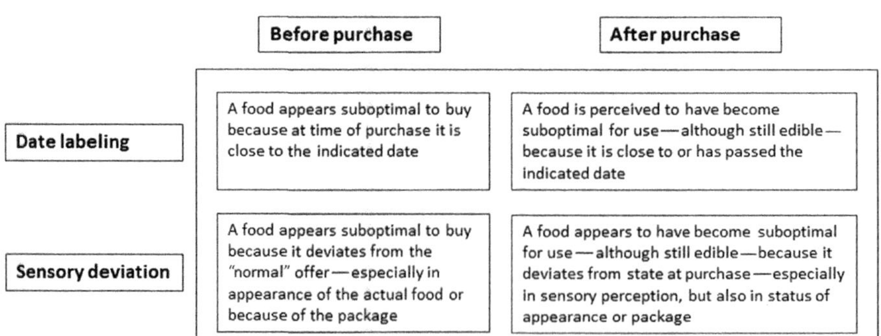

Figure 12.1 Decisions on optimal versus suboptimal food by consumers.
Source: Own.

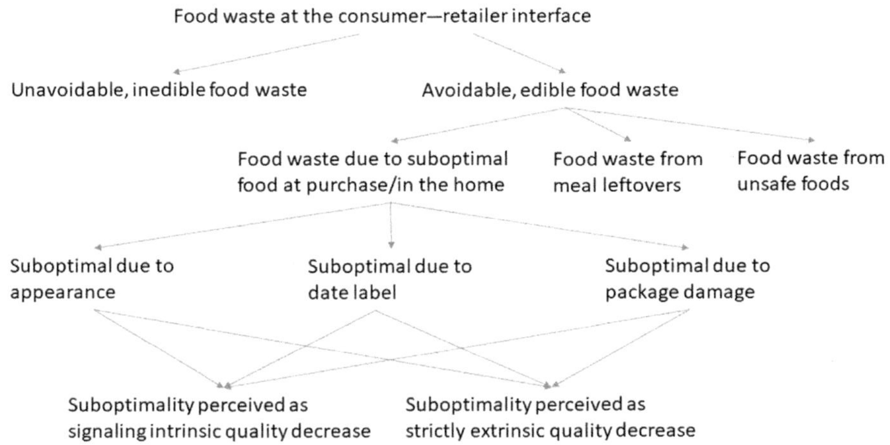

Figure 12.2 Suboptimal food definition and its major categories within the broader context of food waste.
Source: Own. Adapted from Aschemann-Witzel, J., Giménez, A., Ares, G., 2018b. Consumer in-store choice of suboptimal food to avoid food waste: the role of food category, communication and perception of quality dimensions. Food Qual. Preference, 68, 29–39. Available from: https://doi.org/10.1016/j.foodqual.2018.01.020.

The definition of suboptimal food only concerns foods that are not used or cooked yet, and it does not concern leftovers from prepared food and meals. The reason is that optimality refers to whether the food is optimal enough to be used, and use of the food is the consumption in terms of eating it or using it for preparing a meal. For leftovers, the definition does not apply, although leftovers can of course be in more or less good status for being reused. The process of becoming inedible over time happens for leftovers as well, though, as consumers might store leftovers in the fridge or freezer, until it can be discarded more easily, that is, with less bad consciousness (Evans, 2014), as it is not perfectly edible anymore.

Fig. 12.2 sets the definition of suboptimal food into context, showing the most typical categories and the process of becoming suboptimal through the decisions that consumers make. The figure also shows that even though the food is fine and fit for purchase or consumption, both intrinsic and extrinsic quality dimensions might be assumed to be or perceived to be affected. This happens along a continuum between optimality and suboptimality.

12.4 Types of interactions at the consumer–retailer interface

In principal, the interaction between consumers and retailers consists of the retailers offering food physically in the store, and of the retailer communicating about the food and assortment offered. Consumers react to this communication and offer of food by visiting the store, selecting or deselecting food for purchase, and returning for repurchase. Consumers might also communicate—back to the retailer as well as to other consumers—about their experience with the store and about the assortment. In this interaction in the interface between consumers and retailers, a number of relations are suspected to cause food waste.

With regard to the causes of food waste that relates to the consumer–retailer interface, it is useful to first consider the general factors causing food waste at the consumer level. Extensive research on consumers and food waste and a number of reviews have identified and categorized these factors (Aschemann-Witzel et al., 2015; Hebrok and Boks, 2017; Priefer et al., 2016; Quested et al., 2013). The multitude of factors causing food waste at the consumer level can be boiled down to six clusters of factors, of which the first three are tied to the consumer side, and the last three are tied to the context (Aschemann-Witzel et al., 2015). The first consumer side cluster is about consumer *motivation* to avoid food waste. This cluster concerns consumers being aware and knowledgeable about the food waste issue and about its consequences, the extent to which they hold underlying values, beliefs and attitudes such as environmental concern, perception of social unfairness of food wastage, or a dislike of wasting own monetary resources, which all can play a role in shaping a motivation to avoid food waste in one or the other form. The second consumer side cluster is about the existence of goal conflicts between various consumer motives to buy food related to what the food is used for and the role that it plays in a consumer's life. These conflicts lead to potentially necessary *trade-offs* between the goal of not wasting food and other goals, such as food safety, convenience, sensory experience, or health. As a third consumer side cluster, consumers can possess the knowhow, skills, and *capabilities* to purchase, store, handle, and use food in a way that aims to solve the conflicts and trade-offs between different consumer goals. Yet, oftentimes consumers lack such capabilities, resulting in consumer-related food waste.

The first context-side cluster of factors relates to issues that have to do with the consumer's *social influence* and background, such as upbringing, the family, friends, and neighbors, or the further social surrounding of the consumer in question. Consumers want to fulfill the needs of their loved ones, show affection via the food and meals they provide, or signal acknowledgment, respect, or status to guests with the food that they offer on social occasions. At the same time, consumers might also feel embarrassed by not being able to offer sufficient food for guests, but also when rummaging through and purchasing "shabby" suboptimal food, or if observed wasting food. The second cluster of factors encompasses the concrete *purchase situation*, which is heavily determined by how supermarkets or other stores design their offer, organize the store management, or the market practices enacted in the store. The layout of the store, the type of foods offered, the pricing mechanism applied as well as the packaging all influence consumers food choices for optimal or suboptimal food, or the likelihood that they at some point waste some of the food they have purchased (e.g., if no single-size packages are available, or the food did not have the expected shelf-life, etc.). Third and finally, the consumer's context is shaped by the *macroenvironment*, which means that the political directions, the legal requirements, the societal trends, the technological possibilities, environmental issues, and economic prosperity impact the context in which the consumer and the retailer act. As various reviews on the causes of food waste underline (Aschemann-Witzel et al., 2015; Hebrok and Boks, 2017; Priefer et al., 2016; Quested et al., 2013), these factors are interrelated and thus do not exert an influence on consumer level food waste separately of the other factors. Fig. 12.3 visualizes the three consumer-side and the three context-side cluster of factors, and underlines that these are interrelated.

With these clusters of causes of food waste as background knowledge, a number of interactions at the consumer–retailer interface can be pointed out. They are basically based on the idea that the point of purchase of food is a place where *foods, money, and information* are exchanged between consumers and retailers.

With regard to the first interaction, the foods, it is the types of foods, the assortment, and the functionalities of the food products offered that determine which foods consumers can buy, what they can do with the foods, and what they expect to find in the assortment the next time they come back to the store. The producer or processor decision of the date label to be printed on the food (a longer or shorter one?) is tied to the food item, as is the package chosen and the functionalities that the package has (does the package protect the content well, can it be separated into units, easily closed again, stored, completely emptied?), or the appearance and sensory aspects of the food product (does the food have an abnormal shape or color, of which size is it, does it have a long shelf-life?). Consumer purchase responses contribute to how retailers determine which foods are offered. For example, when a product package innovation that can easily be fully emptied is not purchased sufficiently for a longer period of time, the retailer will not reserve further shelf space for the product. Also, if consumers do not buy single-household sizes of foods, then it is not a worthwhile investment for producers. As another example, retailers adapt their esthetic standards required from the supplier based, among others, on their observation of which fruits and vegetables are deselected. Consumer motivation can lead consumers to accept abnormal shapes of fruits, trade-offs with convenience

Figure 12.3 Clusters of factors causing consumer-related food waste.
Source: Own. Adapted from Aschemann-Witzel, J., De Hooge, I.E., Amani, P., Bech-Larsen, T., Oostindjer, M., 2015. Consumer–related food waste: causes and potential for action. Sustainability 7 (6), 6457–6477.

might stop consumers from taking the time and hassle to fully empty the package, and consumer capabilities can determine whether consumers discard items after the date label has passed or determine whether they trust their own judgment of edibility. Social relations can influence consumers' motivation, preferences, and the trade-offs they perceive in food purchase, and the foods offered in the purchase situation comply to current legal requirements on, for example, date labeling, or change with technological possibilities.

With regard to the second interaction, the money exchange, this is an area repeatedly mentioned as a major cause of food waste (Stuart, 2009). It has been discussed that the cost of food makes up only a small share of the household budget in affluent societies as well as that the price levels are rather low for food, often so low it is difficult to imagine the resource input and processing costs in the supply chain are covered at all. The low price is then said to contribute to consumers not perceiving the value of a food item, in particular when the underlying idea is that the traditionally scarce or religiously sacred food should not be wasted or treated disrespectfully (Gjerres and Gaiani, 2013). In addition, pricing mechanisms designed to increase purchase volume are said to potentially cause waste in that consumers tend to buy more than they actually need. Price gradients with lower unit costs for larger packages and multiitem offers such as "buy one get one free" are frequently used strategies in supermarkets. Consumer demand and price-oriented behaviors impact the success of store formats and of marketing strategies in retail. Retailers apply similar demands upstream in the supply chain, where they exert considerable power in the relations to the suppliers. Consumers' psychographic profiles determine their motivation to, for example, be deal prone and seek price promotions or not, and their capabilities might be limited depending on their budget constraints. Economic prosperity, in turn, influences consumers budgets, and social trends have an impact on the importance given to be a "smart shopper."

Coming to the point of information, the third interaction, one has to be aware that not only information in terms of facts and recommendations is exchanged in the consumer–retailer interface, but also that consumers infer something from the foods, the assortment, presentation, and state. Communication is known to function via language and action. For example, retailers offering fruits and vegetables homogeneous in appearance can shape consumers' impressions of what is normal, and placing eggs in or outside the fridge can shape consumers' assumption of how they should store eggs at home. Single households might feel disrespected when they encounter difficulties in finding appropriately small package units, and consumers can feel frustrated and misunderstood when a new package functionality is simply not as easy as the processor conveys. Selling per weight or per unit of, for example, celery heads communicates something about how consumers are expected to select food. Of course, the point of sale is used by retailers to convey a lot of information in text and visuals, with information about the food's characteristics and value on packages, and with in-store promotions or customer brochures. These communication channels are used to provide suggestions on best storage conditions, meal suggestions, as well as the retailer's brand values and corporate social responsibility (CSR) actions, including what retailers do against food waste and for social and environmental sustainability. Consumers' motivation and the priorities when deciding on trade-offs are influenced, as well as their capabilities to deal with food, by whether they received helpful information from packages and in-store communication, and by their perceptions of social norms and societal trends around food based on what they are told and see in the store. Consumers' reactions in terms of purchase, loyalty, and word of mouth determine the directions that retailers take with regard to the presentation of and communicational context of their offer.

12.5 Consumer perception of suboptimal food

When considering how consumers perceive suboptimal food, it is important to underline that optimality and suboptimality are in no way fixed distinctions. Suboptimality is not only relative to the assessment of the *characteristics* of the food (e.g., is it an abnormal shape, or is it not? Is the product "as such" affected, or the appearance and package?) but also relative to the *purpose of use* (e.g., is the shape in any way affecting what I wanted to use it for? Does the date hinder me to store it for later use?). The clusters of factors in Fig. 12.3 can all exert an influence here, but in particular capabilities come into play. A consumer who has the capability to safely assess edibility and who can use a food item in various ways might not perceive a suboptimality as a hindrance to use the product in the same way as a consumer who lacks such capabilities. An apple with a spot might then simply be an item best suited for apple cake, as a possible alternative to being eaten right away.

Furthermore, there is another important influence on the perception of suboptimality, and that is the relative *value perception* of the item. As Fig. 12.3 shows, the cluster of factors called motivation is important, and it encompasses being aware of the issue and impact of food waste, holding values and beliefs that motivate the desire to avoid it, such as benevolence or environmental concern. Such factors can alter consumers' beliefs about suboptimality and can influence consumers' perceptions of a benefit of suboptimal food, as for example when irregularly shaped fruit and vegetables become an instrument for actively taking care of the environment by selecting these on purpose. As value perceptions are linked to the exchange—for example, what do you get in return for what you give—it is also relevant for suboptimal food perceptions which price one has to pay for the item. A package of milk might be optimal when it has the longest date instead of being close to the date. However, when the item closest to the expiry date suddenly is offered at half the price, it becomes more desirable to choose, and in these circumstances it might not be perceived as suboptimal as it was before, given the change in price–value relation.

The role of the price that a consumer has to pay can logically lead to a difference in consumer behavior in the situation before purchase versus the situation at time of potential consumption. In the first case, the consumer can still make a choice about whether or not to pay for the product, if the value received in return is regarded as sufficient enough. In the second case, the product has already been paid for, and the value invested is fixed. What can matter, however, is how much has been paid at time of purchase, as this determines how much value is perceived to be lost, should the item be discarded instead of used. This is the thought underlining the critique of pricing strategies triggering food waste, because one reason might be that an item bought on promotion does not appear as valuable to be saved from the bin as an item for which the full price had been paid. Fig. 12.4 visualizes how characteristics, purposes and value perceptions are interrelated crucial elements of consumer perceptions of suboptimal food.

Some examples of results from research studies can underline and exemplify how consumers perceive suboptimal food. For example, a quantitative experimental survey study in five European countries (De Hooge et al., 2017) showed that there is a clear difference between whether or not a suboptimal food choice situation takes place in the supermarket or the home setting. On average, two out of six choices between optimal and suboptimal products fell on using the suboptimal item first when the choice took place within the household and the consumer already possessed the item in question. On the other hand, only for 0.5 of the 6 choices was a suboptimal item selected when the choice concerned a purchase decision in the supermarket. A clearly external suboptimality such as in the case of the bent cucumber was relatively most often accepted from among the six examples of suboptimality. This might indicate that value perception is hardly affected by the abnormal shape. The contrary case is the example of an apple with a spot, which is hardly chosen in the supermarket situation. Interestingly, it is also least likely chosen to be consumed first within households. This observation might indicate the role of both trade-offs and capabilities—when the primary purpose of use is affected, many consumers might not want to engage in the effort of an alternative use, or simply cannot readily see what they could do instead. Related to that, it was also found that consumers who were more likely to accept the suboptimal items were characterized by greater

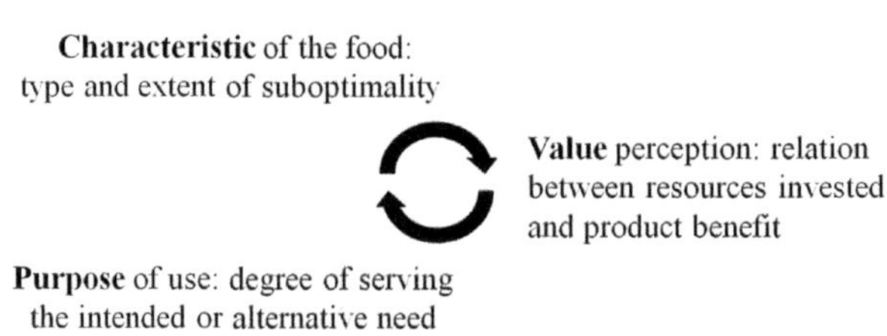

Figure 12.4 Consumer perception of suboptimal food—interactive factors during purchase. Source: Own. Adapted from Aschemann-Witzel, J., Giménez, A., Ares, G., 2018b. Consumer in-store choice of suboptimal food to avoid food waste: the role of food category, communication and perception of quality dimensions. Food Qual. Preference, 68, 29–39. Available from: https://doi.org/10.1016/j.foodqual.2018.01.020; Aschemann-Witzel, J., Jensen, J.H., Jensen, M.H., Kulikovskaja, V., 2017. Consumer behaviour towards price-reduced suboptimal foods in the supermarket and the relation to food waste in households. Appetite 116, 246–258. Available from: https://doi.org/10.1016/j.appet.2017.05.013.

environmental orientation, and by engaging more regularly in shopping and cooking tasks. Overall, the role of different types of suboptimality, product category differences, and consumers product associations underline that consumers give particular attention to the food characteristics that are perceived to signal food safety, risk, and intrinsic quality of the food when making purchase and consumption decisions.

Another research study qualitatively explored consumer consideration when met with suboptimal food in the store (Aschemann-Witzel et al., 2017). In 16 accompanied shopping interviews in a Danish supermarket, consumers continuously voiced their current stream of thought while doing their normal shopping tour in the presence of a student interviewer. Findings show how consumers go back and forth between food characteristics on the one hand and purpose of use at home on the other hand: consumers consider product quality in various dimensions, package unit size, and the current date label of the item, as well as their storage capacity at home, their household's needs, and the plans for meals on that day when making purchase decisions about foods with a price reduction. As the quantitative survey study also showed, consumers appeared willing to choose suboptimal products under certain price reductions, depending on the product category and on the extent of price reduction. Here again, value perception is shown to play a particular role. Due to precisely that reason, it is a common practice of multiple supermarkets to reduce prices of food for example when nearing the date label (Theotokis et al., 2012), an idea that has been taken up in the wake of the societal interest in food waste at the consumer–retailer interface.

In the home situation, consumers have to consider whether or not they want to consume the item now or at a later point, and whether they want to consume the item the way it is, or in any way altered or prepared form, depending on what is among the range of possibilities of use for the individual and item in question. Motivations, trade-offs, and capabilities have a particular role here. A qualitative research study used focus groups in the same five European countries as the quantitative study mentioned before and using a range of techniques to elicit consumers' considerations and perceptions about suboptimal food and food waste in their home. In this study, consumers ($n = 83$ in total) were asked to bring a photo showing a food item they recently had to discard, and to elaborate on the reasons for waste in their home. Interestingly, many of the pictured items did not belong to the main categories often discussed in the literature—that is, fruits, vegetables, bread, dairy products, and meal leftovers—but consisted of packaged foods with a long shelf life such as canned food, snacks, dressings and sauces, or herbs. The most frequent reason mentioned for discarding these items was that they had been lying for a long time in the fridge or in the kitchen closet (sometimes several years), that they had lost all attractiveness for the respondent and that the likelihood of still using them was nil. This observation underlines that food waste in the home is also about foods that were optimal at a point of time in the past, and that may even still be optimal from nutritional and organoleptic perspectives, but that have become suboptimal due to inadequately long storage and loss of attractiveness.

The qualitative research mentioned above also entailed that consumers were given a range of suboptimal foods of different types to discuss jointly, and a task where they had to sort fruit and vegetable images in different stages of suboptimality, to elicit the underlying dimensions that consumers use in perceiving such foods. These foods included among other dairy products of varied best-before dates (milk and yoghurt), vegetables of varied freshness and calibers (carrots, potatoes, apples), and bread of varied freshness. It appeared clearly across the different countries that consumers have much stricter thresholds for determining that a product is acceptable in the store than at home. When purchasing dairy products, respondents are especially observant of the date labeling and tend to systematically purchase the freshest milk (longest date). Milk with a best-before date similar to the purchase date would not be purchased unless a significant price reduction was offered. This was expressed by statements such as "It cannot be that milk with best-before date today is sold at full price" (51-year-old female, Germany). Few would select milk expiring on the day of purchase even at reduced price, but some would if they had a concrete plan, such as to cook porridge on the same day. Similarly, yoghurt would not be purchased past the best-before date at full price.

In a home setting, though, respondents strongly differed with regard to consuming yoghurt past the best-before date. Some would discard the product automatically, some would rely on their senses of sight, smell and taste to test it first, and some would have no hesitation at all for consumption. Several consumers also exposed strategies for usage, such as including yoghurt in baking. Further, consumers displayed very diverse strategies for avoiding waste of bread as, for example, adapting purchase frequency, or ideas for bread storage, including using containers (wooden box, plastic bag), and freezing (whole loaves, in portions or sliced). Many bread-saving recipes were evoked through the different discussions, the most frequently mentioned one being toasting. Respondents also discussed odd-shaped vegetables and fruits of different freshness levels. Reasons for potentially rejecting vegetables were in particular lack of appeal due to poor freshness, apparent decay, and fear of getting sick. Interestingly, in all countries odd-shaped vegetables triggered associations to naturalness, organic production, and higher taste expectations in the consumer's mind. Yet, most would not have picked these in the supermarket due to their difficulty to peel, and because a large share of the item may be thrown away.

Admittedly, consumers stated that discounts on such products would influence purchase decision a lot. In summary, this research provided rich insights on consumers' motivations and practices with regard to food rejection of suboptimal foods. Consumers' perception of suboptimal foods varies across product categories and across situations, and knowledge for evaluating food safety as well as skills for alternative usage of nonstandard products are crucial in the fight against food waste. Important questions remain to understand the deeper roots of why some people better accept suboptimal foods than others—how did they learn these attitudes and how can we make others learn from them? Through upbringing, through school, through the media? "There is a relationship between taste experience and visual experience," declared one of our participants (50-year-old male, Norway). Visually suboptimal foods face the challenge of convincing the consumer to get past their visuals and to give them the chance they deserve to prove their palatability.

12.6 Retailer actions against food waste

Given the role that the consumer–retailer interface plays for food waste, there has been a lot of discussion about the responsibility of supply chain stakeholders for the food waste observed in developed societies (Bloom, 2010; Juul, 2016; Stuart, 2009). Following the core idea of CSR, it is demanded that the businesses in the food sector assume responsibility for issues beyond their profit goal and beyond legal requirements (Carroll, 1999). The EU defines such CSR as "integrating social, environmental, ethical, consumer, and human rights concerns into their business strategy and operations" (EC, 2015, p. 1). A number of issues of particular relevance have been identified for the food sector, such as safety and healthiness, environmental impact, animal welfare, labor rights, and the prosperity of the local community (Forsman-Hugg et al., 2013). Newer conceptualizations of CSR and related concepts have shifted the focus to the idea that CSR might as well be designed and used strategically as well as enacted proactively. This allows aligning profit goals with favorable societal outcomes, and thus creating "shared value" (Porter and Kramer, 2011) and a "business case for sustainability" (Carroll and Shabana, 2010; Schaltegger et al., 2012).

Interestingly, the societal focus on food waste has made it relatively easier than before for retailers to align their commercial goals with actions against food waste in the supply chain. The more society pays attention to the food waste issue and acknowledges the effort, the more can such effort pay off in terms of a favorable attitude towards the company, higher levels of trust, a higher degree of loyalty and ultimately greater sales. Such a relation between CSR activities and commercial success has been repeatedly underlined as a favorable effect that might motivate the business to engage in CSR (Dixon-Fowler et al., 2013; Grewatsch and Kleindienst, 2016). Thus, just as CSR research suggests, there can be a range of advantages when retailers engage in food waste avoidance (Aschemann-Witzel et al., 2017), as listed below:

1. Decreasing food waste reduces cost of disposal
2. Food waste avoidance improves environmental and social sustainability
3. Company image improves where customers and stakeholders appreciate efforts
4. Better employees are attracted, and current employees are more satisfied

Research on the key success factors of commercial as well as noncommercial initiatives against food waste (Aschemann-Witzel et al., 2017) has shown that both "timing" and "business opportunity" are among the key success factors mentioned by actors and experts. This might indicate that the societal focus has gained a momentum that has helped in the starting phase and development of various anti-food waste initiatives. Overall, the food waste initiatives within the supply chain can be categorized into three types of initiatives. First, information and capacity building initiatives that target consumer awareness, knowledge, preferences, and skills. Such initiatives can help to increase consumer acceptance of suboptimal food and their capabilities to use suboptimal foods. Examples of these types of initiatives are information and awareness raising campaigns conducted by nongovernmental organizations (NGOs) or supply chain stakeholder platforms. Secondly, initiatives that redistribute suboptimal food into alternative channels of sale, such as the food banks that have been founded across Europe in the past decade, which allow consumers with lower income to make use of food donated by processors or retailers and otherwise potentially wasted. Thirdly, there are initiatives that work on changing the current status and functioning of the supply chain and of the retail environment. Examples of this are social entrepreneurs who innovate new products based on suboptimal food, as for example "fruit paper" snacks made from fruits otherwise wasted due to suboptimalities, or soups from misshapen vegetables.

Retailers can play a role in all three of these types of initiatives. For example, retailers can support the information and capacity building actions of NGOs or platforms, donate food to food banks, and provide shelf space for new products with the benefit of a good cause towards avoiding food waste in the supply chain. Retailers have in the past years, however, also become active themselves in a number of ways. Examples that have received a lot of media attention are the Danish branch of Norwegian retail chain Rema1000, who had already abolished multiitem offers ("buy one, get one free") in their stores in 2008. They had been in contact with the NGO Stop Wasting Food and this collaboration had triggered the action. However, the marketing manager also explains that he himself had experienced annoyance about the appeal of price promotions, which then led him into wasting some of the food bought. Rema1000 received very positive customer feedback and observed an improvement of their brand perception in the market, and they have also been awarded acknowledged industry prices for their engagement. Even more international coverage was dedicated to the French supermarket Intermarché when they very famously introduced and heavily marketed "inglorious" fruits and vegetables in abnormal shapes in their stores, and the action was very successful in terms of the sales of these items—which they claimed quickly sold out (Aschemann-Witzel et al., 2016).

A mapping study conducted on the types of actions that retailers engage in to avoid food waste provides some more insights into the type of actions that retailers engage in. The study was done in Denmark, a country where food waste at the time of study had been particularly high on the agenda in the media, and where most retailers had already begun to take a standpoint and to introduce actions (Kulikovskaja and Aschemann-Witzel, 2017). The website and CSR report information on the actions against food waste, as well the actions observed in-store were considered. All major retail chains were explored and 18 stores were visited. These visits included some mystery shopping interviews, where a researcher acting as a regular, interested food shopper asked store personnel about their activities. Six different types of actions were found:

1. Price or pricing
2. Product
3. Unit
4. Communication
5. Collaboration
6. In-store management

Price actions (the first type) were particularly prominent and used across all retail chains and store formats to lesser or greater extent. They took the form of price reductions of suboptimal food, changes in price strategies such avoiding multiitem offers, or prolonging the offer in that the second item can be fetched from the store at some other point in time. A second type of action is connected to the product, and consists of changes in the packaging (e.g., shelf-life, ease of emptying) or of offering foods that deviate from optimality. The third is linked to unit sizes. This concerns changing the package sizes (e.g., single-household units) or the possibility to partition the food in serving units, as well as shifting the price from per item to per weight for fruit and vegetables. Furthermore, in the category of communication actions, stores were found to communicate about the pricing actions, about suboptimality and about the potential to avoid food waste by the choice of an item on stickers attached to the suboptimal foods, in in-store posters, or on the website. As a fifth, the collaboration with other actors in the supply chain was identified. For example, retailers might collaborate with food banks or other noncommercial organizations by donating suboptimal foods or by supporting the organizations efforts. Sixth and finally, retailers can also engage in refinement or alteration of their in-store management, improvement of processes and inefficiencies, as well as giving greater focus to the avoidance of food waste. This can be in the form of improved technology or management software or procedures, changing the placing of suboptimal food within the store assortment, or increasing the frequency of personnel checking the status of suboptimality so that items can quickly be donated or offered at a reduced price.

An action that has become particular widespread not only in Denmark but also in retailers in other countries—as already the example of Intermarché shows—is selling suboptimal food at a reduced price alongside the "normal" food items. This increases favorable consumer perceptions of the relation between the resources invested in the production of, the price of, and the value received in return for the suboptimal food. On a day-to-day basis, the most common product examples might be those nearing the date, but fruits and vegetables that have become somewhat unappealing or overripe as well as slightly damaged packages are also among those items. The mapping study revealed that store personnel observed approximately 9 out of 10 items of suboptimal food reduced in price to be sold by the end of the day. Given the increased interest in the topic of food waste in society, many retailers highlight this action now as a contribution to food waste avoidance. Thus, it is an example of an action that belongs to CSR activities, which has gained a strategic importance in positioning the company as being responsible, and which can improve image among customers and stakeholders.

Yet, there are a number of challenges when retailers offer otherwise wasted foods for reduced prices (Aschemann-Witzel et al., 2017; De Hooge et al., 2018). The actions add cost, in particular if the retailer would not have to pay for the wasted food, as contractual agreements might lay the burden of it on suppliers (Eriksson et al., 2017). Another challenge is that low prices encountered in store might affect quality perceptions of the store. Moreover, store management becomes more complicated and tedious if date and appearance of items have to be checked by store personnel. And then, the company might fear that the suboptimal food sales "cannibalize" the sales of the "normal" foods (De Hooge et al., 2018). An example of this encountered in the mapping study mentioned earlier (Kulikovskaja and Aschemann-Witzel, 2017) is that one store was selling bake-off bread at 50% off from 7 p.m. onwards, but moved the time to 8 p.m. once the store manager noticed consumers queuing up to wait for the price reduction to happen.

The type of suboptimality that appears to hold particular potential to be changed by altering consumer perceptions is the shape and color deviations of fruits and vegetables. The claim that suboptimal products are otherwise perfectly edible can be said to hold most apparently when it concerns products with simply a strange shape or abnormal color. Originally, the European Union formalized the norms, or so-called cosmetic specifications, concerning the shape, size, color, skin, and weight for fruits and vegetables (European Union, 2007). After realizing that such specifications might generate food waste across the food supply chain, the European Union abolished cosmetic specifications for 26 of the 36 fruits and vegetables in 2009 (European Union, 2008). Yet, despite this abolishment, suboptimal fruits and vegetables are rarely found in stores, making it impossible for consumers to purchase such suboptimal products. To explore in greater detail how producers, producer organizations, and retailers deal with such suboptimal fruits and vegetables, an interview study was conducted (De Hooge et al., 2018). In this study, 33 German and Dutch producers, producer organizations, and retailers were interviewed to discuss how they dealt with cosmetic specifications and with fruits and vegetables not fulfilling these specifications, and the possibilities and challenges of changing these business practices.

The findings demonstrate that not only the European Union, but also producers, producer organizations, and retailers set standards concerning the appearance of fruits and vegetables (De Hooge et al., 2018). The supply chain actors tend to use such standards to signal being a company that delivers only high-quality products. The consequences of setting such cosmetic specifications for fruits and vegetables are clearly visible across the supply chain. The amount of foods that do not fulfill these specifications differ depending on the type of supply chain actor, varying from about 1% for greenhouse producers and Dutch retailers to 40% for open field producers. In all cases, supply chain actors try to find alternatives for these products before wasting them. Yet, most actors are limited in the alternatives available for suboptimal fruits and vegetables. For example, producers try to export suboptimal fruits and vegetables to alternative market channels abroad or to food processing markets, but otherwise they are limited to options that are already defined as food waste by the literature (e.g., transforming the product to biogas, cattle feed, or manure). Retailers are limited to donating suboptimal foods to charity, but can only do so if the food fulfills the safety food requirements of the national and European laws. Consequently, the majority of suboptimal fruits and vegetables are wasted.

In the interview study it was also examined to which extent there appears to be a business potential in selling suboptimal fruits and vegetables. It appeared that supply chain actors have mixed feelings about producing and marketing suboptimal products. Societal motivations, company image, and CSR motivations may provide reasons for supply chain actors to produce and market suboptimal products, but actors also perceive the current market mechanism and pricing strategies to be problematic. With limited shelf space being available in stores and no increase in consumer demand, the introduction of suboptimal products would actually harm the market for all products. Also, producing and marketing suboptimal products would, in the actors' views, not provide a sustainable solution, as oddly shaped food products actually increase the burden on transportation and logistics. Finally, and perhaps most important of all, all types of actors perceive the consumer to be the underlying barrier for the marketing of suboptimal products: consumers are believed to be the ones who determine what is being sold, and as consumers' first impressions and food purchase decisions are based on product appearances, consumers are thought to be unwilling to purchase suboptimal products.

12.7 Consumer response to retailer actions

Ultimately, consumers need to favorably respond to the retailers efforts, so that these stores continue being engaged in food waste avoidance actions. When consumers do not support or even dislike these actions, the actions will ultimately fail. When consumer responses are more favorable, these responses can either have more direct consequences, such as cost saving, inefficiencies being removed, and higher profit margins, or have more indirect consequences, such as activities being more positively received within society and by stakeholders.

In a case study exploring food waste avoidance initiatives across the supply chain, examples of all three cases (failure, indirect success, direct success) of market and consumer responses were found (Aschemann-Witzel et al., 2017). In the case of a failed attempt, a bakery had observed that they were always discarding quite a large share of the assortment at the end of the day, and decided to deliver only the average sold amount to each subsidiary for a while. However, consumers complained heavily about the reduced assortment and about multiple items being frequently sold out. The interesting observation was that consumers responded as if being personally offended by the bakery's behavior, commenting on whether or not the bakery cared about satisfying their customers. This affected both business and employee satisfaction. Therefore, the bakery returned to the original approach. In the case of Rema1000 in Denmark, the marketing manager admitted that after abolishing the multiitem offers, sales volumes for some products declined. However, the positive feedback from consumers supported the management belief that the action would be worthwhile in the longer run (Aschemann-Witzel et al., 2016), and indeed, the supermarket has gained a very favorable reputation since then. It is difficult to assess how the initiative to remove the multiitem offer is related to the improved image and brand position, given it is an indirect effect, but the retailer appears to be convinced of it being a worthwhile decision. Finally, the case study identified an example in which an idea to avoid food waste at the same time developed into a business success: An ICA (ICA is the name of ta Swedish he supermarket chain) store located in a supermarket in Lund installed an in-store kitchen preparing lunch and dinners from suboptimal food previously wasted. This allowed the supermarket to sell the food that otherwise would have been discarded, and to attract new consumers with relatively low-cost meal options. It also gave the store the opportunity to engage in more risk-taking strategies such as opening a fish and meat counter and broadening their assortment, because they had a "second opportunity" to use the food in the kitchen (Aschemann-Witzel et al., 2016). Thus, it is possible to develop actions of avoiding food waste into a win—win situation where commercial goals are also supported.

It is frequently mentioned that the occurrence of food waste is a multifaceted issue with a diverse set of interactive underlying factors (Hebrok and Boks, 2017; Quested et al., 2013). Therefore, food waste does not only depend on consumers, and consumer responses to supply chain actions are not equal across consumers. Indeed, sociodemographic and psychographic individual factors, as well as social and contextual factors, have been found to relate to food waste. Therefore, it seems logical to take into account consumers' distinct lifestyles (Ganglmair-Wooliscroft and Lawson, 2010) when considering how they would respond to retailer actions.

In a survey across five European countries, a food-related lifestyle measurement was used in combination with statements specific to the topic of food waste (for the findings for Denmark, see Aschemann-Witzel et al., 2018). Through factor and cluster analysis, dimensions of interaction with and preferences for food attributes in day-to-day life were identified, as well as the consumer segments that differed in these dimension. The consumer groups differed in their food involvement, as expressed in the dimension of cooking enjoyment, in food planning, social relationships via meals, food safety concerns, and price orientation. From what these consumers reported about food wastage in their home, their attitude towards the food waste issue, and their experimentally derived choices for optimal versus suboptimal foods, recommendations of which actions likely fit best to which segment can be derived. For example, the consumer segment characterized by a high involvement with food but a spontaneous approach to meal planning might make use of ready-made meals based on suboptimal food from the store, and might use apps suggesting how to creatively use leftovers. A consumer segment already good in planning food purchase and meals can be taken a step further in food waste avoidance with more advanced storage tips. Consumers characterized by a low interest in cooking but quite a price orientation, in turn, are most likely attracted by price reductions of suboptimal food, or by food banks selling donated items.

Price-reduction of suboptimal food is one of the actions that has become rather widespread among retailers. This holds in particular in Denmark, where the mapping study found that it has become rather an industry standard to do so for at least certain food categories, and quite often with colored stickers indicating this to be a food waste avoidance action (Aschemann-Witzel and Kulikovskaja, 2016). An experimental survey (Aschemann-Witzel, 2018) explored in greater depth which factors explain consumer reactions to the foods offered at reduced price. It also tested varying approaches to communicate the benefit of the price reduction, that is, highlighting either the economic savings or the food waste avoidance. Both arguments take different routes, appealing to opposing motivations and value orientations, but both work towards changing the relative value perception of the suboptimal food. The study mimicked the yellow stickers actually used in-store by a range of four supermarkets, but varied the text. Results showed that rather than the actual communicational argument, a decisive factor for consumers was whether or not they were familiar with the supermarket and with the sticker. A gender effect was found, such that men were less likely to react favorably to the food waste messages compared with women. The findings also confirmed the crucial importance of the characteristic of the food in terms of perceived quality, as well as use in terms of likelihood of being consumed entirely. Interestingly, a similarly designed experimental survey study conducted in a very different context, namely the country of Uruguay in which food waste has not yet been discussed to a great extent, showed that offering the price-reduced suboptimal food with communication did indeed have an effect. Suboptimal food communicated with the food waste avoidance argument was more favorably received among respondents of low or of high socioeconomic status. Overall, consumers of low socioeconomic status and men were more likely willing to buy the suboptimal items. The food category in question was crucial for quality perception, and depending on the food and its suboptimality, different dimensions of quality were affected (Aschemann-Witzel et al., 2018a,b).

12.8 Conclusions

The research examples discussed here explore how consumers perceive and choose suboptimal food, which can contribute to food waste avoidance at the consumer–retailer interface. Consumer acceptance of suboptimality is clearly higher in the home versus the supermarket, and greatly differs by product category and type of suboptimality. Consumers individual characteristics also play a role. During either in-store choice or when considering at home usage, consumers negotiate the specific suboptimal food item's characteristic in relation to the household needs and usage context, underlining that consumers indeed assess benefits versus costs. Safety, quality, and potential usage options are of particular importance. Retailers can and already are acting against food waste at the consumer–retailer interface, and six types of actions by supermarkets have been identified, for which price-reduction of suboptimal food is most widespread. Retailers, however, see quite a number of barriers to also selling fruits and vegetables currently not meeting cosmetic standards and wasted in the supply chain early on, and point to a need in consumer expectations. Furthering sales of suboptimal food with accompanying communication is found to increase choice likelihood, but it depends on the context, as for example, consumer group and market. What remains yet underresearched is the relative effectiveness of various approaches to reduce food waste in the consumer–retailer interface, in particular in carefully designed interventions or with actual food waste measures in-store or at within the household.

References

Aschemann-Witzel, J., 2018. Consumer perception and preference for suboptimal food under the emerging practice of expiration date based pricing in supermarkets. Food Qual. Preference 63, 119–128. Available from: https://doi.org/10.1016/j.foodqual.2017.08.007.

Aschemann-Witzel, J., Giménez, A., Ares, G., 2018a. Convenience or price orientation? Consumer characteristics influencing food waste behaviour in the context of an emerging country and the impact on future sustainability of the global food sector. Glob. Environ. Change 49, 85–94. Available from: https://doi.org/10.1016/j.gloenvcha.2018.02.002.

Aschemann-Witzel, J., Giménez, A., Ares, G., 2018b. Consumer in-store choice of suboptimal food to avoid food waste: the role of food category, communication and perception of quality dimensions. Food Qual. Preference 68, 29–39. Available from: https://doi.org/10.1016/j.foodqual.2018.01.020.

Aschemann-Witzel, J., De Hooge, I.E., Amani, P., Bech-Larsen, T., Oostindjer, M., 2015. Consumer–related food waste: causes and potential for action. Sustainability 7 (6), 6457–6477.

Aschemann-Witzel, J., De Hooge, I.E., Normann, A., 2016. Consumer–related food waste: role of food marketing and retailers and potential for action. J. Int. Food Agribusiness Mark. 28 (3), 271–285.

Aschemann-Witzel, J., De Hooge, I.E., Rohm, H., Normann, A., Bossle, M.B., Grønhøj, A., et al., 2017. Key characteristics and success factors of supply chain initiatives tackling consumer–related food waste—a multiple case study. J. Clean. Prod. 155 (2), 33–45. Available from: https://doi.org/10.1016/j.jclepro.2016.11.173.

Aschemann-Witzel, J., Jensen, J.H., Jensen, M.H., Kulikovskaja, V., 2017. Consumer behaviour towards price-reduced suboptimal foods in the supermarket and the relation to food waste in households. Appetite 116, 246–258. Available from: https://doi.org/10.1016/j.appet.2017.05.013.

Aschemann-Witzel, J., Kulikovskaja, V., 2016. Food waste avoidance initiatives in Danish food retail: COSUS WP 6.2 report. Retrieved from <http://bss.au.dk/fileadmin/user_upload/COSUS_WP6.2_Food_waste_avoidance_initiatives_in_Danish_food_retail_Onlineversion.pdf> (accessed 06.10.16.).

Aschemann-Witzel, J., Lengard Almli, V., De Hooge, I.E., Oostindjer, M., 2018. Fine-tuning the fight against food waste by aligning mitigation policies with consumer lifestyles. J. Macromark. 38, 2168–2184.

Bloom, J., 2010. American Wasteland - How America Throws Away Nearly Half of Its Food (And What We Can Do About It). Perseus Books Group.

Carroll, A.B., 1999. Corporate social responsibility: evolution of a definitional construct. Business Soc. 38 (3), 268–295.

Carroll, A.B., Shabana, K.M., 2010. The business case for corporate social responsibility: a review of concepts, research and practice. Int. J. Manage. Rev. 12 (1), 85–105.

De Hooge, I.E., Oostindjer, M., Aschemann-Witzel, J., Normann, A., Loose, S.M., Almli, V. L., 2017. This apple is too ugly for me!. Food Qual. Preference 56, 80–92. Available from: https://doi.org/10.1016/j.foodqual.2016.09.012.

De Hooge, I.E., Van Dulm, E., van Trijp, H.C.M., 2018. Cosmetic specifications in the food waste issue: supply chain considerations and practices concerning suboptimal food products. J. Clean. Prod. 183, 698–709. Available from: https://doi.org/10.1016/j.jclepro.2018.02.132.

Dixon-Fowler, H.R., Slater, D.J., Johnson, J.L., Ellstrand, A.E., Romi, A.M., 2013. Beyond "does it pay to be green?": A meta-analysis of moderators of the CEP-CFP relationship. J. Business Ethics 112 (2), 353–366.

EC, 2010. Prepatory study on food waste across EU 27: Technical Report - 2010 - 054 (No. ISBN: 978-92-79-22138-5). Retrieved from European Commission website: <http://ec.europa.eu/environment/eussd/pdf/bio_foodwaste_report.pdf>.

EC, 2015. Corporate social responsibility (CSR). Retrieved from: <http://ec.europa.eu/growth/industry/corporate-social-responsibility/index_en.htm>.

European Union, 2007. Council Regulation (EC) No 1234/2007.

European Union, 2008. Commission Regulation No. 1221/2008.

Eriksson, M., Ghosh, R., Mattsson, L., Ismatov, A., 2017. Take-back agreements in the perspective of food waste generation at the supplier-retailer interface. Resour., Conserv. Recycl. 122, 83–93. Available from: https://doi.org/10.1016/j.resconrec.2017.02.006.

Evans, D., 2014. Food Waste: Home Consumption, Material Culture and Everyday Life. Bloomsbury Academic, London and New York.

FAO, 2011. Global food losses and food waste - extent, causes and prevention: study conducted for the International Congress SAVE FOOD! at Interpack2011 Düsseldorf, Germany. Swedish Institute for Food and Biotechnology (SIK), Gothenburg, Sweden, FAO, Rome, Italy.

FAO. (2013). Food wastage footprint: Impacts on natural resources - Summary report. Retrieved from: <http://www.fao.org/docrep/018/i3347e/i3347e.pdf>.

Forsman-Hugg, S., Katajajuuri, J.-M., Riipi, I., Mäkelä, J., Järvelä, K., Timonen, P., 2013. Key CSR dimensions for the food chain. British Food Journal 115 (1), 30–47. Available from: https://doi.org/10.1108/00070701311289867.

Fusions, 2015. FUSIONS Definitional Framework for Food Waste: Full report (No. ISBN 978-91-7290-331-9). Fusions research project. Retrieved from: <www.eu-fusions.org/>.

Ganglmair-Wooliscroft, A., Lawson, R., 2010. Subjective well-being of different consumer lifestyle segments. J. Macromark. 31 (2), 172–183. Available from: https://doi.org/10.1177/0276146710393251.

Gjerres, M., Gaiani, S., 2013. Household food waste in Nordic countries: estimations and ethical implications. Etik i praksis. Nordic J. Appl. Ethics 7 (1), 6–23.

Grewatsch, S., Kleindienst, I., 2016. When does it pay to be good? Moderators and mediators in the corporate sustainability- corporate financial relationship: a critical review. J. Business Ethics 1–33.

Grunert, K.G., 2005. Food quality and safety: consumer perception and demand. Eur. Rev. Agric. Econ. 32 (3), 369–391. Available from: https://doi.org/10.1093/eurrag/jbi011.

Grunert, K.G., 2007. How consumers perceive food quality. In: Frewer, L., van Trijp, H.C. M. (Eds.), Understanding Consumers of Food Products. Woodhead, Cambridge, pp. 181–199.

Hebrok, M., Boks, C., 2017. Household food waste: drivers and potential intervention points for design—an extensive review. J. Clean. Prod. 151, 380–392.

Henz, G.P., 2017. Postharvest losses of perishables in Brazil: what do we know so far? Horticult. Bras. 35 (1), 6–13. Available from: https://doi.org/10.1590/s0102-053620170102.

Juul, S., 2016. Food Waste. Retrieved from: <http://www.huffingtonpost.com/selina-juul/>.

Kulikovskaja, V., Aschemann-Witzel, J., 2017. Food waste avoidance actions in food retailing: the case of Denmark. J. Int. Food Agribusiness Mark. 1–18. Available from: https://doi.org/10.1080/08974438.2017.1350244.

Oude Ophuis, P.A., van Trijp, H.C., 1995. Perceived quality: a market driven and consumer oriented approach. Food Qual. Preference 6, 177–183.

Parfitt, J., Barthel, M., Macnaughton, S., 2010. Food waste within food supply chains: quantification and potential for change to 2050. Philos. Trans. R. Soc. B: Biol. Sci. 365 (1554), 3065–3081.

Porter, M.E., Kramer, M.R., 2011. Creating shared value. Harv. Business Rev. 89 (1/2), 62–77.

Priefer, C., Jörissen, J., Bräutigam, K.-R., 2016. Food waste prevention in Europe—a cause-driven approach to identify the most relevant leverage points for action. Resour. Conserv. Recycl. 109, 155–165. Available from: https://doi.org/10.1016/j.resconrec.2016.03.004.

Quested, T.E., Marsh, E., Stunell, D., Parry, A.D., 2013. Spaghetti soup: the complex world of food waste behaviours. SI: Resour. Behav. 79 (0), 43–51.

Raak, N., Symmank, C., Zahn, S., Aschemann-Witzel, J., Rohm, H., 2017. Processing- and product-related causes for food waste and implications for the food supply chain. Waste Manage. 61, 461–472. Available from: https://doi.org/10.1016/j.wasman.2016.12.027.

Schaltegger, S., Lüdeke-Freund, F., Hansen, E.G., 2012. Business cases for sustainability: the role of business model innovation for corporate sustainability. Int. J. Innov. Sustain. Dev. 6 (2), 95.

Stuart, T., 2009. Waste: Uncovering the Global Waste Scandal. Penguin, London.

Theotokis, A., Pramatari, K., Tsiros, M., 2012. Effects of expiration date-based pricing on brand image perceptions. J. Retail. 88 (1), 72–87.

Van Boxstael, S., Devlieghere, F., Berkvens, D., Vermeulen, A., Uyttendaele, M., 2014. Understanding and attitude regarding the shelf life labels and dates on pre-packed food products by Belgian consumers. Food Control 37 (0), 85–92.

The concept of Zero Waste

13

Christine Costello
Assistant Professor, Industrial & Manufacturing Systems Engineering, University of Missouri, Columbia, MO, United States

Chapter Outline

13.1 Introduction and background

In the early 1970s, a chemist, Paul Palmer, PhD, founded Zero Waste Systems (ZWS) Inc. (Palmer, n.d.). ZWS was a small company formed by Dr. Palmer and other chemists and drivers who traveled around the Bay Area, including the nascent Silicon Valley, collecting chemicals associated with electronics manufacturing, for example, acids, solvent mixtures, copper-rich etchant fluids, and finding other businesses that could utilize these materials as inputs (Palmer, n.d.). In the pursuant 40 + years, Zero Waste (ZW) has evolved and grown in a number of directions, but its core idea persists to this day: to render, fundamentally and systematically, the very notion of waste as obsolete.

The concept of ZW represents a paradigm shift that encourages a complete rethinking of current systems. According to the ZW International Alliance, "Zero waste means designing and managing products and processes to systematically avoid and eliminate the volume and toxicity of waste and materials, conserve and recover all resources, and not burn or bury them" (ZWIA, 2009). ZW challenges humanity to close materials loops such that no waste is generated during the production or

Saving Food. DOI: https://doi.org/10.1016/B978-0-12-815357-4.00013-4

consumption of any product or service. As a result, "Implementing ZW will eliminate all discharges to land, water or air that are a threat to planetary, human, animal or plant health" (ZWIA, 2009). Grounded in ecological theory, ZW calls upon humans "to emulate natural cycles, where all discarded materials are designed to become resources for others to use" (ZWIA, 2009). This concept is thus intended to inspire creative design innovations not simply at the waste management stage, but throughout every stage of production and consumption (Zaman, 2015).

Numerous institutions, municipalities, and governances have set ambitious ZW goals. Food waste (FW) is often a particular target as it is typically a large portion of the nonrecyclable municipal solid waste stream (Kim et al., 2011; EPA, 2014). The European Union adopted a set of waste reduction, recycling, and landfill diversion goals under the Circular Economy Package, including a goal to halve food waste by 2030 and to reduce disposal of municipal waste, including food, to a maximum of 10% by 2030 (EC, 2015). Landfilling of FW was banned in Korea in 2005 (Kim et al., 2011) and New York City has set a goal of achieving ZW for landfills by 2030, and is rolling out numerous campaigns to capture FW in particular (OneNYC, n.d.). To develop infrastructure capacity to manage all of the FW NYC is rolling out residential and commercial collections and treatment programs. As of 2017, NYC was collecting organics from one million residents (Balkan, 2017). In particular, a 2013 NYC law required some commercial businesses generating FW to divert their organics, starting with foodservice in hotels with 150 or more rooms, food manufacturers with greater than 25,000 ft^2, food wholesalers with greater than 20,000 ft^2, and stadiums and arenas with 15,000 seats or more (Balkan, 2017; The NYC Council, 2013). Similarly, the state of California passed Mandatory Commercial Organics Recycling legislation in October of 2014, which requires businesses generating more than 8 cubic yards (cy) of organic waste to recycle it on or after April 1, 2016, with a progressive inclusion of organic waste generating businesses to 2 cy by 2021 (California, 2017). An interactive database (maintained by ReFED, a multistakeholder nonprofit collectively dedicated to reducing US FW) is available for researching recent policies related to FW across the United States (ReFED, 2018)

ZW goals have also proved popular among sporting event venues, particularly in colleges and universities (NRDC, 2013). In practice, however, many such venues have operationalized ZW as diverting 90% of event-generated waste from entering landfills. Additionally, among such venues, recycling and composting still remain the most prevalent options for managing diverted waste. Beyond waste diversion, there are currently no requirements in place to delimit usage of the term ZW, and some, including this author, have criticized this laxity and have called for more stringent requirements given the robust standards of ZW philosophy (Costello et al., 2017). In addition to developing the infrastructure to divert waste, use of the term ZW should require a systematic evaluation of materials entering a facility and waste management options using clear metrics to define improved performance. Examples of environmental considerations could include greenhouse gas emissions (GHGs), energy use, nutrient recovery, and social impacts, for example, employment opportunities or disproportionate risk exposure. Importantly, not limiting the inquiry to waste management allows for more proactive options, for example, reducing the types of packaging materials to only those that are recyclable or compostable to increase the probability of fans properly sorting wastes (NRDC, 2013). Costello et al. found that FW was the largest portion of the waste stream during an audit at a college football stadium and that mitigating FW would result in the greatest GHG and energy savings over a range of management options including composting (2017). Management were averse to exploring strategies that would risk running out of food for customers, particularly those seated in the catered boxed seating areas. This reflects an ethos in hospitality management that should be explored more deeply.

Perishable organic materials, such as food, present unique considerations and challenges for ZW thinking, which was originally conceived for the management of chemicals and other materials that could be safely stored for long periods of time without changing form. Perishable organic materials undergo various biogeochemical transformations due to their perishable nature, or within the human body if they are consumed. As a result, the goals of ZW—such as closing materials loops, avoiding or eliminating volume or toxicity, and conserving or recovering resources from these materials flows through society—require a different set of strategies. The existing food production systems impose numerous constraints that are far more challenging than the relatively simple task of separating organic materials from a waste stream to prevent their deposition into a landfill. To truly eliminate FW could require reimagining the basic tenets of the hospitality industry, how residential communities are designed, the distance between croplands and animal production facilities, and the expectation that food will always be plentiful and cheap, as just a few potential examples.

One promising approach for accomplishing this kind of systematic evaluation is life cycle thinking, formalized through life cycle assessment (LCA) (International Standards Organization, 2006). LCA and ZW have similar motivations, which are to holistically and systematically account for the impacts of producing a good or service. LCA is an analytical approach for quantifying the material and energetic flows associated with a product, process, or service over its entire supply chain. This allows for the identification of the most resource-intensive or impactful portions of the supply chain to be identified as targets for improvement. That is, the materials and energy along with the corresponding environmental or social impacts associated with a product are considered in the analysis.

For example, food products typically begin with using diesel-powered machinery to prepare soils, plant seeds, and apply pesticides and fertilizers. Combustion of fossil fuels results in the release of a myriad of molecules, for example, CO_2, nitrous oxides, particulate matter, to the atmosphere that have varying environmental and human health impacts (Union of Concerned Scientists, 2016). In addition, the application of nitrogenous fertilizers to agricultural soils are the leading source of GHGs in the agriculture sector (EPA, 2018) and together with phosphorous are the leading cause of eutrophication in waterbodies (Committee on Environment and Natural Resources, 2010; Dodds et al., 2009). Once applied reactive nitrogen (e.g., anhydrous ammonia, urea) begins to undergo a process called denitrification, a series of biogeochemical reactions involving nitrogen to return to its most stable form, N_2; along this path nitrous oxide forms (N_2O) and is released to the atmosphere. N_2O is a GHG that traps ~300 times more heat than a molecule of carbon dioxide CO_2 (EPA, 2018). Following harvest, which, again requires machinery, agricultural products are then transported using diesel- or gasoline-powered vehicles, processed using machinery, and packaged in plastics made of fossil fuels. Foods are distributed to retail stores, purchased by people who take them to their refrigerators and eventually prepare foods on a stove using electricity or natural gas and, finally if any food is not consumed they are disposed using a waste management technology. In theory all of these processes, including materials and energy associated with mining fossil fuels, are included in the LCA of a food product, though in many cases retail, cooking, and waste disposal are not included (de Vries and de Boer, 2010; González et al., 2011). From a life cycle thinking perspective, when food is wasted, all of these upstream inputs and impacts are wasted as well. Thus far, LCA studies on FW management identify source reduction as the most beneficial option by far (Hamilton et al., 2015), followed by wet then dry animal feed (Kim et al., 2011). Consideration of the entire supply chain can encourage more creative solutions to reduce impact than focusing on waste management alone. In this chapter, a life cycle thinking perspective will be employed wherever possible.

To illustrate difference between perishable organic (i.e., carbon-based biologically derived materials) and nonorganic (i.e., metals) or relatively stable organics (i.e., petroleum-based materials and synthetic chemicals) with regard to ZW, imagine the recovery and recycling of aluminum. Aluminum does not change in form over the course of its supply chain; it may go from ingots to rolled sheets to individual beverage cans, but the material itself is not fundamentally changed. At the end of its useful life aluminum materials can be baled and sit for long periods of time while waiting to be melted down and returned to a useful purpose, perhaps another beverage can. There are numerous upstream uses of energy for mining and smelting resulting in various environmental and human health impacts, but the refined aluminum is stable and truly recyclable. Further, recycling aluminum uses far less energy and materials than virgin aluminum so the advantage is very clear (EPA, 2015).

Now imagine a tomato, plucked from a plant—likely still green, packed into a box, loaded into a truck bed and then onto a rail car, which may or may not have climate control features. Upon arrival into a distribution center, the tomato—perhaps starting to turn red and beginning to soften—is loaded onto a tractor trailer, driven to a grocery store, unpacked—perhaps a few have gotten too soft or have a mold spot and are discarded—and arranged onto a shelf where discerning customers select for the qualities that they have been told indicate a delicious and safe tomato, then the customer transports this tomato to their home to their counter or refrigerator for later consumption. The consumer had imagined using the tomato in a salad to accompany dinner on a Wednesday evening, but on Wednesday morning something came up at work and she ended up grabbing a sandwich from a nearby shop so she could stay until well past dinner. Thursday her boss asked her to go to dinner with a client and by the time she's thought about the tomato again on Saturday it has a mold spot and she's afraid to eat it. Now there is a decision about what to do with this tomato. If the consumer is aware that the tomato is safe to eat, she may simply cut off the molded portion, greatly reducing the volume of food to be managed as waste. In either case, some or all of the tomato must now be managed as a waste.

Waste management options offer different abilities for recovering or recycling the resources embodied in them, particularly when we take a life cycle perspective. But, complete recycling is not entirely possible as the molecular composition of foods are changing in real time. While it is true that some nutrients, such as nitrogen (N), phosphorous (P), minerals, and carbon may be reintroduced to soils to both improve soil health and grow new foods, it is very rare, and sometimes energetically disadvantageous, to return these components of a food to the field from whence they came given that the distance can be hundreds of miles, for example, the distance from New York City to Des Moines, Iowa is about 1100 miles. Further, the amount of material and energy that can be recovered pales in comparison to the upstream input of materials and energy associated with producing the food item (Hamilton et al., 2015). And, finally, it is unclear whether it is physically possible given the geography of lands for cultivation in relation to population centers to produce sufficient human nutrition through the "localization" of food production, at least with the foods, and rates of consumption of those foods, that currently comprise the American diet (Zumkehr and Campbell, 2015; Peters et al., 2009; Dunbar et al., 2016).

13.2 A brief overview of the US agricultural system

To begin to consider the challenges of managing FW so as to close materials flow loops it is important to have a sense of the vast nature of the agricultural system in relation to populations of human and animals in the United States today. In press and conversation about FW management, people will often refer to the benefits of placing composted materials back onto land to recycle nutrients and add carbon back to the soil. Indeed, in an ideal situation, vegetative crops would be grown, fed to people and/or animals, animals would then eaten by people, and in a very progressive system the manure from animals, including humans, would then be composted to remove pathogens and applied to fields to grow more vegetative crops to continue this cycle. In this scenario the materials loop is closed. In the slightly more realistic, but still idealistic scenario, FW and losses would be fed to animals and the manure would again be composted and applied to fields.

The current US agricultural system includes numerous existing geographical constraints that make closing nutrient loops largely impossible. Recovering and transporting nutrients hundreds of miles—back to the fields from which they were taken up by plants and introduced to our food system—is difficult to make economical (Araji et al., 2001) and would almost certainly result in positive net energy and GHG emissions. The US agricultural sector is enormous in the amount of land that it occupies, resources that it consumes, and money that it generates. It is a system that supports ever-expanding urban populations domestically and abroad. The majority of people in the United States live in urban areas and the most populous areas are coastal states. In 2016 Census Bureau Director John H. Thompson stated that "Rural areas cover 97 percent (%) of the nation's land area but contain 19.3% of the population, about 60 million people" (US Census Bureau, 2016). The majority of crop production occurs in interior states and the majority of animal production is highly centralized, often distant from crop fields (Gollehon et al., 2001). The majority of chicken production occurs in the Southeastern states, pigs in the upper Midwest, and cattle in the Midwest and Western Plains states (USDA, 2017). Finally, even if it were possible to return recovered nutrients to the fields in which they were taken up by crops and eventually embodied in foods, the excess nutrients that runoff into the nation's waterways far exceed that which can be recovered through any waste management approach (Steffen et al., 2015; Galloway et al., 2008). It is a debatable question whether these production concerns should be considered within the context of Zero FW, but it is certainly a worthwhile and necessary conversation to be had.

To further illustrate the scale and geography of the US food system and, in particular, the role of commodity crops, that is, corn, soy, wheat, some statistics regarding the land occupation of major crops are provided. In 2017, the major commodity crops planted covered 243,720,000 acres (986,300 km^2) and 203,413,000 acres (823,180 km^2) or 83.5% of the total were located in the Midwest, Northern Plains, and Southern Plains (USDA, 2018). The majority of these crops are consumed in coastal locations where populations are more dense (Hong et al., 2013). The major commodity crops include corn, soybeans, wheat, cotton, oats, barley, rice, sorghum, sugar beets, and sugar cane. Vegetable production occupied 4,492,100 acres (18,180 km^2), fruit production occupied 3,086,900 acres (12,490 km^2), and nut production occupied 2,112,870 acres (8550 km^2) (USDA, 2012 Census Volume 1, Chapter 1: U.S. National Level Data, 2012), or relatively much less land area. Fruits and vegetables utilize less fertilizer input as well. Commodity crops are responsible for the majority of fertilizer inputs to produce the grains and oilseeds that are the starting point for the majority of the foods that we eat, either in processed forms (e.g., breakfast cereals, snack foods, pasta, bread, etc.) or are fed to animals to ultimately be consumed by humans. Humans are more likely to be located in a coastal state, so there is a direct flow of materials from the interior states to the coastal states. There are both real flows, that is, the foods themselves contain nitrogen that is physically transported, and virtual flows, that is, one can take the stance that the fertilizers applied in the interior states occur due to the demand by those living in coastal states (Zhang et al., 2018; Mekonnen and Hoekstra, 2011).

Another complexity introduced by feeding FW directly to animals or reintegrating compost or biosolids to agricultural soils is that, firstly, transporting material comprised of largely of water is energetically costly (Araji et al., 2001) and, secondly, existing agricultural operations, including equipment and labor practices, are not aligned with spreading solid materials. The majority of crops are fertilized using equipment that is designed for gaseous and fluid liquids, not solids. Options are then to change equipment or to use additional treatment technology to consolidate nutrients from the waste streams (Zeng et al., 2006; Yoshino et al., 2003). Additional discussion of utilizing FW as animal feed and options for recovering and utilizing nutrients is presented below.

13.3 Definitions of food waste

Prior to a brief discussion of options within each tier of the hierarchy, it will be necessary to provide some definitions of terms that will be used to characterize FW. Clarity about these terms is important for considering management options, though there remains a great deal of inconsistency (Jakobsen et al., 2016). Many authors and agencies, including the Food and Agriculture Organization, differentiate between food losses and FW. Food loss is define by the FAO as "the decrease in quantity or quality of food" (Food and Agriculture Organization of the United Nations, 2014). FW is considered a part of food loss and refers to "discarding or alternative (nonfood) use of food that is safe and nutritious for human consumption along the entire food supply chain, from primary production to end household consumer level" (Food and Agriculture Organization of the United Nations, 2014). A recent European project, Fusions, proposed the following definition, "Food waste is any food, and inedible parts of food, removed from the food supply chain to be recovered or disposed (including composted, crops ploughed in/not harvested, anaerobic digestion, bio-energy production, co-generation, incineration, disposal to sewer, landfill or discarded to sea)" (FUSIONS, 2016). Finally, the USDA's Economic Research Service defines food loss as "the edible amount of food, post-harvest, that is available for human consumption but is not consumed for any reason" (USDA-ERS, n.d.). In relation to the US FW Challenge, launched on September 16, 2015, a general term of "food loss and waste" was adopted to describe reductions in edible food mass anywhere along the supply chain, though in some instances of food recycling "waste" is stretched to include nonedible parts of food, that is, shells, banana peels, bones (USDA ERS, n.d.).

As can be seen above, distinctions are sometimes drawn regarding whether the food materials are edible to humans or not. Many foods have portions that are typically not considered to be edible, for example, peels, skins, and bones; these portions of food are inedible FW, and also sometimes categorized as unavoidable (Costello et al., 2016; Parfitt et al., 2010). Edible food losses and waste are typically considered avoidable (Parfitt et al., 2010). There is debate about whether foods that have rotted should be considered inedible; in this chapter these foods will be considered edible based on the assertion that these organics could have been consumed or preserved for later consumption. These definitions are made to generate policy and management interventions, which vary along the food supply chain. Definitions are often helpful from the technical perspective as well as a management or treatment technology will differ for many tons of only bones versus collecting and treating FW generated at the household level.

The point at which a food is lost or wasted along the supply chain also has important implications for waste mitigation and management. Indeed, much of the organic materials derived from agricultural products or commodities (i.e., major staple grains and animal carcasses) during food processing supply chains are utilized in secondary markets, for example, offal is utilized in pet food manufacturing, animal bones can be used as fertilizers, and fruit and vegetable losses or wastes hold promise for the production of a variety of valuable materials and chemicals (Galanakis, 2012). Stages of a typical food supply chain, adapted from Parfitt et al. (2010), include:

1. Harvesting and collection from the field: losses occur due to rodents, insects, poor agricultural techniques. Loss reduction is aligned with pest management and optimization of agricultural equipment.
2. Drying: transport and distribution of grains and oilseeds—it is important that harvested grains and oilseeds be properly dried before storage and distribution to avoid spoilage.
3. Storage: lack of infrastructure for storage leading to losses due to pests, disease, spillage, contamination. In developed nations this and the previous two stages are highly studied, infrastructure is quite good, and losses are minimal. In developing nations these stages account for considerable losses of nutritious organic material.
4. Primary processing: cleaning, classification, dehulling (for grains), pounding, grinding, packaging, soaking, drying, sieving, milling. This is the stage in which bulk grains and oils seeds are processed into flours and other feedstocks for food manufacturing. Losses occur due to contamination or poor facilities. In some cases, by-products are recovered and sold into animal food markets.
5. Secondary processing: mixing, cooking, frying, molding, cutting, extrusion, that is, the production of many of the foods found in stores today, for example, breakfast cereals, snack foods, chicken nuggets. Again, losses can occur due to production malfunctions and by-products may be utilized in other markets.
6. Product evaluation: products may not meet company specifications for quality and thus are not allowed to enter the market.
7. Packaging: Losses may occur due to packaging damaging the foods or being insufficient in protection against pests such as rodents or insects.
8. Marketing: publicity, selling, distribution—losses may occur during transport due to insufficient temperature control or protection. If a product is not successfully marketed and consumers are not interested it may expire on the shelves.
9. Preparation: Inedible portions of foods, spoiled foods generated in kitchens that serve food to people generate a relatively pure stream of organic waste to be managed.
10. Postconsumer: Food that is discarded after being purchased or prepared by an individual consumer. This is the area that attracts the most attention and is perhaps the most challenging to manage with regard to final treatment as it is generated by the hands of many, many people thus increasing the risk of contamination with nonorganic or other toxic organic materials.

13.4 The hierarchy of options for managing food losses and wastes

The Environmental Protection Agency has developed a Food Recovery Hierarchy to demonstrate the preferential order for FW management; see Fig. 13.1. The EU and ReFED have also adopted this order of preferential treatment options. ReFED

Figure 13.1 Prioritization order for FW management according to Environmental Protection Agency.

has estimated the economic values of implemented a variety of FW abatement options and the results confirm that the largest potential for economic savings lies in prevention efforts, followed by recycling efforts; see Fig. 13.2. This section will explore each of these tiers of the hierarchy in relation to ZW and life cycle thinking. Sensibly, the most preferred option, and often the most difficult to implement, is Source Reduction, that is, to avoid wasting food in the first place. Pursuit of this option typically involves behavioral or cultural change. The second option in the hierarchy is to Feed Hungry People, which requires that the food be edible and that a distribution network exists. The third option is to Feed Animals, in megaregions, such as the Northeast Corridor the proximity of majority of animal production is quite distant from this supply of FW; further, the methods employed by the vast majority of animal food production in the United States does not align with the wet, heterogeneous FW stream. To be utilized in the most prevalent swine operations the FW must be processed to reduce risks to the health of animals and dried into pellet form. The fourth option, Industrial Uses, includes anaerobic digestion (AD) and industrial uses of fats, oils, and grease. This would also include processes that generate value-added materials from FW streams. These options employ FW to generate energy through methane or through use as biodiesel. Food is generally not digestible on its own; human or animal manures are a common option for successful codigestion (Chen et al., 2008). In some cases, a municipal wastewater treatment plant may have sufficient capacity to accept FW. The fifth option down the hierarchy, despite its popularity, is aerobic composting. And, finally, congruent with ZW philosophy, landfill or incineration is the last option in the hierarchy.

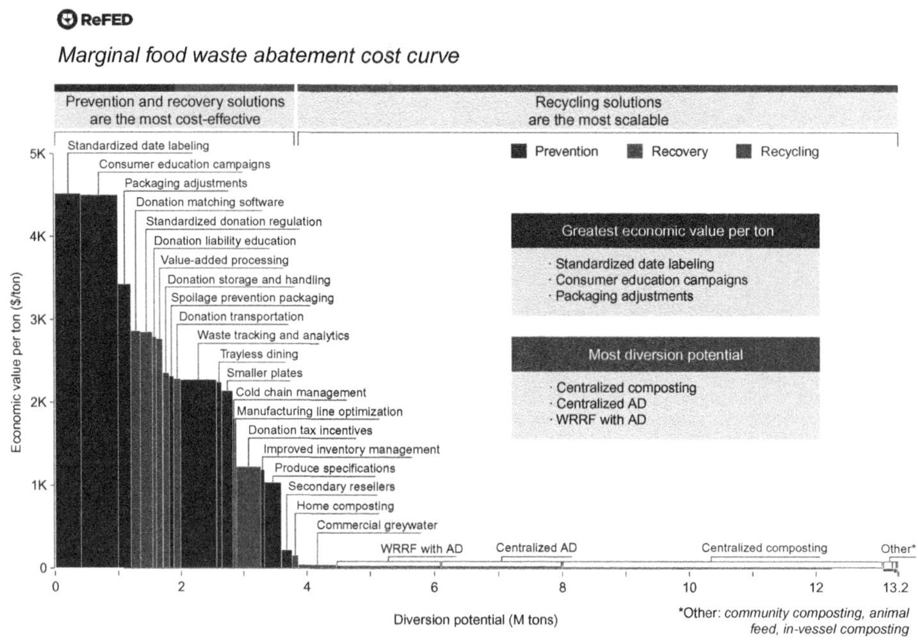

Figure 13.2 Marginal waste abatement curve (Cochran et al., 2018).

13.4.1 Source reduction

Not wasting food at all will have the greatest environmental and materials use benefits (Jakobsen et al., 2016; Hamilton et al., 2015). Strategies and definitions of source reduction depend on the stage in the food supply chain under consideration and are also tied to definitions of FW. If "food" is understood as being edible to humans, then many by-products generated during food processing, for example, peels, trimmings, bones, or recalled food items may not be eligible for source reduction, but may still be utilized to recover materials (e.g., bones used as fertilizer) or energy (e.g., vegetable by-products used as feedstock in an anaerobic digester).

In developed nations there is considerable research, effort, and infrastructure to mitigate losses in the first four stages of the food supply chain and thus loss rates for most grains and oilseeds are fairly small, for example, barley losses are estimated to be less than 3%, while loss rates for fruits and vegetables can be considerably higher ranging from 2% to 23%, average 12%, depending on the type (Parfitt et al., 2010). As will be described below, the potential for capturing food losses or wastes to feed animals, create value-added products, or to produce energy from losses or waste at stages 5 and 9 are much greater than in stages 8–10. This is due to the relative homogeneity, greatly reduced risk of contamination with nonedible materials, and scheduled, larger-in-volume and consistent production rates compared to consumer-generated waste streams. And, similar to stages 1–4 there is already more effort and research to capture these materials than in the consumer waste streams. For these reasons, the remainder of this section will focus more on the research and challenges associated with source reduction in homes, restaurants, and other food-service institutions, like cafeterias, buffets, and stadiums. In many municipalities, FW is the largest category by weight in the waste stream and is thus a good target for reducing waste that is to be managed by a municipality (Zacho and Mosgaard, 2016).

13.4.1.1 Food waste reduction in households

A great deal of the discussion around FW reduction has centered on individual behavioral change. Suggestions for reducing FW include outreach efforts to inform consumers about product expiration dates (Abeliotis et al., 2014) and "ugly" fruits and vegetables and the environmental, social, and personal economic benefits of FW reduction. Other suggestions include offering specific skill-building courses on meal planning, safe storage of foods, and cooking with leftovers (Zacho and Mosgaard, 2016; Graham-Row et al., 2014). It is worth noting that there is often a disconnect between waste managers often tasked with devising communications strategies to reduce waste are most often engineers or scientists lacking the skillsets required to launch such campaigns, which typically require expertise from psychology, sociology, education, communications, and humanities fields (Zacho and Mosgaard, 2016).

While appealing to individuals is an important element of addressing the problem of postconsumer FW, it is also important to consider broader social and physical issues that contribute to the wasting of food. Recognizing the realities of how our communities are structured, portion sizes provided, cultural norms that exist, the demanding nature of work and school, and other societal factors beyond most individuals' reach of control may offer additional approaches to successfully mitigating FW. Southerton et al. (2011) describe three contexts in relation to influencing sustainable consumption practices including:

- The individual, referring to initiatives that focus on education in the hopes of influencing attitudes and ultimately behavior change, for example, reducing the amount of food thrown away at the household level;
- The social, which refers to social norms, cultural conventions, and shared understandings of consumer practices;
- The material, which refers to objects, technologies, and infrastructures that both enable and constrain ways of behaving.

These three contexts allow for a much more comprehensive set of approaches toward reducing FW streams by recognizing that there are many complex factors at play beyond a simple understanding that something is "bad" and thus "should not be done." Evans conducted and in-depth study of 18 households in the United Kingdom to determine individual, social, and material reasons that people wasted food (Evans, 2011, 2012). A complex combination of a desire to provide "proper," that is, healthy, fresh/not frozen or prepared, well-balanced meals made from scratch, combined with the bulk of groceries obtained during a shopping event occurring every 7–10 days at a distant larger store, and demanding and unpredictable work, child (e.g., sports), and other social activities not aligning with the expectations of meal preparation for the week led to FW. Advice focused on meal planning, meal preparation, freezing of leftovers, etc., is good and well-intended, but for many people they simply don't have the time to accommodate the additional labor given the competing demands on their time and a primary goal of ensuring that there is adequate food for themselves and their families. Or even more anecdotal, but highly relatable, children are notoriously picky eaters and may not respond well to improvised meals based on leftovers (Evans, Blaming the consumer—once again: the social and material contexts of everyday FW in some English households, 2011).

At the household level, many of the typical suggestions for people to reduce their FW would translate to additional hours of domestic labor, which many people simply do not have. Based on the calculation of the average annual hours worked (defined by the OECD as the "total number of hours actually worked per year") divided by the average number of people in employment per year, Americans worked 34.4 hours per week (OECD, 2018). In the United States the time spent on household work has decreased considerably from 1965 to 2010 from 54 hours to 43 per week based on the assumption of a two-adult (male and female) household (Bridgman et al., 2012). In 1965 women spent approximately 40 hours per week on domestic labor and 12.8 of those hours were spent on cooking, compared with 26 hours total with 5.9 hours spent on cooking in 2010 (Bridgman et al., 2012). In 1965 men spent about 14 hours per week on domestic labor with 1.8 hours on cooking compared with 17 hours per week with 2.4 hours spent on cooking in 2010 (Bridgman et al., 2012). This amounts to an overall net reduction of 11 hours spent on domestic labor and a reduction of 6.3 hours on time spent cooking per week from 1965 to 2010. It is also important to realize that these statistics assume a two-adult household comprised of healthy adults capable of a full load of work for monetary compensation as well as domestic labor, which is not always the lived reality of individuals. While an increasing number of people are taking up gardening and cooking as hobbies, at the end of the day the provisioning and preparation of food is labor, which is not compensated. For many the additional labor associated with personal FW reduction techniques is simply infeasible in the calculus of time and demands on individual's time in our society.

13.4.1.2 Food waste reduction in hospitality and institutions

Source reduction of FW at hotels, restaurants, and institutions can include customer education as well as service options that mitigate the potential for FW. For example, institutions that serve food via buffets have found that eliminating trays leads to less postconsumer FW (Aramark, 2008). Other approaches for reducing the probability of customers taking more food than they can eat include offering smaller plates (in conjunction with a tray-less service) and staffing the buffet so that customers have to ask for additional portions, which has been shown to reduce waste

(Deutsch, 2018). Steve Finn, vice president of FW reduction at LeanPath, also suggests making FW visible to the consumer (Deutsch, 2018). One way to do this would be to conduct FW audits and share results with consumers (Deutsch, 2018). Campus dining operations at the University of Missouri initiated a program to regularly conduct FW audits and share results with customers and initially saw reductions in waste per customer, which eventually plateaued (Costello et al., 2016).

Beyond trying to influence consumer behavior, ReFED has identified a number of actions that could reduce FW in restaurants and has charted them along two dimensions: profit potential and feasibility; see Fig. 13.3 (Cochran et al., 2018). They identify the highest profit potential in the implementation of waste tracking and analytics and inventory management and production planning (Cochran et al., 2018). These approaches are often implemented in larger organizations such as schools, prisons, and large franchised restaurants. In addition to the other strategies already noted in this section, ReFED identifies the following approaches: portion choices and customized dishes (medium profit potential, highly feasible), produce specifications, and menu design (Cochran et al., 2018). ReFED also identifies a number of options that align with other options on the Food Recovery Hierarchy, such as donation liability education, donation matching partnerships, animal feed, centralized composting or AD, and cooking oil recycling (Cochran et al., 2018). Many of these FW source reduction options involve considerable shifts in management mentality and operations, which can be painful for people at first, but presents

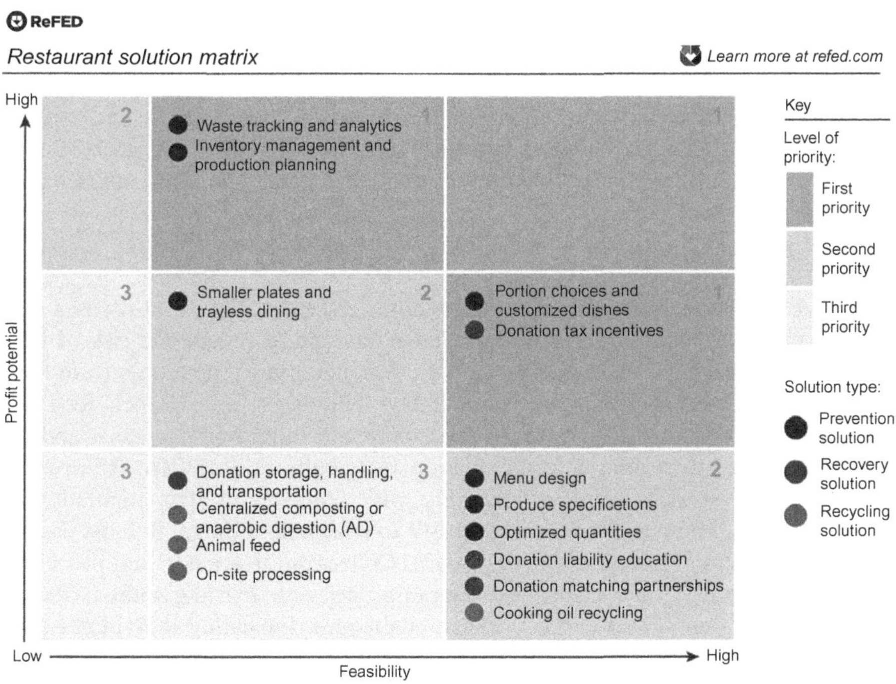

Figure 13.3 ReFED Restaurant Solution Matrix (Cochran et al., 2018).

enormous potential and, ultimately, the largest potential for GHG and energy savings (Costello et al., 2017; Hamilton et al., 2015).

13.4.2 Feed hungry people

In instances where excess food meets the requisite health and safety requirements it can be directed away from landfills and provided to people. Two important pieces of legislation have been enacted by the United States to encourage donations of edible foods to those in need. The Bill Emerson Good Samaritan Food Donation Act, which protects donors and recipients from liability (Government Printing Office, 1996). And, the 2015 Protecting Americans from Tax Hikes Act food donations entitles eligible food business to claim deductions in relation to the value of the donated food (Cochran et al., 2018). Food must be safely packaged and transported with care to maintain food at proper temperatures to prohibit microbial growth, that is, cool foods at 41°F or below, frozen foods at 0°F or below, and hot foods at 135°F or above (Public Health Alliance of Southern California et al., 2018). However, many managers are either unaware of these pieces of legislation, are still too risk-averse to pursue food donations, or don't find the time investment to pursue these options as worthwhile (Cochran et al., 2018).

In larger cities this is a bit easier to accomplish given the increased probability of being able to connect excess food with people within the time constraints associated with ensuring food safety. A variety of apps and other web-based services have been developed around the globe in efforts to connect restaurants, institutions, and retailers with near expiration food products to those that can utilize them at reduced costs or for free (Wong, 2017; Food Cowboy, 2018). These apps provide a variety of services to facilitate connections between food producers and those in need of food and sometimes offer incentivizing services like keeping track of donations for restaurants, hotels, and institutions that can be translated into tax savings.

13.4.3 Feed animals

It is possible to directly feed food scraps to some livestock, but in most cases additional treatment is required to ensure animal safety and to reduce the risk of foodborne illness to humans. As noted previously, residual organic materials from farms or food processing operations are more likely candidates for livestock feed than retailer or postconsumer FW as these feedstocks are more homogeneous and less risky with regard to potential contamination (Mirabella et al., 2014; Rivin et al., 2014). These feedstocks must still comply with federal and state regulations to ensure animal and human safety. Feeding FW to livestock is illegal in most cases in the European Union (Salemdeeb et al., 2017). Collection, treatment, and processing of postconsumer FW into feedstocks commiserate with existing animal feeding infrastructure present additional significant challenges. Educating individuals about proper sorting of FWs and developing the requisite infrastructure and logistics, that is, a separate vehicle fleet for collection of wet FW are nontrivial and significant challenges, as previously noted by Elizabeth Balkan, the NYC Director of Policy for the NYC Department of Sanitation (Balkan, 2017). Following collection, these perishable organic materials must be sorted to remove debris such as plastic and metals and then thermally treated as specified by the laws of the location, described in more detail below (Salemdeeb et al., 2017; Kim et al., 2011).

The US Federal Drug Administration allows for ruminant livestock, for example, cattle, to be fed with "inspected meat products which have been cooked and offered for human food and further heat processed for feed... milk byproducts... and any product whose mammalian protein consists entirely of porcine or equine protein." (Rivin et al., 2014). Postconsumer FW can be fed to livestock, with the exception of bovine brains and spinal cords or bovine spongiform encephalopathy–infected bovine to other bovines, if the material is treated for foodborne pathogens by boiling for 30 minutes (Rivin et al., 2014).

It is also legal in the United States to feed FW to swine so long as the requirements of the Swine Health Protection Act are met, but it is not legal in all states (Rivin et al., 2014). A 2009 amendment to the Swine Health Protection Act allowed for the following exceptions to the rule of 30 minutes of boiling prior to being used as feed for rendered products, bakery waste, candy waste, eggs, domestic dairy products (including milk), fish from the Atlantic Ocean within 200 miles of the continental United States or Canada and fish from inland waters of the United States or Canada that do not flow into the Pacific Ocean, and processed materials [derived in whole or part from the meat of any animal (including fish and poultry)] that have, at minimum, been cooked to a temperature of 167°F (75°C) for at least 30 minutes or has been subjected to an industrial process demonstrated to provide an equivalent level of inactivation of disease organisms, or as approved by the Administrator (Rivin et al., 2014). In 2007, only about 3% of swine producers (representing about 0.12% of swine marketed) utilized garbage as feed (US Government, 2009).

In addition to these collection, pretreatment, and heating requirements important for animal safety, significant technical hurdles remain with regard to integration of postconsumer FW into prevailing livestock operations. Novel production systems that might be more readily adaptable to accepting and feeding wet agricultural/FW are much less numerous and would still require screening and thermal treatment to reduce pathogens and to remove nonfood materials. The predominate swine production methods that supply the vast majority of pork meat in the United States do not readily allow for feeding of wet food materials, such as FW, as the majority of these operations offer dry feeds to swine (Safranski, 2018). Thus, without the possibility of drastic changes to these operations, and potentially supply of bacon, FW must be dried and pelletized to be integrated to existing US swine production infrastructure (Safranski, 2018). Use of foods that might vary in nutrient content can also reduce growth rates for animals, ultimately impacting the economics for the producer. Despite these requisite energy expenditures, recent LCAs on the utilization of FW as animal feeds have found that both dry and wet FW as animal feed are preferable across a variety of environmental impacts to AD and composting; see Table 13.1 (Salemdeeb et al., 2017; Kim et al., 2011).

Table 13.1 Comparative overview of food waste management options in relation to commonly cited metrics for evaluation of environmental performance

Management option	Process energy needs	Direct process greenhouse gas emissions	Electricity generation possible?	Methane generation	Nutrient recovery—N	Nutrient recovery—P
Wet feed to animals	Low	Low—negligible	No	NA	NA	NA
Dry feed to animals	Medium	Low—negligible	No	NA	NA	NA
Landfill w/ CH_4 capture	Low	Low	Yes	Yes	No	Potential if mined in the future
Landfill w/o CH_4 capture	Low	High	No	No	No	
Aerobic composting	Low	Variable	No	No	Yes	Yes
Anaerobic digestion	Medium	Low	Yes	Yes	Yes	Yes
Incineration	High	Low	Yes	No	No	No

Note: NA = Not applicable (1) Swine manure, including urine, can be treated via composting or anaerobic digestion and the resulting materials can be applied as fertilizer.

13.4.4 Industrial uses

Industrial uses is a broad category, which could include production of value-added products from FW (Galanakis, 2012), but is more typically realized through AD and recovery of oils to be utilized as biodiesel. The extraction of value-added products is most likely to occur through the utilization of food processing byproduct streams, rather than postconsumer FW (Mirabella et al., 2014). This is due to the more homogeneous and consistent nature of food processing operations, which is more conducive to subsequent engineered extractions of specific compounds. Examples include phenolic compounds that may be used as antioxidant food additives, essential oils (citrus rinds), proteins, pectin, and a variety of other specialty chemicals (Galanakis, 2012).

AD is a process that relies on microorganisms to break down organic materials into methane gas, and the solids remaining following digestion contain nutrients that can be used as a soil amendment. In addition, AD can be an option for reducing the volume of solid waste that needs to be managed, even if the methane is simply flared and the solids are ultimately disposed of in a landfill. This gas can be recovered and combusted to directly produce heat or within a generator to produce electricity, Table 13.1. AD is often utilized in the management of manures and is often a component of municipal wastewater treatment plants. FW is often codigested with other materials such as human or animal manures to increase chemical stability of the process (Zhang et al., 2014). AD of FW alone is inhibited due to nutrient imbalances including lack of trace elements, excess macronutrients, a carbon to nitrogen ratio outside of ideal ranges, excess lipid concentrations (Zhang et al., 2014). In the case of fruit and vegetable processing byproducts the addition of manure can provide sufficient N and P to the microbial community (Callaghan et al., 2002). AD of FW often requires mechanical and/or thermal pretreatment stages to stabilize various chemical properties of the FW (Zhang et al., 2014). While some technical challenges remain, overall AD is a viable option for recovering nutrients and generating electricity from FW.

13.4.5 Aerobic composting

Composting is the last option in food recovery hierarchies. Composting involves the aerobic microbial decomposition of organic substrates into a stable, humus-like material (EPA, 2015). Composting offers the potential for recycling nutrients and is relatively easy, if land is available and affordable, compared with other, more infrastructure-intensive waste management options; see Table 13.1. In a municipal setting, the collection equipment, that is, vehicle fleet, and logistical challenge collecting FW, and pretreatment to remove debris and shred the materials described above would apply (Cornell Waste Management Institute, 1996). However, as mentioned above, the potential to fully close the loop on the flow of nutrients through the food supply chain is limited due to geographic realities of human populations in relation to where the majority of foods are grown.

Many people chose to compost at home, though successfully keeping a compost pile aerated and well-balanced biogeochemically is fairly complex and studies suggest that home composting results in positive GHG emissions due to the formation of anaerobic pockets that generate methane (Amlinger and Peyr, 2008). Home composting is often not an option in dense urban environments due to the potential for attracting insect and animal pests. Vermicomposting may be a useful option in urban environments due to the more rapid nature of the decomposition and potential for small indoor setups (EPA, n.d.).

Composting at a municipality scale is often a better option as routine maintenance in accordance with optimal pile operations reduces methane formation and rapid breakdown of organics (Amlinger and Peyr, 2008). Larger compositing operations can also achieve the conditions required to break down biobased polymers though rates are still not as rapid as other organics, but which are highly unlikely to breakdown in home-scale compost piles at all (Mohee et al., 2008; Mohee and Unmar, 2007). Common options for composting at a municipal scale include aerated (turned) windrow and aerated static pile composting. In-vessel composting may be another option for municipalities or food-service providers.

Windrow composting is particularly suitable for large volumes of food and yard waste and involves the formation of long rows, "windrows," that are 8 ft tall and 14–16 ft wide (EPA, n.d.). Piles are periodically turned using machinery that can pass over the pile. The pile size is large enough to generate and maintain temperatures up to 140°F, even if the outside of the pile becomes frozen, and small enough in some dimensions to allow oxygen to flow through to the core (EPA, n.d.) again these advantageous features for decomposing biobased plastics.

Aerated static piles are suitable for larger quantity generators of yard trimmings, food scraps, and paper products and cannot readily accommodate animal byproducts or grease from food processing industries (EPA, n.d.). The piles must be aerated using either layers of bulking agents, such as wood chips or shredded newspaper or by using a network of perforated pipes to ensure continual airflow (EPA, n.d.). Both windrow and aerated static pile composting may require measures such as installing a shelter or misting system to maintain optimal moisture rates in the pile (EPA, n.d.).

Finally, in-vessel composting involved feeing organic waste into an enclosed container, which can allow for greater control of the conditions, for example, temperature, moisture, mechanical turning of the pile, and airflow (EPA, n.d.). If managed properly compost can be created in just a few weeks (EPA, n.d.). This may be an option for smaller operations that generate a steady stream of organic waste to install on premise (EPA, n.d.).

In all cases, considerable technical expertise is required to maintain composting operations (Cornell Waste Management Institute, 1996). Utilizing compost can be challenging since it is relatively heavy and has low value as a soil amendment, so it is usually marketed close to the composting facility at fairly low prices (Morris et al., 2014).

13.5 Life cycle assessment and systems analysis of food waste management options

A few studies have systematically compared FW treatment options from a life cycle perspective to determine which options result in the least environmental impacts. Generally speaking, LCA results align with the Food Recovery Hierarchy. Source reduction is the most preferable option for energy and phosphorous savings (Hamilton et al., 2015). Salemdeeb et al. (2017) found that wet and dry animal feed production performed better than AC or composting across 13/14 and 12/14 environmental metrics, respectively. And Kim et al. (2011) also found wet animal feed production to be the most preferable option from a life cycle costing analysis, with dry animal feed production the second. However, and incongruent with basic ZW definition, Kim et al. (2011) found that waste incineration coupled with electricity production performed better than AD or composting. Results of these works demonstrate that selecting an optimal FW management option is likely to be a multicriteria decision problem that will need to be made with respect to stakeholder priorities and existing infrastructure.

13.6 Concluding thoughts

If the philosophy of ZW is to be stringently applied to FW, it challenges humanity to deeply consider our relationships with food and the systems that we have constructed to supply us with plentiful and varied food. And a deep consideration of the societal reasons for why food is wasted, for example, sociology, class/labor, poverty, urban planning, and zoning. In particular, over the past hundred or so years humans have radically altered the land surface and disrupted nutrient cycling through the production and application of synthetic fertilizers and the separation and aggregation of animal production operations (Steffen et al., 2015; Galloway et al., 2008). As urbanization increases, devising approaches to close these loops presents enormous challenges that will require a diversity of tactics to successfully manage. In the near term, focusing on reducing the occurrence of FW and developing infrastructure to separate and recover as much use as possible from FW streams will be beneficial, but it is not the end of the road.

References

Abeliotis, K., Lasaridi, K., Chroni, C., 2014. Attitudes and behavior of Greek households regarding food waste prevention. Waste Manage. Res. 32 (3), 237–240.

Amlinger, F., Peyr, S., 2008. Green house gas emissions from composting and mechanical biological treatment. Waste Manage. Res. 26, 47–60.

Araji, A.A., Abdo, Z.O., Joyce, P., 2001. Efficient use of animal manure on cropland - economic analysis. Bioresour. Technol. 79, 179–191.

Aramark, 2008. The business and cultural acceptance case for trayless dining. Retrieved May 2018, from: <http://www.elon.edu/docs/e-web/bft/sustainability/ARAMARK%20Trayless%20Dining%20July%202008%20FINAL.pdf>.

Balkan, E., May 8, 2017. Director of Policy NYC Department of Sanitation. Zero Waste Food Conference—Reimagining Sustainable Connections Within Food Chains—The New School. (F. Parasecoli, Interviewer). Available from: <https://youtu.be/kh3c-Mvpwc0>.

Bridgman, B., Dugna, A., Lal, M., Osborne, M., Villones, S., 2012. Accounting for Household Production in the National Accounts, 1965-2010. U.S. Bureau of Economic Analysis.

Callaghan, F.J., Wase, D.A.J., Thayanithy, K., Forster, C.F., 2002. Continuous co-digestion of cattle slurry with fruit and vegetable wastes and chicken manure. Biomass Bioenergy. 27, 71−77.

California, May 7, 2017. Mandatory Commercial Organics Recycling (MORe). Retrieved May 2018, from Calrecycle: <http://www.calrecycle.ca.gov/recycle/commercial/organics/>.

Chen, Y., Cheng, J.J., Creamer, K.S., 2008. Inhibition of anaerobic digestion process: A review. Bioresour. Technol. 99 (10), 4044−4064.

Cochran, C., Goulbourne, E., Hunt, C., Veza, A., 2018. Restaurant food waste action guide. Retrieved May 2018, from ReFed: <http://www.refed.com/downloads/Restaurant_Guide_Web.pdf>.

Committee on Environment and Natural Resources, 2010. Scientific assessment of hypoxia in U.S. coastal waters. Interagency Working Group on Harmful Algal Blooms, Hypoxia, and Human Health of the Joint Subcommittee on Ocean Science and Technology. Washington, DC, GPO. Retrieved from: <https://www.coastalscience.noaa.gov/publications/handler.aspx?key = 5690>.

Cornell Waste Management Institute, January 16, 1996. Cornell composting. Retrieved May 2018, from Municipal Solid Waste Composting Fact Sheets: <http://compost.css.cornell.edu/MSWFactSheets/msw.fs.toc.html>.

Costello, C., Birisci, E., McGarvey, R.G., 2016. Food waste in campus dining operations: inventory of pre- and postconsumer mass by food category, and estimation of embodied greenhouse gas emissions. Renew. Agric. Food Syst. 31 (3), 191−201.

Costello, C., McGarvey, R.G., Birisci, E., 2017. Achieving sustainability beyond zero waste: a case study from a college football stadium. Sustainability 9, 1236−1256.

de Vries, M., de Boer, I., 2010. Comparing environmental impacts for livestock products: a review of life cycle assessments. Livestock Sci. 128, 1−11.

Deutsch, J., April 18, 2018. Operations: how to reduce food waste at the buffet. Retrieved May 2018, from Restaurant Business: <http://www.restaurantbusinessonline.com/advice-guy/how-reduce-food-waste-buffet>.

Dodds, W.K., Bouska, W.W., Eitzmann, J.L., Pilger, T.J., Pitts, K.L., Riley, A.J., et al., 2009. Eutrophication of U.S. freshwaters: analysis of potential economic damages. Environ. Sci. Technol. 43 (1), 12−19.

Dunbar, B., Costello, C., McGarvey, R.G., 2016. Robust optimization evaluation of reliance on locally produced foods. Environ. Syst. Decisions 36 (3), 34−41.

EPA, February, 2014. Retrieved from Municipal Solid Waste Generation, Recycling, and Disposal in the United States: Facts and Figures for 2012: <https://www.epa.gov/sites/production/files/2015-09/documents/2012_msw_fs.pdf>.

EPA, 2015. WARM v. 13 Documentation: Composting.

EPA, 2018. Greenhouse gas emissions. Retrieved May 2018, from Inventory of U.S. Greenhouse Gas Emissions and Sinks: 1990−2016: <https://www.epa.gov/ghgemissions/inventory-us-greenhouse-gas-emissions-and-sinks-1990-2016>.

EPA, n.d. Sustainable management of food. Retrieved May 2018, from Types of Composting and Understanding the Process: <https://www.epa.gov/sustainable-management-food/types-composting-and-understanding-process#aeratedturned>.

EC (European Commission), 2015. Retrieved January 25, 2018, from EC-Press Release, Closing the loop: Commission adopts ambitious new Circular Economy Package to boost competitiveness, create jobs and generate sustainable growth: http://europa.eu/rapid/press-release_IP-15-6203_en.htm.

Evans, D., 2011. Blaming the consumer—once again: the social and material contexts of everyday food waste in some English households. Crit. Public Health 21 (4), 429–440.

Evans, D., 2012. Beyond the throwaway society: ordinary domestic practice and a sociological approach to household food waste. Sociology 46 (1), 41–56.

Food and Agriculture Organization of the United Nations, 2014. Retrieved May 2018, from Technical Platform on the Measurement and Reduction of Food Loss and Waste: <http://www.fao.org/platform-food-loss-waste/food-waste/definition/en/>.

Food Cowboy, 2018. Retrieved May 2018, from: <http://www.foodcowboy.com/>.

FUSIONS, 2016. EU FUSIONS. Retrieved May 2018, from Food Waste Definition: <http://www.eu-fusions.org/index.php/about-food-waste/280-food-waste-definition>.

Galanakis, C.M., 2012. Recovery of high added-value components from food wastes: conventional, emerging technologies and commercialized applications. Trends Food Sci. Technol. 26, 68–87.

Galloway, J.N., Townsend, A.R., Erisman, J.W., Bekunda, M., Cai, Z., Freney, J.R., et al., 2008. Transformation of the nitrogen cycle: recent trends, questions, and potential solutions. Science 320 (5878), 889–892.

Gollehon, N., Caswell, M., Ribaudo, M., Kellogg, R., Lander, C., Letson, D., 2001. Confined Animal Production and Manure Nutrients. Resource Economics Division, Economic Research Service, USDA, Washington, DC.

González, A.D., Frostell, B., Carlsson-Kanyama, A., 2011. Protein efficiency per unit energy and per unit greenhouse gas emissions: potential contribution of diet choices to climate change mitigation. Food Policy 36, 562–570.

Government Printing Office, 1996. Retrieved May 2018, from Bill Emerson Good Samaritan Food Donation Act: <https://www.gpo.gov/fdsys/pkg/PLAW-104publ210/pdf/PLAW-104publ210.pdf>.

Graham-Row, E., Jessop, D.C., Sparks, P., 2014. Identifying motivations and barriers to minimising household food waste. Resour., Conserv. Recycl. 84, 15–23.

Hamilton, H.A., Peverill, M.S., Müller, D.B., Brattebø, H., 2015. Assessment of food waste prevention and recycling strategies using a multilayer systems approach. Environ. Sci. Technol. 49, 13937–13945.

Hong, B., Swaney, D.P., Howarth, W.R., 2013. Estimating net anthropogenic nitrogen inputs to U.S. watersheds: comparison of methodologies. Environ. Sci. Technol. 47 (10), 5199–5207.

International Standards Organization, 2006. ISO 14040: environmental management—life cycle assessment—principles and framework.

Jakobsen, B., Brems, E., Stern, K., Voinea, D., Oliveira, P., 2016. Combating food waste: an opportunity for the EU to improve the resource-efficiency of the food supply chain. European Court of Auditors.

Kim, M.-H., Song, Y.-E., Song, H.-B., Kim, J.-W., Hwang, S.-J., 2011. Evaluation of food waste disposal options by LCC analysis from the perspective of global warming: Jungnang case, South Korea. Waste Manage. 31, 2112–2120.

Mekonnen, M.M., Hoekstra, A.Y., 2011. The green, blue and grey water footprint of crops and derived crop products. Hydrol. Earth Syst. Sci. 1577–1600.

Mirabella, N., Castellani, V., Sala, S., 2014. Current options for the valorization of food manufacturing waste: a review. J. Cleaner Prod. 65, 28–41.

Mohee, R., Unmar, G., 2007. Determining biodegradability of plastic materials under controlled and natural composting environments. Waste Manage. 27, 1486–1493.

Mohee, R., Unmar, G.D., Mudhoo, A., Khadoo, P., 2008. Biodegradability of biodegradable/degradable plastic materials under aerobic and anaerobic conditions. Waste Manage. 28, 1624–1629.

Morris, J., Brown, S., Matthews, H.S., Cotton, M., 2014. Evaluation of Climate, Energy, and Soils Impacts of Selected Food Discards Management Systems. Oregon Department of Environmental Quality.

NRDC, August, 2013. Collegiate game changers: how campus sport is going green. Retrieved May 2018, from: <https://www.nrdc.org/resources/collegiate-game-changers-how-campus-sport-going-green>.

OECD, 2018. OECD data. Retrieved May 2018, from Hours Worked: <https://data.oecd.org/emp/hours-worked.htm>.

OneNYC, n.d. OneNYC. Retrieved May 2018, from Vison3 Goal Zero Waste: <https://onenyc.cityofnewyork.us/goals/zero-waste/>.

Palmer, P., n.d. History. Retrieved May 2018, from Zero Waste Institute: <http://zerowasteinstitute.org/?page_id = 202>.

Parfitt, J., Barthel, M., Macnaughton, S., 2010. Review: food waste within food supply chains: quantification and potential for change to 2050. Philos. Trans. R. Soc. 365, 3065–3081.

Peters, C.J., Bills, N.L., Lembo, A.J., Wilkins, J.L., Fick, G.W., 2009. Mapping potential foodsheds in New York State: a spatial model for evaluating the capacity to localize food production. Renew. Agric. Food Syst. 24 (1), 72–84.

Public Health Alliance of Southern California, California Conference of the Directors of Environmental Health, The Center for Climate Change and Health, 2018. Safe Surplus Food Donation Best Management Practices: Guidance for Environmental Health Departments. Retrieved from: <bit.ly/2yXzHAe>.

ReFED, May 11, 2018. ReFED. Retrieved May 2018, from U.S. Food Waste Policy Finder: U.S. Food Waste Policy Finder.

Rivin, J., Miller, Z., Matel, O., 2014. Using food waste as livestock feed. University of Wisconsin, Extension.

Safranski, T., May 17, 2018. Professor. State Swine Breeding Specialist. University of Missouri. (C. Costello, Interviewer).

Salemdeeb, R., Ermagassen, E.K., Kim, M.H., Balmford, A., Al-Tabbaa, A., 2017. Environmental and health impacts of using food waste as animal feed: a comparative analysis of food waste options. J. Cleaner Prod. 140, 871–880.

Southerton, D., McMeekin, A., Evans, D., 2011. International Review of Behaviour Change Initiatives: Climate Change Behaviors Research Programme. Queens Printers of Scotland, Edinburgh, Scottish Government, Social Research.

Steffen, W., Richardson, K., Rockström, J., Cornell, S.E., Fetzer, I., Bennett, E.M., et al., 2015. Planetary boundaries: guiding human development on a changing planet. Science 347, 6223.

The NYC Council, December 30, 2013. Legislation. Retrieved May 2018, from A Local Law to amend the administrative code of the city of New York, in relation to commercial organic waste: <http://legistar.council.nyc.gov/LegislationDetail.aspx?ID = 1482542&GUID = DDD94082-C0E5-4BF9-976B-BBE0CD858F8F>.

US Census Bureau, December 8, 2016. Newsroom. Retrieved May 2018, from New Census Data Show Differences Between Urban and Rural Populations: <https://www.census.gov/newsroom/press-releases/2016/cb16-210.html>.

US Government, April 3, 2009. Federal Register vol. 74, No. 63. Swine Health Protection; Feeding of Processed Product to Swine.

Union of Concerned Scientists, August 30, 2016. Retrieved May 2018, from The Hidden Costs of Fossil Fuels: <https://www.ucsusa.org/clean-energy/coal-and-other-fossil-fuels/hidden-cost-of-fossils#.WxBGBhMvzOQ>.

USDA, 2012. 2012 Census Volume 1, Chapter 1: U.S. National Level Data. Retrieved May 2018, from: <https://www.agcensus.usda.gov/Publications/2012/Full_Report/Volume_1,_Chapter_1_US/>.

USDA, June 15, 2017. USDA Census of Agriculture. Retrieved May 2018, from 2012 Census Ag Atlas Maps: <https://www.agcensus.usda.gov/Publications/2012/Online_Resources/Ag_Atlas_Maps/>.

USDA, 2018. Crop Acreage Data. Retrieved May 2018, from United States Department of Agriculture—Farm Service Agency: <https://www.fsa.usda.gov/news-room/efoia/electronic-reading-room/frequently-requested-information/crop-acreage-data/index>.

USDA ERS, n.d. USDA Office of the Chief Economist. Retrieved May 2018, from: <https://www.usda.gov/oce/foodwaste/faqs.htm>.

Wong, K. (2017). Tackling Food Waste Around the World: Our Top 10 Apps. The Guardian. 6 February 2017.

Yoshino, M., Yao, M., Tsuno, H., Somiya, I., 2003. Removal and recovery of phosphate and ammonium as struvite from supernatant in anaerobic digestion. Water Sci. Technol. 48 (1), 171–178.

Zacho, K.O., Mosgaard, M.A., 2016. Understanding the role of waste prevention in local waste management: a literature review. Waste Manage. Res. 34, 980–994.

Zaman, A.U., 2015. A comprehensive review of the development of zero waste management: lessons learned and guidelines. J. Cleaner Prod. 91, 12–25.

Zeng, L., Mangan, C., Li, X., 2006. Ammonia recovery from anaerobically digested cattle manure by steam stripping. Water Sci. Technol. 54 (8), 137–145.

Zhang, C., Su, H., Baeyens, J., Tan, T., 2014. Reviewing the anaerobic digestion of food waste for biogas production. Renew. Sustain. Energy Rev. 383–392.

Zhang, Y., Liu, Y., Shibata, H., Gu, B., Wang, Y., 2018. Virtual nitrogen factors and nitrogen footprints associated with nitrogen loss and food wastage of China's main food crops. Environ. Res. Lett. 13, 014017.

Zumkehr, A., Campbell, J.E., 2015. The potential for local croplands to meet US food demand. Front. Ecol. Environ. 244–248.

ZWIA, August 12, 2009. Zero waste international alliance. Retrieved May 2018, from ZW Definition: <http://zwia.org/standards/zw-definition/>.

Further reading

USDA-ERS, February 21, 2018. USDA. Economic research service. Retrieved May 2018, from Fertilizer Use and Price: <https://www.ers.usda.gov/data-products/fertilizer-use-and-price/>.

USDA-NASS, May 4, 2018. National agricultural statistics service. Retrieved May 2018, from 2017 Agricultural Statistics Annual: <https://www.nass.usda.gov/Publications/Ag_Statistics/2017/index.php>.

Index

Note: Page numbers followed by "*f*" and "*t*" refer to figures and tables, respectively.

M

N